Genomics and Proteomics: Functional and Computational Aspects

Genomics and Proteomics: Functional and Computational Aspects

Edited by Miguel Rudolph

New York

Published by Syrawood Publishing House,
750 Third Avenue, 9th Floor,
New York, NY 10017, USA
www.syrawoodpublishinghouse.com

Genomics and Proteomics: Functional and Computational Aspects
Edited by Miguel Rudolph

International Standard Book Number: 978-1-68286-765-5 (Hardback)

Cataloging-in-Publication Data

Genomics and proteomics : functional and computational aspects / edited by Miguel Rudolph.
p. cm.
Includes bibliographical references and index.
ISBN 978-1-68286-765-5
1. Genomics. 2. Proteomics. 3. Functional genomics. 4. Genomics--Data processing.
5. Proteomics--Data processing. I. Rudolph, Miguel.
QH447 .G46 2019
572.86--dc23

TABLE OF CONTENTS

PREFACE

The study of the mapping, function, structure, evolution and editing of genomes is studied under genomics. Genes direct the production of proteins through enzymes and messenger molecules. Research in genomics delves into functional genomics, which studies gene transcription, translation and protein-protein interactions. It further strives to understand the functional role of DNA at the level of genes, RNA transcripts and protein products. Proteomics is the in-depth study of proteins. It is an interdisciplinary field which has rapidly advanced due to the progress in genomics. This book is a valuable compilation of topics, ranging from the basic to the most complex advancements in the fields of genomics and proteomics. It includes some of the vital pieces of work being conducted across the world, on various topics related to the functional and computational aspects of these domains. It is a complete source of knowledge on the present status of these important fields.

The information contained in this book is the result of intensive hard work done by researchers in this field. All due efforts have been made to make this book serve as a complete guiding source for students and researchers. The topics in this book have been comprehensively explained to help readers understand the growing trends in the field.

I would like to thank the entire group of writers who made sincere efforts in this book and my family who supported me in my efforts of working on this book. I take this opportunity to thank all those who have been a guiding force throughout my life.

Editor

1

Unraveling the evolution and coevolution of small regulatory RNAs and coding genes in *Listeria*

Franck Cerutti[1], Ludovic Mallet[1], Anaïs Painset[1,7], Claire Hoede[1], Annick Moisan[1], Christophe Bécavin[2,3,4,5], Mélodie Duval[2,3,4], Olivier Dussurget[2,3,4,6], Pascale Cossart[2,3,4], Christine Gaspin[1] and Hélène Chiapello[1*]

Abstract

Background: Small regulatory RNAs (sRNAs) are widely found in bacteria and play key roles in many important physiological and adaptation processes. Studying their evolution and screening for events of coevolution with other genomic features is a powerful way to better understand their origin and assess a common functional or adaptive relationship between them. However, evolution and coevolution of sRNAs with coding genes have been sparsely investigated in bacterial pathogens.

Results: We designed a robust and generic phylogenomics approach that detects correlated evolution between sRNAs and protein-coding genes using their observed and inferred patterns of presence-absence in a set of annotated genomes. We applied this approach on 79 complete genomes of the *Listeria* genus and identified fifty-two accessory sRNAs, of which most were present in the *Listeria* common ancestor and lost during *Listeria* evolution. We detected significant coevolution between 23 sRNA and 52 coding genes and inferred the *Listeria* sRNA-coding genes coevolution network. We characterized a main hub of 12 sRNAs that coevolved with genes encoding cell wall proteins and virulence factors. Among them, an sRNA specific to *L. monocytogenes* species, *rli133,* coevolved with genes involved either in pathogenicity or in interaction with host cells, possibly acting as a direct negative post-transcriptional regulation.

Conclusions: Our approach allowed the identification of candidate sRNAs potentially involved in pathogenicity and host interaction, consistent with recent findings on known pathogenicity actors. We highlight four sRNAs coevolving with seven internalin genes, some of which being important virulence factors in *Listeria*.

Keywords: *Listeria*, sRNA, Phylogenomics, Coevolution network, Regulation, Cell wall, Pathogenicity, Internalin

Background

Small regulatory RNAs are widespread in all kingdoms of life and are recognized as key negative or positive regulators of gene expression [1–3]. They are involved in a wide panel of physiological processes and adaptive responses in bacteria including stress responses, quorum sensing, toxin-antitoxin systems or pathogenicity [4–6]. They generally act post-transcriptionally in *cis* (antisense) or *trans* by base pairing with their target messenger RNA (mRNA) but can also bind specific proteins and modify their activity, as illustrated by CsrB and 6S sRNA [1]. The most extensively studied class of sRNA includes *trans*-encoded sRNAs which regulate their target mRNA by forming short and imperfect duplexes. In silico identification of these duplexes remains a major challenge due to a prohibitively high level of false positive candidates [7–9]. Nevertheless, an improvement in target prediction was shown [10, 11] by focusing on site-specific regions such as the ribosome binding site (RBS), the accessibility of unstructured seed regions in both the sRNA and target mRNA, and the use of comparative genomics of interaction candidates. Altogether, these features argue for a better understanding of sRNA history during bacterial evolution and shed light on how regulatory networks

* Correspondence: helene.chiapello@inra.fr
[1]Université de Toulouse, INRA, UR 875 Unité Mathématiques et Informatique Appliquées de Toulouse, Auzeville, 31326 Castanet-Tolosan, France
Full list of author information is available at the end of the article

involving *trans*-acting-sRNA and target mRNA have emerged and evolved. Unfortunately, little is known about sRNA evolution, sRNA expression control and sRNA-mRNA coevolution within bacteria, and very few studies have been carried out on these topics so far. This can be explained by the lack of sRNA annotation in available genome resources as well as by the low number of well-characterized regulatory sRNAs and the rapid evolution of regulatory sRNAs in bacteria [1].

In the last decade, high throughput sequencing and transcriptome-wide approaches led to a continuous accumulation of complete genomic data in public databases and contributed to the discovery of hundreds of putative and confirmed new sRNAs in many bacteria such as *Escherichia coli* [12], *Salmonella* [13], *Bacillus subtilis* [14, 15] and *Listeria* [5, 6, 16–27], giving rise to large-scale comparative analyses and sRNA evolutionary studies. Existing studies on that topic focused on Gram-negative species, including *Escherichia coli* and related genomes [8, 28–30]. Phyletic analysis of *E.coli* sRNAs [29] led to the first insights into the distribution of sRNAs in gamma-proteobacteria, greatly improving our understanding of the origin of sRNA-mediated regulation and the underlying mechanisms at the source of sRNA acquisition. To our knowledge, such a global evolutionary study has never been performed in Gram-positive bacteria.

Listeria are Gram-positive bacteria that are widespread in the environment and encompass 17 species, two of which are pathogenic: *Listeria monocytogenes*, the human foodborne agent responsible for listeriosis, and *Listeria ivanovii*, an animal pathogen [31]. *L. monocytogenes* has become a model for the study of host-pathogen interactions due to its unique ability to cross host barriers, escape from immune defenses, invade cells and manipulate cellular machineries [32–34]. The comparative analysis of the complete genome sequence of *L. monocytogenes* and the non pathogenic species *L. innocua* in 2001 was the first study to shed light on *Listeria* virulence and its evolution [35]. Following this pioneer work, *Listeria* genomic data grew exponentially and more than 80 complete genomes have been sequenced [36]. Small non-coding RNAs were also extensively studied in *L. monocytogenes* [5, 6, 16–27]. Indeed, 304 non-coding RNAs elements were reported in *L. monocytogenes* EGD-e including 154 sRNAs, 104 antisense and 46 *cis*-encoded [16, 17, 37]. Among these sRNAs, several were shown to be upregulated in bacteria growing in murine intestinal lumen and in human blood, suggesting that they may play a role in adaptation of the bacteria to the niches occupied during infection [1, 5, 21].

Comparative analyses of *Listeria* sRNAs by Kuenne et al. [38] revealed the organization of CRISPR arrays and *cas* genes in 38 complete *L. monocytogenes* genomes. Becavin et al. compared three *L. monocytogenes* species and observed a high conservation of sRNAs compared to protein-coding genes [37]. A comparative transcriptomics approach was also used to compare the expression of non-coding RNAs in *L. monocytogenes* and *L. innocua* species, which revealed conservation across most transcripts, but significant divergence between the species in a subset of non-coding sRNAs [22].

In this article, we present a robust phylogenomics approach that extends and improves existing strategies dedicated to the study of sRNA evolutionary dynamics. We use it to provide the first evolutionary dynamics study of 79 complete genomes of the *Listeria* genus with regards to protein-coding genes, and a selected set of 112 sRNA loci experimentally identified in the pathogenic strain *L.monocytogenes* EGD-e. This dataset includes intergenic trans-encoded sRNAs assumed to target independently expressed and distant mRNAs. We built the core and accessory sRNA and coding genes sets and deduced the ancestral presence-absence states for all *Listeria* genes. Using these patterns, we identified a subset of 23 sRNAs that significantly coevolved with 5′ untranslated regions of coding genes (5'UTRs) and coding DNA sequence (CDS) regions of 52 *Listeria* coding genes. We reconstructed the coevolution network between sRNAs and coding genes and revealed a hub of 12 sRNAs coevolving with genes encoding cell wall proteins and virulence factors. Among them, we focused on *rli133*, an sRNA specific of *L. monocytogenes* species that coevolved with 12 coding genes, six of which exhibited a documented function linked to either virulence or interaction with the host cell, possibly acting as a negative post-transcriptional regulator.

Results

A robust screening strategy for sRNA and coding genes coevolution

We designed an original approach to build a reference phylogenetic tree to infer observed and ancestral evolution patterns and to identify coevolution relationships between pairs of sRNAs and coding-genes. The four main steps of this approach are presented in Fig. 1 and a full description of each step of the workflow is provided in the Methods section. Briefly, the approach starts from a set of annotated genomes and a list of sRNAs to proceed through four main steps: (1) the construction of a reference phylogenetic tree based on orthologous genes; (2) the construction of the presence-absence matrix for sRNAs, 5'UTRs and CDS parts allowing across all the genomes to define core and accessory sets for all elements; (3) the inference of ancestral presence-absence patterns for all variable sRNAs, 5'UTRs and CDS; and (4) the detection of coevolution events between regulatory sRNA and coding genes and construction of a coevolution

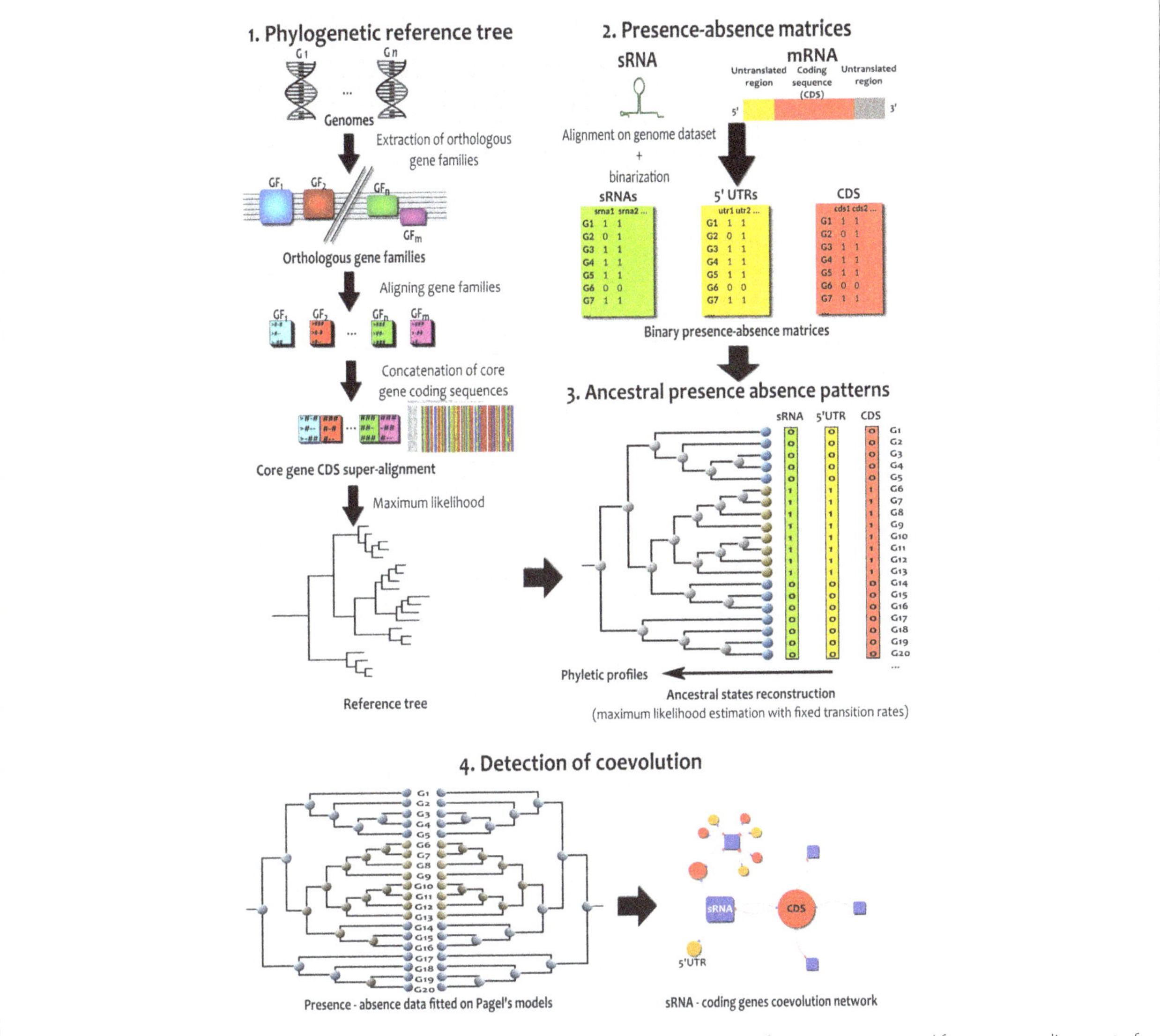

Fig. 1 Strategy and workflow. The strategy consists in 4 steps: (1) Construction of a phylogenetic reference tree computed from a super-alignment of syntenic core genes and a Maximum Likelihood approach (2) Presence-absence matrices computation using alignments of sRNAs, 5'UTRs and CDS (3) Ancestral presence-absence pattern reconstruction for sRNAs, 5'UTRs and CDS based on Markov Model and a Maximum Likelihood approach (4) Detection of coevolution events between sRNAs and 5'UTRs or CDS using both observed and ancestral patterns and construction of the sRNA-coding genes coevolution network

network using both the observed and the reconstructed ancestral presence-absence patterns. The detection of correlated evolution events relies on a phylogenetic-statistical method based on continuous-time Markov modeling of trait evolution developed by M. Pagel [39]. It compares the statistical likelihood of the observed data (in this case, sRNAs, 5'UTRs and CDS presence/absence patterns) under two alternative scenarii: one in which the two features are allowed to evolve independently on the phylogeny, and another where they coevolve together.

This strategy was applied on 112 *L. monocytogenes* EGD-e putative trans sRNAs, all screened on 79 *Listeria* genomes (see Additional file 1: Table S1) obtained from the Listeriomics database [36]. To deal with the remaining paralogs in the dataset, sRNAs exhibiting overlapping positions on the EGD-e reference genome were merged in 15 sRNA loci (see Additional file 2: Table S2 and the Methods section for details).

The *Listeria* phylogenetic reference tree obtained from the 1399 syntenic core coding genes of *Listeria* was robust and consistent with previous studies [40] (see Fig. 2). The four major phylogenetic lineages of *L. monocytogenes* were clearly separated with good Shimodaira Hasegawa (SH) supports (Fig. 2b) [40]. We however observed a few branches of lineage I with lower SH support that correspond to highly conserved genomes, resulting in

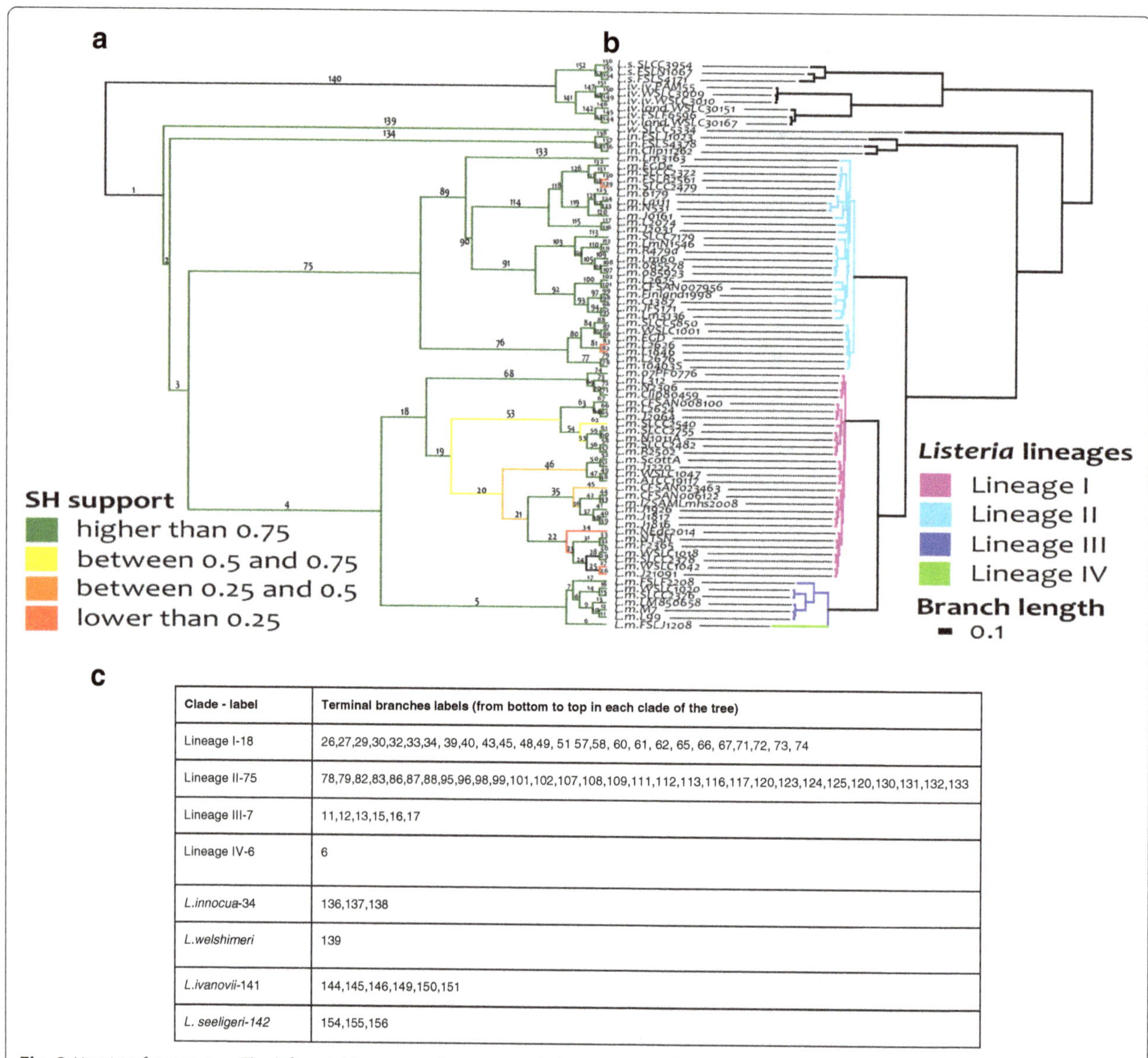

Clade - label	Terminal branches labels (from bottom to top in each clade of the tree)
Lineage I-18	26,27,29,30,32,33,34, 39,40, 43,45, 48,49, 51 57,58, 60, 61, 62, 65, 66, 67,71,72, 73, 74
Lineage II-75	78,79,82,83,86,87,88,95,96,98,99,101,102,107,108,109,111,112,113,116,117,120,123,124,125,120,130,131,132,133
Lineage III-7	11,12,13,15,16,17
Lineage IV-6	6
L.innocua-34	136,137,138
L.welshimeri	139
L.ivanovii-141	144,145,146,149,150,151
L. seeligeri-142	154,155,156

Fig. 2 *Listeria* reference tree. The left part (**a**) presents the tree as a cladogram to visualize Shimodaira–Hasegawa (SH) supports for all branches. Best SH support branches (SH support values >0.75) are indicated in green. Branches with a support value between 0.5 and 0.75 are indicated in yellow and those with a support value between 0.25 and 0.5 are indicated in orange. Low SH support branches are indicated in red (SH support values <0.25). The right part (**b**) represents the tree with estimated branch lengths. The four highlighted clades correspond to the four known Listeria lineages. Branch labels are used in Additional file 4: Table S4. Terminal branch labels of each main clade of the tree are listed in a separated table below (**c**)

short branch lengths and a weaker phylogenetic signal. This reference tree was subsequently used to compare sRNAs and coding gene content of *Listeria* genomes.

sRNA content of *Listeria* genomes

On the 112 *L. monocytogenes* EGD-e sRNAs, 52 (46%) were found to be variable in *Listeria* genomes, i.e., absent in at least one *Listeria* genome, and 60 (54%) were found to be present in all *Listeria* genomes (Additional file 3: Table S3) the later constitute the core *Listeria*-sRNAs. Six of these core sRNAs (rli102, rli119, rli120, rli19-ssrA, rli69 and the rli2-LhrC-2_rli4-LhrC-4_rli7-LhrC-5_rli3-LhrC-3_rli1-LhrC-1 sRNA locus) were kept in the core set despite that their presence could not be confirmed in one or two genomes due to unsequenced regions. Among the core sRNAs, 79% of their occurrences were located in syntenic regions (meaning that both 5′ and 3′ adjacent genes were also found conserved). We then focused on the 52 variable sRNA loci to decipher their evolutionary history in *Listeria* genomes.

Small-RNA presence-absence patterns along the *Listeria* phylogenetic reference tree are shown in Fig. 3. Most sRNAs are present in nearly all genomes, except mostly

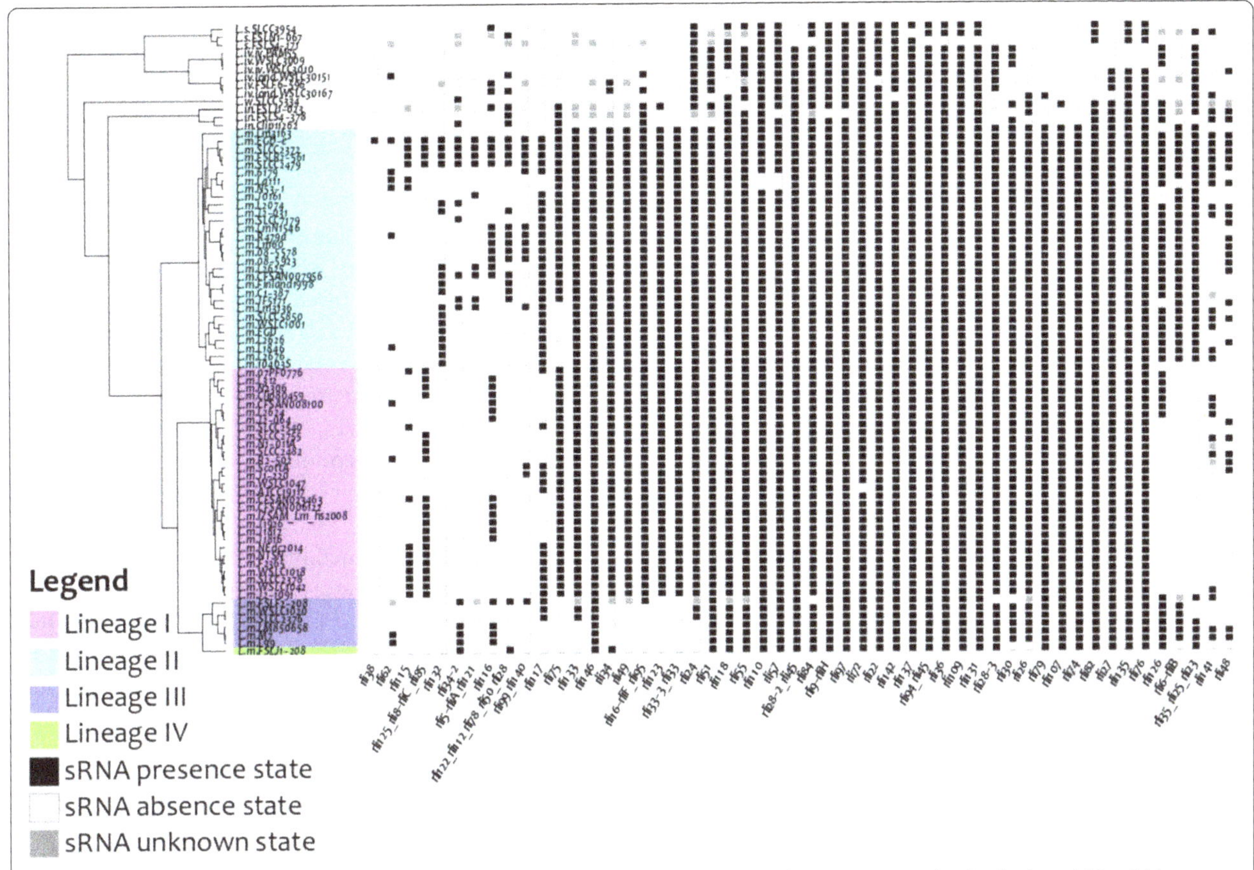

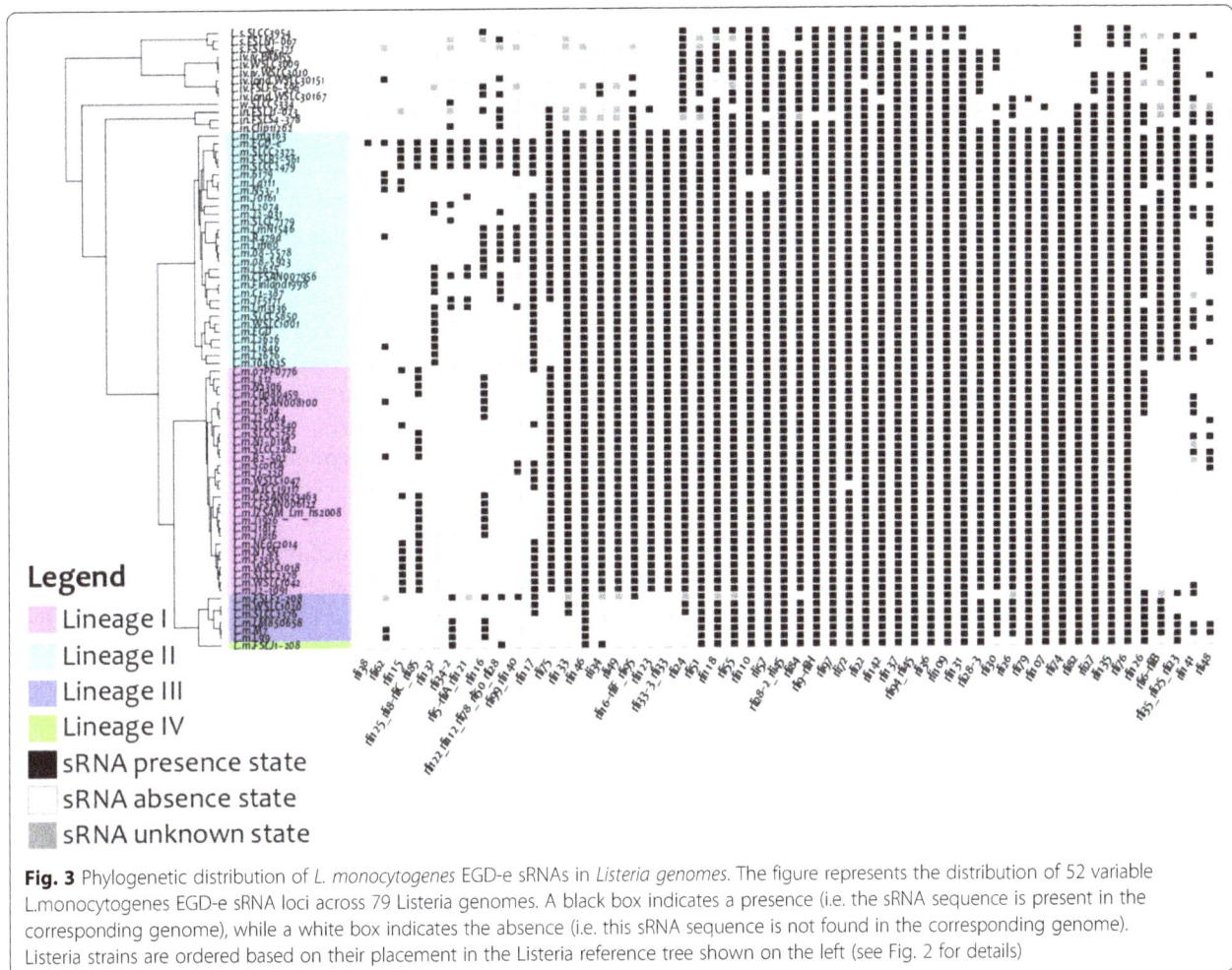

Fig. 3 Phylogenetic distribution of *L. monocytogenes* EGD-e sRNAs in *Listeria genomes*. The figure represents the distribution of 52 variable L.monocytogenes EGD-e sRNA loci across 79 Listeria genomes. A black box indicates a presence (i.e. the sRNA sequence is present in the corresponding genome), while a white box indicates the absence (i.e. this sRNA sequence is not found in the corresponding genome). Listeria strains are ordered based on their placement in the Listeria reference tree shown on the left (see Fig. 2 for details)

non-*L. monocytogenes* species. Small-RNA presence-absence patterns (Fig. 3) also suggest a link between sRNA content and previously defined *Listeria* lineages. For example most of sRNAs present in genomes of lineage I are also present in genomes of lineages II, but not systematically in lineage III and IV. Two sRNAs are found only in lineages I and II of *Listeria monocytogenes* (i.e., *see rli49* and *rli33-3_rli33*). Two other sRNAs are systematically absent of lineage I, while they are present in almost all the other *Listeria* genomes (e.g., *rli6-rliB*, *rli23_rli25_rli35*). *Rli74* is specifically present in all four *L. monocytogenes* lineages and absent in other *Listeria* species. Additionally, several sRNAs exhibit sparse presence-absence patterns probably related to complex evolutionary histories.

Listeria sRNA evolution and coevolution profiles

To investigate sRNA evolutionary histories, we inferred ancestral presence-absence patterns for all 52 variable sRNAs and obtained 44 different phyletic profiles (see Additional file 3: Table S3). Only eight sRNAs shared identical phyletic profiles as following: (1) *rli109, rli131, rli36* and *rli94_rli45*: one loss in branch 154 (*L. seeligeri/str.* FSL S4–171), (2) *rli9-rliH* and *rli97*: one loss in branch 137 (*L. innocua/str.* FSL S4–378) and (3) *rli135* and *rli76*: one loss in branch 147 (common ancestor of three *L. ivanovii* strains). This indicates that most profiles and evolutionary histories are specific to an sRNA, even if some evolutionary events are shared by several sRNAs. On the basis of the reconstructed sRNA gains and losses on the reference phylogenetic tree, we found only three sRNAs with monophyletic patterns (present in the most-recent common ancestor and all its descendants) (*rli146, rli38* and *rli74*) and two sRNAs with polyphyletic patterns (present in some genomes but not in their most-recent common ancestor) (*rli62* and *rli99_rli140*). All the other 47 (90%) sRNAs exhibit more or less complex paraphyletic patterns (present in the most-recent common ancestor and some of its descendants). Additionally, several sRNAs have undergone either a large number of gains (e.g., *rli116*: 10 gains, *rli115*: 5 gains) or a large number of losses (e.g., *rli122_rli112_rli78_rli94_rli50_rli28* locus: 13 losses; *rli141*: 10 losses; *rli26, rli48* and *rli117*: 8 losses) or both (e.g., *rli48*: 8 gains and 8 losses), suggesting that some sRNAs are subject to frequent reshuffle, even at short evolutionary scales.

The three monophyletic profiles indicate a scenario of gene appearance and descent. For instance, *rli146* and *rli74* exhibit the same monophyletic profile, i.e., one acquisition at branch 3 (ancestor of *L. monocytogenes* strains). It was inferred that *rli38* was acquired at branch 132 in *L. monocytogenes* EGD-e. Two different and complex polyphyletic patterns observed for *Rli62* and *rli99_rli140* suggest potential Horizontal Transfer events. All other sRNAs exhibit paraphyletic patterns, suggesting they underwent one or several loss events in the *Listeria* reference tree. Thirty-five out of 47 sRNAs (74%) with paraphyletic patterns are inferred to be present at the tree root, suggesting that the majority of sRNAs were present early, in *Listeria* evolution.

The *Listeria* sRNA-coding gene coevolution network

We used Pagel's model statistical framework [39] (see Methods) and both observed and ancestral presence/absence states to identify significant coevolution relationships between sRNAs and 5'UTRs/CDS regions along the *Listeria* reference tree. We obtained 23 putative sRNAs showing significant coevolutionary relationships with 23 5'UTRs and 39 CDS of 52 coding genes (see complete list in Additional file 4: Table S4).

All results of sRNAs, 5'UTRs and CDS phyletic patterns, coevolution analyses and the resulting coevolution network were made available on a dedicated web server that provides several facilities to browse the results: http://genoweb.toulouse.inra.fr/Listeria_sRNA. The web application was developed with the *Shiny* technology [41] and allows interactive visualization of individual phyletic patterns (i.e., observed and inferred ancestral presence/absence patterns) for all sRNAs and their coevolution partners along the *Listeria* reference tree (see an example in Fig. 4).

The *Listeria* sRNA-coding gene coevolution hub

The inferred sRNA-coding genes coevolving network (see Fig. 4b) reveals interesting features. We observed a hub of 12 sRNAs (*rli107, rli117, rli123, rli133, rli146, rli26, rli30, rli33-3_rli33, rli34, rli49, rli74* and *rli79*) that are connected through common coevolution partners. This cluster includes mainly distant (i.e. distance >40 kb) 5'UTRs and CDS coevolving partners, with the exception of partners of only two sRNAs: *rli30* paired to CDS partners *lmo0501* to *lmo0508,* and *rli74* with its CDS partner *mpl* (*lmo0203).* This cluster includes many genes encoding cell wall proteins, proteins involved in secondary metabolism and virulence factors (see next section and *rli133* case study for details). The other 11 sRNAs are all included in 11 individual clusters that contain either 5'UTR coevolving regions (*rli132, rli116*) exclusively or CDS coevolving regions (*rli28–3, rli99_rli140, rli141*) exclusively or a mix of both 5'UTRs and CDS regions (*rli75, rli5-rliA_rli121, rli34–2, rli115, rli125_rli8-rliC_rli85, rli48*). Interestingly, nine out of 11 of these individual clusters include evolving partners that are close on the genome (< 8 kb). Two individual coevolving groups [*rli28–3/lmo0035*] and [*rli5-rliA_rli121/lmo2309-lmo2407*] were found with distant coevolution partners (distance >800 kb for both clusters). To summarize, our results reveal an sRNA hub including mainly distantly located coevolving 5'UTRs and CDS regions, some of them exhibiting functions related to *Listeria* pathogenicity. Most of the remaining coevolution clusters include pairs of sRNA and 5'UTRs/CDS that exhibit very close genomic positions, i.e. a distance between the start of their sRNA and the start of their 5'UTR/CDS under 8 kb.

We investigated the functional classes of genes coevolving with *Listeria* regulatory sRNAs by using annotations from the Clusters of Orthologous Groups (COGs) database [42] retrieved from the Listeriomics website [36]. The distribution of coding genes in COG categories reveals a significant functional enrichment of coding genes associated with cell wall or membrane biogenesis (see Table 1, Fisher exact test [43], p-value = 0.0131). Interestingly, among coding genes coevolving with *Listeria* sRNAs, we found seven internalin genes (out of an expected 26 in Listeria [36]), two coding genes of the *Listeria* Pathogenicity Island LIPI-1 *mpl* (lmo0203) and *orfX* (lmo0206) [44, 45], one component of the flagellar biosynthesis pathway, eight genes involved in secondary metabolism and bacteriophage genes (see Additional file 4: Table S4 for details).

rli133, an sRNA coevolving with genes known to be involved in pathogenicity

The detailed analysis of *rli133* phyletic pattern (Fig. 5a) reveals an early acquisition event in *L. monocytogenes* common ancestor. *Rli133* cannot be found in other *Listeria* species, indicating that this sRNA is specific to *L. monocytogenes* species. Nevertheless, *rli133* is lost in four strains of lineage III (*L. monocytogenes FSLF2–208, L. monocytogenes LM850658, L. monocytogenes M7, L. monocytogenes L99*) and in the *L. monocytogenes FSLJ1–208* strain of lineage IV. In these five strains, the corresponding intergenic region is missing due to the insertion of two genes. These genes appear to be specific of these five strains and do not have any homolog in public databases (see Additional file 5: Figure S1). In genomes where *rli133* is present, the corresponding sequence is well-conserved and includes few mutation events, i.e., six transitions, two transversions and two indels events corresponding to 12 variables sites out of 126 (9.6%) in the *rli133* alignment (see Additional file 6:

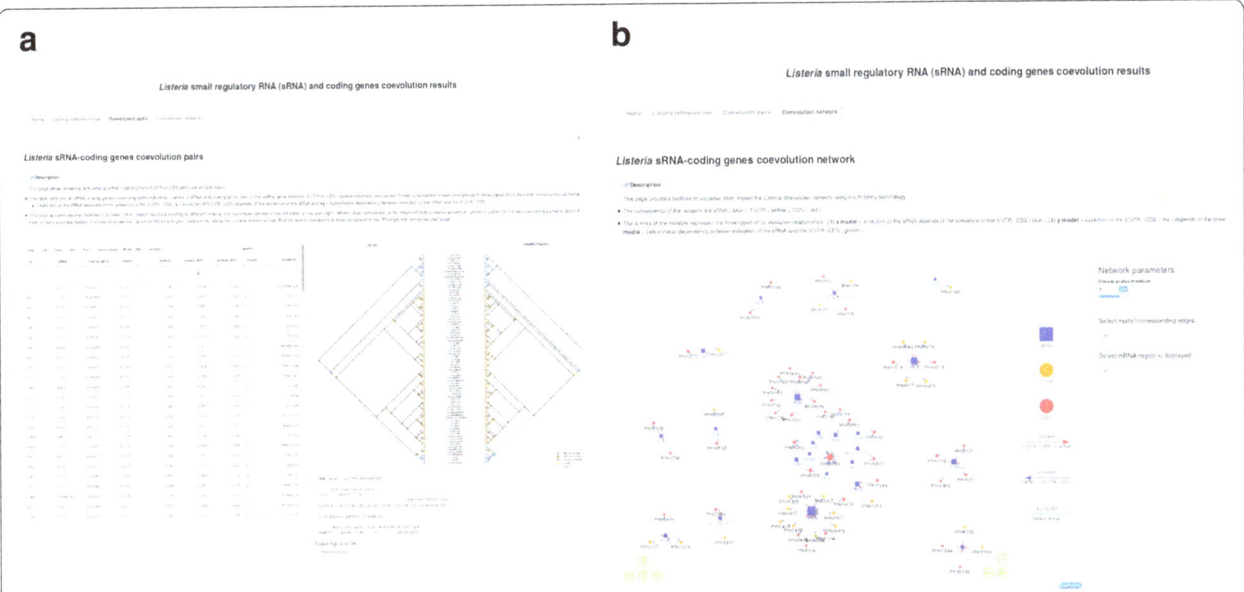

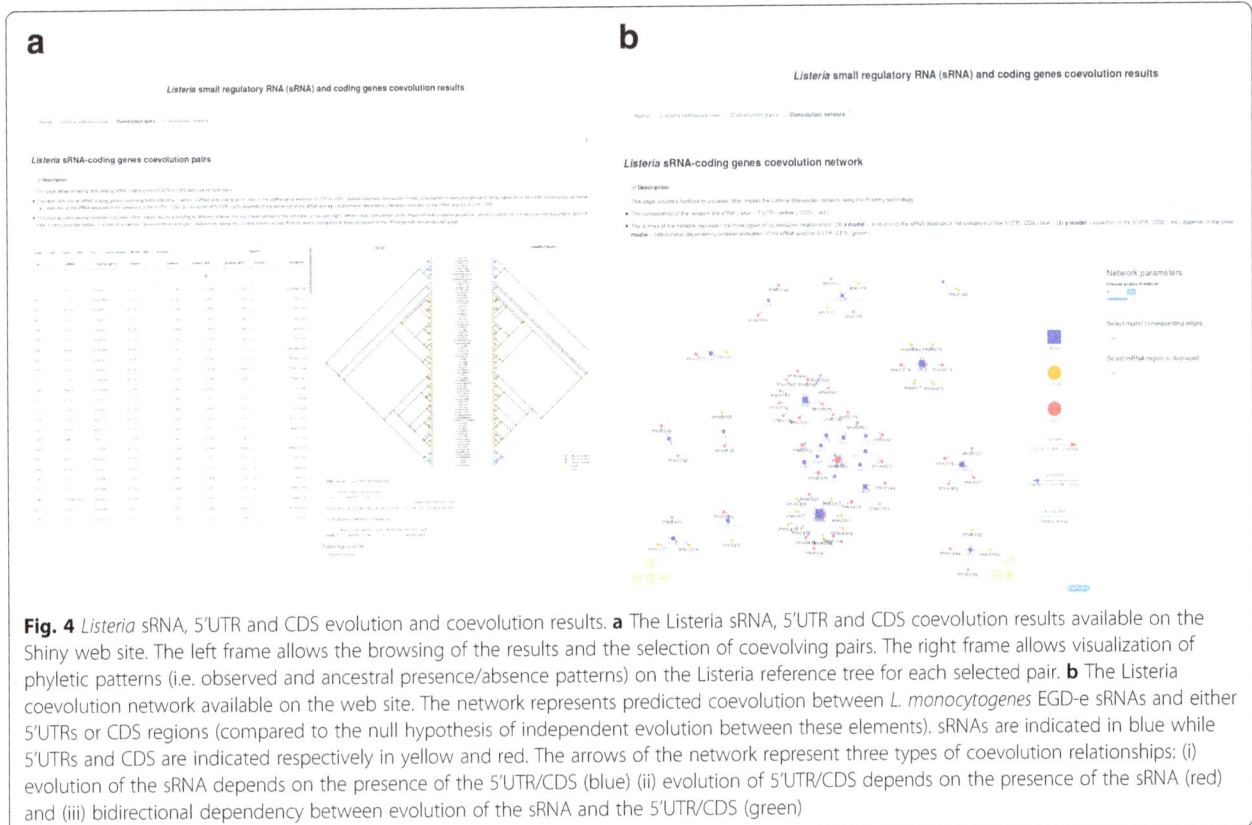

Fig. 4 *Listeria* sRNA, 5′UTR and CDS evolution and coevolution results. **a** The Listeria sRNA, 5′UTR and CDS coevolution results available on the Shiny web site. The left frame allows the browsing of the results and the selection of coevolving pairs. The right frame allows visualization of phyletic patterns (i.e. observed and ancestral presence/absence patterns) on the Listeria reference tree for each selected pair. **b** The Listeria coevolution network available on the web site. The network represents predicted coevolution between *L. monocytogenes* EGD-e sRNAs and either 5′UTRs or CDS regions (compared to the null hypothesis of independent evolution between these elements). sRNAs are indicated in blue while 5′UTRs and CDS are indicated respectively in yellow and red. The arrows of the network represent three types of coevolution relationships: (i) evolution of the sRNA depends on the presence of the 5′UTR/CDS (blue) (ii) evolution of 5′UTR/CDS depends on the presence of the sRNA (red) and (iii) bidirectional dependency between evolution of the sRNA and the 5′UTR/CDS (green)

Figure S2). Considering these mutated sites, *rli133* homologous sequences can be easily separated in two clusters that correspond to *Listeria* lineage I and II. *Rli133* presents coevolutionary relationships with 12 coding genes, eight 5′ UTRs and seven CDS regions. Coevolving gene partners include three internalin genes: *inlI* (*lmo0333*), *inlE* (*lmo0264*) and *inlP* (*lmo2470*). Internalins are important virulence factors [46, 47]. *InlE* may contribute to host tissue colonization [46, 48] and *InlP* has recently been shown to promote placental infection [47]. The role of *inlI* in pathogenicity remains to be determined [49]. Interestingly, we found three other sRNAs potentially coevolving with internalin genes: *rli117* (*lmo0549*, *lmo0263* and *lmo2470*), *rli30* (*lmo2445*) and *rli132* (*lmo2017*). Other genes were found to coevolve with *rli133*, e.g., *lntA*, *lmo0206* and *sepA*. The virulence factor *lntA* (*lmo0438*) targets the chromatin repressor *BAHD1* to activate interferon-stimulated genes in the host cell nucleus [50]. Expression of *LntA* seems to be tightly controlled to subvert immune responses and prevent antibacterial responses [50]. The *orfX* (*lmo0206)* gene is located within the *L. monocytogenes* pathogenic island 1 (LIPI-1), which includes genes required for *Listeria* intracellular lifestyle such as *hly, plcA, plcB and actA* [33] and contributes to bacterial survival in macrophages. *SepA* (*lmo2157*) encodes a protein involved in septum formation and play a role in stress response [51]. To summarize, six out of the twelve coevolution partners of *rli133* exhibit a documented function linked to either pathogenicity, interaction with host cells or stress response [36, 46, 47, 51]. Moreover, *rli133* sRNA was found to be expressed in several transcriptomes during infection, especially in blood and intestine [5, 22, 36]. Coevolution between sRNAs and coding genes may be resulting from the existence of direct or indirect functional links. Direct functional links can be explained by physical interaction through base-

Table 1 Functional enrichment of sRNAs and coding genes coevolution groups

Functional category	*P* value
Amino acid transport and metabolism	0.9055
Carbohydrate transport and metabolism	0.2202
Cell wall/membrane biogenesis	0.0131*
Energy production and conversion	0.8563
Replication, recombination and repair	0.8610
Secondary metabolites biosynthesis, transport and catabolism	0.5206
Signal transduction mechanisms	0.3669
Transcription	0.2648

This table contains p-values obtained with Fisher tests to measure a potential enrichment of a COG functional category in coding genes found to co-evolve with sRNAs (Additional file 4: Table S4). The * indicates a significant (under 0.05) p-value for genes of the category Cell Wall/membrane biogenesis

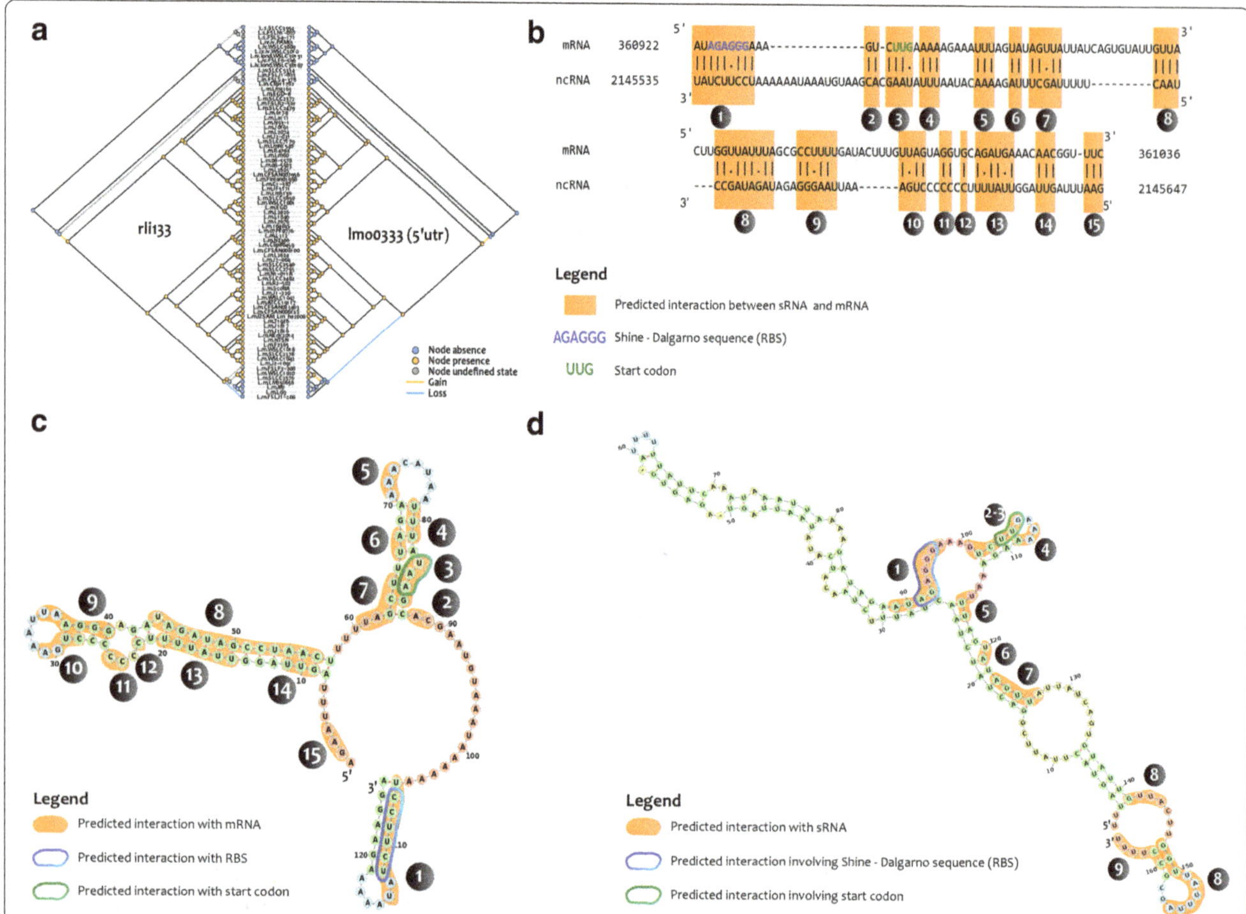

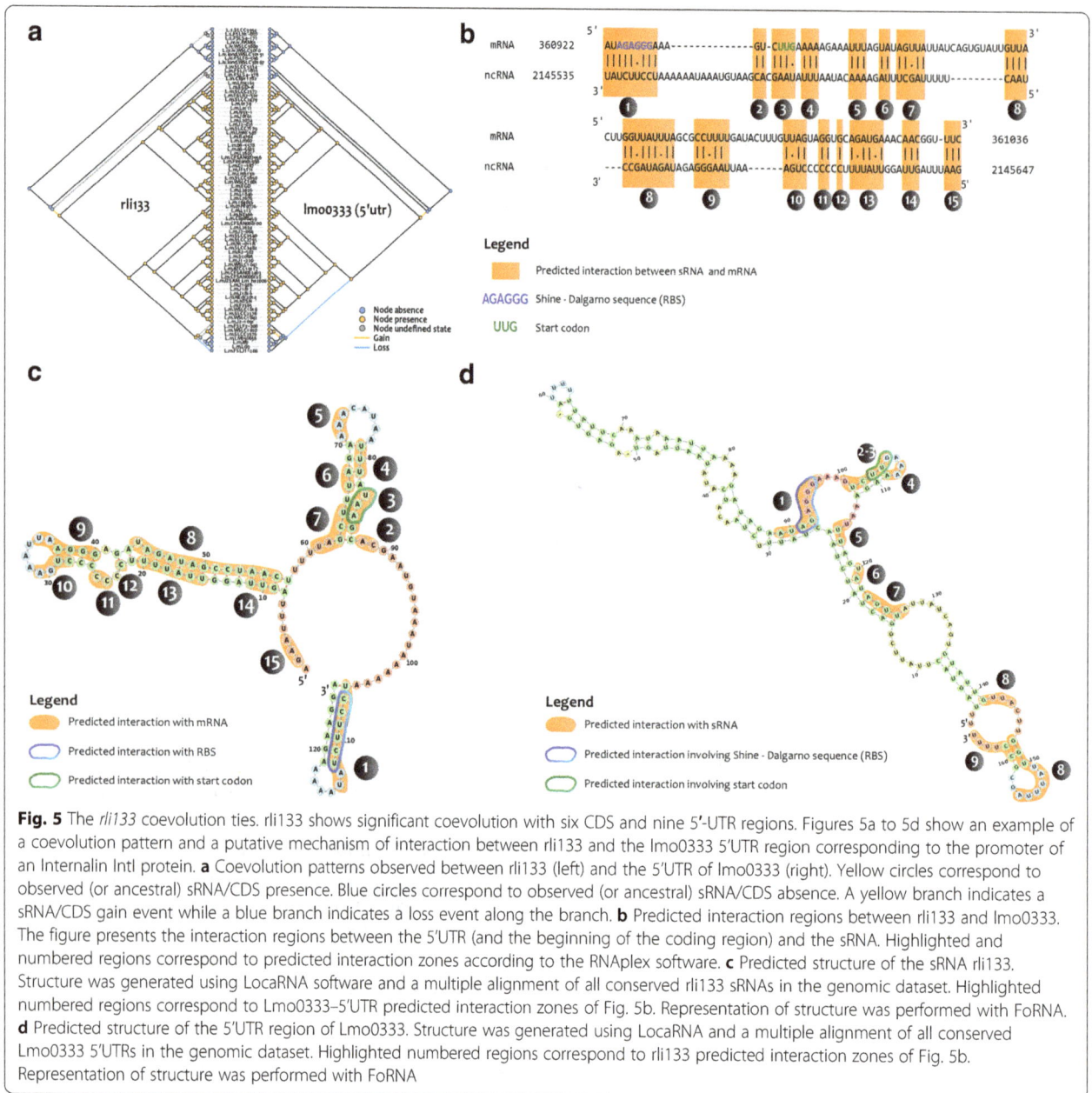

Fig. 5 The *rli133* coevolution ties. rli133 shows significant coevolution with six CDS and nine 5'-UTR regions. Figures 5a to 5d show an example of a coevolution pattern and a putative mechanism of interaction between rli133 and the lmo0333 5'UTR region corresponding to the promoter of an Internalin IntI protein. **a** Coevolution patterns observed between rli133 (left) and the 5'UTR of lmo0333 (right). Yellow circles correspond to observed (or ancestral) sRNA/CDS presence. Blue circles correspond to observed (or ancestral) sRNA/CDS absence. A yellow branch indicates a sRNA/CDS gain event while a blue branch indicates a loss event along the branch. **b** Predicted interaction regions between rli133 and lmo0333. The figure presents the interaction regions between the 5'UTR (and the beginning of the coding region) and the sRNA. Highlighted and numbered regions correspond to predicted interaction zones according to the RNAplex software. **c** Predicted structure of the sRNA rli133. Structure was generated using LocaRNA software and a multiple alignment of all conserved rli133 sRNAs in the genomic dataset. Highlighted numbered regions correspond to Lmo0333-5'UTR predicted interaction zones of Fig. 5b. Representation of structure was performed with FoRNA. **d** Predicted structure of the 5'UTR region of Lmo0333. Structure was generated using LocaRNA and a multiple alignment of all conserved Lmo0333 5'UTRs in the genomic dataset. Highlighted numbered regions correspond to rli133 predicted interaction zones of Fig. 5b. Representation of structure was performed with FoRNA

pairing at specific regions of a sRNA with their target mRNA. To identify possible physical interaction between *rli133* and its coevolution partners, we used several methods to predict structure of both the sRNAs and the 5'UTRs/CDS interacting regions and look for putative interacting zones (see Methods).

We found regions possibly interacting with *rli133* for all the 12 genes coevolving with *rli133*. Nine of these genes were identified to present interacting regions compatible with a negative regulation mechanism: *inlI* (*lmo0333*), *inlE* (*lmo0264*), *inlP* (*lmo2470*), *lntA* (*lmo0438*), *orfX* (*lmo0206*), *lmo0082, lmo0334, lmo0550 and lmo2107.* For the three remaining coevolving genes (*lmo0419, lmo0017* and *lmo2157*) we did not identify a consistent interacting region. As illustrated in Fig. 5b, fifteen interacting regions were predicted between *lmo0333*–5'UTR and *rli133.* Two interacting regions were overlapping in the proximity of the Shine-Dalgarno sequence and the initiation codon (see regions 1 and 3 in Fig. 5b, c and d), which are crucial sites for ribosome recruitment during the initiation of the translation. On the mRNAs, these sequences are mainly found to be accessible in loops or pseudo-loops (Fig. 5d), suggesting that they are constitutively available for translation. Interaction with complementary regions on sRNAs potentially makes them unavailable for ribosome binding during translation initiation, suggesting a potential inhibitory action of *rli133* at posttranscriptional level on these genes. The existence of direct interaction between *rli133* and coevolving genes partners could explain their coevolution.

To conclude, interacting regions corresponding to putative translation inhibition regions targeted by rli133 were identified for nine coevolving genes including *inlE* [46, 48], *inlI* [49], *inlP* [47], *lntA* [50] and *orfX* which were already found to be involved in host-interaction in previous studies.

Discussion

We built a robust workflow that provided new insights on *Listeria* sRNA evolution and coevolution patterns. First, the screening of sRNA presence-absence patterns suggests that 60 out of 112 *L. monocytogenes* EGD-e sRNAs (53%) shape the *Listeria* sRNA core set, in the sense that they were found to be present and conserved in all *Listeria* genomes. These 60 sRNAs were hence inferred present in the common ancestor of *Listeria*, suggesting that they were present early during the evolutionary history of the genus. This is a lower proportion compared to the 60 out of 83 *E. coli* K12 sRNAs (72%) that were found present and conserved in a dataset of 27 complete genomes of *E. coli-Shigel*la [29]. However, this is consistent with the higher number of genomes and the wider evolutionary scale (genus) used in our analysis compared to the species level of the *E.coli-Shigella* study.

The 52 remaining *L. monocytogenes* EGD-e sRNA loci constitute the variable sRNA set that is part of the *Listeria* accessory genome. This number is higher than the 43 accessory *Listeria* sRNAs previously identified by Kuenne et al. [38] in a smaller dataset of 11 genomes restricted to *L. monocytogenes* species. We found a higher proportion of them exhibiting complex paraphyletic distribution compared to the *E. coli/Shigella* study: 47 variable sRNAs out of 52 were shown to have paraphyletic pattern (90%) compared to 25 out of 32 (78%) in the 27 genomes of *E. coli-Shigella* [29]. Only three and two sRNAs have monophyletic and polyphyletic patterns, respectively. This indicates complex and various evolutionary histories underlying diverse origins and a potentially wide panel of sRNA acquisition and loss mechanisms in *Listeria*.

Detection of coevolution and analysis of the *Listeria* sRNA-coding genes coevolving network highlighted many interesting features.

We revealed an evolutionary link between sRNAs and coding genes related to pathogenicity and interaction with the host cell that suggests a key role for these sRNAs to shape *Listeria* virulence and adaptation. More precisely, we identified a hub of 12 sRNAs (*rli26, rli30, rli33-3_rli33 locus, rli34, rli49, rli74, rli79, rli107, rli117, rli123, rli133* and *rli146*) coevolving with many genes encoding cell wall proteins, especially internalins, that are known to be involved in host cell interaction [52], proteins involved in secondary metabolism, stress response and virulence factors. We detected a significant coevolution pattern of four sRNAs *(rli117, rli30, rli132* and *rli133*) and seven internalin genes (*lmo0549, lmo0263, lmo2470, lmo2445, lmo2027, lmo0333 and lmo0264*), indicating a probable key functional role of these sRNAs on these genes, possibly regulatory. To our knowledge, the relationship between small regulatory RNAs and internalin evolution was never observed before and opens several new perspectives concerning the possible impact of sRNAs in *Listeria* evolution and virulence. These results are consistent with previous observations that several internalin genes present long 5'UTRs that may also be post-transcriptionally regulated and that *Listeria* controls many of its virulence genes by a mechanism that involves 5'UTRs [23].

Interestingly, previous studies performed in *E. coli* or *S. enterica* have shown that sRNAs are often found to control the expression of cell wall proteins, particularly in outer membrane [53] or lipopolysaccharide layer synthesis [54]. This is consistent with our result revealing that the 'cell wall or membrane biogenesis' functional category is significantly overrepresented in *Listeria* sRNAs coevolving genes. Namely, we found seven internalin genes and two coding genes of the *Listeria* Pathogenicity Island LIPI-1 (*mpl, orfX*) in the *Listeria* sRNA coevolution partners. These results suggest a possible key regulatory role of some *Listeria* sRNAs on genes involved in host-bacteria interaction and pathogenicity.

We focused on *rli133*, a *L. monocytogenes*-specific sRNA, and identified 6 out of 12 *rli133* coevolution partners exhibiting a function linked to either pathogenicity, interaction with the host cells or stress response. Interacting regions compatible with mechanisms of mRNA translation inhibition were predicted for *rli133* and nine coevolving genes, including *inlE* [46, 48], *inlI* [49], *inlP* [47], *lntA* [50] and *orfX*. These results suggest a possible direct negative regulatory role of *rli133*, which potentially impairs the translation process of some of its coevolving partners. The presence of compatible interacting regions is not a feature specific to genes coevolving with *rli133*, but taking together the observations of coevolution pattern and the presence of a consistent interacting zone argue in favor of a functional link. Moreover, we looked for the presence of the nine 5'UTR-interacting zones of the genes co-evolving with rli133 in 5'UTRs and CDS that do not coevolve with rli133 and found only two similar regions for the inlP (lmo2470) interacting zone: one located in another internalin 5'UTR region (inlP/lmo2027) and one located in the 5'UTR region of the lmo0974 gene that is involved in LPS synthesis and conserved in all the genomes of the dataset. This argues for quite a good specificity of the *rli133* predicted interacting zones. For coevolving partners in which no clear mechanism were highlighted, such as *sepA* (*lmo2157*) [51], a well-known stress response gene involved in septum formation, coevolution patterns may correspond to presence of direct interaction at post-translational level or indirect functional links involving intermediate genes. These results suggest that *rli133* could act as a negative regulator of genes involved in *Listeria* pathogenicity.

Interestingly, rli133 sRNA is missing in the *L. monocytogenes* M7 and *L. monocytogenes* L99 genomes of *L. monocytogenes* lineage III that also have a reduced internalin-coding genes content (respectively 17 and 18 internalins) [55–57]. This suggests a possible link between the presence of *rli133*, the internalin gene content and the regulation of pathogenicity. The situation may be more complicated in other genomes such as the pathogenic strain *L. monocytogenes* J1–208 (lineage IV) identified in goat and whose chromosome contains only 16 internalin-coding genes and no *rli133* sRNA. This strain includes a plasmid (pLMIV) which contains additional internalins that may be involved in another mechanism of regulation of pathogenicity [57]. This indicates that the presence of *rli133* is not an absolute hallmark for pathogenesis and that other, yet unannotated sRNAs may interact with internalin genes in pathogenic strains of lineage IV. Additional genomes and sRNA experimental datasets are needed in this clade to fully understand the role of sRNAs and internalin coding genes in *Listeria* pathogenicity.

The *Listeria* coevolution network also pointed out 11 sRNAs exhibiting correlated evolution, mostly with close 5'UTRs and CDS regions. Screening for distances between coevolving sRNAs and genes indeed revealed two trends concerning gene location: on one hand, genes close to the corresponding sRNA (putative *cis*-regulated genes closer than 8 kb), and on the other hand, genes found at distant locations (putative *trans*-regulated genes with distances higher than 40 kb) (see Additional file 7: Figure S3). One possibility is that some of the closer coevolving sRNAs may correspond to uncharacterized or unannotated 5'UTR regions.

The workflow proposed in this work is resourceful and, to our knowledge, does not have any equivalent in previous work. Our strategy proposes several methodological enhancements and additional analyses compared to the pioneer work of Skippington and Ragan [29]. For instance, our strategy was designed to deal with uncovered regions of draft genomes and paralogy problems (both for sRNAs and coding genes). Moreover, three key steps of our workflow, i.e. the reference tree construction, the inference of ancestral presence-absence states and the detection of coevolution between sRNAs and coding genes, rely on the statistical framework of continuous-time Markov models and maximum likelihood, improving on parsimony approaches that do not provide consistent branch length estimation and may lead to lower precision.

Another key advantage of our approach is its extensively generic character since it can be transposed to any type of organism, any type of functional data and, more generally, to any kind of qualitative trait. For example, the strategy developed may be used to look for a possible coevolution between regulatory or structural RNAs and any type of element or feature such as pathogeny islands, pseudogenes, CRISPRs, phages, insertions sequences, etc. The entire workflow is built on an open source frame that is flexible, optimized and implements parallel and distributed computation, while however remaining computationally demanding.

Several features may be proposed in the future to enhance the proposed strategy. First, as we currently only consider presence, absence and unknown states, it constitutes an oversimplification of the way functional elements are defined, also undermining paralogy for sRNA or coding genes. Consequently, an enhancement of our strategy would be to deal with the occurrence of sRNAs and coding genes for both evolution and coevolution analysis. Second, another useful extension of the current strategy would be to include the analysis of mutation patterns and coevolving sites also for the core sRNAs and coding genes present in all the genomes of the dataset as well. This could be performed by including an additional step in the workflow that relies on a previously published method like the CoMap software [58].

Conclusion

The analysis of the *Listeria* coevolving network sheds light on several sRNAs which might play a role in virulence regulation. Since our approach makes it possible to obtain a list of sRNAs present only in the virulent strains, this study paves the way for new biochemical and biological analyses aimed at identifying and deciphering new factors involved in virulence.

Methods

Listeria genome dataset

Seventy-nine complete public genomes of *Listeria* were obtained from GenBank (release 211). A full description of the 79 *Listeria* genomes is available in Additional file 1: Table S1). The dataset includes 70 complete and 9 draft genomes representing five different *Listeria* species (*L. monocytogenes, L. ivanovii, L. inoccua, L. welshimeri* and *L. seeligeri*). *L. monocytogenes* genomes were the most represented (73 genomes corresponding to 92% of our dataset).

Listeria monocytogenes EGD-e sRNA

A set of 304 experimentally validated sRNAs from *L. monocytogenes* EGD-e was extracted from the Listeriomics database [36]. We focused on the 154 sRNAs annotated as putative trans sRNAs which are known as important regulators of gene expression in bacteria acting on independently expressed targets. A group of 19 sRNAs were excluded because they recently have been found to include small ORFs [6]. Overlapping sRNAs and sRNAs harboring paralogs in the *L. monocytogenes* EGD-e genome were processed using the following procedure. sRNAs were aligned on the *L. monocytogenes* EGD-e genome sequence using BLASTN+ [59]. Overlapping hits were merged considering a minimal overlap length of 15 pb, independently of their orientation. Finally, 112 sRNA loci were used as input sequences in the following analyses, including 15 loci built from merged hits and 97 original sequences.

sRNA and coding gene coevolution strategy

The strategy we developed is implemented in a Snakemake workflow [60] that consists in four main steps (see Fig. 1).

Step 1: Phylogenetic reference tree.

PanOCT, version 3.23 [61], was used to build groups of orthologs from annotated genomes. PanOCT is able to deal with recently diverging paralogs by using neighborhood gene information (synteny). All the parameters were set to default values except for the length ratio to discard shorter protein fragments when a protein is split due to a frameshift or other mechanisms was set to 1.33 as recommended by the authors. Amino-acid sequences of ortholog families were then aligned using ProbCons, version 1.12 [62], and resulting alignments were post-processed using GBLOCKS, version 0.91b [63], using the following parameters: the minimum number of sequences for a conserved position was set to (n/2) + 1, the minimum number of sequences for a flank position to (n/2) + 1 (where n is the total number of sequences in the aligned dataset), the maximum number of contiguous non-conserved positions was set to 20, the minimum length for a block to 5, and gap positions were allowed [8].

The reference tree was built using the syntenic core gene families corresponding to the PanOCT clusters with a single unique ortholog in each genome of our dataset. The corresponding nucleic acid alignments were obtained from all these core families filtered amino-acid alignments and concatenated into a single superalignment to compute a maximum likelihood tree using FastTree2, version 2.1.9 [64]. The following parameters were used for FastTree2: the Generalized Time-Reversible model (GTR) was chosen, the likelihood was reported under the Gamma model using 20 categories of sites, the exhaustive search mode ("-slow" option) was selected to obtain a more accurate reconstruction, NNI and SPR heuristics were used to browse the tree space. Support analyses were performed using Shimodaira Hasegawa test (SH) and 1000 resampling steps of site likelihood.

Step 2: sRNA and coding gene presence-absence matrix.

Presence-absence patterns were inferred from BLAST analyses with different parameters for sRNAs and coding genes. *L. monocytogenes* EGD-e sRNAs were aligned on the genome dataset using BLASTN+, version 2.2.29 [59]. Resulting hits were filtered using two criteria: an e-value $<10^{-2}$ and a coverage related to the query sequence ≥70%. Only sRNAs meeting these two criteria were considered as present in the targeted genomes.

For coding genes, we analyzed separately 5'UTRs and CDS regions. *L. monocytogenes* EGD-e CDS were retrieved from GenBank annotations and aligned against all Listeria genomes using BLASTP+, version 2.2.29 [59]. Resulting hits were filtered using the following criteria: an e-value $<10^{-2}$, a coverage relative to the query sequence ≥70%, an identity rate ≥ 60% and a bitscore ≥50. Only CDS meeting these three criteria were considered as present in the targeted genomes.

L. monocytogenes EGD-e 5'UTRs were retrieved using the following procedure. When available, we used experimental data indicating 5'UTR positions [36] to extract the corresponding DNA sequence. When not available, 5'UTR positions were defined arbitrarily as the 100 nearest 5' nucleotides upstream from each *L. monocytogenes* EGD-e CDS start codon of intergenic region. Only 5'UTRs with a minimum size of 15 bp were kept. 5'UTR sequences were then used as queries for BLASTN+ [59] alignments against all genomes. Resulting BLASTN+ hits were filtered using the following criteria: an e-value $<10^{-2}$, and a minimal identity percentage and coverage adjusted to the 5'UTR sequence lengths as follows. For 5'UTRs with lengths from 15 to 20 bp, the minimum identity percentage was set to 90% and the minimum coverage percentage to 100%. For 20–50 bp long 5'UTRs, both identity percentage and coverage were set to a minimum of 80%. For 50–100 bp and >100 bp long 5'UTRs, the minimal identity percentage was set to 80% and the minimal coverage was set to 50% and 25%, respectively. 5'UTRs meeting these criteria in subject genomes were considered as present.

A binary vector (0/1) corresponding to the absence/presence profile in the whole genome dataset was finally generated using BLASTN+ results and filters defined above. Due to their lack of informative value, sRNAs, 5'UTRs and CDS found in all genomes were not taken into account in subsequent analyses.

To avoid absence mispredictions corresponding to unsequenced regions of draft genomes, absence events of non-coding elements (sRNAs and 5'UTRs) were systematically checked as follows: for queries without hit in a given genome, 5′ and 3′ adjacent genes of the query element were screened for putative homologs in the same genome. In case where homologs were found, the non-coding sequence between the two homolog genes was extracted and screened for stretches of Ns, i.e. assembly gaps. If such stretches were found, the state of the query element was considered as undetermined due to missing DNA region in the considered genome ('?' state assigned). If the region was present but not similar to the query sequence, the query element was consider to be absent ('0' state assigned).

Step 3: Ancestral presence-absence pattern reconstruction.

Presence/absence ancestral states were reconstructed using the recent "Hidden rates model" method proposed by Beaulieu et al. [65]. This method uses Hidden Markov Models (HMM) to reconstruct ancestral character states from observed states and a reference phylogenetic tree. It makes it possible to use different transition rate classes. We used the *'rayDISC'* function of the *'corHMM'* R package version 1.20 [17] and selected the *'ARD'* transition model, i.e. independent transition rates. Internal node states were inferred using maximum likelihood estimation and joint probabilities. The root state probabilities were inferred using the method of Fitzjohn and Maddison [66]. Initial transition rates were estimated using the results of PanOCT orthologs obtained in Step 1 and computed using the *'DiscML'* function of the *'DiscML'* R package, version 1.0.1 [67], and the *'ARD'* transition model (assuming independent transition events, in this case, gain and loss events, for each element). This step results in a matrix containing the binary presence/absence (0/1) pattern for each internal node of the reference tree and for the three analysed features (sRNAs, CDS, 5'UTRs). Finally, gain and loss events were determined as follows: if the feature was absent (present) in a given node but present (absent) in its ancestor, it was considered as lost (gained) along the corresponding branch linking both nodes.

Step 4: Detection of coevolution events.

Our strategy allows the detection of coevolution between a sRNA and a 5'UTR or CDS using a reference phylogenetic tree and both observed and ancestral presence-absence patterns. We used the *'corDISC'* function of the *'corHMM'* R package [65] to identify putative coevolutionary relationships. This function fits Pagel's models of independency and dependency [39] to identify dependent evolution between two binary characters (in this case, the presence or absence of sRNAs, CDS/5'UTRs) and related to a phylogenetic tree. The first model supports an independent relationship between both binary traits: sRNA and 5'UTR/CDS (the null hypothesis). The second kind of model (the alternative hypotheses) supports a dependent relationship between both traits (coevolution). The use of ancestral states along the phylogenetic reference tree makes it possible to evaluate the probable temporal ordering of changes between two x and y presence/absence patterns and to test hypotheses about cause and effect. For this, we used three kinds of dependency models: the x model, meaning that the evolution of the sRNA depends on the presence/absence state of the 5'UTR/CDS, the y model, meaning that the evolution of 5'UTR/CDS depends on the presence/absence state of the sRNA and the xy model, assuming bidirectional dependency between evolution of the sRNA and the 5'UTR/CDS element.

The *'corDISC'* function merges the two x, y traits in a single vector, fits them on a precomputed phylogenetic tree using a specified model and then returns the likelihood of the model. The likelihood of each model was computed and a Likelihood Ratio Test (LRT) was performed. The corresponding p-value was computed. All of the analyses were performed between each variable sRNA, each variable CDS and 5′ UTR. P-values were corrected for multiple testing using the Benjamini-Hochberg (BH) procedure [68]. Finally, we only retained coevolving pairs of sRNA loci and coding gene elements (5'UTR and CDS) with a minimum BH corrected p-value threshold of 0.01.

The coevolution network between sRNAs and coding genes was reconstructed using inferred significant dependency relationships between phyletic patterns of sRNAs and coding gene elements (5'UTR and CDS) of *L. monocytogenes* EGD-e. Graph representations were built using the *'igraph',* version 1.0.1 [69], *'visNetwork',* version 1.0.3 [70], and *'Shiny',* version 1.0.1 [41] R packages.

Gene targets functional enrichment

Functional enrichment test was performed using *L. monocytogenes* EGD-e gene COG ontologies [36] and computed using a Fisher's exact test [43] (*'fisher.test'* function from the *'stats'* R package, version 3.5.0 [71]), with a p-value threshold of 0.05.

Interacting regions prediction

Possible physical interactions between sRNAs and coding genes identified as coevolving partners using our method were predicted using several pieces of software: Ssearch (implementation from Wisconsin Package), version 6.1, IntaRNA, version 2.0.2, RIsearch, version 1.1, RNAcofold, version 2.3.4 and RNAplex, version 2.3.4

[72–76], which are all included in the sRNAtabac resource [77]. We used an extended region including 100 bp before and after the start codon to identify putative interactions between a sRNA region and the extended 5'UTR region of mRNAs (original regions were used for CDS). Only interactions containing a minimum of six successive interacting matches were selected and considered as valid.

Homologous sequences of rli133 previously identified (see Step 2 for details) in *L. monocytogenes* genomes were extracted. Homologous sequences of lmo0333 (inlI) 5'UTR (see Step 2 for details) were extended up to 60 nucleotides after the start codon. *Rli133* and *lmo0333*-extended 5'UTR sequences were processed using LocARNA software, version 1.8.9 [73]. LocARNA is a tool that allows simultaneous folding and alignment of input RNA sequences. LocARNA default alignment accuracy was increased using match probabilities and probabilistic consistency transformation. Additional parameters were used since it is recommended by the authors in the software documentation for aligning up to about 15 sequences of lengths up to a few hundred nucleotides. The weight of base pair match contribution was set to 400. An iterative refinement of the progressive alignment was performed using two iterations. The 2D structure representation of *rli133* and *lmo0333*- extended 5'UTR were computed with FoRNA, version 0.1 [78], using consensus structures of *rli133* and the *lmo0333*-extended 5'UTR associated with corresponding sequences of the reference strain *L. monocytogenes* EGD-e.

Additional files

Additional file 1: Table S1. List of the 79 genomes used in this study. The table includes the list of 79 genomes obtained from Listeriomics and retrieved from the NCBI database. Several fields have been abbreviated for easier reading: 'Se.': strain serotype, 'Li.': Listeria lineage and 'Co.': country where the strain was first isolated. (DOCX 98 kb)

Additional file 2: Table S2. List of the 112 sRNA loci used in this study.The table includes 97 sRNAs obtained from the Listeriomics database and 15 merged regulatory sRNA loci tagged with an * in the table and obtained from the procedure described in Methods. (XLSX 21 kb)

Additional file 3: Table S3. Ancestral presence/absence patterns of L.m. EGD-e regulatory sRNAs. For all 52 L.m. EGD-e variable sRNAs, the table includes the following information according to the Listeria reference tree (see Fig. 2): root presence - absence information (Root state column), tree branches labels where gain (Gains column) and loss events (Losses column) were inferred (labels correspond to branch identifiers indicated in the cladogram of Fig. 2). The undefined column corresponds to tree branch labels with undefined state due to missing data in the corresponding genomes (draft genomes). The pattern_type column corresponds to the three different types of phyletic profiles inferred: monophyletic, polyphyletic or paraphyletic profiles. (DOCX 128 kb)

Additional file 4: Table S4. Listeria sRNAs and coding genes coevolution groups. For each sRNA, the table includes the following informations on the corresponding co-evolving elements: the gene locus tag name ('Element' column), the type of element (CDS or 5'UTR, 'Type' column), the type of dependency model that highlighted the interaction: x = evolution of the sRNA depends on the state of the 5'UTR/CDS, y = evolution of 5'UTR/CDS depends on the state of the sRNA and xy = bidirectional dependency between evolution of the sRNA and the 5'UTR/CDS element (Model column), the distance between the sRNA and the element in nucleotides (Distance column) and the description of the gene/operon function according to Listeriomics database ('Description' column); 'id' = identical content. Coevolution groups that are included in the main network hub are highlighted in gray. (DOCX 142 kb)

Additional file 5: Figure S1. Rli133 genomic context conservation in Listeria. 5' and 3' homolog genes are represented using red arrows. GFXXXX names correspond to PanOCT ortholog clusters identifiers. Blue arrows correspond to two genes inserted in several strains of Listeria lineages III and IV. (PDF 212 kb)

Additional file 6: Figure S2. Multiple alignment of rli133. This figure represents the multiple alignment of rli133 sequences in strains where it is present. Red denotes a fully conserved position. The phylogenetic tree at the left corresponds to a Maximum Likelihood tree computed from the corresponding multiple alignment. (PDF 1413 kb)

Additional file 7: Figure S3. Genomic distance between coevolving sRNAs and CDS. Plain curves show the distance density between sRNAs and 5'UTRs (red) or CDS (blue) engaged in coevolution relationships, considering genome circularity. They are compared to distances between sRNAs and all 5'UTRs or CDS (all) respectively represented by red and blue dotted curves. (PDF 117 kb)

Abbreviations

5' UTR: 5' Untranslated Region of coding genes; CDS: Coding DNA Sequence; COG: Clusters of Orthologous Group; mRNA: messenger RNA; SH: Shimodaira Hasegawa; sRNA: Small regulatory RNA

Acknowledgments

We are grateful to the Genotoul bioinformatics platform, Toulouse Midi-Pyrénées, for providing assistance, computing and storage resources.

Funding

This work received financial support from the French National Research Agency (BacNet Investissement d'Avenir project, 10-BINF-02-01). The funders had no role in study design, data collection and analysis, decision to publish, or preparation of the manuscript.

Authors' contributions

HC and CG designed the study. FC, LM, CH, CG and HC conceived the workflow and FC implemented it. FC, AP, MD, CG and HC performed the data analysis and FC, LM, CH, AP, AM, MD, OD, CG, PC and HC discussed and interpreted the results. PC coordinated the BacNet project. FC, LM, MD, OD, CB, CG, and HC wrote the manuscript. All authors read and approved the final manuscript.

Competing interests

The authors declare that they have no competing interests.

Author details

[1]Université de Toulouse, INRA, UR 875 Unité Mathématiques et Informatique Appliquées de Toulouse, Auzeville, 31326 Castanet-Tolosan, France. [2]Département de Biologie Cellulaire et Infection, Institut Pasteur, Unité des

Interactions Bactéries-Cellules, F-75015 Paris, France. INSERM, U604,F-75015 Paris, France. [4]INRA, USC2020, F-75015 Paris, France. [5]Institut Pasteur – Bioinformatics and Biostatistics Hub – C3BI, USR 3756 IP CNRS, Paris, France. [6]Université Paris Diderot, Sorbonne Paris Cité, F-75013 Paris, France. [7]Present address: Public Health England, 61 Colindale Avenue, London NW9 5EQ, England.

References

1. Gottesman S, Storz G. Bacterial small RNA regulators: versatile roles and rapidly evolving variations. Cold Spring Harb Perspect Biol. 2011;3 doi:10.1101/cshperspect.a003798.
2. Modi SR, Camacho DM, Kohanski MA, Walker GC, Collins JJ. Functional characterization of bacterial sRNAs using a network biology approach. Proc Natl Acad Sci U S A. 2011;108:15522–7.
3. Mandin P, Guillier M. Expanding control in bacteria: interplay between small RNAs and transcriptional regulators to control gene expression. Curr Opin Microbiol. 2013;16:125–32.
4. Caldelari I, Chao Y, Romby P, Vogel J. RNA-mediated regulation in pathogenic bacteria. Cold Spring Harb Perspect Med. 2013;3:a010298.
5. Toledo-Arana A, Dussurget O, Nikitas G, Sesto N, Guet-Revillet H, Balestrino D, et al. The listeria transcriptional landscape from saprophytism to virulence. Nature. 2009;459:950–6.
6. Mellin JR, Cossart P. The non-coding RNA world of the bacterial pathogen listeria monocytogenes. RNA Biol. 2012;9:372–8.
7. Peer A, Margalit H. Accessibility and evolutionary conservation mark bacterial small-rna target-binding regions. J Bacteriol. 2011;193:1690–701.
8. Richter AS, Backofen R. Accessibility and conservation: general features of bacterial small RNA-mRNA interactions? RNA Biol. 2012;9:954–65.
9. Beisel CL, Updegrove TB, Janson BJ, Storz G. Multiple factors dictate target selection by Hfq-binding small RNAs. EMBO J. 2012;31:1961–74.
10. Wright PR, Richter AS, Papenfort K, Mann M, Vogel J, Hess WR, et al. Comparative genomics boosts target prediction for bacterial small RNAs. Proc Natl Acad Sci U S A. 2013;110:E3487–96.
11. Updegrove TB, Shabalina SA, Storz G. How do base-pairing small RNAs evolve? FEMS Microbiol Rev. 2015;39:379–91.
12. Raghavan R, Groisman EA, Ochman H. Genome-wide detection of novel regulatory RNAs in E. Coli. Genome Res. 2011;21:1487–97.
13. Kröger C, Dillon SC, Cameron ADS, Papenfort K, Sivasankaran SK, Hokamp K, et al. The transcriptional landscape and small RNAs of salmonella enterica serovar Typhimurium. Proc Natl Acad Sci U S A. 2012;109:E1277–86.
14. Irnov I, Sharma CM, Vogel J, Winkler WC. Identification of regulatory RNAs in Bacillus Subtilis. Nucleic Acids Res. 2010;38:6637–51.
15. Mars RAT, Nicolas P, Ciccolini M, Reilman E, Reder A, Schaffer M, et al. Small regulatory RNA-induced growth rate heterogeneity of Bacillus Subtilis. PLoS Genet. 2015;11:e1005046.
16. Mellin JR, Koutero M, Dar D, Nahori M-A, Sorek R, Cossart P. Riboswitches. Sequestration of a two-component response regulator by a riboswitch-regulated noncoding RNA. Science. 2014;345:940–3.
17. Mellin JR, Tiensuu T, Bécavin C, Gouin E, Johansson J, Cossart PA. Riboswitch-regulated antisense RNA in listeria monocytogenes. Proc Natl Acad Sci U S A. 2013;110:13132–7.
18. Christiansen JK, Nielsen JS, Ebersbach T, Valentin-Hansen P, Søgaard-Andersen L, Kallipolitis BH. Identification of small Hfq-binding RNAs in listeria monocytogenes. RNA. 2006;12:1383–96.
19. Mandin P, Repoila F, Vergassola M, Geissmann T, Cossart P. Identification of new noncoding RNAs in listeria monocytogenes and prediction of mRNA targets. Nucleic Acids Res. 2007;35:962–74.
20. Oliver HF, Orsi RH, Ponnala L, Keich U, Wang W, Sun Q, et al. Deep RNA sequencing of L. monocytogenes reveals overlapping and extensive stationary phase and sigma B-dependent transcriptomes, including multiple highly transcribed noncoding RNAs. BMC Genomics. 2009;10:641.
21. Mraheil MA, Billion A, Mohamed W, Mukherjee K, Kuenne C, Pischimarov J, et al. The intracellular sRNA transcriptome of listeria monocytogenes during growth in macrophages. Nucleic Acids Res. 2011;39:4235–48.
22. Wurtzel O, Sesto N, Mellin JR, Karunker I, Edelheit S, Bécavin C, et al. Comparative transcriptomics of pathogenic and non-pathogenic listeria species. Mol Syst Biol. 2012;8:583.
23. Loh E, Dussurget O, Gripenland J, Vaitkevicius K, Tiensuu T, Mandin P, et al. A trans-acting riboswitch controls expression of the virulence regulator PrfA in listeria monocytogenes. Cell. 2009;139:770–9.
24. Johansson J, Mandin P, Renzoni A, Chiaruttini C, Springer M, Cossart P, An RNA. Thermosensor controls expression of virulence genes in listeria monocytogenes. Cell. 2002;110:551–61.
25. Sesto N, Koutero M, Cossart P. Bacterial and cellular RNAs at work during listeria infection. Future Microbiol. 2014;9:1025–37.
26. Quereda JJ, Ortega AD, Pucciarelli MG, García-Del Portillo F. The listeria small RNA Rli27 regulates a Cell Wall protein inside eukaryotic cells by targeting a long 5'-UTR variant. PLoS Genet. 2014;10:e1004765.
27. Peng Y-L, Meng Q-L, Qiao J, Xie K, Chen C, Liu T-L, et al. The regulatory roles of ncRNA Rli60 in adaptability of listeria monocytogenes to environmental stress and biofilm formation. Curr Microbiol. 2016;73:77–83.
28. Toffano-Nioche C, Nguyen AN, Kuchly C, Ott A, Gautheret D, Bouloc P, et al. Transcriptomic profiling of the oyster pathogen Vibrio Splendidus opens a window on the evolutionary dynamics of the small RNA repertoire in the vibrio genus. RNA. 2012;18:2201–19.
29. Skippington E, Ragan MA. Evolutionary dynamics of small RNAs in 27 Escherichia Coli and Shigella genomes. Genome Biol Evol. 2012;4:330–45.
30. Peer A, Margalit H. Evolutionary patterns of Escherichia Coli small RNAs and their regulatory interactions. RNA. 2014;20:994–1003.
31. Orsi RH, Wiedmann M. Characteristics and distribution of listeria spp., including listeria species newly described since 2009. Appl Microbiol Biotechnol. 2016;100:5273–87.
32. Hamon M, Bierne H, Cossart P. Listeria monocytogenes: a multifaceted model. Nat Rev Microbiol. 2006;4:423–34.
33. Pizarro-Cerdá J, Cossart P. Subversion of cellular functions by listeria monocytogenes. J Pathol. 2006;208:215–23.
34. Cossart P. Illuminating the landscape of host-pathogen interactions with the bacterium listeria monocytogenes. Proc Natl Acad Sci U S A. 2011;108:19484–91.
35. Glaser P, Frangeul L, Buchrieser C, Rusniok C, Amend A, Baquero F, et al. Comparative genomics of listeria species. Science. 2001;294:849–52.
36. Bécavin C, Koutero M, Tchitchek N, Cerutti F, Lechat P, Maillet N, et al. Listeriomics: an interactive web platform for systems biology of listeria. mSystems. 2017;2 doi:10.1128/mSystems.00186-16.
37. Bécavin C, Bouchier C, Lechat P, Archambaud C, Creno S, Gouin E, et al. Comparison of widely used listeria monocytogenes strains EGD, 10403S, and EGD-e highlights genomic variations underlying differences in pathogenicity. MBio. 2014;5:e00969–14.
38. Kuenne C, Billion A, Mraheil MA, Strittmatter A, Daniel R, Goesmann A, et al. Reassessment of the listeria monocytogenes pan-genome reveals dynamic integration hotspots and mobile genetic elements as major components of the accessory genome. BMC Genomics. 2013;14:47.
39. Pagel M. Detecting correlated evolution on phylogenies: a general method for the comparative analysis of discrete characters. Proc R Soc B Biol Sci. 1994;255:37–45.
40. Orsi RH, den Bakker HC, Wiedmann M. Listeria monocytogenes lineages: genomics, evolution, ecology, and phenotypic characteristics. Int J Med Microbiol. 2011;301:79–96.
41. Beeley C. Web application with R using shiny. Packt Pub Limited; 2013.
42. Tatusov RL. The COG database: a tool for genome-scale analysis of protein functions and evolution. Nucleic Acids Res. 2000;28:33–6.
43. Fisher RA. The logic of inductive inference. J R Stat Soc. 1935;98:39.
44. Vázquez-Boland JA, Domínguez-Bernal G, González-Zorn B, Kreft J, Goebel W. Pathogenicity islands and virulence evolution in listeria. Microbes Infect. 2001;3:571–84.
45. Chatterjee SS, Hossain H, Otten S, Kuenne C, Kuchmina K, Machata S, et al. Intracellular gene expression profile of listeria monocytogenes. Infect Immun. 2006;74:1323–38.
46. Cabanes D, Dehoux P, Dussurget O, Frangeul L, Cossart P. Surface proteins and the pathogenic potential of listeria monocytogenes. Trends Microbiol. 2002;10:238–45.
47. Faralla C, Rizzuto GA, Lowe DE, Kim B, Cooke C, Shiow LR, et al. InlP, a new virulence factor with strong placental tropism. Infect Immun. 2016;84:3584–96.
48. Raffelsbauer D, Bubert A, Engelbrecht F, Scheinpflug J, Simm A, Hess J, et al. The gene cluster inlC2DE of listeria monocytogenes contains additional new internalin genes and is important for virulence in mice. Mol Gen Genet. 1998;260:144–58.

49. Sabet C, Lecuit M, Cabanes D, Cossart P, Bierne HLPXTG. Protein InlJ, a newly identified internalin involved in listeria monocytogenes virulence. Infect Immun. 2005;73:6912–22.
50. Lebreton A, Lakisic G, Job V, Fritsch L, Tham TN, Camejo A, et al. A bacterial protein targets the BAHD1 chromatin complex to stimulate type III interferon response. Science. 2011;331:1319–21.
51. Hain T, Hossain H, Chatterjee SS, Machata S, Volk U, Wagner S, et al. Temporal transcriptomic analysis of the listeria monocytogenes EGD-e σB regulon. BMC Microbiol. 2008;8:20.
52. Bierne H, Sabet C, Personnic N, Cossart P. Internalins: a complex family of leucine-rich repeat-containing proteins in listeria monocytogenes. Microbes Infect. 2007;9:1156–66.
53. Vogel J, Papenfort K. Small non-coding RNAs and the bacterial outer membrane. Curr Opin Microbiol. 2006;9:605–11.
54. Klein G, Raina S. Regulated control of the assembly and diversity of LPS by noncoding sRNAs. Biomed Res Int. 2015;2015:153561.
55. Deng X, Phillippy AM, Li Z, Salzberg SL, Zhang W. Probing the pan-genome of listeria monocytogenes: new insights into intraspecific niche expansion and genomic diversification. BMC Genomics. 2010;11:500.
56. Roberts A, Nightingale K, Jeffers G, Fortes E, Kongo JM, Wiedmann M. Genetic and phenotypic characterization of listeria monocytogenes lineage III. Microbiology. 2006;152:685–93.
57. den Bakker HC, Bowen BM, Rodriguez-Rivera LD, Wiedmann MFSL. J1-208, a virulent uncommon phylogenetic lineage IV listeria monocytogenes strain with a small chromosome size and a putative virulence plasmid carrying internalin-like genes. Appl Environ Microbiol. 2012;78:1876–89.
58. Dutheil J, Galtier N. Detecting groups of coevolving positions in a molecule: a clustering approach. BMC Evol Biol. 2007;7:242.
59. Camacho C, Coulouris G, Avagyan V, Ma N, Papadopoulos J, Bealer K, et al. BLAST+: architecture and applications. BMC Bioinformatics. 2009;10:421.
60. Köster J, Rahmann S. Snakemake–a scalable bioinformatics workflow engine. Bioinformatics. 2012;28:2520–2.
61. Fouts DE, Brinkac L, Beck E, Inman J, Sutton G. PanOCT: automated clustering of orthologs using conserved gene neighborhood for pan-genomic analysis of bacterial strains and closely related species. Nucleic Acids Res. 2012;40:e172.
62. Do CB, Mahabhashyam MSP, Brudno M, Batzoglou S. ProbCons: probabilistic consistency-based multiple sequence alignment. Genome Res. 2005;15:330–40.
63. Castresana J. Selection of conserved blocks from multiple alignments for their use in phylogenetic analysis. Mol Biol Evol. 2000;17:540–52.
64. Price MN, Dehal PS, Arkin AP. FastTree 2–approximately maximum-likelihood trees for large alignments. PLoS One. 2010;5:e9490.
65. Beaulieu JM, O'Meara BC, Donoghue MJ. Identifying hidden rate changes in the evolution of a binary morphological character: the evolution of plant habit in campanulid angiosperms. Syst Biol. 2013;62:725–37.
66. FitzJohn RG, Maddison WP, Otto SP. Estimating trait-dependent speciation and extinction rates from incompletely resolved phylogenies. Syst Biol. 2009;58:595–611.
67. Kim T, Hao W. DiscML: an R package for estimating evolutionary rates of discrete characters using maximum likelihood. BMC Bioinformatics. 2014;15:320.
68. Yekutieli D, Benjamini Y. under dependency. Ann Stat. 2001;29:1165–88.
69. igraph – Network analysis software [Internet]. [cited 7 Feb 2017]. Available: http://igraph.org/.
70. datastorm-open. datastorm-open/visNetwork. In: GitHub [Internet]. [cited 9 Feb 2017]. Available: https://github.com/datastorm-open/visNetwork.
71. R: The R Stats Package [Internet]. [cited 11 Apr 2017]. Available: https://stat.ethz.ch/R-manual/R-devel/library/stats/html/stats-package.html.
72. Wenzel A, Akbasli E, Gorodkin J. RIsearch: fast RNA-RNA interaction search using a simplified nearest-neighbor energy model. Bioinformatics. 2012;28: 2738–46.
73. Gruber AR, Lorenz R, Bernhart SH, Neuböck R, Hofacker IL. The Vienna RNA websuite. Nucleic Acids Res. 2008;36:W70–4.
74. Ropelewski AJ, Nicholas HB, Deerfield DW. Mathematically complete nucleotide and protein sequence searching using Ssearch. Curr Protoc Bioinformatics. 2004;
75. Wright PR, Georg J, Mann M, Sorescu DA, Richter AS, Lott S, et al. CopraRNA and IntaRNA: predicting small RNA targets, networks and interaction domains. Nucleic Acids Res. 2014;42:W119–23.
76. Womble DDGCG. The Wisconsin package of sequence analysis programs. Methods Mol Biol. 2000;132:3–22.
77. Toulouse APO-M-. I. sRNA-TaBac | Home [Internet]. [cited 11 Apr 2017]. Available: http://srnatabac.toulouse.inra.fr:8080/.
78. Kerpedjiev P, Hammer S, Hofacker IL. Forna (force-directed RNA): simple and effective online RNA secondary structure diagrams. Bioinformatics. 2015;31:3377–9.

2

Genomic analysis of endemic clones of toxigenic and non-toxigenic *Corynebacterium diphtheriae* in Belarus during and after the major epidemic in 1990s

Steffen Grosse-Kock[1], Valentina Kolodkina[2], Edward C. Schwalbe[1], Jochen Blom[3], Andreas Burkovski[4], Paul A. Hoskisson[5], Sylvain Brisse[6], Darren Smith[1], Iain C. Sutcliffe[1], Leonid Titov[2] and Vartul Sangal[1*]

Abstract

Background: Diphtheria remains a major public health concern with multiple recent outbreaks around the world. Moreover, invasive non-toxigenic strains have emerged globally causing severe infections. A diphtheria epidemic in the former Soviet Union in the 1990s resulted in ~5000 deaths. In this study, we analysed the genome sequences of a collection of 93 *C. diphtheriae* strains collected during and after this outbreak (1996 – 2014) in a former Soviet State, Belarus to understand the evolutionary dynamics and virulence capacities of these strains.

Results: *C. diphtheriae* strains from Belarus belong to ten sequence types (STs). Two major clones, non-toxigenic ST5 and toxigenic ST8, encompassed 76% of the isolates that are associated with sore throat and diphtheria in patients, respectively. Core genomic diversity is limited within outbreak-associated ST8 with relatively higher mutation rates (8.9×10^{-7} substitutions per strain per year) than ST5 (5.6×10^{-7} substitutions per strain per year) where most of the diversity was introduced by recombination. A variation in the virulence gene repertoire including the presence of *tox* gene is likely responsible for pathogenic differences between different strains. However, strains with similar virulence potential can cause disease in some individuals and remain asymptomatic in others. Eight synonymous single nucleotide polymorphisms were observed between the *tox* genes of the vaccine strain PW8 and other toxigenic strains of ST8, ST25, ST28, ST41 and non-toxigenic *tox* gene-bearing (NTTB) ST40 strains. A single nucleotide deletion at position 52 in the *tox* gene resulted in the frameshift in ST40 isolates, converting them into NTTB strains.

Conclusions: Non-toxigenic *C. diphtheriae* ST5 and toxigenic ST8 strains have been endemic in Belarus both during and after the epidemic in 1990s. A high vaccine coverage has effectively controlled diphtheria in Belarus; however, non-toxigenic strains continue to circulate in the population. Recombination is an important evolutionary force in shaping the genomic diversity in *C. diphtheriae*. However, the relative role of recombination and mutations in diversification varies between different clones.

Keywords: *Corynebacterium diphtheriae*, Toxigenic, Non-toxigenic, Epidemic, Endemic, Vaccine, Virulence, Diphtheria, Sore throat

* Correspondence: vartul.sangal@northumbria.ac.uk
[1]Faculty of Health and Life Sciences, Northumbria University, Newcastle upon Tyne, UK
Full list of author information is available at the end of the article

Background

Diphtheria is a toxin-mediated disease caused by toxigenic strains of *Corynebacterium diphtheriae* which is characterised by the presence of an inflammatory pseudomembrane in the upper respiratory tract, resulting in breathing difficulties with fatal outcomes [1]. Historically, *C. diphtheriae* isolates have been typed phenotypically into four biovars (belfanti, gravis, intermedius and mitis) although genetic approaches have questioned the basis of biovar separation [2]. Diphtheria toxin, which is the most prominent virulence factor of *C. diphtheriae*, inhibits protein synthesis by catalysing NAD^+-dependent ADP-ribosylation of elongation factor 2, thus inducing apoptosis, resulting in the cell death [3]. The *tox* gene is regulated by a metalloregulatory transcriptional regulator DtxR which induces the toxin production under low iron conditions [4]. The cell death caused by the toxin likely makes the host iron sources available to the pathogen [1]. The toxoid vaccine induces a strong IgG antibody response that neutralises the diphtheria toxin [5] and has approximately 97% efficacy [6]. However, diphtheria remains endemic to many countries [7] and multiple diphtheria outbreaks have been reported across the globe [8–11].

Non-toxigenic *C. diphtheriae* strains are also causing significant invasive infections such as endocarditis, septic arthritis and osteomyelitis [12–14]. These strains lack the *tox* gene, which is present on lysogenising corynephages in toxigenic strains [15]. In addition, non-toxigenic *tox* gene-bearing strains (NTTB) of *C. diphtheriae* are also circulating in the population [16]. The *tox* gene is a pseudogene in NTTB strains due to frameshift mutations but these strains may be able to genetically revert to active toxin production [16].

The major post-vaccine epidemic in the former Soviet Union in the 1990s caused >157,000 cases with approximately 5000 deaths [17]. Belarus, a former Soviet state, reported a significant shift in the distribution of *C. diphtheriae* ribotypes following the epidemic period (between 2000 and 2001) with an increase in the number of infections caused by non-toxigenic strains [18]. This potentially suggests a change in the evolutionary dynamics of *C. diphtheriae* strains. Therefore, to gain a deeper understanding of the population genetics, evolutionary dynamics and virulence capacities, we have sequenced the genomes of a collection of 93 representative toxigenic and non-toxigenic *C. diphtheriae* strains from Belarus isolated between 1996 and 2014 (Additional file 1: Table S1).

Results

Major endemic clones of toxigenic and non-toxigenic *C. diphtheriae* in Belarus

A total of 4382 *C. diphtheriae* isolates were collected in Belarus from 1996 to 2014 (Additional file 2: Table S2). Toxigenic strains accounted for approximately 47% of the total isolates in 1996, which fell to zero in 2011. Only non-toxigenic strains have been isolated since 2011 in Belarus. Two to nine isolates from each year were selected for genomic analyses with a proportional representation of toxigenic and non-toxigenic strains (Additional file 2: Table S2). In total, 93 *C. diphtheriae* isolates were selected including one isolate from 1979. These were isolated from all six provinces of Belarus (Brest, Gomel, Grodno, Minsk, Mogilev and Vitebsk) from asymptomatic carriers ($n = 22$) and patients that presented with diphtheria ($n = 26$) or sore throat ($n = 45$). As a part of the clinical diagnosis, these strains were assigned to biovars belfanti, gravis or mitis. The genomes of these strains were sequenced on an Illumina MiSeq instrument and the size of assemblies varied between 2.3-2.6 Mb. Further information on *C. diphtheriae* isolates and genomes assemblies is provided in Additional file 1 (Table S1).

For comparative analyses, genome sequences of two reference strains CCUG 2706A, a strain of rare biovar intermedius, and CCUG 5865, a distinct sequence type (ST)-106 isolate of biovar belfanti, and 22 previously published *C. diphtheriae* strains were also included (Additional file 1: Table S1). The core genome was calculated using EDGAR [19] that is consisted of 1267 genes. A maximum-likelihood (ML) tree from nucleotide sequence alignment of the core genome separated two lineages, one including 116 strains of all four biovars and the second lineage with a single biovar belfanti isolate, CCUG 5865 (Fig. 1). These results support the conclusion of a multilocus sequence typing (MLST) study showing two lineages within *C. diphtheriae* [20].

The majority (76%) of *C. diphtheriae* isolates from Belarus formed two groups within lineage 1, ST5 (37 isolates, 39.8%) and ST8 (34 isolates, 36.6%; Fig. 1). ST5 is a non-toxigenic clone while most isolates in ST8 are toxigenic. Toxigenic ST8 isolates were responsible for the epidemic in the former Soviet Union in the 1990s [17, 20] and this study reveals that these strains were also circulating after the epidemic period (Additional file 1: Table S1).

Two non-toxigenic isolates were ST32 that are known to cause severe pharyngitis and tonsillitis among patients in Europe [21, 22]. Other minor groups include toxigenic ST25 (4 isolates) and ST28-ST41 (4 isolates), non-toxigenic ST76 (5 isolates) and NTTB ST40 (5 isolates). Two strains, one toxigenic ST53 and one nontoxigenic ST123, are singletons.

Spatio-temporal distribution of *C. diphtheriae* clones

To investigate the reported shift in the major genotypes between 2000 and 2001 in Belarus [18], we analysed the temporal distribution of *C. diphtheriae* strains in Belarus

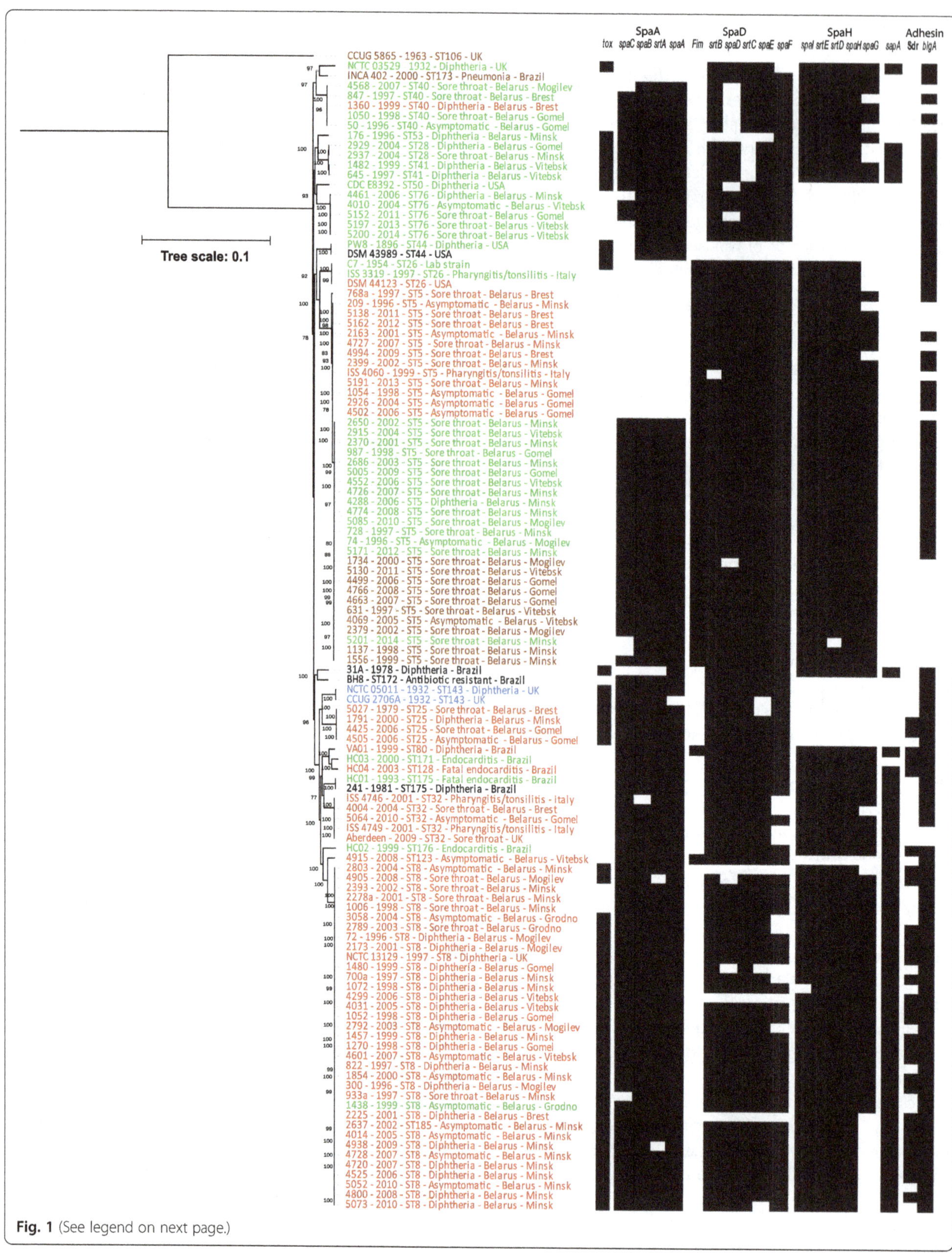

Fig. 1 (See legend on next page.)

(See figure on previous page.)
Fig. 1 A maximum-likelihood tree derived from concatenated nucleotide sequenced alignment of the core genome. The scale bar represents nucleotide substitutions per nucleotide site. The strain designations of isolates of biovar belfanti, gravis, intermedius and mitis are presented in brown, red, blue and green colour, respectively. The presence of virulence genes is mapped on the tree in black whereas a white box shows the absence of genes

predating 2001 (epidemic period) and since January 2001 (post-epidemic period; Fig. 2). ST5 and ST8 strains have been prevalent in most of the Belarussian provinces in both the epidemic and post-epidemic periods and we did not observe any shift in the distribution of strains in these STs (χ^2 test, $p > 0.05$; Additional file 2: Table S3) to correlate with the replacement of biovar gravis by mitis and change in the distribution of ribotypes [18]. Ribotyping is error prone as the resolution and reproducibility are dependent on multiple factors including the restriction enzymes and stringency of the hybridisation conditions [23]. We have previously shown that biovar designations are not necessarily reliable and are not supported by genomic diversity [24] which is strengthened

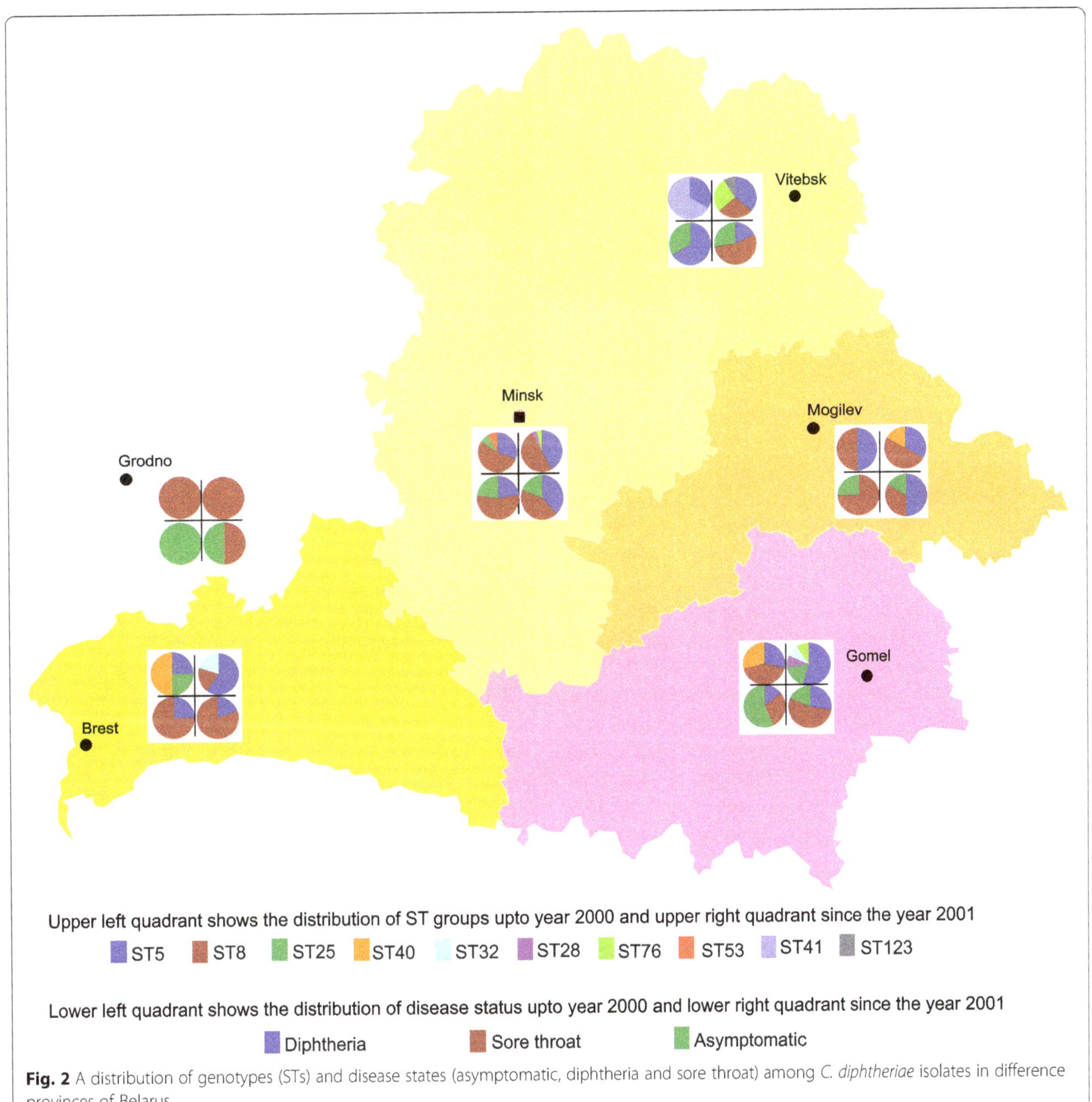

Fig. 2 A distribution of genotypes (STs) and disease states (asymptomatic, diphtheria and sore throat) among *C. diphtheriae* isolates in difference provinces of Belarus

by the fact that ST5 includes strains assigned to biovars belfanti, gravis and mitis (Fig. 1). Although biovars belfanti and mitis isolates formed a subgroup (ST5-B; Additional file 3: Figure S1) within ST5, the dataset does not indicate any replacement of gravis isolates by mitis as the strains belonging to these biovars were isolated both during and after the epidemic period. These findings further suggest that genetic approaches should be adopted over biotyping for studying *C. diphtheriae* epidemiology.

At the provincial level, all three isolates were ST8 in Grodno province in both the periods (Fig. 2). In some provinces, certain clones were observed either in the epidemic (ST8 in Gomel, ST25 in Brest and Minsk, ST40 in Brest and Gomel, and ST41 in Vitebsk) or in the post-epidemic period only (ST8 in Brest and Vitebsk, ST25 in Gomel, and ST40 in Mogilov; Fig. 2). Although ST25 and ST40 strains are rare, they seem to be maintaining a reservoir as they appeared in some provinces in the epidemic period and in other provinces in the post-epidemic period. Some clones were only observed in the post-epidemic period; for example, ST28 in Gomel and Minsk, and ST32 in Brest and Gomel. It is possible that these strains were introduced in Belarus after 2001. Alternatively, these rare strains may have also been circulating prior to year 2000 but a larger sample size from the epidemic period needs be analysed to detect them.

Asymptomatic carriage and disease status

Of the 93 isolates from Belarus, 22 (23.7%) were isolated from asymptomatic carriers, 45 (48.4%) from patients with a sore throat and only 26 (28.0%) were from diphtheria patients (Additional file 1: Table S1; Fig. 2). The diseases status is clearly associated with different *C. diphtheriae* clones (χ^2 test, $p < 0.001$; Additional file 2: Table S4). The majority of the ST5 strains (29/37 isolates; 78.4%) caused sore throat and 7 strains (18.9%) were associated with asymptomatic carriage. Although all ST5 isolates are non-toxigenic, one isolate caused diphtheria-like symptoms in a patient. 29.4% (10/34 isolates) ST8 isolates were carriage-associated, 52.9% (18/34) caused diphtheria and 17.7% (6/34) caused sore throat. ST8 isolates are toxigenic except for three that are non-toxigenic isolated from the patients with sore throat.

Similarly, isolates of minor *C. diphtheriae* clones, toxigenic ST25 and non-toxigenic ST40 and ST76, were isolated from healthy carriers as well as from patients with sore throat, diphtheria or diphtheria-like symptoms (Figs. 1 & 2). Non-toxigenic ST32 isolates were either asymptomatic or caused sore throat whereas isolates of the ST28-ST41 group caused sore throat or diphtheria in patients. Interestingly, strains of the same ST have the ability to asymptomatically inhabit the human respiratory tract or to cause sore throat and diphtheria or diphtheria-like symptoms (non-toxigenic strains), regardless of their toxigenicity.

Clonal expansion of major *C. diphtheriae* clones

To understand the mechanism of clonal expansion in *C. diphtheriae*, we focused on the major clones, non-toxigenic ST5 and toxigenic ST8. Overall, 94,033 single nucleotide polymorphisms (SNPs) were observed within the core genomic alignment (1,226,854 bp) of 117 isolates. 3577 SNPs were present within ST5 whereas only 426 SNPs were present among ST8 isolates. The concatenated core genomic alignment was analysed by Gubbins [25] that indicated higher diversity being introduced by recombination than point mutations at the internal branch level which is shared by all the isolates within each group (Additional file 3: Figure S2). Gubbins identifies the regions introduced by recombination in the whole genomic alignments and calculates relative frequencies of recombination and mutations in clonal diversification. *C. diphtheriae* genomes analysed in this study are draft assemblies with some gaps. We did not attempt to predict genome-wide recombination rates and only focused on identification of regions introduced by recombination in the core genome of *C. diphtheriae*.

The regions predicted to be acquired through recombination were removed from the core genomic alignment, resulting in an alignment of 806,921 bp for ST5 (414 SNPs) and 861,883 bp for ST8 isolates (263 SNPs). Therefore, ST5 isolates acquired more diversity among the core genes through recombination than ST8.The level of temporal signal slightly varied between ST5 and ST8 after stripping the imported regions (Additional file 3: Figure S3). The correlation between the root-to-tip distances and strain isolation dates was relatively stronger within ST8 ($R^2 = 0.501$) than ST5 ($R^2 = 0.310$) with core genomic clock-rates of 8.9×10^{-7} (95% highest posterior density interval $5.6 \times 10^{-7} - 1.2 \times 10^{-6}$) and 5.6×10^{-7} (95% highest posterior density interval $3.7 - 7.7 \times 10^{-7}$) substitutions per strain per year, respectively. Therefore, point mutations are slightly more frequent in ST8 than ST5.

Virulence potential of *C. diphtheriae* clones

The genome sequences are quite conserved within each clone which is reflected in the CDS BLAST maps of ST5 and ST8 (Additional file 3: Figure S4A-B). 1807 genes (81-87% genes in individual isolates) were shared by all ST5 isolates and 1770 genes (78-86% genes) were common within ST8. A ML-tree from the accessory genome retrieved similar groupings as the core genome but indicated minor variations in the gene content that may result in functional variations between individuals of a clone (Additional file 3: Figure S5). The strains assigned to subgroup ST5-A and ST5-C possessed only two pilus gene clusters, SpaD and SpaH, whereas an additional SpaA cluster was present among the strains in ST5-B (Fig. 1; Additional file 3: Figure S1). Pilus gene clusters are borne by genomic islands [26] and ST5-B may have horizontally acquired the SpaA gene cluster from other *C. diphtheriae* strains. In addition, some genes in the pilus gene clusters have lost their function due to frameshift mutations; for example, the *spaC* gene in the SpaA gene cluster of strains 5201 and 1137. A gene encoding BigA-like adhesin which is known to mediate adhesion to epithelial cells [27] was possessed by some ST5 isolates, both in subgroups ST5-A and ST5-B (Fig. 1). Therefore, recombination and gain or loss of gene functions are introducing functional variations among isolates within a single clone [21].

Most of the ST8 isolates are equipped with all three Spa gene clusters, except for three strains that lacked the SpaD cluster. Similar to ST5, some genes in different *spa* clusters were pseudogenes. All ST8 isolates possessed gene encoding BigA-like adhesin (DIP2014) and an additional gene, *sapA* (DIP2066), encoding a surface-anchored pilus protein which is absent in ST5 isolates. In addition, most of the ST8 isolates carried another gene (DIP2093) encoding an adhesin of the Sdr family (Fig. 1). Similarly, the numbers and organisation of *spa* clusters and the presence/absence of *sapA* and adhesin genes varied both within and between other clones (Fig. 1). As reported in the previous sections, individual isolates of the same clone can cause different pathologies in different individuals. It is possible that gain or loss of the gene functions in pilus gene clusters or other virulence genes is partially contributing to the degree of disease; however, such a correlation is not obvious as some isolates from asymptomatic carriers have the *tox* gene along with all the above-mentioned virulence genes.

The key virulence factor in *C. diphtheriae* is the *tox* gene which is present among ST8 (except for three isolates), ST25 and ST28-ST41 isolates in Belarus. Interestingly, ST40 isolates were NTTB strains where a deletion of a nucleotide at position 52 in the *tox* gene resulted in the frameshift. A total of eight SNPs were observed between the *tox* gene of vaccine strain PW8 and other toxigenic strains in the dataset but all of them are synonymous (Fig. 3), suggesting that the impact of the vaccine will be similar on all toxigenic isolates.

Discussion

C. diphtheriae is genetically diverse with >11 distinct groups identified by the analysis of MLST data [2]. Most of the isolates from Belarus belong to two major clones, ST5 and ST8, with the remaining isolates distributed to eight other STs (Fig. 1). ST5 and ST8 strains from Belarus vary in their virulence gene repertoire and differ in their ability to cause disease (Fig. 1). ST5 isolates lack the *tox* gene, *sapA* and Sdr-like adhesin with additional absence of the SpaA gene cluster among subgroup ST5-A and ST5-C isolates. SpaA pili are responsible for adhesion to pharyngeal epithelial cells and SpaD and SpaH pili interact with laryngeal and lung epithelial cells [28, 29].

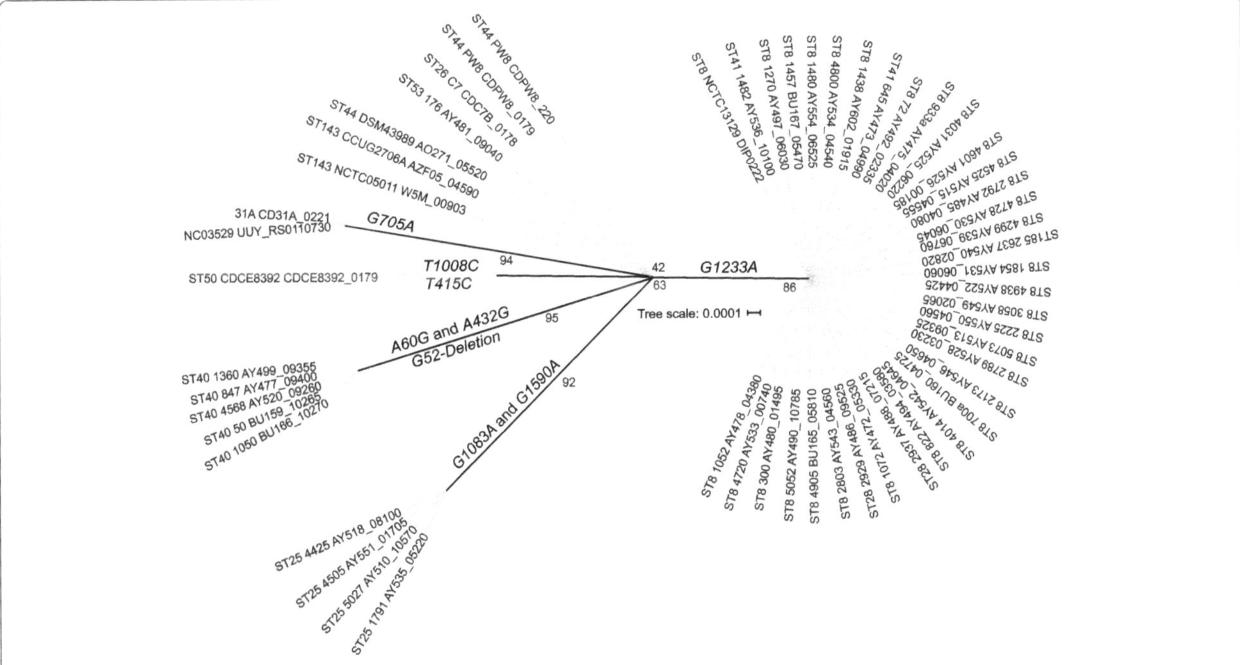

Fig. 3 A ML tree from the nucleotide sequence alignment of the *tox* gene. The scale bar represents the number of nucleotide substitutions per site. SNPs separating the clones from the vaccine strain PW8 are mapped on the branches

Sdr-like adhesin (DIP2093) also helps the pathogen in interacting with host cells and biofilm formation [30, 31]. Therefore, ST8 isolates may have greater abilities to adhere and invade host cells in comparison to ST5 isolates. However, regardless of the virulence potential, *C. diphtheriae* strains can cause disease in some individuals, while others remain asymptomatic. These asymptomatic carriers may serve as reservoir for dissemination of the pathogen to the community [32].

The toxin is responsible for the cell death which is produced under low iron conditions [3, 4]. Iron is essential for growth of all organisms and pathogenic bacteria often rely on the host for iron supply [33]. Most of the genes involved in iron uptake and transport including Irp6A-C (DIP0108-DIP0110), DIP0582-0586, HmuT-V (DIP0626-0628) and DIP1059-1062 are conserved in *C. diphtheriae* with minor exceptions. However, ChtC-CirA (DIP0522-DIP0523), ChtAB (DIP1519-DIP1520) and HtaA-C (DIP0624, DIP0625 and DIP0629) that are involved in uptake of hemoglobin-haptoglobin complexes [34], are only present in 48-73 strains. Interestingly, the majority of ST8 isolates possess these genes, suggesting that they are better equipped to utilise iron from the host cells than ST5 isolates.

MLST studies analysing *C. diphtheriae* strains from Russia and Poland also revealed the presence of diverse strains in these neighbouring countries during the epidemic period [14, 20]. However, ST8 isolates were apparently more prevalent in Poland in the post-epidemic period [14]. Consistent with the present study, all ST8 isolates from Russia belonged to biovar gravis and were toxigenic and ST5 isolates were non-toxigenic [20]. Invasive ST8 isolates in Poland were also biovar gravis but they were non-toxigenic [14]. A high diphtheria vaccination coverage in Poland probably protected the population from the epidemic in the neighbouring Soviet States in the 1990s [14, 35]. A consistency between *C. diphtheriae* ribotypes and grouping from other typing approaches has been previously reported [14, 20, 36]. Ribotyping information was available for 50 of the 93 strains from Belarus and we looked at their distribution within *C. diphtheriae* clones. In agreement with previous findings, epidemic ST8 clone encompassed ribotypes D1 and D4; however, one isolate each of ribotypes D6 and D7 were also present in this group (Additional file 1: Table S1) [37–39]. All D10 ribotype isolates were confined to ST5, whereas all isolates within ST25 were ribotype D6. Ribotypes of ST40 isolates were unclear except for a single isolate which was identified as D4. ST41 included one isolate each of ribotype D7 and D8 and one D7 isolate fell within ST53. ST76 isolate was identified to be a new (unassigned) ribotype. Therefore, ribotyping is generally concordant with the MLST and genomic groupings, with some exceptions.

The diversity at the clustered regularly interspaced short palindromic repeat (CRISPR) loci has been used to characterise *C. diphtheriae* outbreaks [37–39]. 20 *C. diphtheriae* isolates from Belarus belonging to ribotype D4 were divided into three spoligotypes based on the diversity at both the DRA and DRB CRISPR loci [39]. We have previously highlighted the extensive diversity at CRISPR loci between different *C. diphtheriae* strains based on the direct repeat and spacers sequences extracted from the genome sequences [40]. In this study, 16 combinations of direct-repeats and spacers are observed at the DRA locus among ST8 isolates while this locus is absent in one strain (Additional file 2: Table S5). The DRB locus was more diverse among these isolates with 20 direct-repeat and spacer combinations, resulting in a total of 30 combined profiles among 35 ST8 isolates (Additional file 2: Table S3). These findings are consistent with the previous studies revealing 45 combined spoligotypes among epidemic *C. diphtheriae* isolates in Russia [37, 38]. The DRB locus was absent among ST5 isolates and they were subdivided into 21 CRISPR types based on the diversity at the DRA locus (Additional file 2: Table S5).

Interestingly, the evolutionary dynamics of the non-toxigenic clone ST5 varied from the toxigenic clone ST8, with recombination being more prevalent within ST5, particularly in subgroup ST5-B (Additional file 3: Figure S2). *C. diphtheriae* inhabits the human upper respiratory tract which is also a niche for a variety of other bacteria [41], providing opportunities for recombination. Indeed, recombination frequencies are high among bacteria in the upper respiratory tract [42]. This variation in the relative role of recombination and point mutation in diversification of toxigenic and non-toxigenic strains is interesting and confirms that different lineages of the same species may have different recombination and mutation rates [43]. It is also possible that vaccine-induced immune response may be influencing the evolutionary dynamics by applying a selective pressure to toxigenic ST8 isolates. Vaccination was found to affect the evolution in *Bordetella pertussis* where the molecular clock rate was associated with the vaccination coverage in different countries [44]. It will be interesting if more studies characterising genomic variations in the collection of other toxigenic and non-toxigenic *C. diphtheriae* clones also observe similar differences in the evolutionary dynamics.

A high coverage of diphtheria vaccine in Belarus has significantly reduced the number of diphtheria cases and no new cases have been reported to WHO since 2011 [7] (Additional file 3: Figure S6). However, infections caused by non-toxigenic strains continue to emerge in most Belarusian provinces (Additional file 1: Table S1). IgG antibodies induced by the vaccine neutralise the toxin; however, it is unclear if the vaccine eliminates the organism or not. The toxoid vaccine for botulism reduced free neurotoxin in cows as well as the number of *Clostridium botulinum* spores in the faeces [45]. It is possible that neutralising antibodies eliminate the pathological effects of the toxin that potentially allow for the development of an adaptive immune response to limit growth of the bacteria. Several membrane-associated and secreted proteins have been detected in diphtheria vaccines by highly sensitive mass-spectrometry (Möller and Burkovski, unpublished data). These proteins may stimulate production of antibodies against additional targets on the cell surface. Therefore, the vaccine may be more effective against toxigenic strains but may also target non-toxigenic strains.

Conclusions

In conclusion, the diphtheria vaccine remains effective against toxigenic strains and has largely controlled diphtheria in Belarus after the major epidemic in the 1990s. This study describes the diversity among *C. diphtheriae* strains that were circulating in the period from 1996 to 2014 and demonstrates variation in the evolutionary dynamics between the two prevalent *C. diphtheriae* clones in Belarus. The variation in the presence of virulence genes and gain or loss of gene function is likely responsible for the differences in virulence characteristics, not only between different clones but also between isolates within a single clone. Regardless of their virulence potential, *C. diphtheriae* strains can asymptomatically colonise some individuals which exacerbates the threat of dissemination to the wider community.

Methods

Bacterial strains

The details of 93 *C. diphtheriae* isolates from Belarus are presented in Additional file 1 (Table S1). Two reference strains CCUG 2706A and CCUG 5865 were obtained from the CCUG Culture Collection, Göteborg, Sweden.

Genome sequencing

All 95 isolates (93 isolates from Belarus and two reference strains) were cultured on Brain-Heart Infusion agar overnight at 37 °C and a single colony was used to inoculate a 5 ml Brain-Heart Infusion broth. DNA was extracted from 2 ml overnight culture incubated at 37 °C for 16 h in a shaking incubator using the UltraClean® Microbial DNA Isolation Kit (MoBio). The genomes were sequenced on an Illumina MiSeq instrument and the reads were assembled using the CLC Genomic Workbench (Qiagen) or SPAdes 3.9.0 [46]. Genome assemblies were submitted to the GenBank for annotation by the NCBI Prokaryotic Genome Annotation Pipeline [47]. The genomes sequences of 22 previously published *C. diphtheriae* strains were obtained from the GenBank (Additional file 1: Table S1).

Comparative genomic and phylogenetic analyses

A comparative analysis on the complete dataset of 117 genomes was performed using EDGAR [19]. Orthologs of virulence genes including *sapA* (DIP2066), adhesin of Sdr family (DIP2093) and BigA-like adhesin (CDC7B_1983; DIP2014) were searched within the dataset using EDGAR. The CDS BLAST maps were generated using the CGView Comparison Tool [48] for isolates in ST5 and ST8 using concatenated sequence of ISS 4060 and NCTC 13129 as the reference, respectively. MLST profiles were extracted from the genome sequences using MLST 1.8 [49]. Single locus variants to known sequence types were treated as the same clone (ST).

The nucleotide sequences of the concatenated core genome were aligned using MUSCLE [50] and poorly aligned regions were removed by GBLOCKS [51]. A ML tree from the core genomic alignment was generated using GTR + I + G4 model according to Bayesian Information Criterion (BIC) with 100,000 SH-like approximate likelihood ratio tests (SH-aLRT) and 100,000 ultrafast bootstrap iterations using IQ-Tree [52]. A ML tree from the binary data of the presence or absence of genes in the accessory genome was generated using GTR2 + FO + ASC + R5 model with 1000 SH-aLRT and 1000 ultrafast bootstrap iterations. Both the trees were re-rooted using the strain CCUG 5865. ML tree were generated separately from the core genome of ST5 isolates using TIM + I model and from the nucleotide sequence alignment of *tox* gene using the HKY model, each with 100,000 SH-aLRT and 100,000 ultrafast bootstrap iterations. All phylogenetic trees were visualized using iTOL [53].

Spatio-temporal association of *C. diphtheriae* clones

χ^2 tests were performed to analyse differential distribution of *C. diphtheriae* strains in ST5 and ST8 clones in epidemic and post epidemic periods and their association with the disease status using the package SPSS v24 (IBM).

Evolutionary analyses

The regions introduced by recombination in the core genome were identified using Gubbins [25] and were masked from the alignment using the script maskrc-svg.py (provided by Kwong, J. and Seemann, T; https://github.com/kwongj/maskrc-svg). ML trees were constructed using IQ-Tree [52] with the best-fit substitution models and 100,000 SH-aLRT and 100,000 ultrafast bootstrap iterations from the core genomic alignment of Belarussian ST5 and ST8 isolated after stripping the masked regions. These phylogenetic trees were analysed by tempEST v 1.5 with the sampling dates and best-fit root criteria to detect temporal signal [54]. The clock-rates were calculated from these alignments using BEAST [55]. The HKY substitution model was used with Coalescent Bayesian Skyline and 10,000,000 MCMC chain length and 10,000 burn-in iterations. The trace file was analysed using Tracer v1.6 [56].

Additional files

Additional file 1: Table S1. Details of *C. diphtheriae* strains analysed in this study. (XLSX 21 kb)

Additional file 2: Table S2. Number and toxigenicity of isolates collected between 1996 and 2014 in Belarus and those selected for genomic analyses in this study; **Table S3.** Distribution of isolates from epidemic (≤ year 2000) and postepidemic period (≥ year 2001) in major groups. All minor groups are pooled together for statistical analysis; **Table S4.** Distribution of isolates from asymptomatic carriage, diphtheria and sore throat patients in major groups. All minor groups are pooled together for statistical analysis; **Table S5.** Allelic variation in the direct repeat and spacer sequences among the CRISPR loci of ST8 and ST5 isolates. (PDF 396 kb)

Additional file 3: Figure S1. A ML tree from core genomic alignment of ST5 strains; **Figure S2.** Gubbins analysis of recombination in the core genome of *C. diphtheriae*. Predicted recombination events on internal branches are shown in red and those occurred at terminal branches are shown in blue; **Figure S3.** A plot of root-to-tip divergence (Y-axis) against the sampling dates (X-axis); A. within ST5 and B. within ST8; **Figure S4. A.** CDS BLAST map of ST5 using strain ISS 4060 as the reference. **B.** CDS BLAST map of ST8 using strain NCTC 13129 as the reference; **Figure S5.** A ML tree from the binary data of the presence or absence of genes in the accessory genome. The scale bar with a distance of 0.1 represents the difference of 341.7 genes; **Figure S6.** A plot showing the average global vaccine coverage, reported vaccination in Belarus and the reported number of diphtheria cases between 1992 and 2015. (PDF 5948 kb)

Abbreviations

BLAST: Basic local alignment search tool; CDS: Coding DNA Sequence; IgG: Immunoglobulin G; MCMC: Markov chain Monte Carlo; ML tree: Maximum-Likelihood tree; MLST: Multilocus sequence typing; NTTB: Non-toxigenic *tox* gene-bearing strains; SH-aLRT: SH-like approximate likelihood ratio test; ST: Sequence type

Acknowledgements

The authors would like to thank the NU-OMICS facility of Northumbria University, and Nicholas J. Loman and MicrobesNG team from University of Birmingham for assistance in genome sequencing. We thank Jimmy Gibson for IT assistance and Melanie Pache for assistance in analysing CRISPR loci. SG-K was supported by OSZ-Lise-Meitner, Berlin under the EU exchange program to the laboratory of VS.

Funding

No Funding.

Authors' contributions

VS and LT conceived and designed the study. VK, LT, AB and PAH contributed representative strains for sequencing and phenotypic data. SG-K and VS cultured the strains and extracted DNA for sequencing. DS sequenced the strains. SG-K, JB, ICS, ECS, SB and VS analysed the sequence data. VS drafted the manuscript. All the authors provided intellectual input in finalising the manuscript. All authors have read and approved the final version of this manuscript.

Competing interests

The authors declare that they have no competing interests.

Author details

[1]Faculty of Health and Life Sciences, Northumbria University, Newcastle upon Tyne, UK. [2]Republican Research and Practical Centre for Epidemiology and Microbiology, Minsk, Republic of Belarus. [3]Justus-Liebig-Universität, Gießen, Germany. [4]Friedrich-Alexander-Universität Erlangen-Nürnberg, Erlangen, Germany. [5]Strathclyde Institute of Pharmacy and Biomedical Sciences, University of Strathclyde, Glasgow, UK. [6]Institut Pasteur, Biodiversity and Epidemiology of Bacterial Pathogens, Paris, France.

References

1. Hadfield TL, McEvoy P, Polotsky Y, Tzinserling VA, Yakovlev AA. The pathology of diphtheria. J Infect Dis. 2000;181(Suppl 1):S116–20.
2. Sangal V, Hoskisson PA. Evolution, epidemiology and diversity of *Corynebacterium diphtheriae*: new perspectives on an old foe. Infect Genet Evol. 2016;43:364–70.
3. Murphy JR. Mechanism of diphtheria toxin catalytic domain delivery to the eukaryotic cell cytosol and the cellular factors that directly participate in the process. Toxins. 2011;3(3):294–308.
4. Schmitt MP, Holmes RK. Iron-dependent regulation of diphtheria toxin and siderophore expression by the cloned *Corynebacterium diphtheriae* repressor gene *dtxR* in *C. diphtheriae* C7 strains. Infect Immun. 1991;59(6):1899–904.
5. Malito E, Rappouli R. History of diphtheria vaccine development. In: Burkovski A, editor. *Corynebacterium diphtheriae* and related toxigenic species. Heidelberg: Springer; 2014. p. 225–38.
6. Bisgard KM, Rhodes P, Hardy IR, Litkina IL, Filatov NN, Monisov AA, Wharton M. Diphtheria toxoid vaccine effectiveness: a case-control study in Russia. J Infect Dis. 2000;181(Suppl 1):S184–7.
7. Diphtheria reported cases. [http://apps.who.int/immunization_monitoring/globalsummary/timeseries/tsincidencediphtheria.html]. Accessed 30 Mar 2017.
8. Besa NC, Coldiron ME, Bakri A, Raji A, Nsuami MJ, Rousseau C, Hurtado N, Porten K. Diphtheria outbreak with high mortality in northeastern Nigeria. Epidemiol Infect. 2014;142(4):797–802.
9. Parande MV, Parande AM, Lakkannavar SL, Kholkute SD, Roy S. Diphtheria outbreak in rural North Karnataka, India. *JMM Case Rep*. 2014;1(3):1–3.
10. Nanthavong N, Black AP, Nouanthong P, Souvannaso C, Vilivong K, Muller CP, Goossens S, Quet F, Buisson Y. Diphtheria in Lao PDR: insufficient coverage or ineffective vaccine? PLoS One. 2015;10(4):e0121749.
11. Santos LS, Sant'anna LO, Ramos JN, Ladeira EM, Stavracakis-Peixoto R, Borges LL, Santos CS, Napoleao F, Camello TC, Pereira GA, et al. Diphtheria outbreak in Maranhao, Brazil: microbiological, clinical and epidemiological aspects. Epidemiol Infect. 2015;143(4):791–8.
12. Romney MG, Roscoe DL, Bernard K, Lai S, Efstratiou A, Clarke AM. Emergence of an invasive clone of nontoxigenic *Corynebacterium diphtheriae* in the urban poor population of Vancouver, Canada. J Clin Microbiol. 2006;44(5):1625–9.
13. Edwards B, Hunt AC, Hoskisson PA. Recent cases of non-toxigenic *Corynebacterium diphtheriae* in Scotland: justification for continued surveillance. J Med Microbiol. 2011;60(Pt 4):561–2.
14. Farfour E, Badell E, Zasada A, Hotzel H, Tomaso H, Guillot S, Guiso N. Characterization and comparison of invasive *Corynebacterium diphtheriae* isolates from France and Poland. J Clin Microbiol. 2012;50(1):173–5.
15. Sangal V, Hoskisson PA. Corynephages: infections of the infectors. In: Burkovski A, editor. *Corynebacterium diphtheriae* and related toxigenic species. Heidelberg: Springer; 2014. p. 67–82.
16. Zakikhany K, Neal S, Efstratiou A. Emergence and molecular characterisation of non-toxigenic *tox* gene-bearing *Corynebacterium diphtheriae* biovar mitis in the United Kingdom, 2003-2012. Euro Surveill. 2014;19(22)

17. Dittmann S, Wharton M, Vitek C, Ciotti M, Galazka A, Guichard S, Hardy I, Kartoglu U, Koyama S, Kreysler J, et al. Successful control of epidemic diphtheria in the states of the former Union of Soviet Socialist Republics: lessons learned. J Infect Dis. 2000;181(Suppl 1):S10–22.
18. Kolodkina V, Titov L, Sharapa T, Grimont F, Grimont PA, Efstratiou A. Molecular epidemiology of *C. diphtheriae* strains during different phases of the diphtheria epidemic in Belarus. BMC Infect Dis. 2006;6:129.
19. Blom J, Kreis J, Spanig S, Juhre T, Bertelli C, Ernst C, Goesmann A. EDGAR 2. 0: an enhanced software platform for comparative gene content analyses. Nucleic Acids Res. 2016;44(W1):W22–8.
20. Bolt F, Cassiday P, Tondella ML, Dezoysa A, Efstratiou A, Sing A, Zasada A, Bernard K, Guiso N, Badell E, et al. Multilocus sequence typing identifies evidence for recombination and two distinct lineages of *Corynebacterium diphtheriae*. J Clin Microbiol. 2010;48(11):4177–85.
21. Sangal V, Blom J, Sutcliffe IC, von Hunolstein C, Burkovski A, Hoskisson PA. Adherence and invasive properties of *Corynebacterium diphtheriae* strains correlates with the predicted membrane-associated and secreted proteome. BMC Genomics. 2015;16(1):765.
22. von Hunolstein C, Alfarone G, Scopetti F, Pataracchia M, La Valle R, Franchi F, Pacciani L, Manera A, Giammanco A, Farinelli S, et al. Molecular epidemiology and characteristics of *Corynebacterium diphtheriae* and *Corynebacterium ulcerans* strains isolated in Italy during the 1990s. J Med Microbiol. 2003;52(Pt 2):181–8.
23. Bouchet V, Huot H, Goldstein R. Molecular genetic basis of ribotyping. Clin Microbiol Rev. 2008;21(2):262–73.
24. Sangal V, Burkovski A, Hunt AC, Edwards B, Blom J, Hoskisson PA. A lack of genetic basis for biovar differentiation in clinically important *Corynebacterium diphtheriae* from whole genome sequencing. Infect Genet Evol. 2014;21:54–7.
25. Croucher NJ, Page AJ, Connor TR, Delaney AJ, Keane JA, Bentley SD, Parkhill J, Harris SR. Rapid phylogenetic analysis of large samples of recombinant bacterial whole genome sequences using Gubbins. Nucleic Acids Res. 2015; 43(3):e15.
26. Trost E, Blom J, Soares Sde C, Huang IH, Al-Dilaimi A, Schroder J, Jaenicke S, Dorella FA, Rocha FS, Miyoshi A, et al. Pangenomic study of *Corynebacterium diphtheriae* that provides insights into the genomic diversity of pathogenic isolates from cases of classical diphtheria, endocarditis, and pneumonia. J Bacteriol. 2012;194(12):3199–215.
27. Czibener C, Merwaiss F, Guaimas F, Del Giudice MG, Serantes DA, Spera JM, Ugalde JE. BigA is a novel adhesin of *Brucella* that mediates adhesion to epithelial cells. Cell Microbiol. 2016;18(4):500–13.
28. Mandlik A, Swierczynski A, Das A, Ton-That H. *Corynebacterium diphtheriae* employs specific minor pilins to target human pharyngeal epithelial cells. Mol Microbiol. 2007;64(1):111–24.
29. Reardon-Robinson ME, Ton-That H. Assembly and function of *Corynebacterium diphtheriae* pili. In: Burkovski A, editor. *Corynebacterium diphtheriae* and related toxigenic species. Heidelberg: Springer; 2014. p. 123–41.
30. Peixoto RS, Antunes CA, Louredo LS, Viana VG, Santos CSD, Fuentes Ribeiro d, Silva J, Hirata R Jr, Hacker E, Mattos-Guaraldi AL, Burkovski A. Functional characterization of the collagen-binding protein DIP2093 and its influence on host-pathogen interaction and arthritogenic potential of *Corynebacterium diphtheriae*. Microbiology. 2017;163(5):692–701.
31. Feuillie C, Formosa-Dague C, Hays LM, Vervaeck O, Derclaye S, Brennan MP, Foster TJ, Geoghegan JA, Dufrene YF. Molecular interactions and inhibition of the staphylococcal biofilm-forming protein SdrC. Proc Natl Acad Sci U S A. 2017;114(14):3738–43.
32. Adler NR, Mahony A, Friedman ND. Diphtheria: forgotten, but not gone. Intern Med J. 2013;43(2):206–10.
33. Braun V, Hantke K. Recent insights into iron import by bacteria. Curr Opin Chem Biol. 2011;15(2):328–34.
34. Allen CE, Schmitt MP. Utilization of host iron sources by *Corynebacterium diphtheriae*: multiple hemoglobin-binding proteins are essential for the use of iron from the hemoglobin-haptoglobin complex. J Bacteriol. 2015;197(3):553–62.
35. Walory J, Grzesiowski J, Hryniewicz W. The prevalence of diphtheria immunity in healthy population in Poland. Epidemiol Infect. 2001;126(2):225–30.
36. Titov L, Kolodkina V, Dronina A, Grimont F, Grimont PA, Lejay-Collin M, de Zoysa A, Andronescu C, Diaconescu A, Marin B, et al. Genotypic and phenotypic characteristics of *Corynebacterium diphtheriae* strains isolated from patients in Belarus during an epidemic period. J Clin Microbiol. 2003; 41(3):1285–8.
37. Mokrousov I, Limeschenko E, Vyazovaya A, Narvskaya O. *Corynebacterium diphtheriae* spoligotyping based on combined use of two CRISPR loci. Biotechnol J. 2007;2(7):901–6.
38. Mokrousov I, Narvskaya O, Limeschenko E, Vyazovaya A. Efficient discrimination within a *Corynebacterium diphtheriae* epidemic clonal group by a novel macroarray-based method. J Clin Microbiol. 2005;43(4):1662–8.
39. Mokrousov I, Vyazovaya A, Kolodkina V, Limeschenko E, Titov L, Narvskaya O. Novel macroarray-based method of *Corynebacterium diphtheriae* genotyping: evaluation in a field study in Belarus. Eur J Clin Microbiol Infect Dis. 2009;28(6):701–3.
40. Sangal V, Fineran PC, Hoskisson PA. Novel configurations of type I and II CRISPR-Cas systems in *Corynebacterium diphtheriae*. Microbiology. 2013; 159(10):2118–26.
41. Depner M, Ege MJ, Cox MJ, Dwyer S, Walker AW, Birzele LT, Genuneit J, Horak E, Braun-Fahrlander C, Danielewicz H, et al. Bacterial microbiota of the upper respiratory tract and childhood asthma. J Allergy Clin Immunol. 2017; 139(3):826–34. e813
42. Marks LR, Reddinger RM, Hakansson AP. High levels of genetic recombination during nasopharyngeal carriage and biofilm formation in *Streptococcus pneumoniae*. MBio. 2012;3(5):e00200–12.
43. den Bakker HC, Didelot X, Fortes ED, Nightingale KK, Wiedmann M. Lineage specific recombination rates and microevolution in *Listeria monocytogenes*. BMC Evol Biol. 2008;8(277)
44. Xu Y, Liu B, Gröndahl-Yli-Hannuksila K, Tan Y, Feng L, Kallonen T, Wang L, Peng D, He Q, Wang L, et al. Whole-genome sequencing reveals the effect of vaccination on the evolution of *Bordetella pertussis*. Sci Rep. 2015;5:12888.
45. Kruger M, Skau M, Shehata AA, Schrodl W. Efficacy of *Clostridium botulinum* types C and D toxoid vaccination in Danish cows. Anaerobe. 2013;23: 97–101.
46. Bankevich A, Nurk S, Antipov D, Gurevich AA, Dvorkin M, Kulikov AS, Lesin VM, Nikolenko SI, Pham S, Prjibelski AD, et al. SPAdes: a new genome assembly algorithm and its applications to single-cell sequencing. J Comput Biol. 2012;19(5):455–77.
47. Tatusova T, DiCuccio M, Badretdin A, Chetvernin V, Ciufo S, Li W. Prokaryotic genome annotation pipeline. In: The NCBI handbook. 2nd ed. Bethesda, US: National Center for Biotechnology Information; 2013.
48. Grant JR, Arantes AS, Stothard P. Comparing thousands of circular genomes using the CGView comparison tool. BMC Genomics. 2012;13(1):202.
49. Larsen MV, Cosentino S, Rasmussen S, Friis C, Hasman H, Marvig RL, Jelsbak L, Sicheritz-Ponten T, Ussery DW, Aarestrup FM, et al. Multilocus sequence typing of total-genome-sequenced bacteria. J Clin Microbiol. 2012;50(4): 1355–61.
50. Edgar RC. MUSCLE: multiple sequence alignment with high accuracy and high throughput. Nucleic Acids Res. 2004;32(5):1792–7.
51. Talavera G, Castresana J. Improvement of phylogenies after removing divergent and ambiguously aligned blocks from protein sequence alignments. Syst Biol. 2007;56(4):564–77.
52. Nguyen LT, Schmidt HA, von Haeseler A, Minh BQ. IQ-TREE: a fast and effective stochastic algorithm for estimating maximum-likelihood phylogenies. Mol Biol Evol. 2015;32(1):268–74.
53. Letunic I, Bork P. Interactive tree of life (iTOL) v3: an online tool for the display and annotation of phylogenetic and other trees. Nucleic Acids Res. 2016;44(W1):W242–5.
54. Rambaut A, Lam TT, Max Carvalho L, Pybus OG. Exploring the temporal structure of heterochronous sequences using TempEst (formerly path-O-gen). Virus Evol. 2016;2(1):vew007.
55. Drummond AJ, Suchard MA, Xie D, Rambaut A. Bayesian phylogenetics with BEAUti and the BEAST 1.7. Mol Biol Evol. 2012;29(8):1969–73.
56. Rambaut A, Suchard MA, Xie D, Drummond AJ: Tracer v1.6. Available from http://tree.bio.ed.ac.uk/software/tracer/. 2014.

3

Phylogenetic and recombination analysis of the herpesvirus genus varicellovirus

Aaron W. Kolb[1], Andrew C. Lewin[2], Ralph Moeller Trane[1], Gillian J. McLellan[1,2,3] and Curtis R. Brandt[1,3,4*]

Abstract

Background: The varicelloviruses comprise a genus within the alphaherpesvirus subfamily, and infect both humans and other mammals. Recently, next-generation sequencing has been used to generate genomic sequences of several members of the *Varicellovirus* genus. Here, currently available varicellovirus genomic sequences were used for phylogenetic, recombination, and genetic distance analysis.

Results: A phylogenetic network including genomic sequences of individual species, was generated and suggested a potential restriction between the ungulate and non-ungulate viruses. Intraspecies genetic distances were higher in the ungulate viruses (pseudorabies virus (SuHV-1) 1.65%, *bovine herpes virus type 1* (BHV-1) 0.81%, *equine herpes virus type 1* (EHV-1) 0.79%, *equine herpes virus type 4* (EHV-4) 0.16%) than non-ungulate viruses (*feline herpes virus type 1* (FHV-1) 0. 0089%, *canine herpes virus type 1* (CHV-1) 0.005%, varicella-zoster virus (VZV) 0.136%). The G + C content of the ungulate viruses was also higher (SuHV-1 73.6%, BHV-1 72.6%, EHV-1 56.6%, EHV-4 50.5%) compared to the non-ungulate viruses (FHV-1 45.8%, CHV-1 31.6%, VZV 45.8%), which suggests a possible link between G + C content and intraspecies genetic diversity. Varicellovirus clade nomenclature is variable across different species, and we propose a standardization based on genomic genetic distance. A recent study reported no recombination between sequenced FHV-1 strains, however in the present study, both splitstree, bootscan, and PHI analysis indicated recombination. We also found that the recently sequenced Brazilian CHV-1 strain BTU-1 may contain a genetic signal in the UL50 gene from an unknown varicellovirus.

Conclusion: Together, the data contribute to a greater understanding of varicellovirus genomics, and we also suggest a new clade nomenclature scheme based on genetic distances.

Keywords: Varicellovirus, Genome, Herpes, Veterinary, Phylogeny, Virus, Recombination

Background

The *Varicellovirus* genus is part of the larger alphaherpesvirus subfamily which includes herpes simplex viruses 1 and 2, as well as Marek's disease virus. Like other alphaherpesviruses, varicelloviruses typically infect epithelial surfaces, and most appear to be neurotropic, establishing latency in neurons [1–9]. The first varicellovirus to be clinically described as a unique disease was varicella zoster virus (VZV), the causative agent of chickenpox and shingles in 1767 [10]. Numerous other varicelloviruses have been identified, including pseudorabies virus/SuHV-1 (Aujeszky's disease) in pigs, BHV-1 (*bovine herpes virus type 1*; infectious bovine rhinotracheitis; IBR), EHV-1 (*equine herpes virus type 1*; epidemic abortion and myeloencephalopathy in horses), EHV-4 (*equine herpes virus type 4*; equine rhinopneumonitis), FHV-1 (*feline herpes virus type 1*; feline rhinotracheitis), and CHV-1 (*canine herpes virus type 1*; fading puppy syndrome). These viruses have significant impact on livestock and companion animals. Due to high transmissibility and virulence, pseudorabies virus and EHV-1 are both classified by the United States Department of Agriculture (USDA) as reportable diseases [11]. Vaccines have been developed against several varicelloviruses, including VZV, pseudorabies virus, BHV-1, EHV-1, EHV-4, FHV-1, CHV-1, which have been effective at reducing morbidity and mortality [12–22]. Pseudorabies vaccination and eradication efforts in the United States have been effective, with the

* Correspondence: crbrandt@wisc.edu
[1]Department of Ophthalmology and Visual Sciences, School of Medicine and Public Health, University of Wisconsin-Madison, 550 Bardeen Laboratories, 1300 University Ave., Madison, WI 53706, USA
[3]McPherson Eye Research Institute, University of Wisconsin-Madison, Madison, WI, USA
Full list of author information is available at the end of the article

country declared disease free in 2004. The USA and Canada have also enacted BHV-1 control programs, and several European countries have successfully eradicated the disease [23]. Despite vaccination and control efforts, many of these diseases continue to negatively affect humans and animals worldwide.

The first complete sequence of a varicellovirus, (VZV) was reported in 1986 [24], followed by several others [25–27] . The advent of next-generation sequencing (NGS) has revolutionized genomics, and has allowed additional varicellovirus species and sub-strains to be sequenced. The first comprehensive computational multigene phylogenetic analysis of the three main herpes virus subfamilies was a major step forward in cementing the basic structure for *Alphherpesvirinae,* including varicelloviruses [28]. As increasing numbers of viral strains have been sequenced, full genome phylogenetic and recombination analysis of VZV, SuHV-1, EHV-1, and EHV-4 have been performed [29–32].

The genetic code is degenerate, resulting in most amino acids being encoded by multiple codons. The usage of some codons and not others for an amino acid is often not random, and is called codon usage bias [33]. Codon usage and mutational bias analysis has been examined in several viruses, including phages, canine parvovirus, Japanese encephalitis virus, rabies, Zika virus, herpesviruses, and other vertebrate DNA viruses [34–40]. Shackelton et al. [39], showed that codon usage bias is strongly linked with genomic G + C content. The previous analysis of codon usage in herpesviruses [38] found strong codon bias in the SuHV-1 and BHV-5 viruses, both high G + C viruses. While the earlier herpesvirus study [38] included several varicelloviruses, additional viruses have now been sequenced, and a more inclusive analysis is now possible.

The goal of the current study was to perform a genome based comprehensive phylogenetic, genetic distance, and recombination analysis of the varicellovirus subfamily. Unique findings reported here are a phylogenetic stricture between ungulate herpesviruses and the remaining species, a possible link between genomic G + C content and intraspecies distance, the identification of recombination amongst FHV-1 strains, and results suggesting that the Brazilian CHV-1 strain BTU-1 may be a recombinant between CHV-1 and an unknown varicellovirus. We also propose a *Varicellovirus* genus clade nomenclature standardization based on genetic distance.

Methods

Genomic multiple sequence alignments

For phylogenetic and distance analysis, currently available *Varicellovirus* genus genomic sequences were downloaded from NCBI, and are cataloged in Additional file 1: Table S1. The first generated alignment was of the *Varicellovirus* genus as a whole, using one strain from each of the viral species, as well as an outgroup, *anatid herpes virus type 1* (AnHV-1). AnHV-1 was chosen as an outgroup because the AnHV-1 is an alphaherpesvirus, and the genome is annotated in a similar fashion to the varicelloviruses. The varicelloviruses and the AnHV-1 outgroup virus have similar genome annotation and gene synteny, with the exception of pseudorabies virus. In pseudorabies virus, the UL27 to UL44 genes are inverted. Prior to genome alignment, the UL27-UL44 genome segment of pseudorabies virus was inverted by reverse complemention in order to generate a gene order similar to the other varicellovirueses. MAFFT v2.66 [41, 42] was utilized to generate the alignment using the FFT-NS-1 method. The subsequent genomic multiple sequence alignment was manually inspected for quality. Areas of the alignment that appeared to be of poor quality were realigned in Mega 6 [43] using ClustalW.

Additional intraspecies genomic alignments of BHV-1, CHV-1, EHV-1, EHV-4, FHV-1, SuHV-1, and VZV, were generated with and without outgroups using MAFFT v2.66. The outgroups for the intraspecies alignments were chosen based on low genetic distance, in other words, the closest known relative. Thus, for EHV-1 analysis, EHV-8 was chosen as the outgroup, with the remaining analyzed species/outgroup combinations being; BHV-1/outgroup BHV-5, EHV-4/outgroup EHV-1, SuHV-1/outgroup BHV-1, FHV-1/outgroup CHV-1, CHV-1/outgroup FHV-1, and VZV/outgroup CeHV-9. All of the genomic alignments generated in thus study are available for download at http://sites.ophth.wisc.edu/brandt/.

Genetic distance and genomic G + C content calculations

The mean maximum likelihood (ML) distances for each alignment were calculated using the Mega 6 package. For the genetic distance analysis, pairwise gap deletion rather than complete deletion of gaps was used, as complete deletion of alignment gaps may exclude valuable phylogenetic data, and could result in an underestimation of distance. To calculate overall genome G + C content, an online calculator found at http://www.endmemo.com/bio/gc.php was used.

Phylogenetic and recombination analysis

For phylogenetic maximum likelihood and network analysis, intraspecies genomic alignments of BHV-1, CHV-1, EHV-1, EHV-4, FHV-1, SuHV-1, VZV, and total sequenced varicelloviruses were generated using duck enteritis virus (AnHV-1) as the outgroup as described above. Maximum likelihood phylogenetic analysis was performed on the genomic alignments containing an outgroup using the RAxMLGUI package [44] with the GTRCAT + I model and 500 bootstrap replicates.

Phylogenetic networks for BHV-1, CHV-1, EHV-1, EHV-4, FHV-1, SuHV-1, VZV, and total sequenced varicelloviruses were generated with Splitstree 4 [45] using multiple sequence alignments and AnHV-1 as the outgroup. Splitstree was also used to calculate the pairwise homoplasy index (PHI) statistical test [46] for recombination. Jmodeltest2 [47] was used to identify the optimal substitution model settings for each individual phylogenetic network. RDP4 [48] recombination analysis performed genomic multiple sequence alignments without outgroups using the Jin and Nei substitution model with the following parameters: 1500 bp window, 750 bp step size, and 200 bootstrap replicates.

Determining shared clade cut-off values between BHV-1, EHV-1, and SuHV-1

Shared clade Cut-off values between BHV-1, EHV-1, and SuHV-1 were determined by first generating a histogram of pairwise *p*-distances and corresponding frequencies for each virus species, similar to the study performed by Grau-Roma et el with *porcine circovirus type 2* (PCV2) [49].The pairwise *p*-distances for each species were calculated using Mega 6, and multiple sequence alignments without outgroups. Histograms of frequency vs *p*-distance for BHV-1, EHV-1, and SuHV-1, and subsequent data were generated using R (version 3.4.2 using the ggplot2 package). An initial shared cut-off of 0.01 was established by examining the upper and lower bounds of the two main groups in the three histograms. This intial cut-off was further evaluated, using a variance analysis framework, where variance between and within groups was examined. For each potential cut-off value, we calculated the following quantities for the *p*-distances for each virus:

For each potential cut-off value, we calculate the following quantities:

$SS_{\text{between}} = \Sigma f_{ij} \cdot (\text{group mean } j - \text{overall mean}),^2$

$SS_{\text{within}} = \Sigma f_{ij} \cdot (\text{p-distance } i - \text{overall}),^2$

where.

group mean_j = mean *p*-distance in group_j, $j = 1, 2$,

overall mean = overall mean *p*-distance,

p-distance i = the ith *p*-distance,

f_{ij} = the frequency of the ith *p*-distance in the jth group.

Finally, we calculate ratio

$$F = \frac{SS_{between}}{SS_{within}/(N_j - 2)}$$

where N_j is the total number of observations in group j, or in terms of frequencies, $N_j = \Sigma_i f_{ij}$. We next wanted to determine the cut-off which maximized the quantity across the groups, by first plotting the F values for each of the three graphs (Figure S1). We next restricted the cut-off values where the *p*-values for the different viruses was divided into two groups. For example, this means that values larger than 0.012 were discarded as no such distances were found in the BHV-1 virus group. The values within each graph were rescaled 0 to 1 in order to make each virus of equal value, and the sum of the curves maximized. To examine how the F measure corresponded to the frequency distributions, the value of rescaled F was overlayed. The point at which the sum of the rescaled F values attains it's maximum, was chosen as the cut-off value.

Codon usage analysis

To investigate codon usage in the *Varicellovirus* genus as a whole, the effective number of codons (ENC) was calculated for the US1 (ICP22), UL30 DNA polymerase, and glycoprotein H genes from each varicellovirus species. These genes represent one member of the α (immediate-early), β (early), and γ (late) gene classes. The ENC value is a measure of how much the codon usage of a gene deviates from the equal usage of synonymous codons [50]. ENC values range from 20 to 61, with 20 indicating maximum bias, with one codon used from each synonymous codon group, to 61 indicating no codon usage bias. The ENC values for the varicellovirus US1, UL30, and gH genes were calculated using DnaSP (v5) [51]. In addition to the ENC values, GC3s values were calculated using DnaSP, while the GC1 and GC2 values were obtained using in the online calculator http://genomes.urv.es/CAIcal/. The GC12 values were calculated using Microsoft Excel. ENC-GC3s plots were generated using SigmaPlot v.11 . In the ENC-GC3s plots, if the plotted values are located on or near the standard curve, then codon usage is constrained only by G + C mutation bias. However, the greater the plotted values deviate from the standard curve, the more additional factors such as natural selection may influence the bias.

Neutral evolution plots (GC12s vs GC3s) were generated to examine the contribution of mutational pressure and natural selection. Sigmaplot v.11 was used to generate the plots, as well as for the linear regression statistical analysis.

Results and discussion

Nomenclature standardization

The nomenclature designation for varicellovirus species strains and intraspecies clades is somewhat variable, with some, such as bovine herpesvirus 1 strains given a BHV-1.1 or 1.2 designation [52], while VZV clades are given a simple numeral designations (1–6) [29]. As such, a clade nomenclature standardization across varicelloviruses may be useful, and we are proposing a common nomenclature system based on genetic distances. The International Committee on Taxonomy of Viruses (ICTV) does not provide guidelines in defining taxonomic clades below species level [53], and it is within the purview of interested groups to do so. Genetic distances have been previously used as a basis for nomenclature systems in H5N1 avian influenza [54] and porcine circovirus type 2 [55, 56]. The genetic distance based nomenclature system we are proposing would preserve classic BHV-1 1.1 and 1.2 clade designations, as well the varicella zoster virus (VZV) numerical designations. Within each species, clade/group distances greater than 1% would be designated by a 1.1, 1.2, ... numbering as seen in BHV-1 [52]. Between clade/group distances of less than 1% would result in a numerical format (i.e. 1, 2, 3...) as has been consistently used with VZV [29]. Under this system, it would be possible to have two distant clades given a 1.x designation, with less distant subclades designated numerically.

The 1% cut-off was determined by calculating a shared value for the BHV-1, EHV-1, and SuHV-1 viruses. BHV-1, EHV-1, and SuHV-1 were chosen, as these three viruses have the highest levels of intraspecies genetic distance of the varicelloviruses (detailed in the sections below). An initial cut-off of 0.01 was chosen, based on a shared *p*-distance value that divides the observed *p*-distances into two groups simultaneously for all three viruses (Fig. 1a, b, and c). Additional evaluation of the initial cut-off was performed using a variance analysis framework, where variation between groups and within groups was examined. Figure 1e, f, and g show the individual rescaled *F* curves (gray dotted line) for each of the three viruses, as well as the sum of the curves in black. The *F* values were rescaled so as not to weight one virus more than the rest so the curves do not directly overlap. The sum of the rescaled *F* curves shows a peak at 0.01, which validates the initial cut-off. The values of the unscaled, and rescaled *F* values are shown in Additional file 1: Tables S2, and S3 respectively.

Phylogenetic network analysis of varicelloviruses

To investigate phylogenetic relationships between the sequenced varicelloviruses, the genomes of each species, along with the AnHV-1 outgroup genome were aligned. Both a maximum likelihood (ML) based tree (Fig. 2a) and phylogenetic network (Fig. 2b) were constructed. The ML whole genome based tree showed that CeHV-9 and VZV occupy a basal position in the genus, and the ungulate viruses share a node with bootstrap support of 100% (Fig. 2a). This result is fairly unremarkable and is similar to analysis performed using smaller sets of genes [57]. To assess the phylogenetic dissonance in the dataset, a phylogenetic network was also generated, and shows a similar basic topology with the ML tree. The network however suggests a

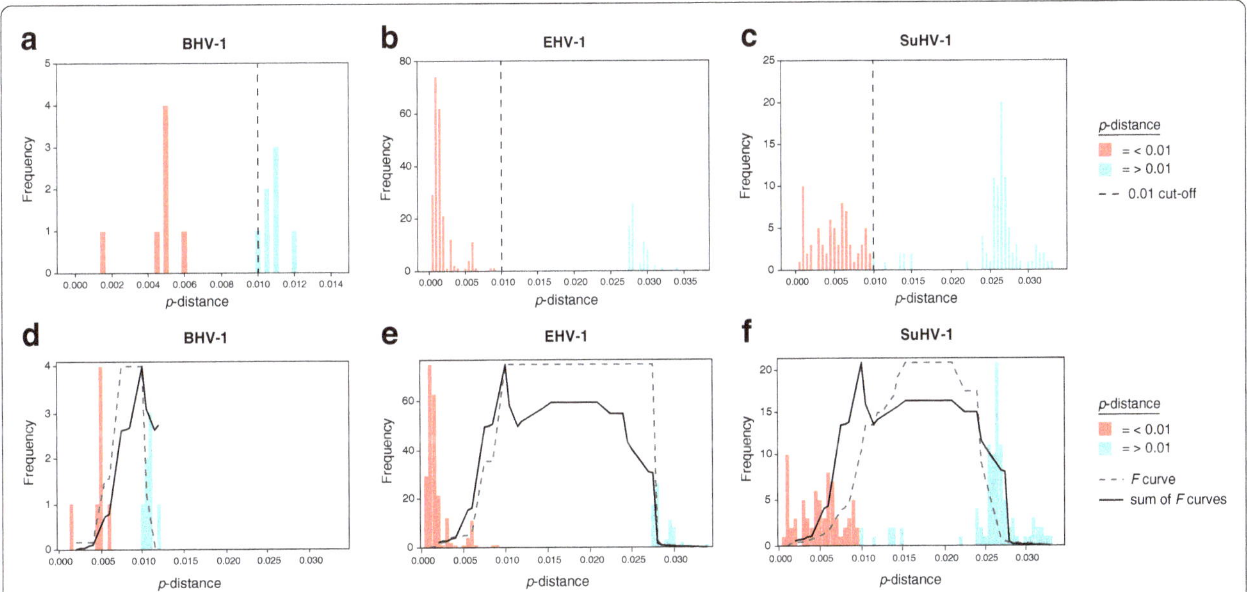

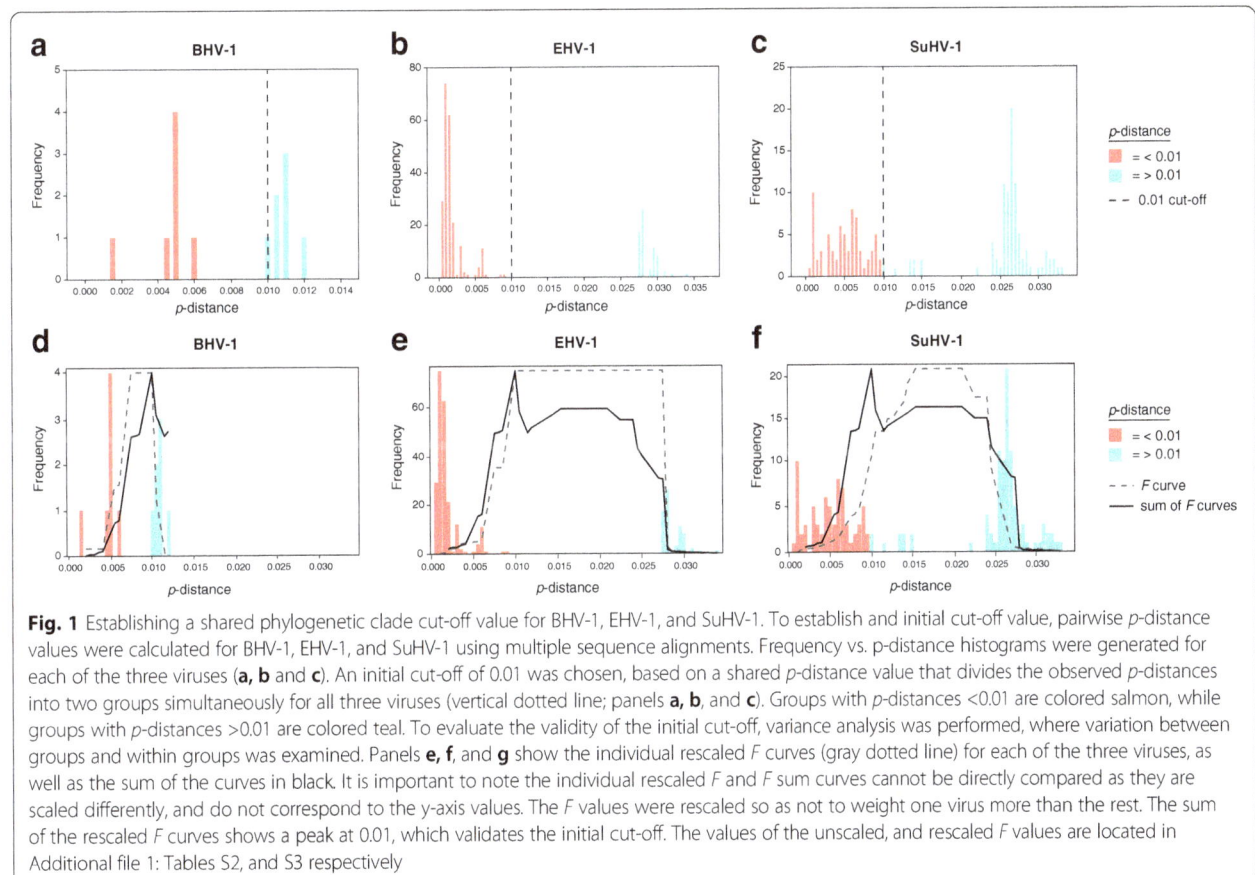

Fig. 1 Establishing a shared phylogenetic clade cut-off value for BHV-1, EHV-1, and SuHV-1. To establish and initial cut-off value, pairwise *p*-distance values were calculated for BHV-1, EHV-1, and SuHV-1 using multiple sequence alignments. Frequency vs. p-distance histograms were generated for each of the three viruses (**a, b** and **c**). An initial cut-off of 0.01 was chosen, based on a shared *p*-distance value that divides the observed *p*-distances into two groups simultaneously for all three viruses (vertical dotted line; panels **a, b**, and **c**). Groups with *p*-distances <0.01 are colored salmon, while groups with *p*-distances >0.01 are colored teal. To evaluate the validity of the initial cut-off, variance analysis was performed, where variation between groups and within groups was examined. Panels **e, f**, and **g** show the individual rescaled *F* curves (gray dotted line) for each of the three viruses, as well as the sum of the curves in black. It is important to note the individual rescaled *F* and *F* sum curves cannot be directly compared as they are scaled differently, and do not correspond to the y-axis values. The *F* values were rescaled so as not to weight one virus more than the rest. The sum of the rescaled *F* curves shows a peak at 0.01, which validates the initial cut-off. The values of the unscaled, and rescaled *F* values are located in Additional file 1: Tables S2, and S3 respectively

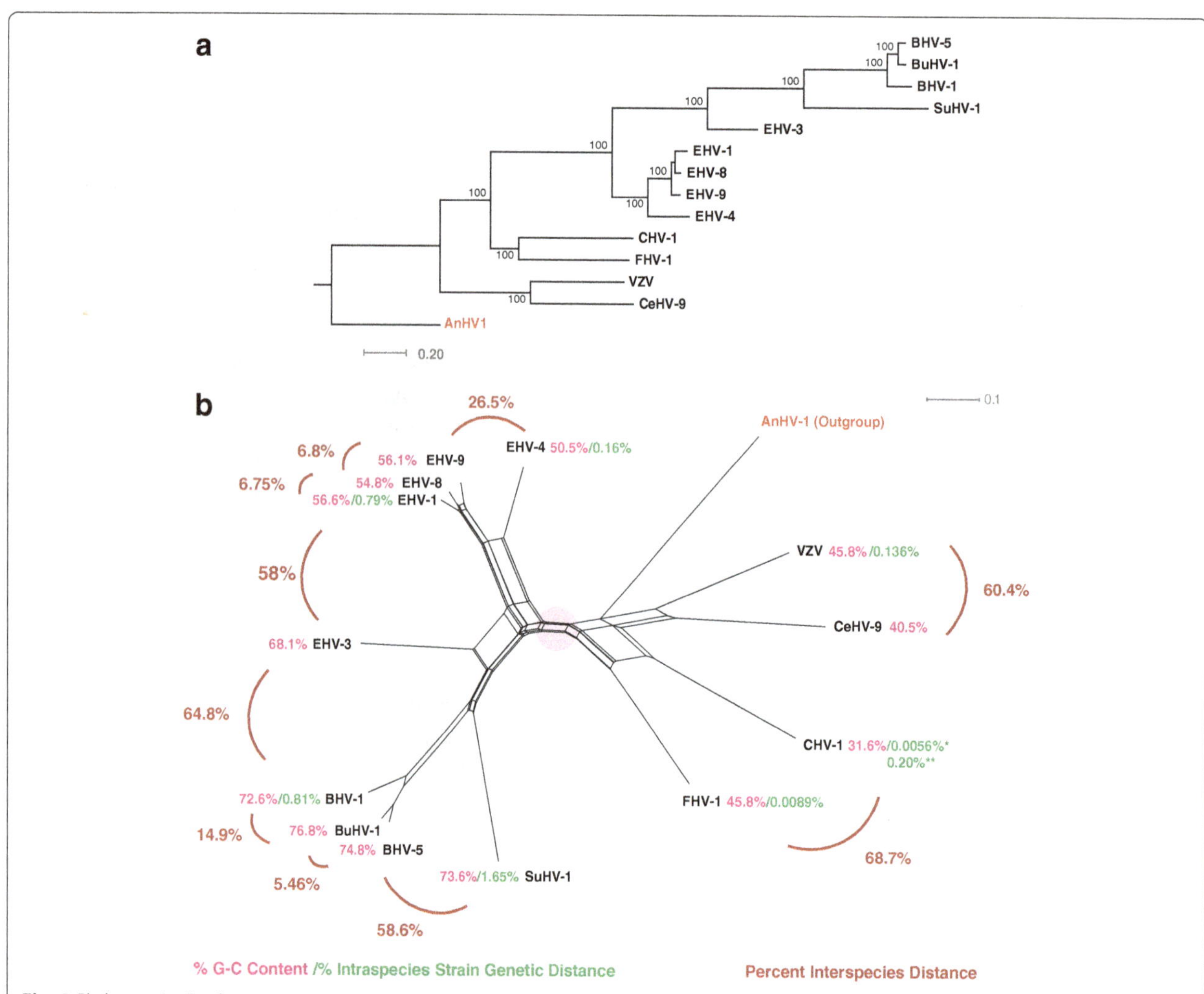

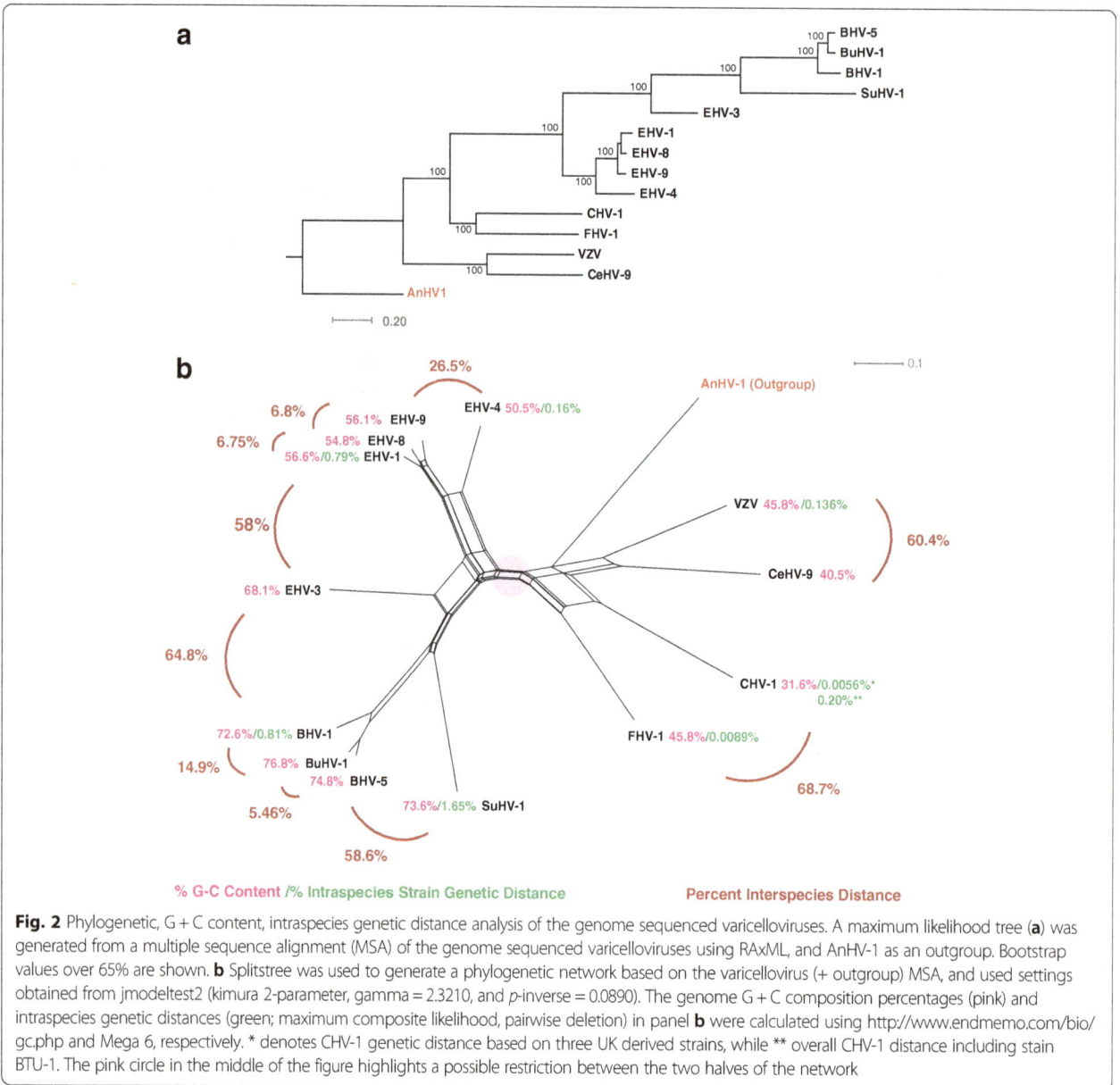

Fig. 2 Phylogenetic, G + C content, intraspecies genetic distance analysis of the genome sequenced varicelloviruses. A maximum likelihood tree (**a**) was generated from a multiple sequence alignment (MSA) of the genome sequenced varicelloviruses using RAxML, and AnHV-1 as an outgroup. Bootstrap values over 65% are shown. **b** Splitstree was used to generate a phylogenetic network based on the varicellovirus (+ outgroup) MSA, and used settings obtained from jmodeltest2 (kimura 2-parameter, gamma = 2.3210, and *p*-inverse = 0.0890). The genome G + C composition percentages (pink) and intraspecies genetic distances (green; maximum composite likelihood, pairwise deletion) in panel **b** were calculated using http://www.endmemo.com/bio/gc.php and Mega 6, respectively. * denotes CHV-1 genetic distance based on three UK derived strains, while ** overall CHV-1 distance including stain BTU-1. The pink circle in the middle of the figure highlights a possible restriction between the two halves of the network

stricture between the ungulates and non-ungulates, and is denoted by a pink circle (Fig. 2b). The stricture could be the result of low amounts of recombination between the two sides of the network, however, it may represent a bottleneck, or may be simply due to divergent phylogenetic signals. To determine if there was recombination within the network, the PHI statistical test for recombination was performed. The PHI test indicated statistically significant signals amongst the ungulate virus portion of the network, as well the non-ungulate portion, however, analysis of the network as a whole (minus outgroup) was not significant (Table 1). This lack of a significant result is likely due to the high amount of genetic distance within the dataset. The genomic distances between virus species are also shown in Fig. 2b to aid in data interpretation.

Varicellovirus G + C content and interstrain genetic distance

The G + C content of each of the varicellovirus species was analyzed, with results shown in both Fig. 2b, and Table 2. All of the ungulate viruses had a G + C content above 50%, ranging from 50.5% in EHV-4 to 74.8% in BHV-5. The primate and carnivore viruses had G+C contents under 50%, and ranged from 31.6% in CHV-1 to 45.8% in both VZV and FHV-1. For each varicellovirus, where multiple strains have been sequenced, the overall intraspecies genetic distance for each species was calculated (Fig. 1b and Table 1). The results suggest a possible link between G + C content and intraspecies genetic distance with higher G + C content and genetic distance in the ungulate viruses (SuHV-1 = 1.65%, BHV-

Table 1 Pairwise homoplasty index (PHI) statistic test p-values for recombination in the varicelloviruses

Virus	PHI p-value
VZV	< 0.001
CHV-1	0.3082
FHV-1	< 0.001
EHV-4	< 0.001
EHV-1	< 0.001
SuHV-1	< 0.001
BHV-1	< 0.001
Varicellovirus Genus	1.00
Ungulate Viruses	< 0.001
Primate and Carnivore Viruses	< 0.001

1 = 0.81%, EHV-1 = 0.79%, and EHV-4 = 0.16%), and lower values in the carnivore and VZV viruses (VZV = 0.136%, CHV-1 = 0.0056/0.020%, and FHV-1 = 0.0089%). It should be noted that for CHV-1, two distance values are given, and this is discussed below.

It is unclear if the higher genomic G + C content of the ungulate viruses is the result of genetic drift or evolutionary pressure. The observation that G + C content in varicelloviruses may be linked to intraspecies genetic distance may not be unprecedented, as G + C content appears to correlate with substitution rates in *Arabadopsis* [58]. It is highly unlikely that G + C content is the main driver of varicellovirus intraspecies genetic distance, however, it may be possible that G + C content is able to influence distance. Additional factors may influence intraspecies genetic variability in varicelloviruses, such as the propensity of the host to form large herds, transmissibility, and the number reactivation events in the life of the host. We also cannot eliminate the possibility that the genomic G + C content and intraspecies distance link is an artifact due to small sample size.

Table 2 Varicellovirus G + C content and intraspecies strain genetic distance

Virus	G + C %	Intrastrain distance %
BHV-1	72.6	0.77
BuHV-1	76.8%	NA
BHV-5	74.8	NA
EHV-1(Combined)	56.6	0.74
EHV-1 (Wild)	56.6	0.60
EHV-1 (Domestic)	56.6	0.14
EHV-3	68.1	NA
EHV-4	50.5	0.14
EHV-8	54.4	NA
EHV-9	56.1	NA
SuHV-1	73.6	1.23
FHV-1	45.8	0.004
CHV-1 (UK strains)	31.6	0.005
CHV-1 (Overall)	31.6	0.20
CeHV-9	40.5	NA
VZV	45.8	0.136

Codon usage and mutational bias in the *Varicellovirus* genus

The observation of varying G + C content across the *Varicellovirus* genus lead us to investigate codon usage and mutational bias. Codon usage and mutational bias has been previously examined in other viruses such as canine parvovirus [35], Japanese encephalitis virus [37], and Zika virus [36]. For the present analysis, the effective codon number values of three genes, US1 (α), UL30 polymerase (β), and UL22 (glycoprotein H; γ) were calculated for each of the varicellovirus species (Table 3). These three genes were chosen to be representative of each kinetic class; α (immediate-early), β (early), and γ (late). ENC values range from 20 to 61, with 61 indicating no bias and 20 indicating extreme bias. The values show greater bias in all three genes from viruses that are either A-T or G-C rich (Table 3), for example CHV-1, SuHV-1, BHV-1, BHV-5, and BuHV-1. Next, ENC values in the context of mutational pressure were assessed by plotting the ENC values against the G + C content in the synonymous third codon position (GC3s), found in Fig. 3a, b, and c. The ENC plots for US1, UL30, and UL22 show that these three *Varicellovirus* genus genes are largely constrained by G + C mutation bias, as most data points are located close to the standard curve. Some of the data points are located farther away from the plot, such as SuHV-1 US1 (Fig. 3a), and EHV-4

Table 3 Effective codon number (ENC) values for the varicellovirus US1 (α), UL30 (β), and UL22 (γ) genes

Virus	US1	UL30	UL22
VZV	58.325	53.455	56.306
CeHV-9	55.196	46.375	48.746
CHV-1	44.246	37.827	34.744
FHV-1	57.692	54.214	55.561
EHV-4	48.973	55.309	58.247
EHV-1	37.591	50.632	57.499
EHV-8	42.716	53.786	57.220
EHV-9	37.448	50.471	57.474
EHV-3	29.358	34.598	38.156
SuHV-1	28.755	28.550	28.835
BHV-1	30.05	33.977	35.195
BHV-5	35.436	30.899	30.029
BuHV-1	29.317	29.396	29.430

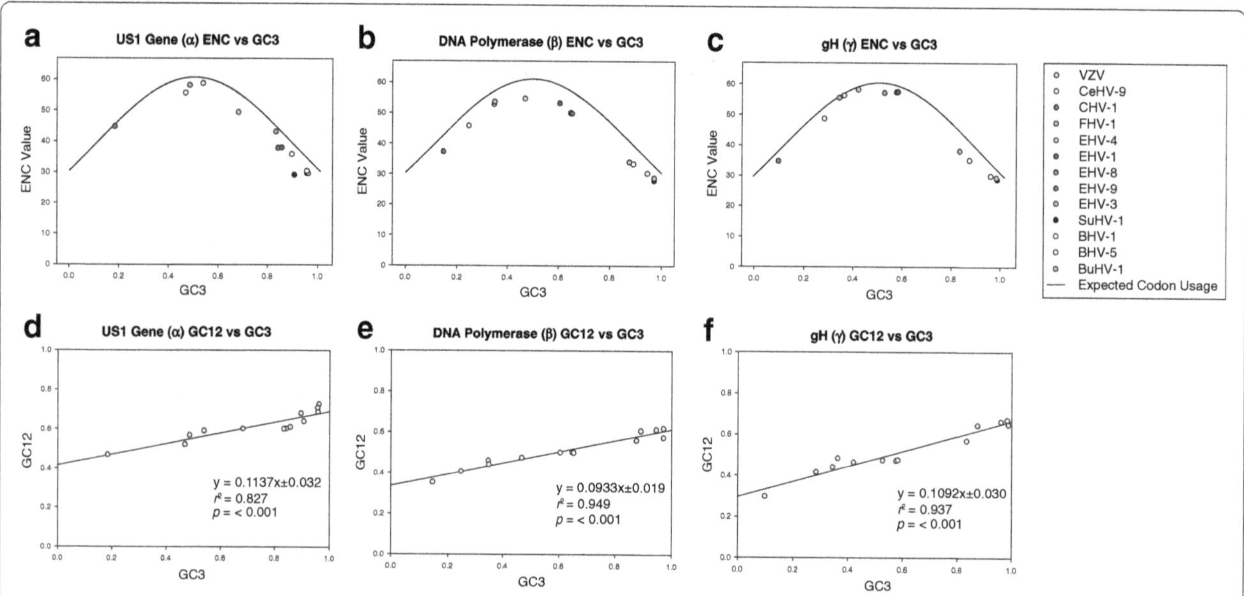

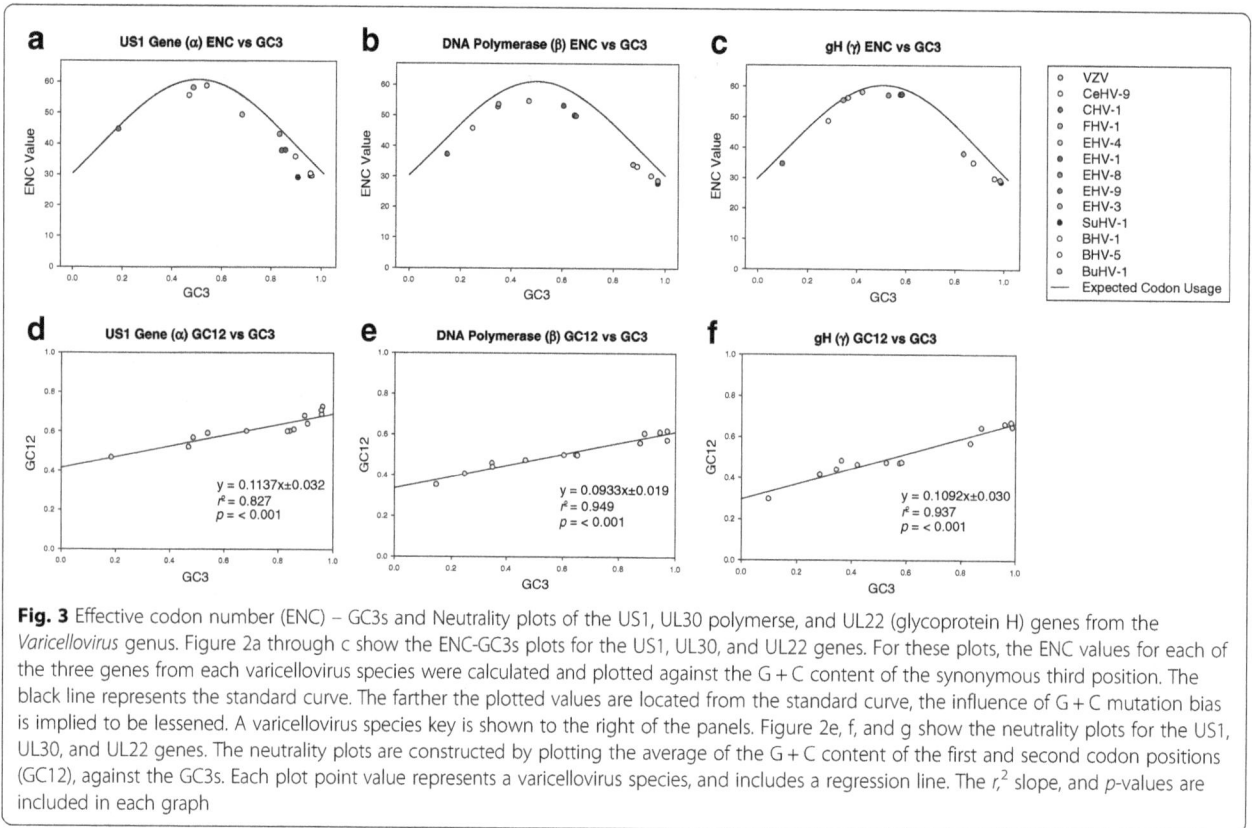

Fig. 3 Effective codon number (ENC) – GC3s and Neutrality plots of the US1, UL30 polymerse, and UL22 (glycoprotein H) genes from the *Varicellovirus* genus. Figure 2a through c show the ENC-GC3s plots for the US1, UL30, and UL22 genes. For these plots, the ENC values for each of the three genes from each varicellovirus species were calculated and plotted against the G + C content of the synonymous third position. The black line represents the standard curve. The farther the plotted values are located from the standard curve, the influence of G + C mutation bias is implied to be lessened. A varicellovirus species key is shown to the right of the panels. Figure 2e, f, and g show the neutrality plots for the US1, UL30, and UL22 genes. The neutrality plots are constructed by plotting the average of the G + C content of the first and second codon positions (GC12), against the GC3s. Each plot point value represents a varicellovirus species, and includes a regression line. The r_i^2 slope, and p-values are included in each graph

(Fig. 3b), which suggest additional pressures influencing bias, possibly including natural selection.

Neutrality plots (GC12 vs GC3s) for US1, UL30, and UL22 were generated to further examine mutation and natural selection biases (Fig. 3c, d, and e). The neutrality plots showed significant results for US1 (r^2 = 0.827; p = < 0.001), UL30 (r^2 = 0.949; p = < 0.001), and UL22 (r^2 = 0.937; p = < 0.001), which confirmed mutational bias. The slopes for all three of the neutrality plots were shallow (US1 = 0.1137, UL30 = 0.0933, and UL22 = 0.1092), which indicated that mutation bias influenced codon usage only 11.37%, 9.33%, and 10.92% for each of these genes, respectively. ENC and neutrality plots appear to result in somewhat different conclusions in investigating codon usage and mutational pressure. Care should be taken in interpretation, as the analysis is genus wide, and not in a single species. G+C constrained mutation bias in varicelloviruses was confirmed as has been previously shown in vertebrate DNA viruses [39], however additional factors such as natural selection are likely to play major role as well.

Bovine herpesvirus 1

BHV-1 has been traditionally divided into three subtypes; BHV-1.1, BHV-1.2a, and BHV-1.2b, with the 1.1 strains generally associated with IBR, and 1.2 with venereal disease phenotypes [59, 60]. It must be noted that strains of either type can cause respiratory and venereal disease phenotypes [61, 62], which suggests that genetic criteria may be a more reliable way to group BHV-1 strains than by clinical phenotype. To examine BHV-1 phylogeny, recombination, and genetic distance between clades, maximum likelihood trees and phylogenetic networks were generated, recombination bootscan analysis was conducted, and inter- as well as intra-clade distances were calculated. The ML tree and the phylogenetic network both recover two main clades (Fig. 4). Genetic distance analysis showed that the distance between the two groups was 1.12%, fulfilling the criteria for designating the clades 1.1 and 1.2. Thus, the BHV-1 clades retain the 1.x organization they had previously under our proposed nomenclature criteria. The genomic sequence analysis of BHV-1 recovered two main clades designated 1.1 and 1.2, however subclades within 1.2 were not detected. It is possible that as additional BHV-1 strains are sequenced, evidence of 1.2. subclades may become apparent. The overall genetic distance within BHV-1.1 was 0.60%, and within BHV-1.2 was 0.43% (Fig. 4b). Bootscan recombination analysis using BHV-1.2 strain B589 revealed extensive recombination from the remaining BHV-1.2 strains, but no significant recombinant signals from the BHV-1.1 viruses were detected (Fig. 4c). PHI recombination test analysis suggests that there is recombination in the dataset (p = <0.001; Table 1).

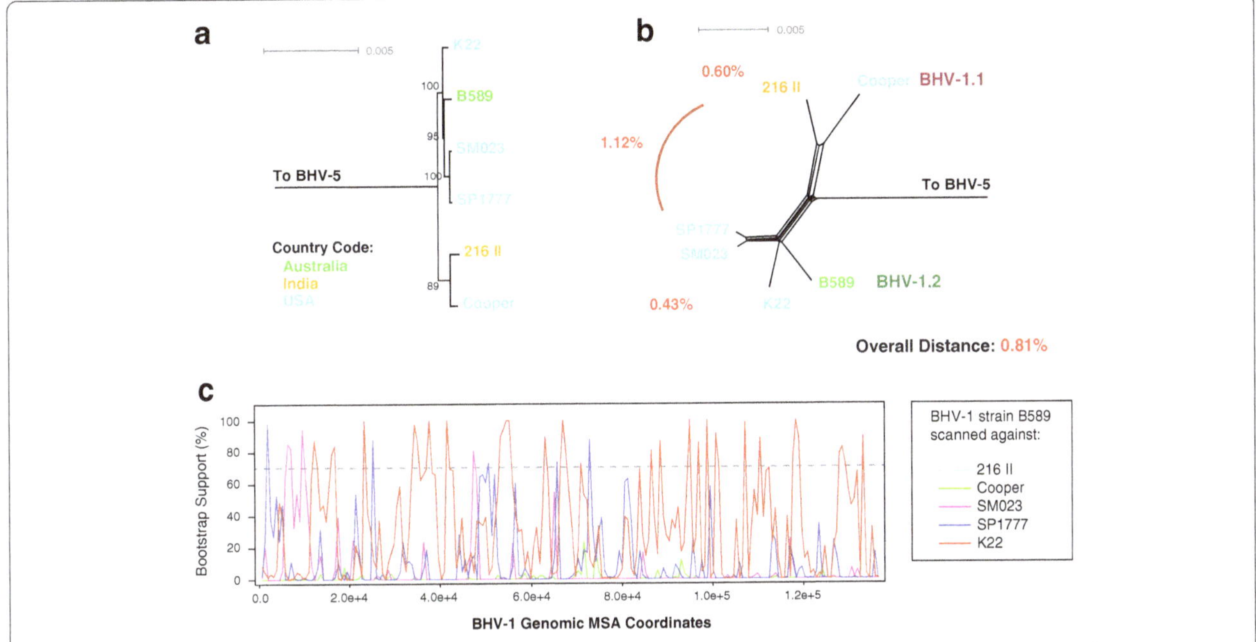

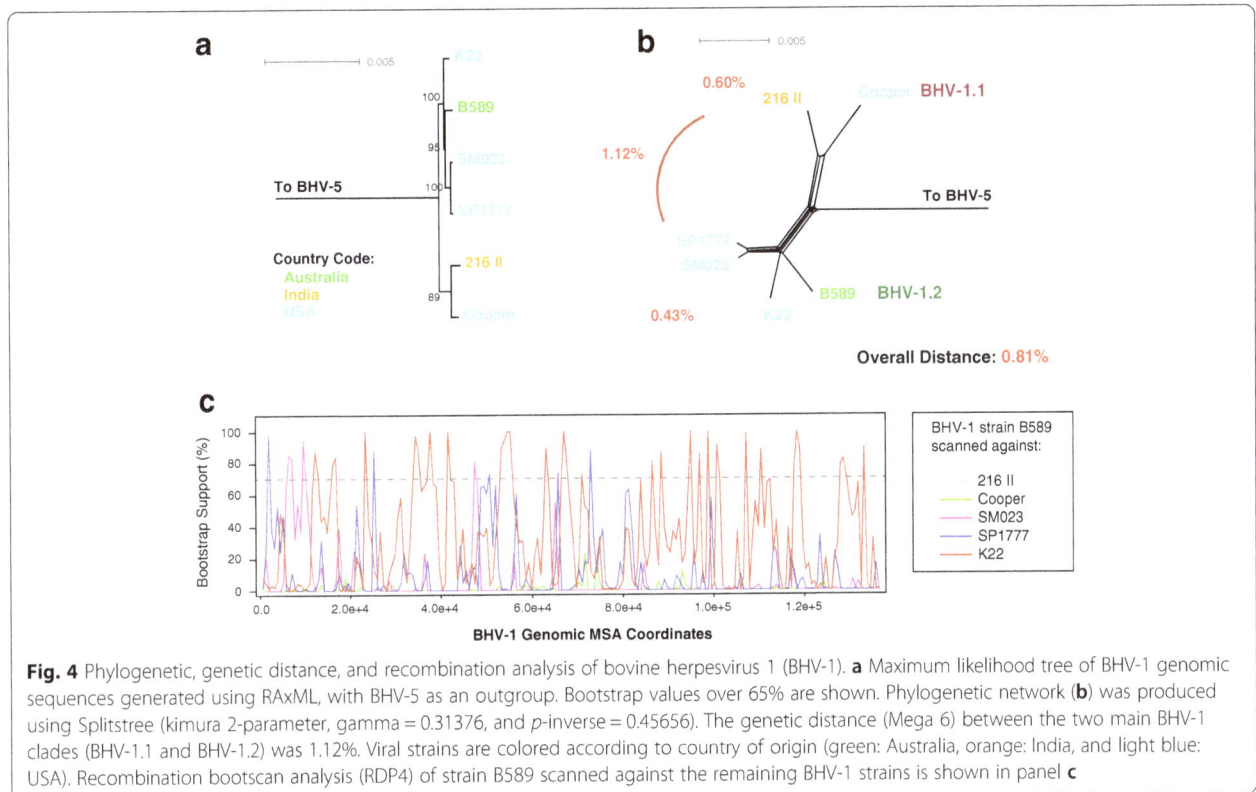

Fig. 4 Phylogenetic, genetic distance, and recombination analysis of bovine herpesvirus 1 (BHV-1). **a** Maximum likelihood tree of BHV-1 genomic sequences generated using RAxML, with BHV-5 as an outgroup. Bootstrap values over 65% are shown. Phylogenetic network (**b**) was produced using Splitstree (kimura 2-parameter, gamma = 0.31376, and *p*-inverse = 0.45656). The genetic distance (Mega 6) between the two main BHV-1 clades (BHV-1.1 and BHV-1.2) was 1.12%. Viral strains are colored according to country of origin (green: Australia, orange: India, and light blue: USA). Recombination bootscan analysis (RDP4) of strain B589 scanned against the remaining BHV-1 strains is shown in panel **c**

Pseudorabies virus/SuHV-1

The phylogenetic structure of SuHV-1 strains was examined next. The genome based ML tree and phylogenetic network both recovered two basic clades, Chinese and European/American, which have been previously identified [63] (Fig. 5a and b). The phylogenetic network however, showed that the Italian domestic dog isolated strain ADV32751 was separated somewhat from the remaining European/American strains (Fig. 5b). Distance analysis showed that the genetic distance between the main Chinese and European/American clades was 2.76% (Fig. 5b). This comparatively large genetic distance between the two groups appears to be consistent with what is thought to be two independent domestication events, one in China [64], and another in modern day Turkey roughly 9000 years before present [65]. Additional calculations showed that the distance between strain ADV32751 and the remaining European/American strains was 1.44% (Fig. 5b). Based on the results of the distance calculations, we suggest that pseudorabies virus be designated as SuHV-1.1 (Chinese), SuHV-1.2 (main European/American), and provisional SuHV-1.3 (strain ADV32751) (Fig. 5b). It is unclear if the ADV32751 strain contains mutations which could have enhanced transmission to a domestic dog. Additionally, given the distance value with respect to the SuHV-1.2 viruses, the chance of genetic contributions from a European wild boar strain should not be excluded. Within the European/American SuHV-1.2 clade, two additional groupings were detected, and designated 1 and 2 based on the genetic distance (0.37%; Fig. 5b). A bootscan using SuHV-1.2 strain NIA3 against the remaining strains showed little to no recombination signals from either SuHV-1.1 or 1.3 (Fig. 5c). The lack of recombination signals between the SuHV-1.2 and 1.1 subclades is not unexpected due to geographic distances, however a recent report showed that the Chinese pseudorabies virus strain SC contained genomic contributions from the vaccine strain Bartha [30]. The PHI recombination test of all the SuHV-1 strains showed (Table 1) statistically significant recombination within the dataset ($p = <0.001$).

EHV-1

The ML tree (Fig. 6a) shows a split between the wild and domestic horse derived EHV-1. The phylogenetic network also confirms this split (Fig. 6b). Genetic distance calculations resulted in 2.92% distance between the wild and domestic EHV-1 clades, and we suggest designating these EHV-1.1 (wild equine) and EHV-1.2 (domestic horse). The distance within the wild horse EHV-1.1 clade was higher than EHV-1.2, at 0.61% vs. 0.18% respectively. An expansion of the EHV-1.2 clade is shown in Fig. 5c, and shows three provisional clades, with clades 1 and 2 being 0.135% distant, and clades 2

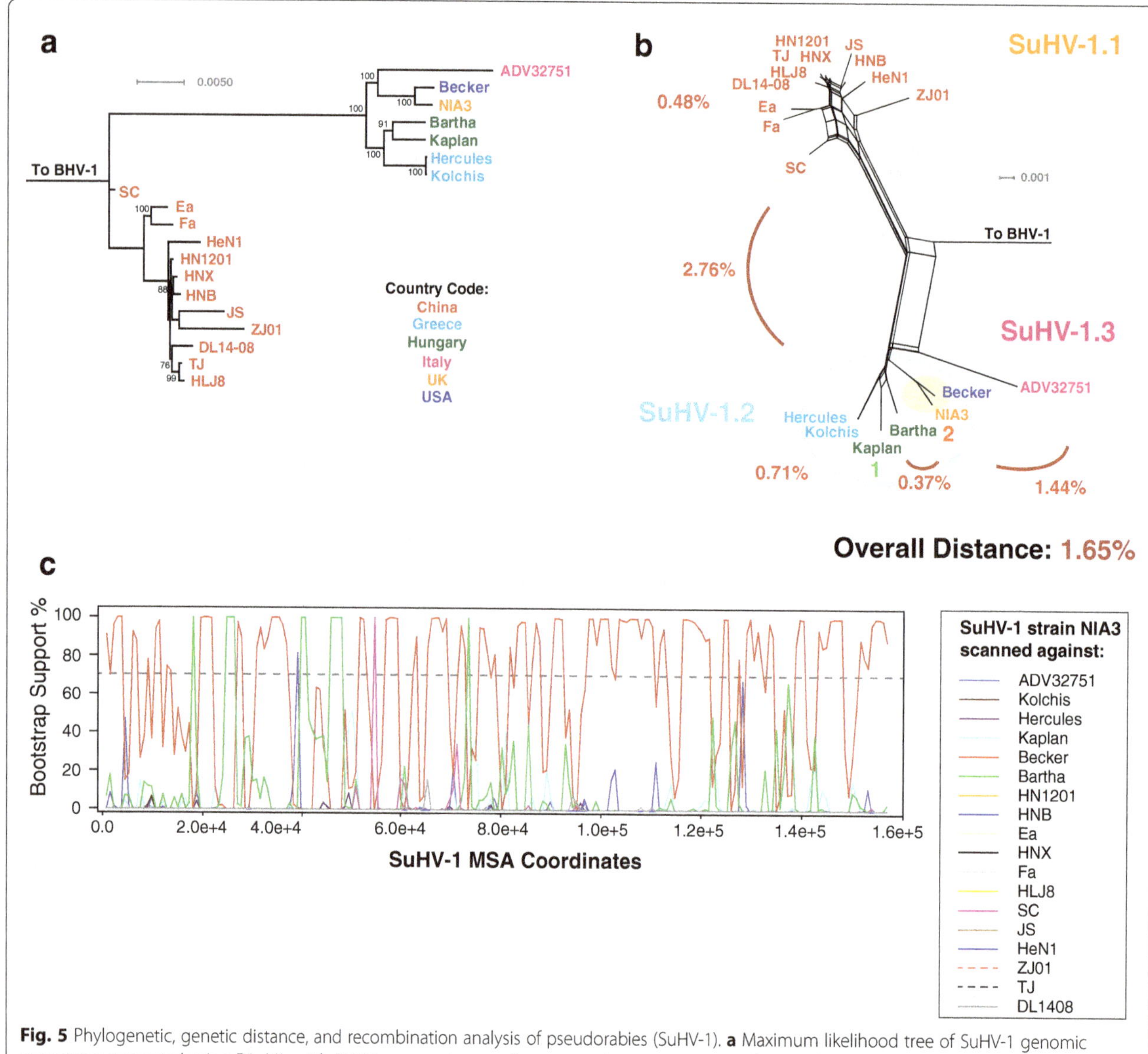

Fig. 5 Phylogenetic, genetic distance, and recombination analysis of pseudorabies (SuHV-1). **a** Maximum likelihood tree of SuHV-1 genomic sequences generated using RAxML, with BHV-1 as an outgroup. Bootstrap values over 65% are shown. Phylogenetic network (**b**) was produced using Splitstree (kimura 2-parameter, gamma = 0.1520, and *p*-inverse = 0. 0.240). The genetic distance (Mega 6) between the Chinese (SuHV-1.1) and European/American (SuHV-1.2) clades was 2.76%. Recombination bootscan analysis (RDP4) of strain NAI3 scanned against the remaining SuHV-1 strains is shown in panel **c**. Viral strains in panels A and B are colored according to country of origin (**a**)

and 3 0.137% distant. Two strains, NY03 and 5586, are separated from the remaining EVH-1.2 viruses and may represent a separate clade, however additional strains need to be islolated before this can be determined. EHV-1.2 strains NMKT04 and V592 occupy a position between clades 1 and 2 may be interclade recombinants. It is notable that EHV-1.2 strains do not appear to correlate to geographic origin, and may reflect the cosmopolitan nature of common breeds such as the Thoroughbred. Bootstrap recombination analysis scanning EHV-1.2 group 3 strain Va02 against the remaining strains showed no recombination from EHV-1.1 stains (Fig. 6c). When the EHV-1 strains were examined using the PHI recombination test (Table 1), statistically significant recombination was detected ($p = <0.001$). Wild equine derived EHV-1 strains cause severe infections, often neurological in both equine and non-equine captive animals [66–70], and some domestic horse EHV-1.2 strains can also cause myeloencephalopathy [71, 72]. A SNP (D752) within the polymerase gene of domestic horse viruses has been shown to influence the neurological disease phenotype, and is shared among wild equine strains [73, 74]. The genomic phylogenetic analysis (Fig. 5) of the domestic horse (EHV-1.2) strains did not sort the strains based on disease phenotype, i.e. neuropathology vs abortion (Fig. 6c). This finding along with data showing that

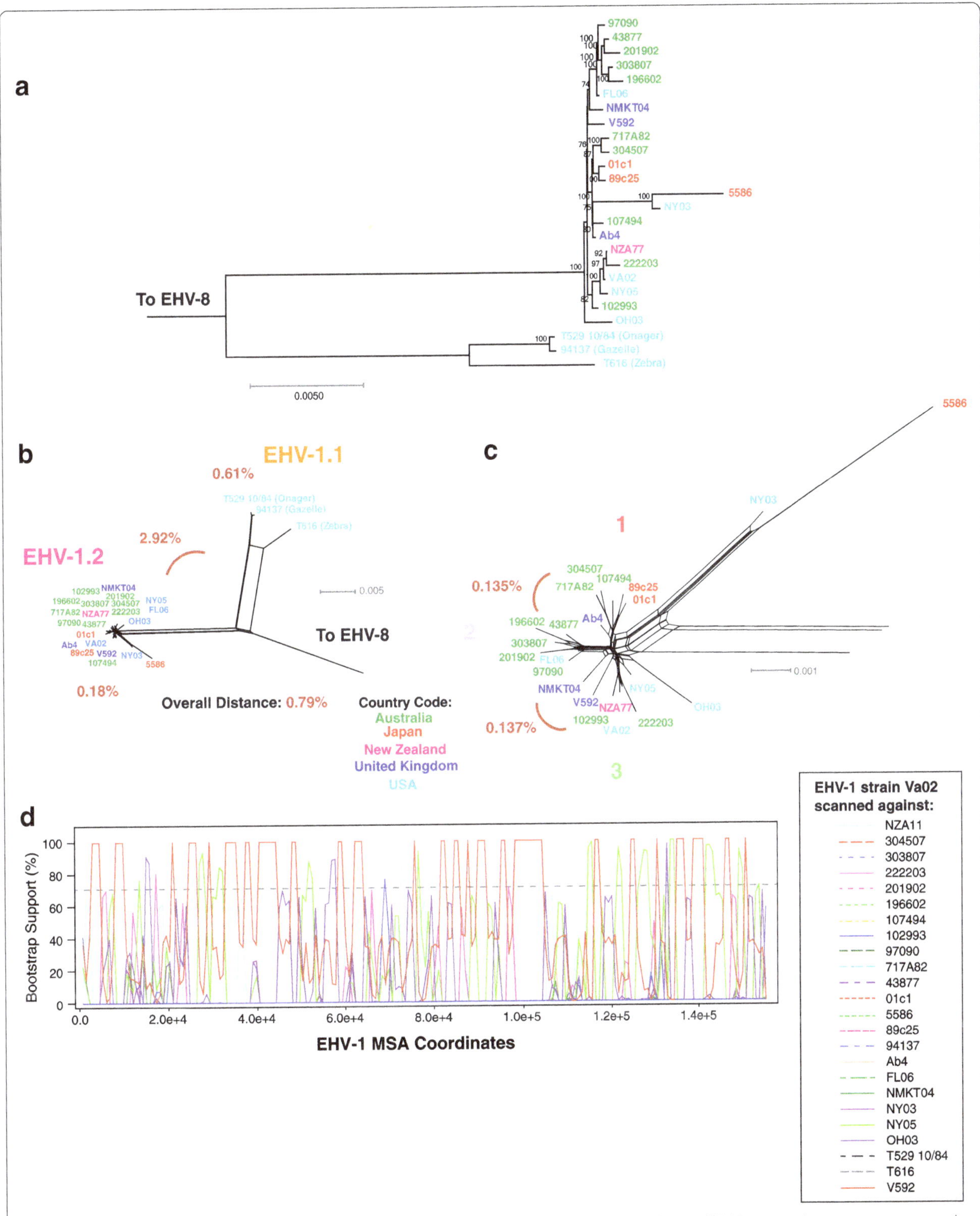

Fig. 6 Phylogenetic, genetic distance, and recombination analysis of EHV-1. **a** Maximum likelihood tree of EHV-1 genomic sequences generated using RAxML, with EHV-8 as an outgroup. Bootstrap values over 65% are shown. Phylogenetic network (**b**) was produced using Splitstree (kimura 2-parameter, gamma = 1.0360, and *p*-inverse = 0.4940). The genetic distance (Mega 6) between the wild equine (EHV-1.1) and domestic horse (EHV-1.2) clades was 2.92%. (Panel **c**) A zoom of the domestic horse (EHV-1.2) strains from panel B shows one main grouping (**a**) and provisional B and C groups. Recombination bootscan analysis (RDP4) of strain Va02 scanned against the remaining EHV-1 strains is shown in Panel **d**. No recombination signals were detected from the EHV-1.1 strains. Viral strains in panels **a** and **b** are colored according to country of origin (Panel **c**)

non-D752 strains [75] can cause encephalitis strongly suggests multiple genes contribute to EHV-1 disease phenotypes, as has been observed in HSV-1 [76–79].

EHV-4

Equine herpesvirus type 4 is an important equine disease, and causes rhinopneumonitis most commonly in foals [80], however it is not a reportable disease as is EHV-1. The phylogenetic ML tree (Fig. 7a) and phylogenetic network (Fig. 7b) of the available EHV-4 showed a split into two main groups (clade 1 and 2) as has been shown previously [32]. The genetic distance between the two clades was 0.23%. Both the ML tree and phylogenetic network showed that the EHV-4 viruses, like EHV-1, do not sort according to geographic origin and this is likely the result of the modern movement of common breeds globally. Recombination bootscan analysis (Fig. 7c) scanning EHV-4 group A strain 12-I-203 against the rest showed extensive recombination from both groups 1 and 2. Additional PHI recombination test analysis (Table 1) detected significant amounts of recombination in EHV-4 ($p = <0.001$).

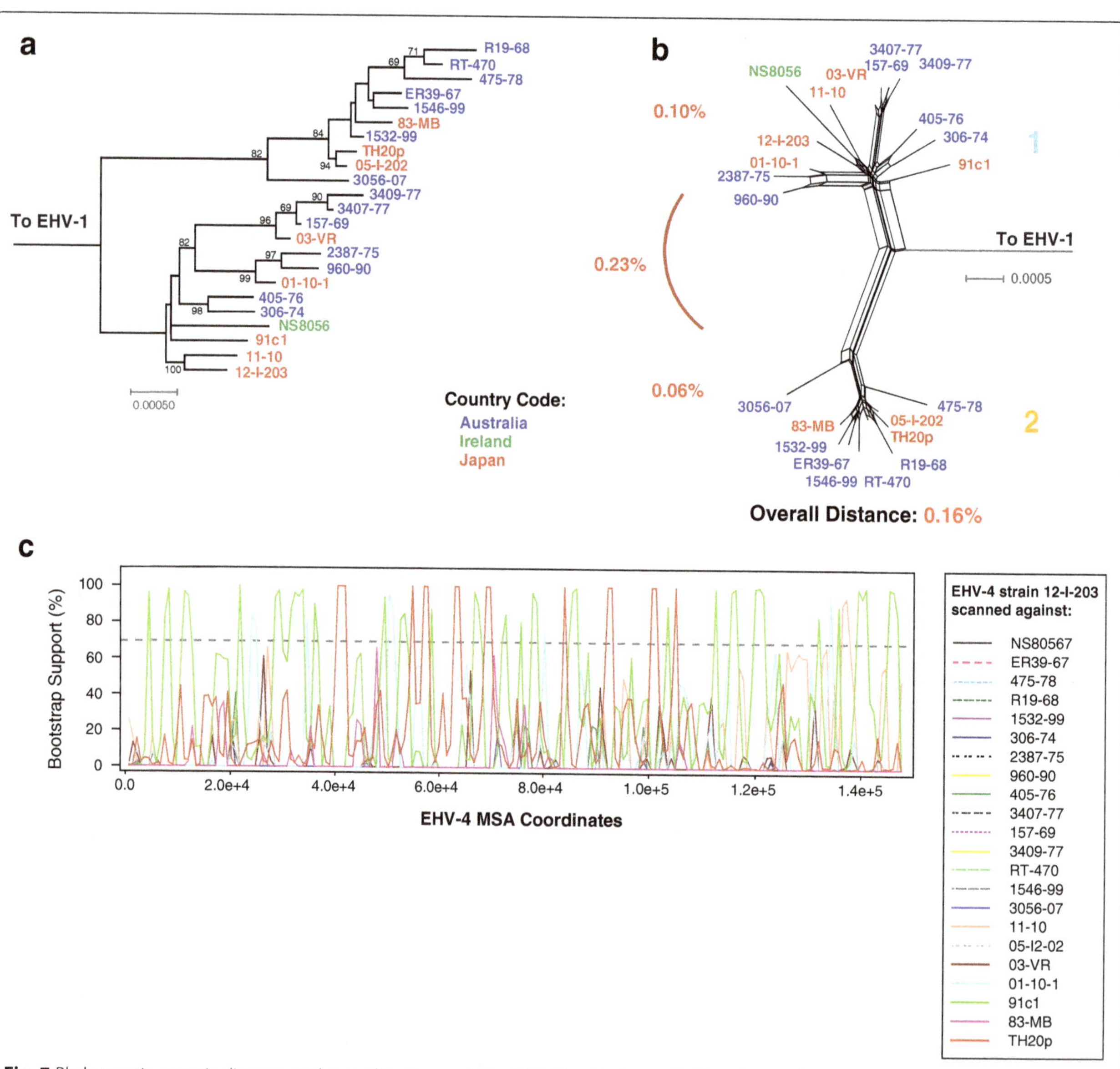

Fig. 7 Phylogenetic, genetic distance, and recombination analysis of EHV-4. **a** Maximum likelihood tree of EHV-4 genomic sequences generated using RAxML, with EHV-1 as an outgroup. Bootstrap values over 65% are shown. Phylogenetic network (**b**) was produced using Splitstree (kimura 2-parameter, gamma = 0.8230, and *p*-inverse = 0.2770). The genetic distance (Mega 6) between group A and group B was 0.23%. Recombination bootscan analysis (RDP4) of strain 12-I-203 scanned against the remaining EHV-4 strains is shown in panel **c**. Viral strains in panels **a** and **b** are colored according to country of origin (Panel **a**)

FHV-1

Feline herpes virus type 1 (FHV-1) is thought to be main cause of corneal ulceration [81] in cats, and can also contribute to upper respiratory disease [82]. Recently, several FHV-1 genomes from Australia were sequenced and genomic analysis did not reportedly detect recombination [83]. As part of our analysis of varicellovirus phylogenetic relationships, we analyzed the available FHV-1 genomic sequences. The ML tree and phylogenetic network (Fig. 8a and b) suggested some strain grouping, however, because the overall interstrain genetic distance is low (0.0089%), we did not designate any clades. Reticulations within the phylogenetic network implied the presence of recombination between the FHV-1 strains. Bootscan recombination analysis (Fig. 8c), as well the PHI recombination test (Table 1; $p = <0.001$) detected recombination signals. The difficulty in detecting recombination is likely due to the low interstrain genetic distance (0.0089%), and it is possible that as additional strains are sequenced, more recombination may be more readily identified. It would be surprising if recombination was not detected in FHV-1, as herpesviruses have been shown to be highly recombinagenic [29, 31, 84, 85].

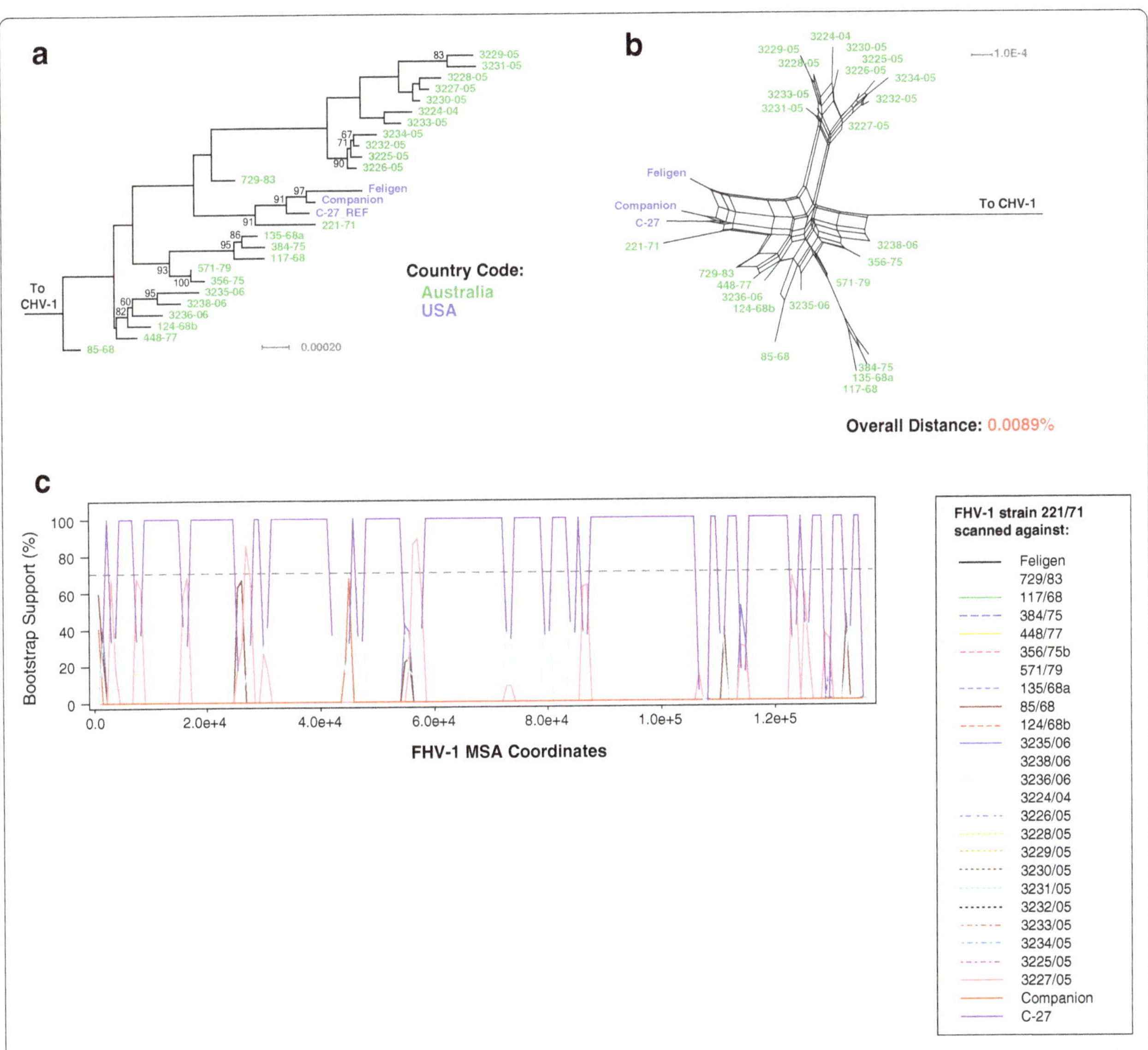

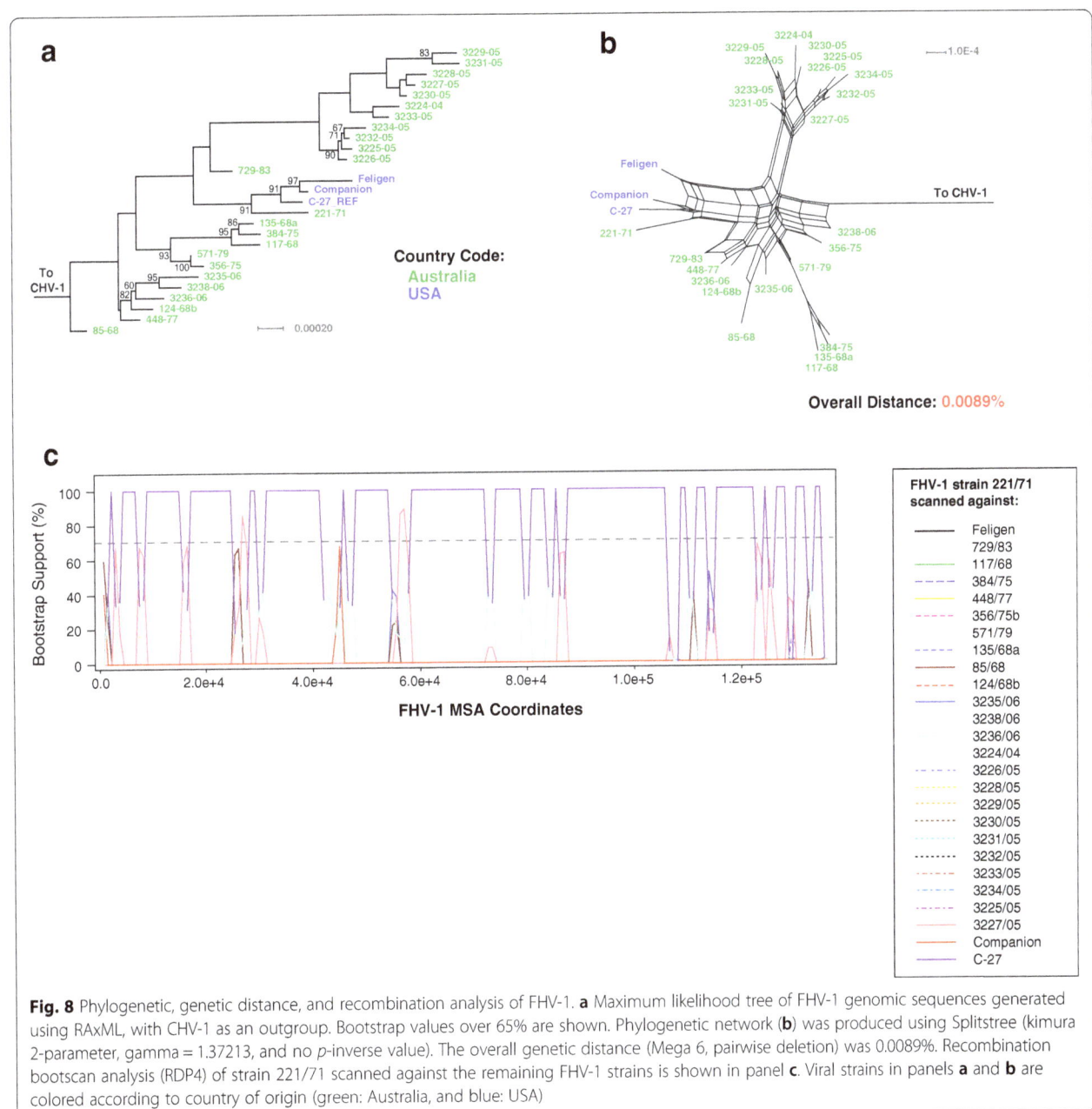

Fig. 8 Phylogenetic, genetic distance, and recombination analysis of FHV-1. **a** Maximum likelihood tree of FHV-1 genomic sequences generated using RAxML, with CHV-1 as an outgroup. Bootstrap values over 65% are shown. Phylogenetic network (**b**) was produced using Splitstree (kimura 2-parameter, gamma = 1.37213, and no *p*-inverse value). The overall genetic distance (Mega 6, pairwise deletion) was 0.0089%. Recombination bootscan analysis (RDP4) of strain 221/71 scanned against the remaining FHV-1 strains is shown in panel **c**. Viral strains in panels **a** and **b** are colored according to country of origin (green: Australia, and blue: USA)

CHV-1

Until very recently, only three CHV-1 strains, collected between 1985 and 2000 from the UK had been sequenced [86], however a short time ago, a CHV-1 strain from Brazil was deposited into GenBank. The overall genetic distance between the three UK strains was very low, at 0.005% (Table 1; Fig. 9a). The CHV-1 strain from Brazil (strain BTU-1) was 0.34% distant from the three UK viruses, with an overall interstrain distance of 0.20% for the four viruses (Table 2; Fig. 9a). We performed a similarity plot using the MSA (multiple sequence alignment) without an outgroup, and found a deep trough at approximately 9 kb from the left end of the genome (Fig. 9b). The similarity trough corresponded to the UL50 deoxyuridine triphosphatase gene. Distance analysis comparing the UL50 protein sequences from the four CHV-1 viruses showed that the Brazilian BTU-1 strain UL50 was 12.2% distant compared to the UK

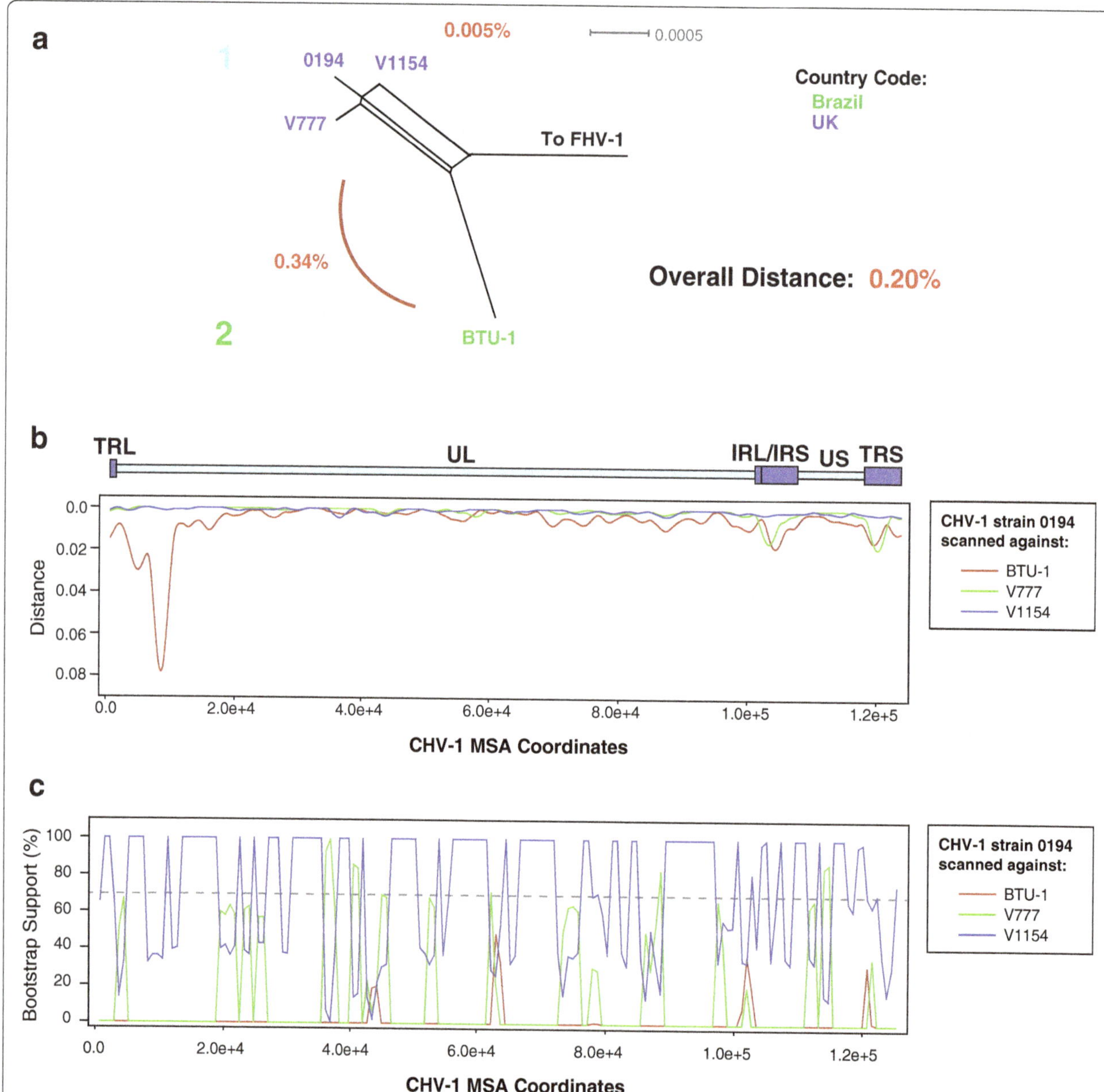

Fig. 9 Phylogenetic, genetic distance, and recombination analysis of CHV-1. **a** Phylogenetic network was produced using Splitstree (kimura 2-parameter, gamma = 0.75049, and no *p*-inverse value). The genetic distance (Mega 6) between group A and group B was 0.34% (Panel **a**). A genomic similarity plot (RDP4, Panel B) was generated by scanning UK strain 0194 against the remaining CHV-1 strains. A schematic of the CHV-1 genome has been placed above the Simplot graph, with TRL = terminal repeat long, UL = unique long coding region, IRL = internal repeat long, IRS = internal repeat short, US = unique short coding region, and TRS = terminal repeat short. Recombination bootscan analysis (RDP4) of strain 0194 scanned against the remaining EHV-4 strains is shown in panel **c**. Viral strains in the phylogenetic network (**a**) are colored according to country of origin (green: Brazil, and blue: UK)

Table 4 UL47 to UL54 maximum likelihood based protein distances of CHV-1 strain BTU-1 compared to UK derived viruses

Protein	Percent distance
UL47	0.78
UL48	0.23
UL49	0.12
UL49A	0
UL50	12.2
UL51	2.1
UL52	1.63
UL53	1.2
UL54	0

strains (Table 4). Further blast searches (data not shown) determined that even though the BTU-1 strain UL50 protein sequence was 12.2% distant, it appeared closest to the remaining CHV-1 strains, rather than FHV-1. Bootscan recombination analysis using the UK derived 0194 strain as a reference only detected recombination signals from the other UK viruses, and none from BTU-1 (Fig. 9c). Curiously, the PHI recombination test did not detect statistically significant recombination (Table 1; $p = 0.3082$), and may be due to the small size of the dataset. Based on the data, we hypothesize that the BTU-1 strain may be the result of a recombination event between canine herpesvirus 1, and an unknown varicellovirus. It would be unlikely that positive selection would only affect a single gene in the virus to such a large extent (12.2% distance), however the possibility cannot be eliminated. Because the UL50 sequence most closely resembles the remaining CHV-1 viruses, it is likely that the unknown virus originated from an animal of the Caniformia suborder, which includes Brazilian species such as the maned wolf, bush dog, pampas fox, tayra, striped hog-nosed skunk, and crab-eating racoon.

Varicella zoster virus

The Varicella-Zoster virus (VZV) causes chickenpox as well as shingles, and the phylogeny of VZV clades has been extensively studied [29]. We treated our analysis of VZV as an update, as additional strains have been sequenced and uploaded into GenBank. A ML tree and phylogenetic network using CeHV-9 (simian varicella virus) as the outgroup were constructed and are found in Fig. 10. The phylogenetic network suggests six clades, which are denoted numerically as described previously

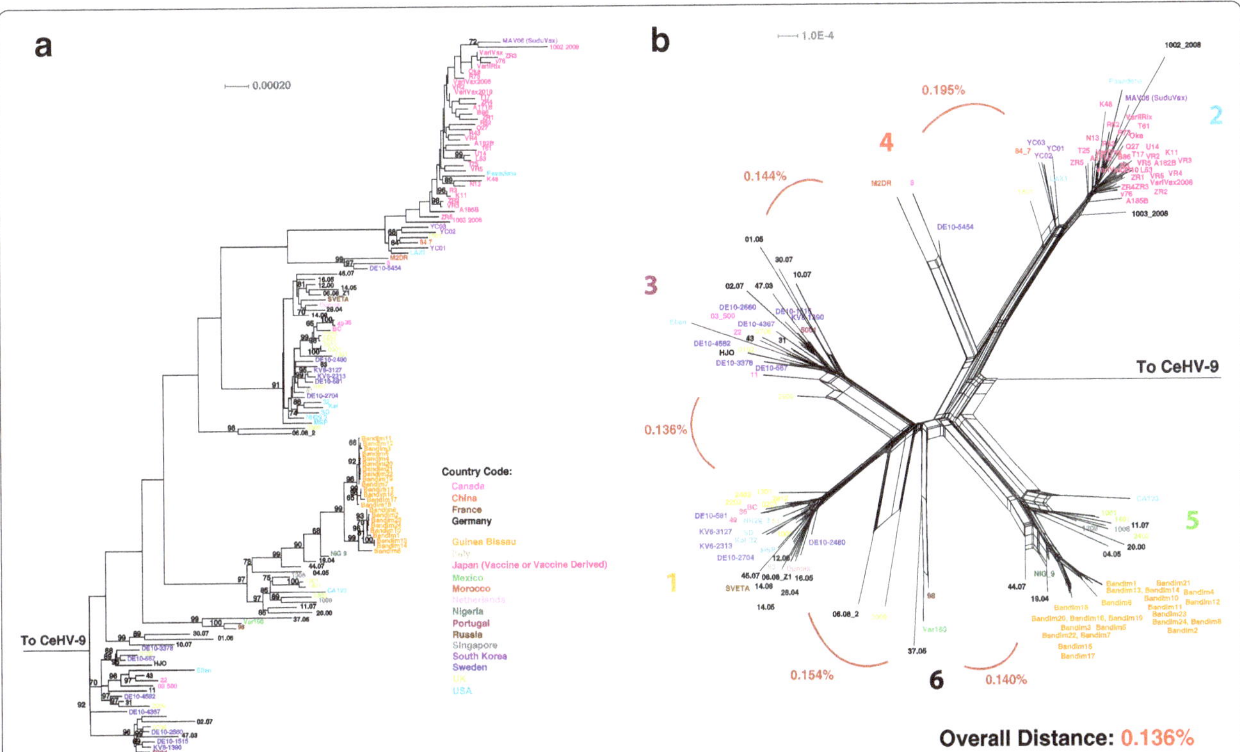

Fig. 10 Phylogenetic, genetic distance, and recombination analysis of VZV. **a** Maximum likelihood tree of VZV genomic sequences generated using RAxML, with CeHV-9 as an outgroup. Bootstrap values over 65% are shown. A Phylogenetic network (**b**) was produced using Splitstree (kimura 2-parameter, gamma = 0.50800, and *p*-inverse = 0.19700). There are seven clades A-G, and the overall genetic distance (Mega 6, pairwise deletion) was 0.136%. Viral strains in panels **a** and **b** are colored according to country of origin (Panel **a**)

[29]. (Fig. 9b). PHI recombination analysis confirmed statistically significant recombination among the VZV strains (Table 1; $p = <0.001\%$).

Conclusions

In conclusion, this is the first genome based phylogenetic study of the entire *Varicellovirus* genus. In this study, we present a number of unique findings including results suggesting that a phylogenetic stricture exists between the ungulate viruses and the primate and carnivore viruses, a possible link between genome G + C content and intraspecies strain genetic diversity, the detection of recombination in all of the varicellovirus species including FHV-1, and that the Brazilian CHV-1 strain BTU may contain a genetic signal from an unknown varicellovirus in the UL50 gene. We also propose a clade nomenclature standardization for varicelloviruses. This work helps to deepen the understanding of varicellovirus genomics and evolution.

Additional file

Additional file 1: Table S1. Accession numbers for the sequences used in the present study. The species, isolation host, strain designation, genome size, country of isolation, isolation source, collection date, and accession number for each sequence is provided. **Table S2.** Unscaled *F* values. The the only reported *F* values were those which included cut-offs that result in two groups for all viruses. **Table S3.** Rescaled *F* values. The rescaled values of *F* simply indicates the relative size of *F*. The total value is on a scale from 0 to 3, with 3 meaning that the separation is as good as possible for all three viruses. (ZIP 42 kb)

Abbreviations

AnHV-1: *Anatid herpes virus type 1*; BHV-1: *Bovine herpes virus type 1*; BHV-5: *Bovine herpes virus type 5*; BuHV-1: *Bubaline herpes virus type 1*; CeHV-9: *Cercopithecine herpes virus type 9*; CHV-1: *Canine herpes virus type 1*; EHV-1: *Equine herpes virus type 1*; EHV-3: *Equine herpes virus type 3*; EHV-4: *Equine herpes virus type 4*; EHV-8: *Equine herpes virus type 8*; EHV-9: *Equine herpes virus type 9*; ENC: Effective number of codons; FHV-1: *Feline herpesvirus type 1*; ICTV: International Committee on Taxonomy of Viruses; ML: Maximum likelihood; MSA: Multiple sequence alignment; NGS: Next-generation sequencing; PCV2: *Porcine circovirus type 2*; PHI: Pairwise homoplasty index; SuHV-1: Pseudorabies virus; USDA: United States Department of Agriculture; VZV: Varicella zoster virus

Acknowledgements

We would like to acknowledge Dr. Ellison Bentley for her general support.

Funding

This study was supported by a Core Grant for Vision Research (EYP30016665), an unrestricted grant to the Department of Ophthalmology and Visual Sciences from Research to Prevent Blindness, Inc. and the Clinical and Translational Science Award (CTSA) program, through the NIH National Center for Advancing Translational Sciences (NCATS), grant UL1TR000427. The funding sources had no role in study design, data collection, data analysis, data interpretation, or the writing of the manuscript.

Author's contributions

AK, AL, RMT and CRB conceived and designed the experiments. AK, AL, and RMT performed the experiments. AK, AL, RMT and CRB analyzed the data. AK, AL, RMT, GM, and CRB contributed to the writing of the manuscript. The authors have read, and approved this manuscript.

Competing interests

The authors declare that they have no competing interests.

Author details

[1]Department of Ophthalmology and Visual Sciences, School of Medicine and Public Health, University of Wisconsin-Madison, 550 Bardeen Laboratories, 1300 University Ave., Madison, WI 53706, USA. [2]Department of Surgical Sciences, School of Veterinary Medicine, University of Wisconsin-Madison, Madison, WI, USA. [3]McPherson Eye Research Institute, University of Wisconsin-Madison, Madison, WI, USA. [4]Medical Microbiology and Immunology, School of Medicine and Public Health, University of Wisconsin-Madison, Madison, WI, USA.

References

1. Cardoso TC, Ferreira HL, Okamura LH, Oliveira BR, Rosa AC, Gameiro R, Flores EF. Comparative analysis of the replication of bovine herpesvirus 1 (BHV1) and BHV5 in bovine-derived neuron-like cells. Arch Virol. 2015;160(11): 2683–91.
2. Charlton KM, Mitchell D, Girard A, Corner AH. Meningoencephalomyelitis in horses associated with equine herpesvirus 1 infection. Vet Pathol. 1976; 13(1):59–68.
3. Verheyen K, Newton JR, Wood JL, Birch-Machin I, Hannant D, Humberstone RW. Possible case of EHV-4 ataxia in warmblood mare. Vet Rec. 1998; 143(16):456.
4. Card JP, Rinaman L, Schwaber JS, Miselis RR, Whealy ME, Robbins AK, Enquist LW. Neurotropic properties of pseudorabies virus: uptake and transneuronal passage in the rat central nervous system. J Neurosci. 1990;10(6): 1974–94.
5. Steiner I, Kennedy PG, Pachner AR. The neurotropic herpes viruses: herpes simplex and varicella-zoster. Lancet Neurol. 2007;6(11):1015–28.
6. Fukushi H, Tomita T, Taniguchi A, Ochiai Y, Kirisawa R, Matsumura T, Yanai T, Masegi T, Yamaguchi T, Hirai K. Gazelle herpesvirus 1: a new neurotropic herpesvirus immunologically related to equine herpesvirus 1. Virology. 1997; 227(1):34–44.
7. Ledbetter EC, Marfurt CF, Dubielzig RR. Metaherpetic corneal disease in a dog associated with partial limbal stem cell deficiency and neurotrophic keratitis. Vet Ophthalmol. 2013;16(4):282–8.
8. Hora AS, Tonietti PO, Guerra JM, Leme MC, Pena HF, Maiorka PC, Brandao PE. Felid herpesvirus 1 as a causative agent of severe nonsuppurative meningoencephalitis in a domestic cat. J Clin Microbiol. 2013;51(2):676–9.
9. Gray WL. Simian varicella in old world monkeys. Comp Med. 2008;58(1):22–30.
10. Heberden W. On the chicken-pox. Med Trans Coll Phys. 1767;1:428.
11. USDA: U.S. National List of Reportable Animal Diseases (NLRAD)- National Animal Health Reporting System (NAHRS) reportable disease list. 2017.
12. Takahashi M, Otsuka T, Okuno Y, Asano Y, Yazaki T. Live vaccine used to prevent the spread of varicella in children in hospital. Lancet. 1974;2(7892):1288–90.
13. Osterrieder N, Neubauer A, Brandmuller C, Kaaden OR, O'Callaghan DJ. The equine herpesvirus 1 IR6 protein influences virus growth at elevated temperature and is a major determinant of virulence. Virology. 1996;226(2):243–51.
14. Kit S, Kit M, Pirtle EC. Attenuated properties of thymidine kinase-negative deletion mutant of pseudorabies virus. Am J Vet Res. 1985;46(6):1359–67.
15. Kit S, Sheppard M, Ichimura H, Kit M. Second-generation pseudorabies virus vaccine with deletions in thymidine kinase and glycoprotein genes. Am J Vet Res. 1987;48(5):780–93.

16. Moormann RJ, de Rover T, Briaire J, Peeters BP, Gielkens AL, van Oirschot JT. Inactivation of the thymidine kinase gene of a gI deletion mutant of pseudorabies virus generates a safe but still highly immunogenic vaccine strain. J Gen Virol. 1990;71(Pt 7):1591–5.
17. Kaashoek MJ, Moerman A, Madic J, Rijsewijk FA, Quak J, Gielkens AL, van Oirschot JT. A conventionally attenuated glycoprotein E-negative strain of bovine herpesvirus type 1 is an efficacious and safe vaccine. Vaccine. 1994;12(5):439–44.
18. Galeota JA, Flores EF, Kit S, Kit M, Osorio FA. A quantitative study of the efficacy of a deletion mutant bovine herpesvirus-1 differential vaccine in reducing the establishment of latency by wildtype virus. Vaccine. 1997;15(2):123–8.
19. Heldens JG, Hannant D, Cullinane AA, Prendergast MJ, Mumford JA, Nelly M, Kydd JH, Weststrate MW, van den Hoven R. Clinical and virological evaluation of the efficacy of an inactivated EHV1 and EHV4 whole virus vaccine (Duvaxyn EHV1,4). Vaccination/challenge experiments in foals and pregnant mares. Vaccine. 2001; 19(30):4307–17.
20. Jas D, Aeberle C, Lacombe V, Guiot AL, Poulet H. Onset of immunity in kittens after vaccination with a non-adjuvanted vaccine against feline panleucopenia, feline calicivirus and feline herpesvirus. Vet J. 2009; 182(1):86–93.
21. Lappin MR, Sebring RW, Porter M, Radecki SJ, Veir J. Effects of a single dose of an intranasal feline herpesvirus 1, calicivirus, and panleukopenia vaccine on clinical signs and virus shedding after challenge with virulent feline herpesvirus 1. J Feline Med Surg. 2006;8(3):158–63.
22. Poulet H, Guigal PM, Soulier M, Leroy V, Fayet G, Minke J, Chappuis Merial G. Protection of puppies against canine herpesvirus by vaccination of the dams. Vet Rec. 2001;148(22):691–5.
23. Raaperi K, Orro T, Viltrop A. Epidemiology and control of bovine herpesvirus 1 infection in Europe. Vet J. 2014;201(3):249–56.
24. Davison AJ, Scott JE. The complete DNA sequence of varicella-zoster virus. J Gen Virol. 1986;67(Pt 9):1759–816.
25. Telford EA, Watson MS, McBride K, Davison AJ. The DNA sequence of equine herpesvirus-1. Virology. 1992;189(1):304–16.
26. Telford EA, Watson MS, Perry J, Cullinane AA, Davison AJ. The DNA sequence of equine herpesvirus-4. J Gen Virol. 1998;79(Pt 5):1197–203.
27. Delhon G, Moraes MP, Lu Z, Afonso CL, Flores EF, Weiblen R, Kutish GF, Rock DL. Genome of bovine herpesvirus 5. J Virol. 2003;77(19):10339–47.
28. McGeoch DJ, Dolan A, Ralph AC. Toward a comprehensive phylogeny for mammalian and avian herpesviruses. J Virol. 2000;74(22):10401–6.
29. Norberg P, Depledge DP, Kundu S, Atkinson C, Brown J, Haque T, Hussaini Y, MacMahon E, Molyneaux P, Papaevangelou V, et al. Recombination of globally circulating varicella-zoster virus. J Virol. 2015;89(14):7133–46.
30. Ye C, Guo JC, Gao JC, Wang TY, Zhao K, Chang XB, Wang Q, Peng JM, Tian ZJ, Cai XH, et al. Genomic analyses reveal that partial sequence of an earlier pseudorabies virus in China is originated from a Bartha-vaccine-like strain. Virology. 2016;491:56–63.
31. Vaz PK, Horsington J, Hartley CA, Browning GF, Ficorilli NP, Studdert MJ, Gilkerson JR, Devlin JM. Evidence of widespread natural recombination among field isolates of equine herpesvirus 4 but not among field isolates of equine herpesvirus 1. J Gen Virol. 2016;97(3):747–55.
32. Izume S, Kirisawa R, Ohya K, Ohnuma A, Kimura T, Omatsu T, Katayama Y, Mizutani T, Fukushi H. The full genome sequences of 8 equine herpesvirus type 4 isolates from horses in Japan. J Vet Med Sci. 2017;79(1):206–12.
33. Marin A, Bertranpetit J, Oliver JL, Medina JR. Variation in G + C-content and codon choice: differences among synonymous codon groups in vertebrate genes. Nucleic Acids Res. 1989;17(15):6181–9.
34. Sau K, Gupta SK, Sau S, Ghosh TC. Synonymous codon usage bias in 16 *Staphylococcus aureus* phages: implication in phage therapy. Virus Res. 2005; 113(2):123–31.
35. Li G, Ji S, Zhai X, Zhang Y, Liu J, Zhu M, Zhou J, Su S. Evolutionary and genetic analysis of the VP2 gene of canine parvovirus. BMC Genomics. 2017;18(1):534.
36. Butt AM, Nasrullah I, Qamar R, Tong Y. Evolution of codon usage in Zika virus genomes is host and vector specific. Emerg Microbes Infect. 2016;5(10):e107.
37. Singh NK, Tyagi A, Kaur R, Verma R, Gupta PK. Characterization of codon usage pattern and influencing factors in Japanese encephalitis virus. Virus Res. 2016;221:58–65.
38. Roychoudhury S, Mukherjee D. A detailed comparative analysis on the overall codon usage pattern in herpesviruses. Virus Res. 2010;148(1–2):31–43.
39. Shackelton LA, Parrish CR, Holmes EC. Evolutionary basis of codon usage and nucleotide composition bias in vertebrate DNA viruses. J Mol Evol. 2006;62(5):551–63.
40. He W, Zhang H, Zhang Y, Wang R, Lu S, Ji Y, Liu C, Yuan P, Su S. Codon usage bias in the N gene of rabies virus. Infect Genet Evol. 2017;54:458–65.
41. Katoh K, Misawa K, Kuma K, Miyata T. MAFFT: a novel method for rapid multiple sequence alignment based on fast Fourier transform. Nucleic Acids Res. 2002;30(14):3059–66.
42. Katoh K, Toh H. Parallelization of the MAFFT multiple sequence alignment program. Bioinformatics. 2010;26(15):1899–900.
43. Tamura K, Stecher G, Peterson D, Filipski A, Kumar S. MEGA6: molecular evolutionary genetics analysis version 6.0. Mol Biol Evol. 2013;30(12):2725–9.
44. Berger SA, Krompass D, Stamatakis A. Performance, accuracy, and web server for evolutionary placement of short sequence reads under maximum likelihood. Syst Biol. 2011;60(3):291–302.
45. Huson DH, Bryant D. Application of phylogenetic networks in evolutionary studies. Mol Biol Evol. 2006;23(2):254–67.
46. Bruen TC, Philippe H, Bryant D. A simple and robust statistical test for detecting the presence of recombination. Genetics. 2006;172(4):2665–81.
47. Darriba D, Taboada GL, Doallo R, Posada D. jModelTest 2: more models, new heuristics and parallel computing. Nat Methods. 9(8):772.
48. Martin DP, Lemey P, Lott M, Moulton V, Posada D, Lefeuvre P. RDP3: a flexible and fast computer program for analyzing recombination. Bioinformatics. 2010;26(19):2462–3.
49. Grau-Roma L, Crisci E, Sibila M, Lopez-Soria S, Nofrarias M, Cortey M, Fraile L, Olvera A, Segales J. A proposal on porcine circovirus type 2 (PCV2) genotype definition and their relation with postweaning multisystemic wasting syndrome (PMWS) occurrence. Vet Microbiol. 2008;128(1–2):23–35.
50. Wright F. The 'effective number of codons' used in a gene. Gene. 1990; 87(1):23–9.
51. Rozas J. DNA sequence polymorphism analysis using DnaSP. Methods Mol Biol. 2009;537:337–50.
52. Engels M, Giuliani C, Wild P, Beck TM, Loepfe E, Wyler R. The genome of bovine herpesvirus 1 (BHV-1) strains exhibiting a neuropathogenic potential compared to known BHV-1 strains by restriction site mapping and cross-hybridization. Virus Res. 1986;6(1):57–73.
53. ICTV Taxonomy [https://talk.ictvonline.org/taxonomy/w/ictv-taxonomy].
54. WHO. Updated unified nomenclature system for the highly pathogenic H5N1 avian influenza viruses. In: Global influenza surveillance and response system (GISRS): World Health Organization; 2011. http://www.who.int/influenza/gisrs_laboratory/h5n1_nomenclature/en/.
55. Xiao CT, Halbur PG, Opriessnig T. Global molecular genetic analysis of porcine circovirus type 2 (PCV2) sequences confirms the presence of four main PCV2 genotypes and reveals a rapid increase of PCV2d. J Gen Virol. 2015;96(Pt 7):1830–41.
56. Segales J, Olvera A, Grau-Roma L, Charreyre C, Nauwynck H, Larsen L, Dupont K, McCullough K, Ellis J, Krakowka S, et al. PCV-2 genotype definition and nomenclature. Vet Rec. 2008;162(26):867–8.
57. Grose C. Pangaea and the out-of-africa model of varicella-zoster virus evolution and phylogeography. J Virol. 2012;86(18):9558–65.
58. DeRose-Wilson LJ, Gaut BS. Transcription-related mutations and GC content drive variation in nucleotide substitution rates across the genomes of *Arabidopsis thaliana* and *Arabidopsis lyrata*. BMC Evol Biol. 2007;7:66.
59. Miller JM, Whetstone CA, Van der Maaten MJ. Abortifacient property of bovine herpesvirus type 1 isolates that represent three subtypes determined by restriction endonuclease analysis of viral DNA. Am J Vet Res. 1991;52(3):458–61.
60. D'Arce RC, Almeida RS, Silva TC, Franco AC, Spilki F, Roehe PM, Arns CW. Restriction endonuclease and monoclonal antibody analysis of Brazilian isolates of bovine herpesviruses types 1 and 5. Vet Microbiol. 2002;88(4):315–24.
61. Muylkens B, Thiry J, Kirten P, Schynts F, Thiry E. Bovine herpesvirus 1 infection and infectious bovine rhinotracheitis. Vet Res. 2007;38(2):181–209.
62. Rijsewijk FA, Kaashoek MJ, Langeveld JP, Meloen R, Judek J, Bienkowska-Szewczyk K, Maris-Veldhuis MA, van Oirschot JT. Epitopes on glycoprotein C of bovine herpesvirus-1 (BHV-1) that allow differentiation between BHV-1.1 and BHV-1.2 strains. J Gen Virol. 1999;80(Pt 6):1477–83.
63. Wang X, CX W, Song XR, Chen HC, Liu ZF. Comparison of pseudorabies virus China reference strain with emerging variants reveals independent virus evolution within specific geographic regions. Virology. 2017;506:92–8.
64. Cucchi HT-BA, Yuan J, Dobney K, et al. J Archeol Sci. 2011;38:11–22.
65. Ervynck A, Dobney K, Hongo H, Meadow R. Born free? New evidence for the status of Sus Scrofa at Neolithic Çayönü Tepesi (southeastern Anatolia, Turkey). Paléorient. 2001;27:47–73.

66. Abdelgawad A, Azab W, Damiani AM, Baumgartner K, Will H, Osterrieder N, Greenwood AD. Zebra-borne equine herpesvirus type 1 (EHV-1) infection in non-African captive mammals. Vet Microbiol. 2014;169(1–2):102–6.
67. Greenwood AD, Tsangaras K, Ho SY, Szentiks CA, Nikolin VM, Ma G, Damiani A, East ML, Lawrenz A, Hofer H, et al. A potentially fatal mix of herpes in zoos. Curr Biol. 2012;22(18):1727–31.
68. Kennedy MARE, Diderrich V, Richman L, Allen GP, Potgieter LND. Encephalitis associated with a variant of equine herpesvirus 1 in a Thomson's gazelle (Gazella Thomsoni). J Zoo Wildl Med. 1996;27:533–8.
69. Montali RJ, Allen GP, Bryans JT, Phillips LG, Bush M. Equine herpesvirus type 1 abortion in an onager and suspected herpesvirus myelitis in a zebra. J Am Vet Med Assoc. 1985;187(11):1248–9.
70. Blunden AS, Smith KC, Whitwell KE, Dunn KA. Systemic infection by equid herpesvirus-1 in a Grevy's zebra stallion (Equus Grevyi) with particular reference to genital pathology. J Comp Pathol. 1998;119(4):485–93.
71. Goehring LS, van Winden SC, van Maanen C, Sloet van Oldruitenborgh-Ossterbaan MM: Equine herpesvirus type 1-associated myeloencephalopathy in The Netherlands: a four-year retrospective study (1999–2003). J Vet Intern Med 2006, 20(3):601-607.
72. Allen GP, Bolin DC, Bryant U, Carter CN, Giles RC, Harrison LR, Hong CB, Jackson CB, Poonacha K, Wharton R, et al. Prevalence of latent, neuropathogenic equine herpesvirus-1 in the thoroughbred broodmare population of central Kentucky. Equine Vet J. 2008;40(2):105–10.
73. Goodman LB, Loregian A, Perkins GA, Nugent J, Buckles EL, Mercorelli B, Kydd JH, Palu G, Smith KC, Osterrieder N, et al. A point mutation in a herpesvirus polymerase determines neuropathogenicity. PLoS Pathog. 2007;3(11):e160.
74. Guo X, Izume S, Okada A, Ohya K, Kimura T, Fukushi H. Full genome sequences of zebra-borne equine herpesvirus type 1 isolated from zebra, onager and Thomson's gazelle. J Vet Med Sci. 2014;76(9):1309–12.
75. Nugent J, Birch-Machin I, Smith KC, Mumford JA, Swann Z, Newton JR, Bowden RJ, Allen GP, Davis-Poynter N. Analysis of equid herpesvirus 1 strain variation reveals a point mutation of the DNA polymerase strongly associated with neuropathogenic versus nonneuropathogenic disease outbreaks. J Virol. 2006;80(8):4047–60.
76. Javier RT, Sedarati F, Stevens JG. Two avirulent herpes simplex viruses generate lethal recombinants *in vivo*. Science. 1986;234:746–8.
77. Kintner RL, Allan RW, Brandt CR. Recombinants are isolated at high frequency following in vivo mixed ocular infection with two avirulent herpes simplex virus type 1 strains. Arch Virol. 1995;140(2):231–44.
78. Kolb AW, Lee K, Larsen I, Craven M, Brandt CR. Quantitative trait locus based virulence determinant mapping of the HSV-1 genome in murine ocular infection: genes involved in viral regulatory and innate immune networks contribute to virulence. PLoS Pathog. 2016;12(3):e1005499.
79. Lee K, Kolb AW, Larsen I, Craven M, Brandt CR. Mapping murine corneal neovascularization and weight loss virulence determinants in the HSV-1 genome and the detection of an epistatic interaction between the UL and IRS/US regions. J Virol. 2016;90:8115–31.
80. Matsumura T, Sugiura T, Imagawa H, Fukunaga Y, Kamada M. Epizootiological aspects of type 1 and type 4 equine herpesvirus infections among horse populations. J Vet Med Sci. 1992;54(2):207–11.
81. Hartley C. Aetiology of corneal ulcers assume FHV-1 unless proven otherwise. J Feline Med Surg. 2010;12(1):24–35.
82. Gaskell R, Dawson S, Radford A, Thiry E. Feline herpesvirus. Vet Res. 2007; 38(2):337–54.
83. Vaz PK, Job N, Horsington J, Ficorilli N, Studdert MJ, Hartley CA, Gilkerson JR, Browning GF, Devlin JM. Low genetic diversity among historical and contemporary clinical isolates of felid herpesvirus 1. BMC Genomics. 2016;17:704.
84. Lee K, Kolb AW, Sverchkov Y, Cuellar JA, Craven M, Brandt CR. Recombination analysis of herpes simplex virus type 1 reveals a bias towards GC content and the inverted repeat regions. J Virol. 2015;89:7214–23.
85. Schynts F, Meurens F, Detry B, Vanderplasschen A, Thiry E. Rise and survival of bovine herpesvirus 1 recombinants after primary infection and reactivation from latency. J Virol. 2003;77(23):12535–42.
86. Papageorgiou KV, Suarez NM, Wilkie GS, McDonald M, Graham EM, Davison AJ. Genome sequence of canine herpesvirus. PLoS One. 2016;11(5): e0156015.

4

Genetic diversity and association mapping of mineral element concentrations in spinach leaves

Jun Qin[1*], Ainong Shi[1*], Beiquan Mou[2], Michael A. Grusak[3], Yuejin Weng[1], Waltram Ravelombola[1], Gehendra Bhattarai[1], Lingdi Dong[1] and Wei Yang[1]

Abstract

Background: Spinach is a useful source of dietary vitamins and mineral elements. Breeding new spinach cultivars with high nutritional value is one of the main goals in spinach breeding programs worldwide, and identification of single nucleotide polymorphism (SNP) markers for mineral element concentrations is necessary to support spinach molecular breeding. The purpose of this study was to conduct a genome-wide association study (GWAS) and to identify SNP markers associated with mineral elements in the USDA-GRIN spinach germplasm collection.

Results: A total of 14 mineral elements: boron (B), calcium (Ca), cobalt (Co), copper (Cu), iron (Fe), potassium (K), magnesium (Mg), manganese (Mn), molybdenum (Mo), sodium (Na), nickel (Ni), phosphorus (P), sulfur (S), and zinc (Zn) were evaluated in 292 spinach accessions originally collected from 29 countries. Significant genetic variations were found among the tested genotypes as evidenced by the 2 to 42 times difference in mineral concentrations. A total of 2402 SNPs identified from genotyping by sequencing (GBS) approach were used for genetic diversity and GWAS. Six statistical methods were used for association analysis. Forty-five SNP markers were identified to be strongly associated with the concentrations of 13 mineral elements. Only two weakly associated SNP markers were associated with K concentration. Co-localized SNPs for different elemental concentrations were discovered in this research. Three SNP markers, AYZV02017731_40, AYZV02094133_57, and AYZV02281036_185 were identified to be associated with concentrations of four mineral components, Co, Mn, S, and Zn. There is a high validating correlation coefficient with $r > 0.7$ among concentrations of the four elements. Thirty-one spinach accessions, which rank in the top three highest concentrations in each of the 14 mineral elements, were identified as potential parents for spinach breeding programs in the future.

Conclusions: The 45 SNP markers strongly associated with the concentrations of the 13 mineral elements: B, Ca, Co, Cu, Fe, Mg, Mn, Mo, Na, Ni, P, S, and Zn could be used in breeding programs to improve the nutritional quality of spinach through marker-assisted selection (MAS). The 31 spinach accessions with high concentrations of one to several mineral elements can be used as potential parents for spinach breeding programs.

Keywords: Genome-wide association study (GWAS), Genotyping by sequencing (GBS), Mineral elements, Single nucleotide polymorphism (SNP), *Spinacia oleracea* L., Spinach

* Correspondence: junqin@uark.edu; ashi@uark.edu
[1]Department of Horticulture, University of Arkansas, Fayetteville, AR 72701, USA
Full list of author information is available at the end of the article

Background

Spinach (*Spinacia oleracea* L., 2n = 2× = 12) is an economically important vegetable crop worldwide with an estimated annual value of $11.8 billion. The United States (US) is the second largest producer of spinach after China with over 550,000 tons harvested, valued at over $300 million annually since 2009 [1, 2]. In addition to its economic importance, spinach is one of the rising vegetable crops in the US in terms of per capita consumption and is considered a healthy vegetable for humans as it is a source of vitamins and mineral nutrients, as well as several health-promoting phytochemicals [3, 4].

Minerals originate in the earth and cannot be made by living organisms [5]. Mineral elements are present in different forms in nature and some of these elements are essential for the body to perform different functions [6]. Most of them mediate vital biochemical reactions by acting as a cofactor or catalyst for many enzymes. They also act as centers of building stabilizing structures such as enzymes and proteins. The five major minerals in the human body are calcium (Ca), phosphorus (P), potassium (K), sodium (Na), and magnesium (Mg) [6, 7]. All of the remaining elements in a human body are called "trace elements". The trace elements that have a specific biochemical function in the human body are iron (Fe), cobalt (Co), copper (Cu), zinc (Zn), manganese (Mg), molybdenum (Mo), iodine (I), and selenium (Se) [8]. Spinach is a dietary source of Ca, Cu, Fe, K, Mg, Mn, P, Zn, folate, vitamins and dietary fiber [9]. Therefore, breeding new spinach varieties with high nutritional components including the mineral elements is one of the main goals in spinach breeding programs worldwide.

Molecular plant breeding has been the foundation for twenty-first-century crop improvement [10]. Marker-assisted selection (MAS) has been successfully used to incorporate specific genes/alleles in plant breeding [11–13]. Single nucleotide polymorphism (SNP) with its high abundance, cost efficiency, and high-throughput scoring, has become a powerful tool in genome mapping, association studies, diversity analysis, and tagging of important genes in plant genomics [14–17]. Therefore, identification of SNP markers for mineral elements will be useful in spinach MAS breeding programs.

Genotyping by sequencing (GBS) is one of the next-generation sequencing platforms that utilizes a simple highly-multiplexed system for constructing reduced representation libraries which reduces sample handling, requires fewer PCR and purification steps, no size fractionation and uses inexpensive barcoding [18, 19]. As a cost-effective tool for MAS, GBS has been used to facilitate genome-wide association studies (GWAS), genetic linkage analysis, molecular marker discovery, and studies of genomic diversity or selection [18, 20, 21]. As the GBS method has no requirement for a priori knowledge of the species genomes, it has been shown to be robust across a range of species and SNP discovery and genotyping are completed together [22, 23]. The spinach genome assembly (PacBio Assembly) (980 Mbp) has been reported on January 14, 2014 [24, 25], but it has not been publically available yet. The spinach genome Spinach-1.0.1 is available to the public at http://www.ncbi.nlm.nih.gov/Traces/wgs/?val=AYZV02 and also at "The *Beta vulgaris* Resource" website with the page at http://bvseq.molgen.mpg.de/Genome/Download/Spinach/, representing approximately half of the spinach genome [26, 27]. We used the AYZV02 as the reference of spinach genome sequences for short reads assembly and SNP discovery in each spinach sample in this study.

To date, several association studies for different phenotypic traits of spinach have been reported, such as oxalate concentration [28], leafminer (*Liriomyza* spp.) resistance [29], Verticillium wilt resistance [30], Stemphylium leaf spot resistance [31], and leaf traits [32]. However, no genetic studies have been conducted to evaluate the genetic diversity of mineral elements and no research has been reported regarding mineral elements using association mapping in spinach natural populations to date. Therefore, the objectives of this study were to perform genetic diversity analysis and GWAS for spinach mineral elements in the USDA spinach germplasm collection, and to identify SNP markers associated with the 14 mineral elements: B, Ca, Co, Cu, Fe, K, Mg, Mn, Mo, Na, Ni, P, S, and Zn. The results will provide information on how to use spinach germplasm accessions with high mineral concentrations and new molecular markers for spinach breeding programs.

Results

Phenotypic variation in the mineral concentrations in spinach USDA-GRIN germplasm

The 14 mineral elements in spinach were analyzed for their mean (average) concentration, range, standard deviation (stdev), and coefficient of variation (CV) for each mineral element (Table 1). Potassium had the highest concentration with greater than 79,500 ppm; Mg the second with 8697 ppm; P, Ca, and S were greater than 4200 ppm; Na and Zn greater than 100 ppm; Mn, Fe, and B greater than 10 ppm; and Cu, Mo, Ni, and Co were less than 7 ppm, with Ni and Co with less than 1 ppm.

All elements showed broad concentration ranges across the germplasm accessions. K, P, Fe, Mg, B, Ca, S, and Cu had about two to five times difference (Maximum/Minimum); Na, Co, Zn, and Mn showed approximately eight to fourteen times difference; while Ni (18 times) and Mo (42 times) exhibited the largest differences (Table 1). The standard deviation and stdev error

Table 1 The average, range, standard deviation, and coefficient of variation of the 14 mineral compounds in spinach

Mineral compound	No. sample	Average (ppm)	Minimum (ppm)	Maximum (ppm)	Range (ppm)	Standard deviation	Stdev error	CV(%)
K	275	79,576.06	47,170.00	108,935.36	61,765.37	11,903.06	43.28	14.96
Mg	292	8697.34	4921.23	14,293.27	9372.04	1443.39	4.94	16.60
P	292	6748.91	4004.33	10,210.77	6206.44	952.55	3.26	14.11
Ca	292	6400.55	3043.91	11,544.80	8500.89	1511.40	5.18	23.61
S	292	4281.57	1899.60	7462.84	5563.23	782.93	2.68	18.29
Na	292	615.76	186.94	1484.01	1297.07	214.52	0.73	34.84
Zn	292	102.50	31.43	386.82	355.38	78.97	0.27	77.04
Mn	292	83.20	29.78	415.76	385.97	79.00	0.27	94.95
Fe	291	74.34	50.43	138.58	88.15	13.57	0.05	18.25
B	269	30.47	17.07	52.36	35.29	5.93	0.02	19.47
Cu	292	6.86	2.22	10.97	8.75	1.25	0.00	18.16
Mo	292	2.36	0.20	8.62	8.42	1.29	0.00	54.96
Ni	288	0.72	0.23	4.23	4.00	0.48	0.00	66.29
Co	238	0.15	0.05	0.46	0.42	0.08	0.00	54.01

showed similar trends with their averages. K had the highest average, and it also had the highest standard deviation and stdev error. The coefficient of variation (CV), also known as relative standard deviation (RSD), is a standardized measure of dispersion of a probability distribution or frequency distribution, where CV = mean/Stdev × 100. All 14 mineral elements had a large CV value of greater than 14%. Among them, Co, Mn, Mo, Ni, and Zn had greater than 50% CV value, indicated that the dispersion in the variable was greater (Table 1), showing there were large variations in these mineral elements across the spinach germplasm accessions. Analyzed from the distributions of mineral concentration values (Fig. 1), Co, Mn, Ni, and Zn showed a skewed distribution toward the lower range, and others had near-normal distributions.

The correlation among the 14 mineral elements was analyzed by JMP Genomics 7. The correlation coefficients among Co, Mn, S, and Zn were greater than 0.7, indicating strong associations (correlations) among the four mineral elements. In addition, the pairwise correlation coefficients of Fe and S, Fe and Co, Fe and Zn, Cu and Mo, K and P were greater than 0.4, indicating significant associations between each pair (Table 2). The two dimension plot of Biplot can be used to visualize the analysis of two–way data. As shown in Fig. 2, the lines that extend from the center and connect to each mineral element trait are considered as index vectors. The angles of the index vectors indicate that correlations existed between the indices (each mineral element trait). The cosine of these angles indicates the genetic correlation. An angle that is less than 90° indicates a positive correlation, and an angle that is greater than 90° indicates a negative correlation. If the angle is close to 0° or 180°, the two indices were highly correlated. The smaller angle between Co, Mn, S, and Zn, than other mineral elements, indicates their higher correlation than the other elements (Fig. 2).

Genetic diversity analysis of spinach germplasm

The population structure of the 292 spinach accessions was initially inferred using STRUCTURE 2.3.4 [33] and the peak of delta K was observed at K = 4, indicating the presence of the four main populations (clusters, Q1-Q4) in the 292 spinach accessions (Fig. 3a). The classification of accessions into populations based on the model-based structure was shown in Fig. 3 and Additional file 1: Table S1. In total, 247 accessions (84.6%) were assigned to one of the four populations (Q1, Q2, Q3, and Q4). Population 1, 2, 3, and 4 (Q1, Q2, Q3, and Q4) consisted of 33 (11.3%), 26 (8.9%), 109 (37.3%), and 79 (27.0%) accessions, respectively. The remaining 45 accessions (15.4%) were categorized as having admixed ancestry, including two, three, and four population admixed among Q1, Q2, Q3, and Q4 (Additional file 1: Table S1).

The genetic diversity was analyzed using the Maximum Likelihood (ML) method by MEGA 6 [34]. Several phylogenetic trees were drawn based on interpretation of results. We defined Q1, Q2, Q3, and Q4 as the clusters and used the same colors as the population structure Q1 (red), Q2 (green), and Q3 (blue), and Q4 (yellow) from the STRUCTURE 2.3.4 (Fig. 3b) to draw the subtrees of the phylogenetic trees in MEGA 6. Two phylogenetic trees were included: (1) without taxon names assigned in order to compare the populations from STRUCTURE (Fig. 3c), and (2) the ring phylogenetic tree (Additional file 2: Figure S1). The phylogenetic trees from MEGA 6 (Fig. 3c and Additional file 2: Figure S1), were well consistent with the

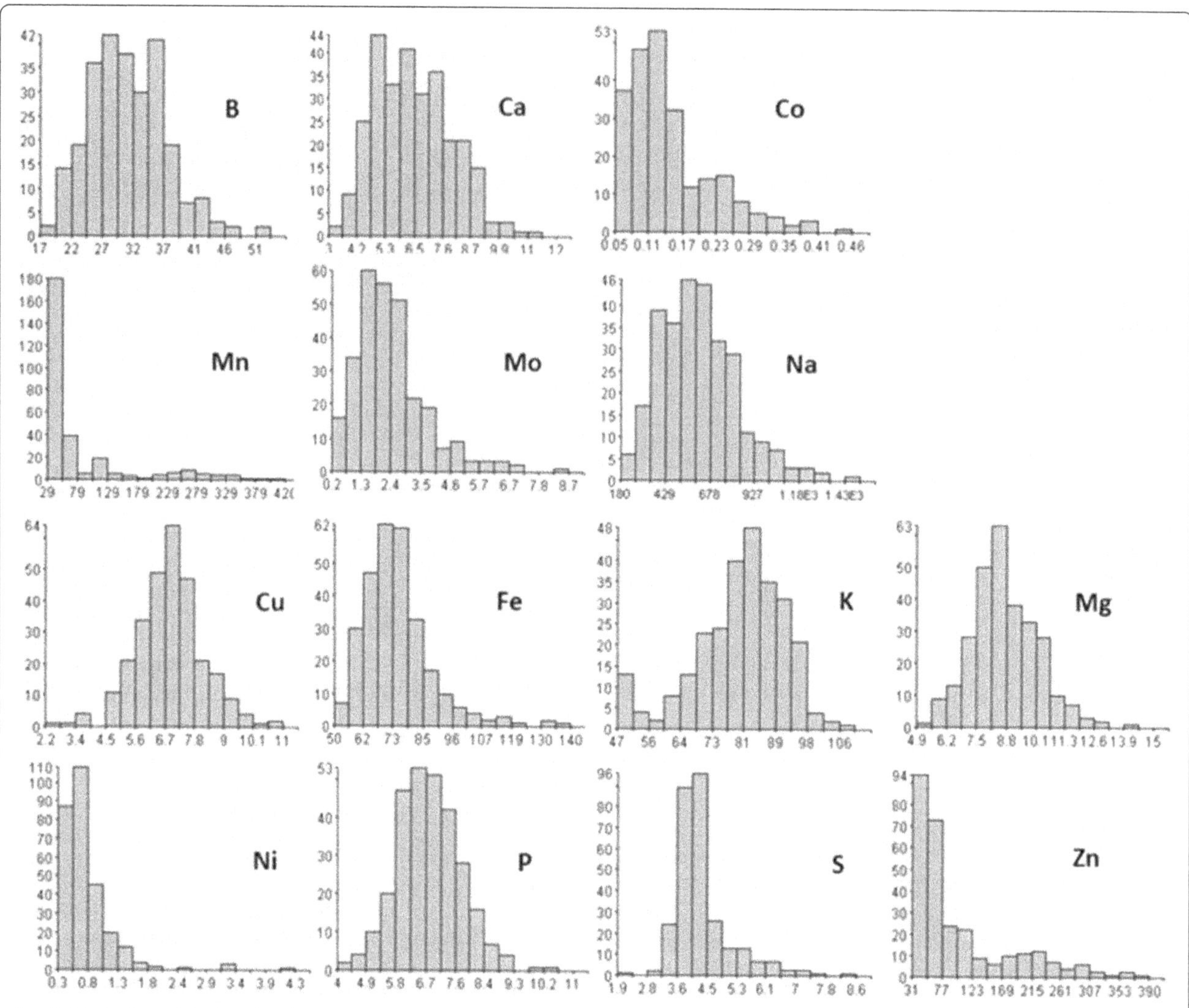

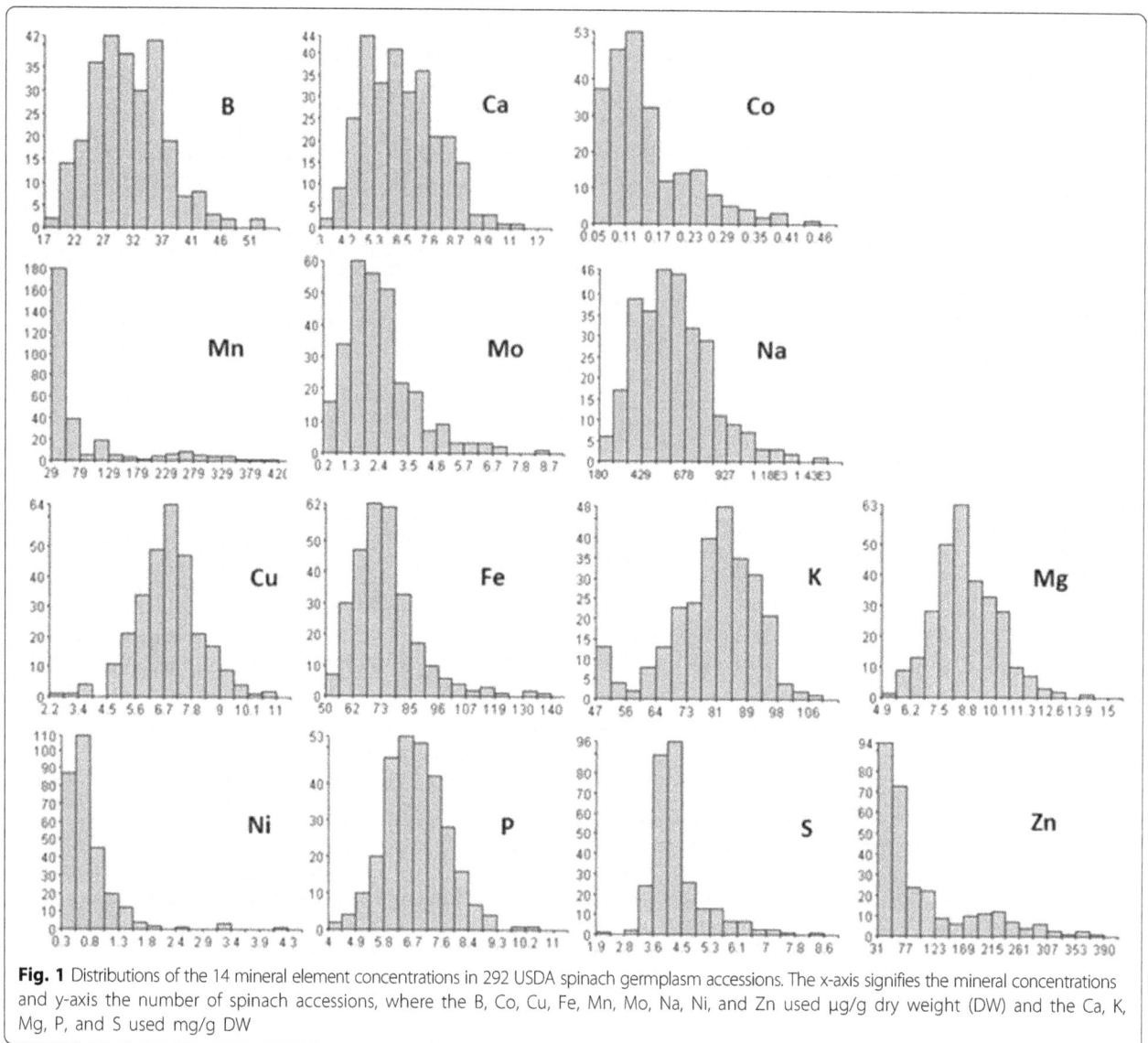

Fig. 1 Distributions of the 14 mineral element concentrations in 292 USDA spinach germplasm accessions. The x-axis signifies the mineral concentrations and y-axis the number of spinach accessions, where the B, Co, Cu, Fe, Mn, Mo, Na, Ni, and Zn used μg/g dry weight (DW) and the Ca, K, Mg, P, and S used mg/g DW

structure populations (Q1-Q4) from STRUCTURE 2.3.4 (Fig. 3a and b), indicating that there were four well-differentiated genetic populations and admixture in the spinach panel.

Association mapping and SNP marker identification

Six methods were used for association analysis of the 14 mineral elements among the 292 spinach accessions using 2401 SNPs. The six methods included: (1) SMR_Qgene: single marker regression using the QGene 4.3.10 [35], (2) SMR_tassel: single marker regression without structure and without kinship using TASSEL 5, (3) GLM_tassel: general linear model using TASSEL 5, (4) MLM_tassel: mixed linear model methods using TASSEL 5, (5) cMLM_gapit: compressed mixed linear model methods using GAPIT, and (6) EcMLM_gapit: enriched compressed mixed linear model methods using GAPIT. The selection standardization in this research is based on LOD values: SMR_QGene, SMR_tassel, and GLM_tassel > = 2.5, and one of MLM (either MLM_tassel, CMLM_gapit, or EcMLM_gapit) > = 2.5.

Based on the criteria above, a total of 45 SNPs were identified to be strongly associated with the 13 mineral elements except for K (Table 3). Among the 45 SNP markers, four were associated with B; one with Ca; seven with Co; two with Cu; six with Fe; four with Mg; five with Mn; one with Mo; five with Na; one with Ni; one with P; seven with S; and five with Zn. In addition, two SNP markers, AYZV02123399_305 and AYZV02147304_372, were detected to be associated with K having a LOD value > = 2.0 at SMR_QGene and SMR_tassel, LOD > = 2.5 at GLM_tassel, and one of MLM models (either MLM_tassel, CMLM_gapit, or EcMLM_gapit). Of these identified markers, four SNPs showed pleiotropic effects:

Table 2 Correlation coefficient among fourteen mineral components in spinach

Correlation	B	Ca	Co	Cu	Fe	K	Mg	Mn	Mo	Na	Ni	P	S	Zn
B	1	0.09	0.12	0.22	0.11	0.09	−0.15	0.07	0.23	0.03	0.07	−0.01	0.15	0.23
Ca	0.09	1	−0.36	0.25	−0.07	0.01	0.37	−0.35	0.34	0.24	0.10	−0.27	−0.32	−0.37
Co	0.12	−0.36	1	0.14	0.44	−0.35	−0.21	0.72	−0.31	−0.04	0.02	−0.14	0.70	0.74
Cu	0.22	0.25	0.14	1	0.30	0.15	−0.16	0.02	0.44	−0.06	0.12	0.03	0.17	0.13
Fe	0.11	−0.07	0.44	0.30	1	−0.04	0.01	0.36	−0.02	−0.03	0.40	0.19	0.44	0.41
K	0.09	0.01	−0.35	0.15	−0.04	1	0.10	−0.36	0.31	−0.01	0.06	0.43	−0.32	−0.35
Mg	−0.15	0.37	−0.21	−0.16	0.01	0.10	1	−0.12	−0.22	0.35	−0.03	0.25	−0.12	−0.30
Mn	0.07	−0.35	0.72	0.02	0.36	−0.36	−0.12	1	−0.35	−0.02	−0.03	−0.09	0.76	0.81
Mo	0.23	0.34	−0.31	0.44	−0.02	0.31	−0.22	−0.35	1	−0.19	0.13	0.11	−0.34	−0.22
Na	0.03	0.24	−0.04	−0.06	−0.03	−0.01	0.35	−0.02	−0.19	1	0.00	−0.06	0.05	−0.16
Ni	0.07	0.10	0.02	0.12	0.40	0.06	−0.03	−0.03	0.13	0.00	1	0.08	−0.03	0.00
P	−0.01	−0.27	−0.14	0.03	0.19	0.43	0.25	−0.09	0.11	−0.06	0.08	1	−0.07	−0.11
S	0.15	−0.32	0.70	0.17	0.44	−0.32	−0.12	0.76	−0.34	0.05	−0.03	−0.07	1	0.72
Zn	0.23	−0.37	0.74	0.13	0.41	−0.35	−0.30	0.81	−0.22	−0.16	0.00	−0.11	0.72	1

AYZV02017731_35 was associated with both Co and S; AYZV02057049_393 with both Fe and S; AYZV02073631_255 with both Co and S; and AYZV02225779_181 with both Fe and S, indicating each of the four markers can be used to select two mineral components in MAS spinach breeding. Three SNP markers, AYZV02017731_40, AYZV02094133_57, and AYZV02281036_185 were identified to be associated with concentrations of four mineral components, Co, Mn, S, and Zn based on LOD values: SMR_Qgene (Table 4). T-test result validated the three SNP markers were significantly associated with the four mineral elements (Table 5), suggesting that it may be possible to select high contents of the four elements, Co, Mn, S, and Zn at the same time through MAS using these markers in breeding.

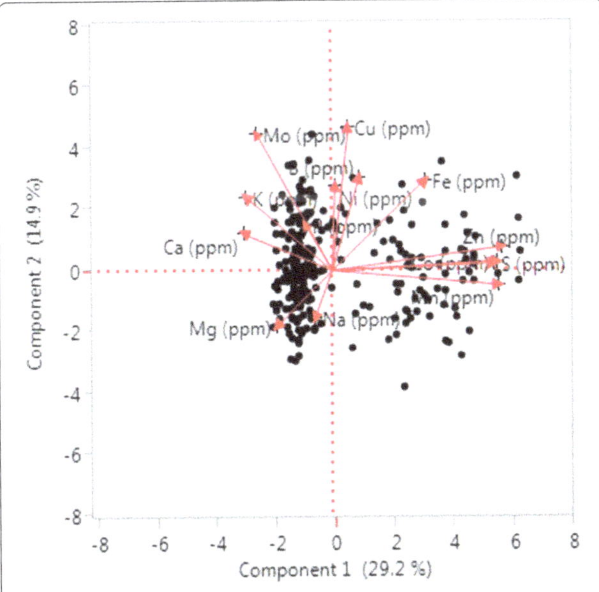

Fig. 2 The two dimension plot of Biplot for 14 mineral element concentrations in 292 spinach germplasm accessions

R-square (Rsq) of the detected markers using the six different methods were listed in Table 3 with a large range from 1.84% of the AYZV02144992_13 marker, associated with Mg in MLM model, to 15.3% of the AYZV02136507_242, associated with Mn in SMR from Tassel. According to the LOD value, the larger Rsq value of a marker, the stronger association is the marker which makes more contribution to the trait. In all detected SNP markers, AYZV02136507_242 had the greatest Rsq value with 10.6% Rsq in SMR model from QGene, 15.3% Rsq in SMR, 13.8% Rsq in GLM, and 12.9% Rsq in GLM from Tassel for Mn element, indicating that the AYZV02136507_242 is strongly associated with Mn from this study.

Evaluation and genetic diversity analysis of the top three spinach germplasm accessions for each mineral

We identified the top 3 accessions with the highest concentrations in each of the 14 mineral elements. First, we ordered the 292 spinach accessions based on their mineral concentrations by each individually from the highest to lowest values and gave the order number from 1 to 292 plus the mineral name for each mineral element. For B, as an example, we ordered the B concentration from the highest to lowest and gave each accession with an order ID from B1 to B292 such as NSL6557 had the highest B value with 52.36 ppm and was given B1, and NSL6083 the lowest B value with 17.07 ppm given B292 (Additional file 1: Table S1). By combining the 14 mineral elements, each accession was given a mineral concentration order ID including the 14 mineral names and

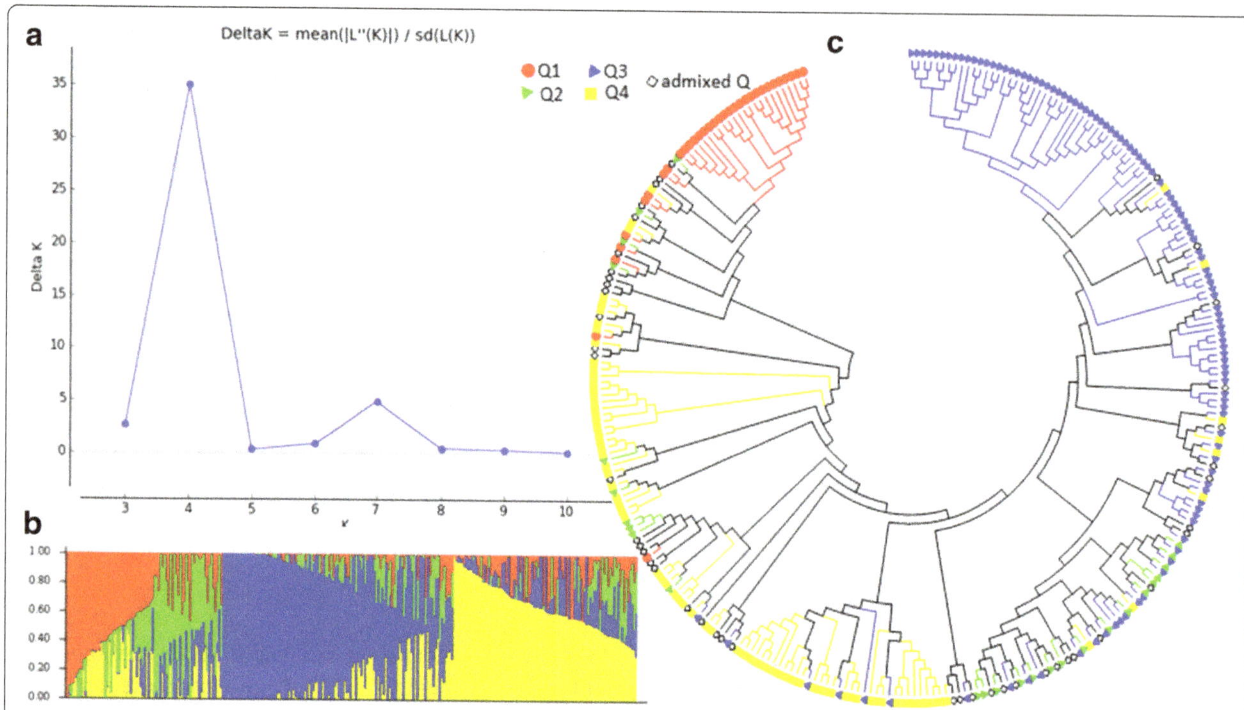

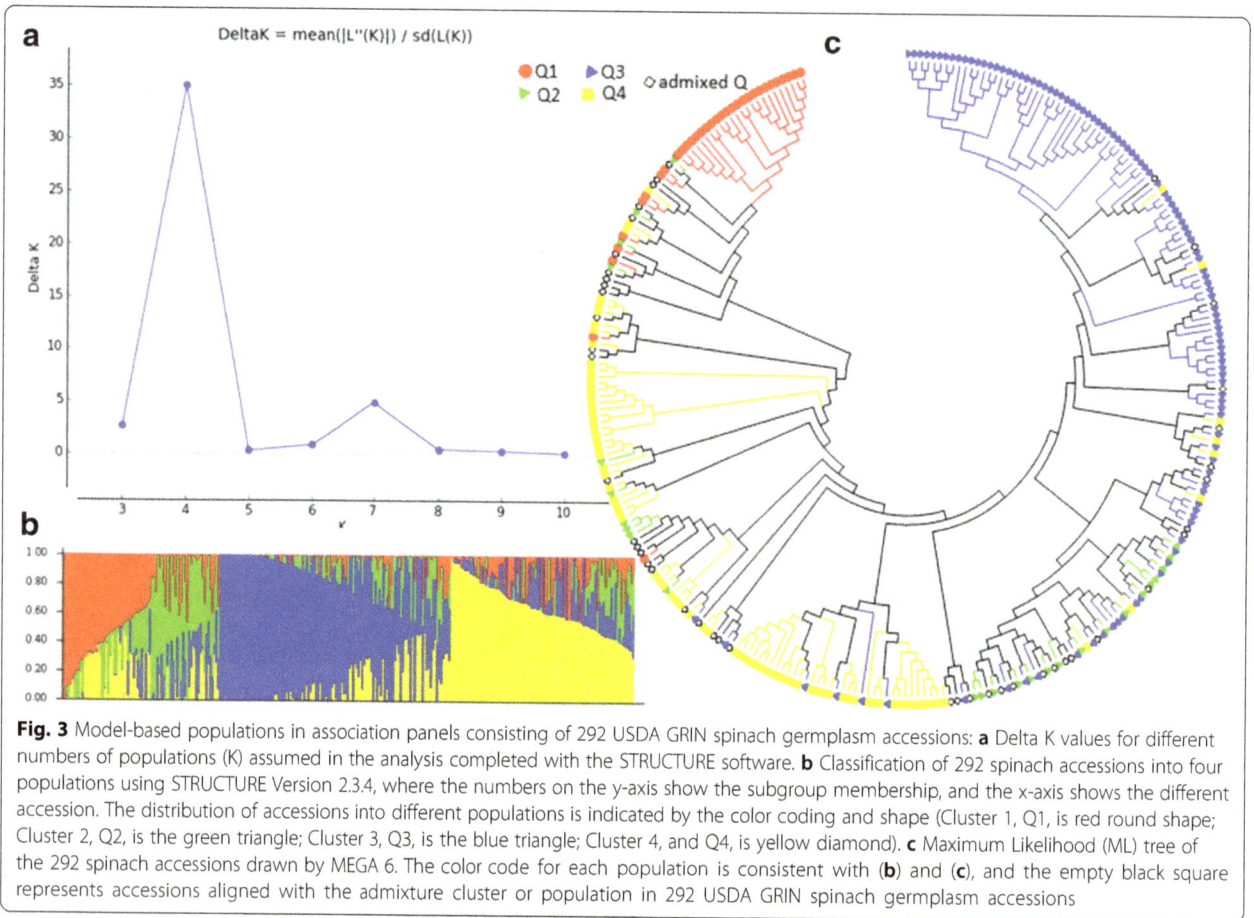

Fig. 3 Model-based populations in association panels consisting of 292 USDA GRIN spinach germplasm accessions: **a** Delta K values for different numbers of populations (K) assumed in the analysis completed with the STRUCTURE software. **b** Classification of 292 spinach accessions into four populations using STRUCTURE Version 2.3.4, where the numbers on the y-axis show the subgroup membership, and the x-axis shows the different accession. The distribution of accessions into different populations is indicated by the color coding and shape (Cluster 1, Q1, is red round shape; Cluster 2, Q2, is the green triangle; Cluster 3, Q3, is the blue triangle; Cluster 4, and Q4, is yellow diamond). **c** Maximum Likelihood (ML) tree of the 292 spinach accessions drawn by MEGA 6. The color code for each population is consistent with (**b**) and (**c**), and the empty black square represents accessions aligned with the admixture cluster or population in 292 USDA GRIN spinach germplasm accessions

their order numbers. For example, the accession NSL6557 was given the designation B1Ca63Co125Cu41-Fe133K284Mg34Mn67Mo36Na161Ni41P19S72Zn100, which means a B rank of 1, Ca rank of 63, Co rank of 125, Cu rank of 41, Fe rank of 133, etc. for this accession (Additional file 1: Table S2). After ranking all 14 mineral elements, 31 spinach accessions had been chosen because they had at least one mineral element (out of 14) ranked in the top three among the 292 spinach accessions (Additional file 1: Table S2), indicating that the 31 accessions were good mineral element resources for spinach breeding to improve mineral concentrations. Eight out of these 31 spinach accessions had more than two mineral elements ranked in the top 3 highest (PI604777, PI604786, PI175595, PI604782, PI360895, PI339547, PI176372, and PI169671). Three out of the eight accessions had three mineral elements ranked in the top 3 highest. PI604777 was ranked as No. 1 in S, No. 2 in Fe, and No. 3 in Co; PI604786 as No. 2 in Zn, No. 2 in Mn, and No. 3 in Cu; and PI175595 was No. 1 in Ni, No. 3 in P, and No. 3 in K (Additional file 1: Table S2).

The 31 spinach accessions were collected from 10 countries plus one unknown location: ten from Turkey, two from Afghanistan, India, Iran, Japan, Macedonia, and Netherlands, respectively, and one from Mongolia, Belgium, China, Ethiopia, Hungary, Italy, Nepal, and the US, respectively (Additional file 1: Table S2), The genetic diversity analysis was performed for the 31 spinach accessions and the phylogenetic tree was drawn using MEGA 6 (Fig. 4). Based on genetic distances among the 31 genotypes, there were three clusters: Cluster I, consisted of 24 accessions; Cluster II had only two accessions; and Cluster III, included five accessions. The cluster I can be further divided into five sub-clusters (groups): I-1 with seven accessions, I-2, six accessions, I-3, three accessions, I-4, five accessions, and I-5, only two accessions. In addition, PI604788 is an outlier (Fig. 4). The genetic diversity and phylogenetic analysis will provide the information for breeders to choose these spinach accessions as parents in breeding programs.

Discussion

Application of marker-assisted selection in the genetic improvement of spinach breeding

In this study, we conducted a comprehensive GWAS to identify genetic loci with SNP markers associated with the 14 mineral elements (B, Ca, Co, Cu, Fe, K, Mg, Mn,

Table 3 The information of the significant SNP markers associated with 14 mineral compounds among the 292 spinach accessions using six statistical models, SMR_Qgene, SMR_tassel, GLM_tassel, MLM_tassel, CMLM_gapit, and EcMLM_gapit

Trait		LOD value (−LOG(p))						R-squre (%)				MAF
		Qgene	Tassel			GAPIT		Qgene	Tassel			
	Marker	SMR	SMR	GLM	MLM	cMLM	EcMLM	SMR	SMR	GLM	MLM	
B	AYZV02030222_447	5.15	4.59	3.75	2.94	3.23	3.18	8.40	8.12	6.36	5.40	14.37
B	AYZV02101102_298	2.67	2.59	3.08	3.32	3.65	3.61	4.50	4.47	5.06	5.92	11.75
B	AYZV02164196_389	3.61	4.02	3.15	2.69	3.45	3.36	6.00	7.51	5.75	5.23	20.34
B	AYZV02164196_391	3.31	3.48	2.66	2.13	2.66	2.58	5.50	6.54	4.88	4.10	20.34
Ca	AYZV02244039_336	2.53	3.13	3.04	2.70	0.17	0.40	3.90	5.79	5.65	5.36	27.32
Co	AYZV02017731_35	5.64	5.75	4.16	3.77	2.99	2.38	10.30	9.80	6.26	6.15	9.92
Co	AYZV02073631_257	3.86	2.70	2.62	1.68	2.73	3.52	7.20	5.35	4.76	3.35	6.75
Co	AYZV02165009_136	3.28	3.41	3.03	2.67	2.90	2.14	6.20	5.37	4.31	4.03	9.07
Co	AYZV02217527_245	3.74	4.15	3.17	2.15	4.34	4.53	7.00	6.58	4.44	3.15	2.74
Co	AYZV02217549_245	3.74	3.65	2.53	1.83	2.59	2.69	7.00	6.96	4.43	3.45	3.38
Co	AYZV02220844_249	3.04	3.60	3.49	3.13	3.53	2.06	5.70	6.99	6.18	6.05	10.97
Co	AYZV02221073_190	3.41	4.18	3.19	2.96	2.28	2.10	6.40	7.15	4.81	4.80	23.63
Cu	AYZV02147304_383	3.01	3.27	3.62	3.01	3.27	3.30	4.60	5.22	5.62	4.98	14.09
Cu	AYZV02277499_79	2.75	3.45	4.67	3.66	3.21	3.33	4.20	5.76	7.49	6.43	43.30
Fe	AYZV02057049_393	4.68	4.61	4.90	4.41	0.10	0.12	7.10	7.33	7.54	7.35	11.90
Fe	AYZV02201149_106	2.85	4.08	5.48	4.74	0.44	0.10	4.40	7.43	9.36	9.37	23.45
Fe	AYZV02201149_51	2.65	3.01	3.59	3.21	0.45	0.05	4.10	5.58	6.27	6.27	24.14
Fe	AYZV02207926_4318	3.88	4.45	3.35	3.05	1.20	1.12	6.00	7.47	5.37	5.52	18.79
Fe	AYZV02225779_181	5.87	5.67	6.38	5.74	3.39	3.38	8.90	8.97	9.71	9.56	5.00
Fe	AYZV02296000_81	3.46	4.53	4.28	3.89	0.92	0.21	5.30	7.65	7.03	6.76	23.45
Mg	AYZV02113550_76	3.09	3.66	3.33	2.64	3.12	2.00	4.70	5.74	4.95	4.13	41.41
Mg	AYZV02144992_13	5.38	3.98	3.17	0.96	1.80	2.78	8.10	6.93	5.45	1.84	18.56
Mg	AYZV02176946_248	3.98	5.05	4.50	3.48	1.29	1.52	6.10	7.84	6.89	5.93	12.71
Mg	AYZV02297745_849	4.21	5.92	6.03	3.32	1.02	3.75	6.40	8.07	8.06	4.62	39.52
Mn	AYZV02026295_260	4.79	4.68	3.79	2.76	1.22	1.11	7.30	7.26	5.87	4.52	5.33
Mn	AYZV02136507_242	7.08	9.20	8.32	6.68	0.91	0.52	10.60	15.30	13.77	12.90	43.81
Mn	AYZV02136507_248	5.18	6.49	5.66	4.81	0.32	0.02	7.90	11.26	9.76	9.20	41.41
Mn	AYZV02245160_278	3.28	3.36	2.82	3.03	0.08	0.56	5.00	5.72	4.77	5.43	17.18
Mn	AYZV02281036_185	3.12	3.45	3.01	2.88	2.97	1.95	4.80	5.69	4.95	5.14	8.59
Mo	AYZV02092600_720	3.15	3.77	4.04	3.88	3.45	3.76	4.80	4.82	5.05	5.10	4.81
Na	AYZV02103789_734	3.52	3.65	2.87	2.50	2.10	2.06	5.40	4.93	3.61	3.17	13.57
Na	AYZV02156839_631	3.08	3.11	2.86	2.68	0.02	0.07	4.70	5.30	4.77	4.93	23.54
Na	AYZV02178194_42	3.38	4.11	2.70	2.08	2.57	1.84	5.20	6.23	3.70	2.84	28.87
Na	AYZV02209560_3	3.77	4.59	3.42	2.84	3.16	3.31	5.80	6.29	4.40	3.72	13.40
Na	AYZV02225745_194	3.05	3.21	2.59	2.24	1.93	2.51	4.70	4.48	3.42	2.91	18.56
Ni	AYZV02051025_93	3.40	6.42	5.77	4.87	0.30	0.31	5.30	10.33	9.23	8.71	28.92
P	AYZV02105368_125	2.88	3.40	3.54	2.82	2.91	2.80	4.40	5.82	6.05	5.09	11.00
S	AYZV02017731_35	3.84	4.33	3.39	2.84	2.08	1.38	5.90	5.92	4.39	3.71	10.14
S	AYZV02057049_393	3.78	3.88	4.21	3.64	0.29	0.07	5.80	6.18	6.55	6.19	11.86
S	AYZV02073631_255	3.58	2.69	2.83	2.52	3.04	2.82	5.50	4.40	4.53	4.28	6.70
S	AYZV02073631_257	3.65	2.79	2.88	2.57	3.34	3.05	5.60	4.54	4.59	4.35	6.36

Table 3 The information of the significant SNP markers associated with 14 mineral compounds among the 292 spinach accessions using six statistical models, SMR_Qgene, SMR_tassel, GLM_tassel, MLM_tassel, CMLM_gapit, and EcMLM_gapit *(Continued)*

Trait		LOD value (–LOG(p))						R-squre (%)				MAF
		Qgene	Tassel			GAPIT		Qgene	Tassel			
	Marker	SMR	SMR	GLM	MLM	cMLM	EcMLM	SMR	SMR	GLM	MLM	
S	AYZV02094133_57	5.36	6.48	6.15	4.96	3.06	2.46	8.10	9.70	8.95	7.87	16.32
S	AYZV02225779_181	4.28	4.12	4.45	3.87	0.01	0.03	6.50	6.57	6.92	6.37	4.98
S	AYZV02248538_132	2.75	3.60	3.49	3.74	1.22	1.08	4.20	5.93	5.61	6.35	9.28
Zn	AYZV02066684_19177	3.71	3.24	3.19	2.72	0.66	0.57	5.70	5.46	4.97	5.44	14.95
Zn	AYZV02151846_237	3.06	5.43	3.09	2.17	2.53	1.71	4.70	8.58	4.29	3.49	36.08
Zn	AYZV02201848_2665	5.11	5.16	4.30	3.96	2.69	2.72	7.70	8.14	6.34	6.59	11.34
Zn	AYZV02212287_92	2.54	3.92	4.91	3.15	1.66	1.46	3.90	6.57	7.41	5.26	17.01
Zn	AYZV02212288_92	3.30	4.13	5.27	3.69	2.53	2.32	5.10	7.19	8.27	6.19	19.59
K	AYZV02123399_305	2.07	2.25	2.77	2.61	2.53	2.80	3.40	4.05	4.94	4.84	25.00
K	AYZV02147304_372	2.33	2.21	2.48	2.53	2.09	1.69	3.80	3.90	4.36	4.47	18.43

Mo, Na, Ni, P, S, and Zn) in 292 spinach accessions of the USDA collection. A total of 45 SNP markers were identified to be associated with the 13 mineral elements except K, based on six different association mapping models. Similar mineral element association mapping research has been reported in other crops, such as pea [36] and rice [37–40] using RIL or diversity populations. Ma et al. (2017) [36] conducted a comprehensive QTL mapping study and identified genetic loci associated with mineral element concentrations (B, Ca, Fe, K, Mg, Mn, Mo, P, S and Zn) in pea seeds using a RIL population.

Co-localization of SNPs for different element concentrations has been discovered in this research. AYZV02017731_35 and AYZV02073631_257 were related to Co and S, AYZV02057049_393 was related to both Fe and S, and AYZV02225779_181 was related to Fe and S (Table 3). Co-localization of QTLs for different element concentrations in seeds has previously been reported in rice [37, 38, 41]. Genetically, the phenomenon of co-localization may be caused by pleiotropy of a single gene product being involved in the transport and/or physiological processing of multiple elements. Another possibility is the presence of clustered genes that are tightly associated together and responsible for the accumulation of different elements [42].

Accordingly, the high positive correlations among Co, S, Mn, and Zn ranged from 0.70 to 0.81 in this study (Table 2). The positive correlations suggest that high Co,

Table 4 The information of three significant SNP markers associated with four mineral compounds among the 292 spinach accessions using six statistical models, SMR_Qgene, SMR_tassel, GLM_tassel, MLM_tassel, CMLM_gapit, and EcMLM_gapit. Co, Mn, S, and Zn based on T-tesing at $P = 0.05$

Trait		LOD value (–LOG(p))						R-squre (%)				MAF
		Qgene	Tassel			GAPIT		Qgene	Tassel			
	Marker	SMR	SMR	GLM	MLM	cMLM	EcMLM	SMR	SMR	GLM	MLM	
Co	AYZV02017730_40	3.74	3.83	2.35	1.98	2.26	1.55	7.00	6.12	3.16	2.78	7.17
Mn	AYZV02017730_40	2.06	2.65	2.10	1.15	0.99	0.66	3.20	3.28	2.46	1.17	6.53
S	AYZV02017730_40	3.25	3.95	3.09	2.42	2.22	1.47	5.00	5.17	3.83	3.01	6.53
Zn	AYZV02017730_40	3.35	3.93	2.63	2.04	1.80	1.19	5.10	5.15	2.98	2.51	6.53
Co	AYZV02094133_57	2.07	2.53	1.46	1.27	1.08	0.65	3.90	4.18	1.93	1.78	16.67
Mn	AYZV02094133_57	4.04	4.50	3.49	2.16	0.33	0.63	6.20	6.56	4.88	2.98	16.32
S	AYZV02094133_57	5.36	6.48	6.15	4.96	3.06	2.46	8.10	9.70	8.95	7.87	16.32
Zn	AYZV02094133_57	1.62	2.30	1.99	1.13	0.49	0.17	2.50	3.04	2.35	1.28	16.32
Co	AYZV02281036_185	2.29	1.54	1.12	1.06	2.37	1.41	4.30	3.17	2.12	2.23	9.07
Mn	AYZV02281036_185	3.12	3.45	3.01	2.88	2.97	1.95	4.80	5.69	4.95	5.14	8.59
S	AYZV02281036_185	2.35	2.00	1.70	1.75	1.71	1.35	3.60	3.34	2.78	3.03	8.59
Zn	AYZV02281036_185	2.86	2.85	2.60	2.47	3.19	2.76	4.40	4.73	3.97	4.22	8.59

Table 5 Three SNP markers significantly associated with two to four of the four mineral compound, Co, Mn, S, and Zn based on T-tesing at P = 0.05

SNP	Co			Mn			S			Zn		
	Allele	Significant	LSM	Allele	Significant	LSM	Allele	Significant	LSM	Allele	Significant	LSM
AYZV02017730_40				GG	A	320.375	GG	A	6539.647	GG	A	287.886
AYZV02017730_40				CC	B	235.260	CC	B	5550.364	CC	B	189.567
AYZV02094133_57				CC	A	302.734	CC	A	6461.113			
AYZV02094133_57				AA	B	256.411	AA	B	5835.309			
AYZV02281036_185	AA	A	0.428	AA	A	283.587	AA	A	6397.044	AA	A	306.576
AYZV02281036_185	CC	B	0.384	CC	B	219.221	CC	B	5847.326	CC	B	240.304
AYZV02281036_185	AC	AB	0.366	AC	AB	361.036	AC	AB	6037.149	AC	AB	215.736

S, Mn and Zn concentrations could be possible in individual spinach accessions. For the four mineral elements (Co, S, Mn and Zn), eleven highly significantly associated SNP markers were identified in this study (Table 3).

These markers related to mineral elements have low LOD and small R-squared values because the mineral elements are controlled by multiple genes with minor effects. How to apply these SNP markers in spinach germplasm evaluation and breeding is a challenge for spinach breeders. Next step of research is going to validate these SNP markers using KASP SNP genotyping: 1) To validate the stability of these SNP markers across multiple locations and years; 2) To conduct an additional association mapping study on another natural population panel in order to identify potential overlaps with what we are reporting; 3) To develop bi-parental populations to validate and identify SNP markers linked to mineral elements and try to find SNP markers with major effect; 4) To use genomic selection approach to select the minor effect alleles to improve element compositions in spinach cultivars.

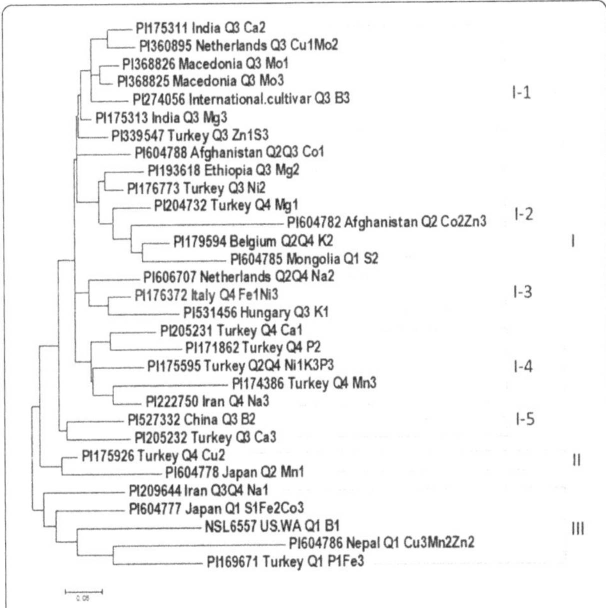

Fig. 4 The ring phylogenetic tree created by the Maximum Likelihood (ML) method from MEGA 6 in 31 spinach germplasm accessions that had at least one mineral element ranked in the top three highest concentration among the 292 spinach accessions for 14 mineral elements

This comprehensive spinach mineral nutrient study provides a foundation of SNP markers to improve mineral content in spinach cultivar development. The future release of a spinach reference genome sequence will enable a complete analysis of these trait loci.

Mineral elements related to nutrition and human healthy food in spinach

The mineral elements are essential and indispensable for growth and health, having a direct or indirect effect on the metabolism and physiological processes of humans and plants as well. Deficiencies or insufficient intake of minerals may lead to several dysfunctions and diseases in humans [43]. There is a growing interest in the mineral and phytochemical composition of foods and diets, and especially in leafy vegetables. Recent studies with Mexican, Central American, and African green leafy vegetables, including *Cnidoscolus aconitifolius*, *Crotalaria longirostrata*, *Solanum scabrum*, *Gynandropsis gyandra*, and several leafy *Amaranthus* species, have highlighted the contributions that these vegetables can provide to one's daily intake of essential nutrients and health-beneficial compounds [44, 45].

Mineral elements in spinach have been reported by other studies employing a large number of techniques [46–51]. On a moisture-free basis, the highest levels of K, Ca, Na, P, Mg, S, Zn, Cu, Co, and B were found in spinach compared to parsley, dill, and mint [52]. In this research, 14 mineral elements were detected in 292 spinach accessions, including macro-elements (K, Mg, P, Ca, and S) and micro-elements (Na, Zn, Mn, B, Fe, Cu, Mo, Co, and Ni). These findings provide a new range of spinach mineral values for dieticians and nutrition scientists to consider when calculating nutrient intakes of spinach-containing diets, or when developing healthful menu recommendations.

Na, K, Mg, and Ca are the four main and essential electrolytes for humans, Na is responsible for controlling the total amount of water in the body. It is also important for regulating blood volume and maintaining muscle and nerve function. Mg is the most abundant intracellular divalent cation. It is an essential cofactor for a multitude of enzymatic reactions that are important for the generation of energy from ATP and for physiologic processes, including neuromuscular function and maintenance of cardiovascular tone [53]. Ca is the major component of bone and assists in tooth development [54]. K is an important component of cell and body fluids that helps to control heart rate and blood pressure (http://www.nutrition-and-you.com/spinach.html).

Among the 292 spinach accessions from this research, the highest levels of Na, K, Mg, and Ca were found to have 1484.01 ppm, 108,935.36 ppm, 14,293.27 ppm, and 10,210.76 ppm in PI209644 from Iraq, PI531456 from Hungary, PI204732 from Turkey, and PI205231 from Turkey, respectively (Additional file 1: Table S1). In addition, the top three spinach germplasm accessions in each mineral element were also listed and provide information on how to use these high mineral accessions in breeding programs. The PI accessions containing beneficial SNPs associated with mineral elements will be highly valuable for the spinach breeders to use for the development of cultivars with high mineral element concentrations through MAS and GS (genomic selection). The significant genetic variations among genotypes as evidenced by the 2 to 42 times difference in mineral concentration (Table 1) suggest that the genetic improvement of mineral traits is feasible in spinach. The co-localization of SNP markers and the positive correlations in concentrations for many mineral elements make it possible to pyramid high concentrations of multiple elements into a single cultivar in a spinach breeding program.

Conclusions

A total of 14 mineral elements: boron (B), calcium (Ca), cobalt (Co), copper (Cu), iron (Fe), potassium (K), magnesium (Mg), manganese (Mn), molybdenum (Mo), sodium (Na), nickel (Ni), phosphorus (P), sulfur (S), and zinc (Zn) were evaluated in 292 spinach accessions originally collected from 29 countries. The 45 SNP markers strongly associated with the concentrations of the 13 mineral elements: B, Ca, Co, Cu, Fe, Mg, Mn, Mo, Na, Ni, P, S, and Zn using six statics methods, including single marker regression using Q-gene (SMR), single marker regression using Tassel (SMR), general linear model using Tassel (GLM), mixed linear model using Tassel (MLM), compressed mixed linear model using Gapit (cMLM), and enriched compressed mixed linear model using Gapit (EcMLM). Three SNP markers, AYZV02017731_40, AYZV02094133_57, and AYZV02281036_185 were identified to be associated with concentrations of four mineral components, Co, Mn, S, and Zn. The markers could be used in breeding programs to improve the nutritional quality of spinach through marker-assisted selection (MAS). Thirty-one spinach accessions with high concentrations of one to several mineral elements can be used as potential parents for spinach breeding programs.

Methods

Plant materials

A total of 292 accessions of spinach (*S. oleracea*) USDA-GRIN (US Department of Agriculture, Germplasm Resources Information Network) germplasm originally collected from 29 countries plus 18 unknown locations were used for genetic diversity and association analysis of mineral elements in this study (Additional file 1: Table S1). All seeds were kindly provided by David Brenner at USDA-ARS (Agricultural Research Service) and Iowa State University at Ames, IA, US.

Leaf mineral concentration evaluation

Concentrations of 14 mineral components: B, Ca, Co, Cu, Fe, K, Mg, Mn, Mo, Na, Ni, P, S, and Zn were evaluated in 292 USDA spinach germplasm accessions (Additional file 1: Table S1). The phenotypic data of the nine among the 14 elements in spinach germplasm accessions, including Ca, Cu, Fe, Mg, Mn, Mo, Ni, P, and Zn, have been published partially in the USDA GRIN website at http://www.ars-grin.gov/cgi-bin/npgs/html/eval.pl?492376.

Accessions were grown in 1-l black plastic pots filled with a 2:1 (vol: vol) mixture of commercially available soil (Metro-Mix 360; Scotts-Sierra Horticultural Products Co., Marysville, Ohio, USA) and vermiculite (Strong-Lite Medium Vermiculite; Sun Gro Horticulture Co, Seneca, Illinois, USA). There were six plants of each accession in a pot, with pots randomly distributed within a growth chamber (model PGW36; Controlled Environments Ltd., Winnipeg, Manitoba, Canada). Plants were grown on a 12 h. photoperiod of 300 μmol $m^{-2} s^{-1}$ photosynthetically active radiation (incandescent and fluorescent lamps) with a 20 ± 0.5 °C / 15 ± 0.5 °C day/ night temperature regime. Relative humidity was maintained at 50% ± 5%. Pots were initially irrigated with deionized water, and after emergence, plants were subirrigated daily with a nutrient solution containing the concentrations of mineral salts: 1.2 mM KNO_3, 0.8 mM $Ca(NO_3)_2$, 0.8 mM NH_4NO_3, 0.2 mM $MgSO_4$, 0.3 mM KH_2PO_4, 25 μM $CaCl_2$, 25 μM H_3BO_3, 2 μm $MnSO_4$, 2 μM $ZnSO_4$, 0.5 μM $CuSO_4$, 0.5 μM H_2MoO_4, 0.1 μM

$NiSO_4$, and 10 μM Fe(III)-N, N′-ethylenebis[2-(2-hydroxyphenyl)-glycine] (Sprint 138; Becker-Underwood, Inc., Ames, Iowa, USA).

Plants were harvested at 4–5 weeks after planting when they had 5–6 fully expanded leaves. Harvested material included both mature and immature leaves (leaf blades and petioles) from the six plants of each accession. Soil contamination of the samples was minimized by cutting plants 0.5 cm above the soil surface. Leaves were dried in paper bags at 65–70 °C for a minimum of 48 h, and after cooling the pooled leaves from the 6 plants were ground to a uniform powder using a coffee grinder with stainless steel blades (model IDS 55; Mr. Coffee, Boca Raton, Florida, USA). Two 0.25 g (dry weight) subsamples of each accession were wet digested in borosilicate glass tubes using ultra-pure nitric and perchloric acids, as previously described [55]. Digestages were resuspended in ultra-pure nitric acid and analyzed for concentrations of B, Ca, Co, Cu, Fe, K, Mg, Mn, Mo, Na, Ni, P, S, and Zn using inductively coupled plasma optical emission spectrometry (CIROS ICP Model FCE12; Spectro, Kleve, Germany). Tomato leaf standards (SRM 1573A; National Institute of Standards and Technology, Gaithersburg, Maryland, USA) were digested and analyzed as quality control along with each run of 50 spinach samples to verify the reliability of the procedures and analytical measurements. Results are reported on a dry weight basis as the average of the two subsamples in ppm (parts per million; equivalent to micrograms per gram).

DNA extraction, GBS, and SNP discovery

Genomic DNA was extracted from freeze-dried fresh leaves of spinach plants using the CTAB (hexadecyltrimethyl ammonium bromide) method [56]. DNA sequencing was done by next generation sequencing technologies using GBS [18, 20] and GBS was conducted by HiSeq 2000 in Beijing Genome Institute (BGI). SOAP family software (http://soap.genomics.org.cn/) was used for sequence assembly, mapping and SNP discovery of GBS data. The GBS data averaged 3.26 M short-read and 283.74 Mbp data-points for each spinach sample. The short reads of the GBS data were aligned to spinach genome reference AYZV02 (http://www.ncbi.nlm.nih.gov/Traces/wgs/?val=AYZV02) by using SOAPaligner/soap2 (http://soap.genomics.org.cn/), while SOAPsnp v. 1.05 was used for SNP calling [57, 58]. Approximately a half million SNPs were discovered from the original GBS data from BGI among the 292 spinach germplasm accessions. The spinach accessions and SNPs were filtered before conducting genetic diversity and association analyses. If an accession had greater than 35% missing SNP data, the accession was removed from the panel. The SNP data were filtered by setting the parameters of minor allele frequency (MLF) > 2%, missing data <20%, and heterozygous genotype <10%. After filtering, 2402 SNPs among 292 spinach accessions were used for genetic diversity and association analysis.

Phenotypic data analysis

Phenotypic data of the 14 mineral elements in spinach were analyzed using Microsoft Excel 2016 for the average, range, standard deviation, and coefficient of variation (CV) and the distributions of the 14 mineral elements were drawn using QGene [35]. The correlation coefficients of the 14 mineral elements were calculated using JMP Genomics 7 software (SAS Institute, Cary, NC, USA). In JMP Genomics 7, the dendrogram construction was done using multivariate methods to do hierarchical clustering. After being clustered, the multivariate principal component analysis (PCA) was used to create biplot on covariance.

Genetic diversity and population structure analysis

The model-based program STRUCTURE 2 [33] was used to infer population structure. In order to identify the number of populations (K) capturing the major population structure in the tested spinach association panel, the burn-in period was set at 50,000 with the Markov Chain Monte Carlo iterations and the run length was set at 10,000 in an admixture model and correlated allele frequencies independent for each run [59]. Ten runs were performed for each simulated value of K, ranging from 1 to 10. The delta K was calculated using the formula provided by Evanno et al. (2005) [60]. The optimal K was determined with Structure Harvester [61]. After the optimal K was determined, a Q-matrix was generated; this was used in Tassel 5 for association analysis of mineral elements. Each spinach accession was also assigned to a cluster (Q) based on a probability for that accession in a cluster, using a cut-off probability of 0.50. Based on the optimal K, a Bar plot with 'Sort by Q' was obtained to visualize the population structure of the spinach association panel.

Genetic diversity was also assessed and the phylogeny trees were drawn using MEGA 6 [34] based on the Maximum Likelihood tree method with the following parameters [17]. Test of Phylogeny: Bootstrap Method, No. of Bootstrap Replications: 500, Model/Method: General Time Reversible model, Rates among Sites: Gamma distributed with Invariant sites (G + I), Number of Discrete Gamma Categories: 5, Gaps/Missing Data Treatment: Use all sites, ML Heuristic Method: Subtree-Pruning-Regrafting-Extensive (SPR level 5), Initial Tree for ML: Make initial tree automatically (Neighbor Joining), and Branch Swap Filter: Moderate. During the drawing of the phylogeny trees, the population structure and the cluster information were imported to MEGA 6 for combined analysis of genetic diversity. For sub-tree of each Q (cluster), the shape of 'Node/Subtree Marker' and the 'Branch Line' was drawn with the same color as in the figure of the Bar plot of the population clusters from the STRUCTURE analysis.

Association analysis

Association analysis was conducted with the single marker regression (SMR) without structure and kinship, the general linear model (GLM), and the mixed linear model (MLM) methods as described in TASSEL 5 [62] (http://www.maizegenetics.net/tassel) and the analysis was also performed with compressed mixed linear model (cMLM) [63] and enriched compressed mixed linear model (EcMLM) [64] implemented in the GAPIT R package [65]. The QGene 4.3.10 was also used to conduct SMR for all SNPs [35], although QGene was developed for QTL mapping, it can also be used in association analysis through SMR. The effect of SNP markers were also conducted by T-test using JMP Genomics and Microsoft Excel 2016.

Additional files

Additional file 1: Table S1. Spinach PI accession number, name, origin/country, cluster assigned in this study, taxon name, and 14 mineral element concentrations in 292 germplasm accessions. **Table S2.** Spinach PI accession number, taxon name, origin country, cluster assigned in this study, mineral element, mineral ID ranked in top three, and 14 mineral element concentrations in 292 germplasm accessions. (XLSX 105 kb)

Additional file 2: Figure S1. The ring phylogenetic tree combining structure populations (Q1 to Q4) from STRUCTURE 2 and the Maximum Likelihood (ML) method from MEGA 6. The spinach accession number, the accession original country, and the structure population (cluster) were merged together into one taxon name as each spinach accession ID in the combined tree drawn by MEGA 6. The colored shape and branch of each cluster matched the structure population (red round shape for Q1, green triangle for Q2, blue triangle for Q3, yellow diamond for Q4, and the black square with the black branch for the admixture in 292 USDA GRIN spinach germplasm accessions. (XLSX 149 kb)

Abbreviations

B: Boron; Ca: Calcium; CMLM: Compressed mixed linear model using Gapit; Co: Cobalt; Cu: Copper; EcMLM: Enriched compressed mixed linear model using Gapit; Fe: Iron; GBS: Genotyping by sequencing; GLM: General linear model; GWAS: genome-wide association study; K: Potassium; MAS: Marker-assisted selection; Mg: Magnesium; MLF: Minor allele frequency; MLM: Mixed linear model; Mn: Manganese; Mo: Molybdenum; Na: Sodium; Ni: Nickel; P: Phosphorus; S: Sulfur; SMR: Single marker regression; SNP: Single nucleotide polymorphism; Zn: Zinc

Acknowledgements

Not applicable.

Funding

(1) by USDA-ARS GRIN GERMPLASM EVALUATION PROPOSAL for National Plant Germplasm System (NPGS) by Crop Germplasm Committee (CGC) with Project Number: 58-5030-6-076 for proving spinach material collection and DNA extraction and sequencing; (2) USDA Specialty Crop Multistate Program (SCMP) for DAN extraction and sequencing; and (3) by the USDA National Institute of Food and Agriculture Hatch project accession number 1002423 for providing the project director and technician salary.

Authors' contributions

JQ, AS, BM, MG, YW, WR, GB, LD, and WY carried out phenotyping and genotyping. AS and JQ analyzed the data. JQ and AS composed the draft of the manuscript. AS and MG directed and managed this research. BM and MG reviewed and edited the manuscript. All authors have read, made corrections and approved the final manuscript.

Competing interests

The authors declare that they have no competing interests.

Author details

[1]Department of Horticulture, University of Arkansas, Fayetteville, AR 72701, USA. [2]Crop Improvement and Protection Research Unit, USDA-ARS, Salinas, CA 93905, USA. [3]USDA-ARS Red River Valley Agricultural Research Center, Fargo, ND 58102, USA.

References

1. Correll J, Bluhm B, Feng C, Lamour K, Du Toit L, Koike S. Spinach: better management of downy mildew and white rust through genomics. Eur J Plant Pathol. 2011;129(2):193–205.
2. NASS U: NASS-National Agricultural Statistics Service. URL http://www nass usda gov/(accessed 27 august 2015) Varunsatian, S, Watanabe, K, Hayakawa, S, Nakamura 2015.
3. Decoteau DR: Vegetable crops: Pearson College div; 2000.
4. Morelock TE, Correll JC. Spinach. Vegetables I. 2008:189–218.
5. Higdon J: An evidence-based approach to vitamins and minerals health benefits and intake recommendations: Thieme medical publishers, Inc.; 2003.
6. Prashanth L, Kattapagari KK, Chitturi RT, Baddam VRR, Prasad LK. A review on role of essential trace elements in health and disease. Journal of Dr NTR University of Health Sciences. 2015;4(2):75.
7. Berdanier CD, Dwyer JT, Heber D: Handbook of nutrition and food: CRC press; 2013.
8. Berdanier CD, Dwyer JT, Heber D: Handbook of nutrition and food: CRC press; 2016.
9. Randhawa MA, Khan AA, Javed MS, Sajid MW. Chapter 18-green leafy vegetables: a health promoting source. Handbook of Fertility Academic Press, San Diego, CA. 2015:205–20.
10. Moose SP, Mumm RH. Molecular plant breeding as the foundation for 21st century crop improvement. Plant Physiol. 2008;147(3):969–77.
11. Collard B, Jahufer M, Brouwer J, Pang E. An introduction to markers, quantitative trait loci (QTL) mapping and marker-assisted selection for crop improvement: the basic concepts. Euphytica. 2005;142(1–2):169–96.
12. Collard BC, Mackill DJ. Marker-assisted selection: an approach for precision plant breeding in the twenty-first century. Philosophical Transactions of the Royal Society B: Biological Sciences. 2008;363(1491):557–72.
13. Xu Y, Crouch JH. Marker-assisted selection in plant breeding: from publications to practice. Crop Sci. 2008;48(2):391–407.
14. Lehne B, Lewis CM, Schlitt T. From SNPs to genes: disease association at the gene level. PLoS One. 2011;6(6):e20133.
15. Taranto F, D'Agostino N, Greco B, Cardi T, Tripodi P. Genome-wide SNP discovery and population structure analysis in pepper (Capsicum Annuum) using genotyping by sequencing. BMC Genomics. 2016;17(1):943.
16. Li P, Guo M, Wang C, Liu X, Zou Q. An overview of SNP interactions in genome-wide association studies. Briefings in Functional Genomics. 2015; 14(2):143–55.

17. Shi A, Buckley B, Mou B, Motes D, Morris JB, Ma J, Xiong H, Qin J, Yang W, Chitwood J. Association analysis of cowpea bacterial blight resistance in USDA cowpea germplasm. Euphytica. 2016;208(1):143–55.
18. Elshire RJ, Glaubitz JC, Sun Q, Poland JA, Kawamoto K, Buckler ES, Mitchell SE. A robust, simple genotyping-by-sequencing (GBS) approach for high diversity species. PLoS One. 2011;6(5):e19379.
19. Davey JW, Hohenlohe PA, Etter PD, Boone JQ, Catchen JM, Blaxter ML. Genome-wide genetic marker discovery and genotyping using next-generation sequencing. Nat Rev Genet. 2011;12(7):499–510.
20. Sonah H, Bastien M, Iquira E, Tardivel A, Légaré G, Boyle B, Normandeau É, Laroche J, Larose S, Jean M. An improved genotyping by sequencing (GBS) approach offering increased versatility and efficiency of SNP discovery and genotyping. PLoS One. 2013;8(1):e54603.
21. He J, Zhao X, Laroche A, Lu Z-X, Liu H, Li Z. Genotyping-by-sequencing (GBS), an ultimate marker-assisted selection (MAS) tool to accelerate plant breeding. Front Plant Sci. 2014;5:484.
22. Poland JA, Rife TW. Genotyping-by-sequencing for plant breeding and genetics. The Plant Genome. 2012;5(3):92–102.
23. Narum SR, Buerkle CA, Davey JW, Miller MR, Hohenlohe PA. Genotyping-by-sequencing in ecological and conservation genomics. Mol Ecol. 2013;22(11): 2841–7.
24. Van Deynze A. A de novo draft assembly of spinach using Pacific biosciences technology. In. 2014:10–5.
25. Van Deynze A, Ashrafi H, Hickey L, Peluso P, Rank D, Chin J, Rapicavoli N, Drake J, Garvin T, Schatz M. Using spinach to compare technologies for whole genome assemblies. In. 2015:10–4.
26. Dohm JC, Minoche AE, Holtgräwe D, Capella-Gutiérrez S, Zakrzewski F, Tafer H, Rupp O, Sörensen TR, Stracke R, Reinhardt R. The genome of the recently domesticated crop plant sugar beet (Beta Vulgaris). Nature. 2014;505(7484): 546–9.
27. Minoche AE, Dohm JC, Schneider J, Holtgräwe D, Viehöver P, Montfort M, Sörensen TR, Weisshaar B, Himmelbauer H. Exploiting single-molecule transcript sequencing for eukaryotic gene prediction. Genome Biol. 2015; 16(1):184.
28. Shi A, Mou B, Correll JC. Association analysis for oxalate concentration in spinach. Euphytica. 2016;212(1):17–28.
29. Shi A, Mou B. Genetic diversity and association analysis of leafminer (Liriomyza Langei) resistance in spinach (Spinacia Oleracea). Genome. 2016; 59(8):581–8.
30. Shi A, Mou B, Correll J, Motes D, Weng Y, Qin J, Yang W: SNP association analysis of resistance to Verticillium wilt ('Verticillium dahliae'Kleb.) in spinach. *Australian Journal of Crop Science* 2016, 10(8):1188.
31. Shi A, Mou B, Correll J, Koike ST, Motes D, Qin J, Weng Y, Yang W. Association analysis and identification of SNP markers for Stemphylium leaf spot (Stemphylium botryosum f. sp. spinacia) resistance in spinach (*Spinacia oleracea*). American Journal of Plant Sciences. 2016;**7**(12):1600.
32. Ma J, Shi A, Mou B, Evans M, Clark JR, Motes D, Correll JC, Xiong H, Qin J, Chitwood J. Association mapping of leaf traits in spinach (Spinacia Oleracea L.). Plant Breed. 2016;135(3):399–404.
33. Pritchard JK, Stephens M, Donnelly P. Inference of population structure using multilocus genotype data. Genetics. 2000;155(2):945–59.
34. Tamura K, Stecher G, Peterson D, Filipski A, Kumar S. MEGA6: molecular evolutionary genetics analysis version 6.0. Mol Biol Evol. 2013;30(12):2725–9.
35. Joehanes R, Nelson JC. QGene 4.0, an extensible java QTL-analysis platform. Bioinformatics. 2008;24(23):2788–9.
36. Ma Y, Coyne CJ, Grusak MA, Mazourek M, Cheng P, Main D, McGee RJ. Genome-wide SNP identification, linkage map construction and QTL mapping for seed mineral concentrations and contents in pea (*Pisum sativum* L.). BMC plant biology, 2017. 17(1):43.
37. Anuradha K, Agarwal S, Rao YV, Rao K, Viraktamath B, Sarla N. Mapping QTLs and candidate genes for iron and zinc concentrations in unpolished rice of Madhukar× Swarna RILs. Gene. 2012;508(2):233–40.
38. Garcia-Oliveira AL, Tan L, Fu Y, Sun C. Genetic identification of quantitative trait loci for contents of mineral nutrients in rice grain. J Integr Plant Biol. 2009;51(1):84–92.
39. Stangoulis JCR, Huynh B-L, Welch RM, Choi E-Y, Graham RD. Quantitative trait loci for phytate in rice grain and their relationship with grain micronutrient content. Euphytica. 2007;154(3):289–94.
40. Huang Y, Sun C, Min J, Chen Y, Tong C, Bao J. Association mapping of quantitative trait loci for mineral element contents in whole grain Rice (Oryza Sativa L.). J Agric Food Chem. 2015;63(50):10885–92.
41. Norton GJ, Deacon CM, Xiong L, Huang S, Meharg AA, Price AH. Genetic mapping of the rice ionome in leaves and grain: identification of QTLs for 17 elements including arsenic, cadmium, iron and selenium. Plant Soil. 2010; 329(1–2):139–53.
42. Du J, Zeng D, Wang B, Qian Q, Zheng S, Ling H-Q. Environmental effects on mineral accumulation in rice grains and identification of ecological specific QTLs. Environ Geochem Health. 2013;35(2):161–70.
43. Sautter C, Poletti S, Zhang P, Gruissem W. Biofortification of essential nutritional compounds and trace elements in rice and cassava. Proc Nutr Soc. 2006;65(02):153–9.
44. Jiménez-Aguilar DM, Grusak MA. Evaluation of minerals, phytochemical compounds and antioxidant activity of Mexican, central American, and African green leafy vegetables. Plant Foods Hum Nutr. 2015;70(4):357–64.
45. Jiménez-Aguilar DM, Grusak MA. Minerals, vitamin C, phenolics, flavonoids and antioxidant activity of Amaranthus leafy vegetables. J Food Compos Anal. 2017;
46. Lin S-w. A comparative study of the determination of phosphorus by electrothermal atomic absorption spectrometry and solution spectrophotometry. Anal Chim Acta. 1984;158:199–206.
47. Stephen SC, Littlejohn D, Ottaway JM. Evaluation of a slurry technique for the determination of lead in spinach by electrothermal atomic-absorption spectrometry. Analyst. 1985;110(9):1147–51.
48. Bhattacharjee S, Dasgupta P, Paul AR, Ghosal S, Padhi KK, Pandey LP. Mineral element composition of spinach. J Sci Food Agric. 1998;77(4):456–8.
49. Alegría A, Barbera R, Farré R. Atomic-absorption spectrophotometric determination of nickel in foods. Journal of Micronutrient Analysis. 1988;4(3): 229–39.
50. Yan D, Schwedt G. Simultaneous ion chromatography of inorganic anions together with some organic anions and alkaline earth metal cations using chelating agents as eluents. J Chromatogr A. 1990;516(2):383–93.
51. Mittal R, Allawadhi KL, Sood BS, Singh N, Kumar A, Kumar P. Determination of potassium and calcium in vegetables by x-ray fluorescence spectrometry. X-Ray Spectrom. 1993;22(6):413–7.
52. Rahmatollah R, Mahbobeh R. Mineral contents of some plants used in Iran. Pharm Res. 2010;2(4):267.
53. Saris N-EL, Mervaala E, Karppanen H, Khawaja JA, Lewenstam A. Magnesium: an update on physiological, clinical and analytical aspects. Clin Chim Acta. 2000;294(1):1–26.
54. Brody T. Protein. Brody T Nutritional biochemistry. 1994;2
55. Sankaran RP, Grusak MA. Whole shoot mineral partitioning and accumulation in pea (Pisum Sativum). Front Plant Sci. 2014;5(April):1–8.
56. Kisha T, Sneller C, Diers B. Relationship between genetic distance among parents and genetic variance in populations of soybean. Crop Sci. 1997; 37(4):1317–25.
57. Li H. A statistical framework for SNP calling, mutation discovery, association mapping and population genetical parameter estimation from sequencing data. Bioinformatics. 2011;27(21):2987–93.
58. Li H, Handsaker B, Wysoker A, Fennell T, Ruan J, Homer N, Marth G, Abecasis G, Durbin R. The sequence alignment/map format and SAMtools. Bioinformatics. 2009;25(16):2078–9.
59. Lv J, Qi J, Shi Q, Shen D, Zhang S, Shao G, Li H, Sun Z, Weng Y, Shang Y: Genetic diversity and population structure of cucumber (*Cucumis sativus* L.). *PLoS One* 2012, **7**(10):e46919.
60. Evanno G, Regnaut S, Goudet J. Detecting the number of clusters of individuals using the software STRUCTURE: a simulation study. Mol Ecol. 2005;14(8):2611–20.
61. Earl DA. STRUCTURE HARVESTER: a website and program for visualizing STRUCTURE output and implementing the Evanno method. Conserv Genet Resour. 2012;4(2):359–61.
62. Bradbury PJ, Zhang Z, Kroon DE, Casstevens TM, Ramdoss Y, Buckler ES. TASSEL: software for association mapping of complex traits in diverse samples. Bioinformatics. 2007;23(19):2633–5.
63. Zhang Z, Ersoz E, Lai C-Q, Todhunter RJ, Tiwari HK, Gore MA, Bradbury PJ, Yu J, Arnett DK, Ordovas JM. Mixed linear model approach adapted for genome-wide association studies. Nat Genet. 2010;42(4):355–60.
64. Huang X, Wei X, Sang T, Zhao Q, Feng Q, Zhao Y, Li C, Zhu C, Lu T, Zhang Z. Genome-wide association studies of 14 agronomic traits in rice landraces. Nat Genet. 2010;42(11):961–7.
65. Lipka AE, Tian F, Wang Q, Peiffer J, Li M, Bradbury PJ, Gore MA, Buckler ES, Zhang Z. GAPIT: genome association and prediction integrated tool. Bioinformatics. 2012;28(18):2397–9.

5

Genome-wide analysis of the *Solanum tuberosum* (potato) trehalose-6-phosphate synthase (TPS) gene family: evolution and differential expression during development and stress

Yingchun Xu[1], Yanjie Wang[1], Neil Mattson[2], Liu Yang[3] and Qijiang Jin[1*]

Abstract

Background: Trehalose-6-phosphate synthase (TPS) serves important functions in plant desiccation tolerance and response to environmental stimuli. At present, a comprehensive analysis, i.e. functional classification, molecular evolution, and expression patterns of this gene family are still lacking in *Solanum tuberosum* (potato).

Results: In this study, a comprehensive analysis of the *TPS* gene family was conducted in potato. A total of eight putative potato *TPS* genes (*StTPSs*) were identified by searching the latest potato genome sequence. The amino acid identity among eight StTPSs varied from 59.91 to 89.54%. Analysis of d_N/d_S ratios suggested that regions in the TPP (trehalose-6-phosphate phosphatase) domains evolved faster than the TPS domains. Although the sequence of the eight *StTPSs* showed high similarity (2571-2796 bp), their gene length is highly differentiated (3189-8406 bp). Many of the regulatory elements possibly related to phytohormones, abiotic stress and development were identified in different *TPS* genes. Based on the phylogenetic tree constructed using *TPS* genes of potato, and four other *Solanaceae* plants, *TPS* genes could be categorized into 6 distinct groups. Analysis revealed that purifying selection most likely played a major role during the evolution of this family. Amino acid changes detected in specific branches of the phylogenetic tree suggests relaxed constraints might have contributed to functional divergence among groups. Moreover, *StTPSs* were found to exhibit tissue and treatment specific expression patterns upon analysis of transcriptome data, and performing qRT-PCR.

Conclusions: This study provides a reference for genome-wide identification of the potato *TPS* gene family and sets a framework for further functional studies of this important gene family in development and stress response.

Keywords: *Solanum tuberosum*, Trehalose-6-phosphate synthase, gene family, expression profiling

Background

Trehalose is a non-reducing disaccharide and known as a quantitatively important compatible solute in distinct organisms, for example, bacteria, fungi, algae, and plants [1–3]. Recent accumulating evidence has caused great interest in trehalose, due to its role as a potential signal metabolite and a cell stabilizer in plants. Trehalose is believed to interact with pathogens and herbivorous insects in plants as well as protect plants from various environmental stresses, i.e. heat, cold, desiccation, freezing, hypoxia and oxidative stress [4, 5]. A striking example is in "resurrection plants", e.g. *Selaginella lepidophylla*, *Myrothamnus flabellifolius* and *Sporobolus spp.*, which survive under extreme desiccation, where up to 99% of their water has been removed. The protective effects of trehalose can be explained by water replacement hypothesis or the glass transition hypothesis [4]. Under water deficiency, resurrection plants accumulate massive

* Correspondence: jqj@njau.edu.cn
[1]College of Horticulture, Nanjing Agricultural University, Nanjing 210095, China
Full list of author information is available at the end of the article

amounts of trehalose reaching levels up to 10 –20% of the dry weight [6] which enable them to persist in metabolic stasis for several years until re-watered. However, it is interesting to note that trehalose levels are much lower in crops plants.

It is well established that an important enzyme, trehalose-6-P synthase (TPS), catalyzes the conversion of Glc-6-P and UDP-Glc into trehalose-6-P (T-6-P) [7]. T-6-P is then catalyzed by T-6-P phosphatase (TPP) and releases trehalose. Both plants and yeast (*Saccharomyces cerevisiae*) share a similar biosynthesis pathway [7, 8]. So far, TPS proteins have been purified from several organisms, including *S. lepidophylla* [9, 10], yeast [11], *Mycobacterium smegmatis* [12], and *Mycobacterium tuberculosis* [13]. Among these organisms, the biosynthesis of trehalose in *Escherichia coli* and *Saccharomyces cerevisiae* has been well studied. It was found that the TPS are specific for either UDP-glucose or GDP-glucose as the glucosyl donor [8, 14]. Further studies indicated that T-6-P could restrict glucose influx by its interaction with sugar kinase activities and glucose transport [15].

In spite of low trehalose content in plants, recent evidence showed that expression or overexpression of *TPS* genes in some plants, i.e. tobacco, could lead to pounced changes on growth performance and morphology under drought stress [16, 17]. In *Selaginella*, studies suggest involvement of a functional TPS (SlTPS1) in regulating plant response to heat and salt stresses [18, 19]. In fact, it has become clear that overexpression or expression of *TPS* genes conferred biotic and abiotic stress tolerance of transgenic plants [7, 17, 20, 21]. Despite this, there is no evidence that the enhanced tolerance in these plants is associated with changes of trehalose content [21]. In wheat and cotton, water deficiency only triggers a slight increase in trehalose content [41]. Whether these observed effects on stress tolerance in these transgenic lines were attributed to small changes in trehalose levels [7] has so far been poorly described.

A large number of putative trehalose synthesis genes have been identified and characterized in a wide range of plants [22–24]. In *Arabidopsis*, studies identified 11 *TPS* gene family members, defined by the presence of conserved TPS and TPP domains and can be categorized into two main subfamilies [25]. It is now well accepted that the plant *TPS* gene family is a large gene family with multiple copies, and known to participate in a great array of biological processes [24, 26, 27]. Other functions have been attributed to *TPS* genes. For example, TPS has been found to act as a sucrose signal for trehalose in stress response. Notably, the *Arabidopsis TPS* gene, *AtTPS2*, was demonstrated to be a regulator of glucose, abscisic acid and stress signaling [28]. Further, T6P is also recognized as a regulator of sugar metabolism in plants [29–31]. T6P was found to inhibit the effect of SNF1-related protein kinase1, which is a central integrator of stress and metabolic signals, to regulate plant growing tissues [30]. However, their actual functions in higher plants are largely unknown, particularly those involved in important signaling pathways. Identification and characterization *TPS* gene family members is particularly relevant for understanding the role of TPS in plants, both for genetic diversity to obtain a broader understanding of the function of TPS, and as a potential gene resource for improving crop plant defense against biotic and abiotic stresses.

Potato is an important food and economic crop globally. *TPS* genes in potato have not been well characterized previously. This study investigated the distribution of *TPS* genes from whole genome-wide resources, genetic structure of *TPS* genes in potato genomes, and expression patterns of the gene family members in different tissues or under various stresses. The evolutionary characterization of the *TPS* gene family in potato, and four other *Solanaceae* plants including tomato, pepper, tobacco and petunia were also examined. These results contribute to a better understanding of potato *TPS* gene family, and facilitate further functional studies of them.

Results and discussion

Identification of the potato *TPS* gene family members

The potato is an important dicotyledonous source of human food. Compared with other important crops, i.e. rice, there is relatively less genetic research on potato. Trehalose protects bioactive substances and cell structures of cells against various environmental stresses [2, 3, 18, 20, 21]. Trehalose-6-phosphate synthases (TPSs), important enzymes in the biosynthesis of trehalose, have emerged recently as key players in protecting plants from heat, nutrient, osmotic, and dehydration stress, as well as toxic chemicals. Of particular interest, TPSs might function as regulatory molecules in linking trehalose metabolism to glucose transport and glycolysis [3]. However, very little research has focused on the identity and function of potato *TPS* genes. In this work, the latest version of the potato genome was downloaded to identify genes encoding TPS using HMMER (v3.1) [32] with HMMs of TPS and TPP domains. While the initial screen identified 11 ORFs predicted to encode putative TPS proteins, only 8 contain both TPS and TPP domain and were identified as TPS protein (Table 1). Previous studies have identified 11 *TPS* genes in *Arabidopsis*, 11 in rice, 12 in *Populus trichocarpa* and 13 in soybean [33]. Our data suggest a loss in *TPS* genes in potato as compared to the *TPS* gene family in *Arabidopsis*, rice, soybean, and *Populus trichocarpa*. To determine the genomic distribution of *StTPS* genes, we noted their position on each chromosome based on the information

Table 1 List of the *StTPS* genes identified in this study

Gene ID	Gene Accession Number	CDS (bp)	Deduced Polypeptide				Predicted Subcellular localization
			Length (aa)	MW (kDa)	pI	GRAVY	
StTPS1	PGSC0003DMG400022778	2574	857	97.354	5.610	-0.218	Cytoplasmic
StTPS2	PGSC0003DMG400028467	2553	850	96.178	5.700	-0.229	Cytoplasmic
StTPS3	PGSC0003DMG400023995	2574	857	96.528	5.520	-0.188	Cytoplasmic
StTPS4	PGSC0003DMG400004114	2760	919	103.894	5.940	-0.131	Nuclear
StTPS5	PGSC0003DMG400027449	2625	874	98.541	6.390	-0.239	Cytoplasmic
StTPS6	PGSC0003DMG400017276	2517	838	94.877	5.630	-0.280	Cytoplasmic
StTPS7	PGSC0003DMG401017546	2589	862	97.255	5.900	-0.172	Cytoplasmic
StTPS8	PGSC0003DMG400010556	2799	932	105.815	5.500	-0.211	Plasma Membrane

obtained from the genome database (Fig. 1). It was found that *StTPS* genes were dispersed in seven potato chromosomes. Eight potato *TPS* genes were designated as *StTPS1–StTPS8* according to their order on the chromosomes. Synteny blocks were analyzed among the potato chromosomes for further investigation of the possible evolutionary mechanism of *StTPS* genes in potato (Fig. 1). It is interesting to note that none of the *StTPS* gene pairs were observed within a synteny block, which indicating that *StTPS* genes were duplicated by other modes but not segmental, tandem and proximal, and *StTPS* gene family might expand after two previously reported whole-genome duplication events [34]. The eight predicted full length TPS proteins varied from 857 to 932 amino acid residues and the relative molecular mass ranged from 94.877 to 105.815 kDa, with isoelectric points in the range of 5.520 to 6.390. Subcellular localization prediction suggested that most of the StTPSs might be located in cytoplasm, while only a few might be located in plasma membrane or nucleus (Table 1). Subcellular localization of a gene product is closely related to its functional involvement.

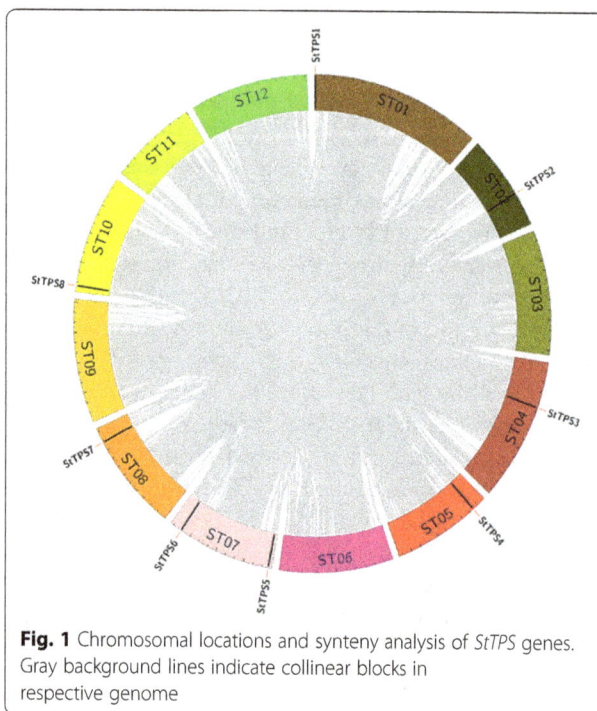

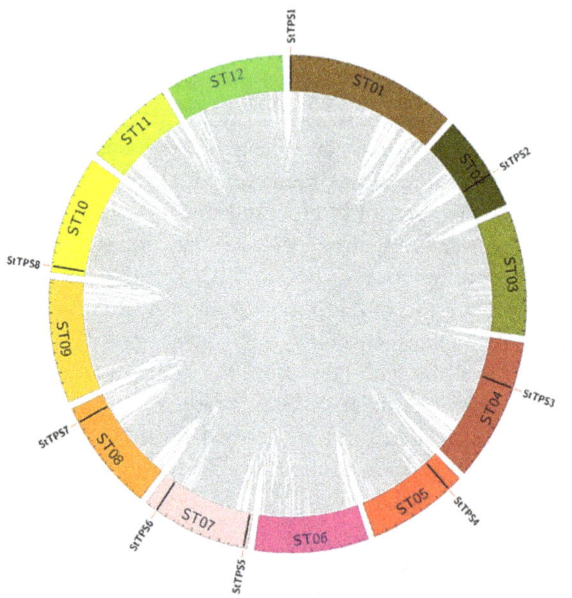

Fig. 1 Chromosomal locations and synteny analysis of *StTPS* genes. Gray background lines indicate collinear blocks in respective genome

Multiple sequence alignment

To clarify the characteristics of *TPS* gene family in potato, multiple sequence alignment of amino acid sequences was performed using Clustalx (Additional file 1: Figure S1). The results showed that the catalytic centers in StTPSs are highly conserved, implying the corresponding genes encode active TPS enzymes. The amino acid identity among eight StTPSs ranged from 59.91 to 89.54%, with the highest identity between StTPS3 and StTPS4 (Fig. 2a), and the lowest identity between StTPS5 and StTPS8. Relatively high divergence was observed in some regions of the amino acid sequences outside of the TPS and TPP domain. The average identity of amino acid sequence of TPS and TPP domains were about 70%, while it was only 60% of the sequences outside domains. It seems likely that these non-conserved regions may contribute largely to functional distinction.

We then analyzed substitution rate ratios of the synonymous substitution rate (d_S) versus the non-synonymous substitution rate (d_N) of *StTPSs*, as this could measure selection pressure on amino acid substitutions, and reflecting whether Darwinian positive selection was involved in driving gene divergence after duplication. Results in Fig. 2b showed that all the estimated d_N/d_S values of different domains and regions outside domains were substantially less than 1. Generally, d_N/d_S ratio >1 indicates positive selection, a ratio <1 indicates negative or purifying selection and a ratio = 1 indicates neutral evolution [35, 36]. This suggested that potato *TPS* gene family might have undergone purifying selection.

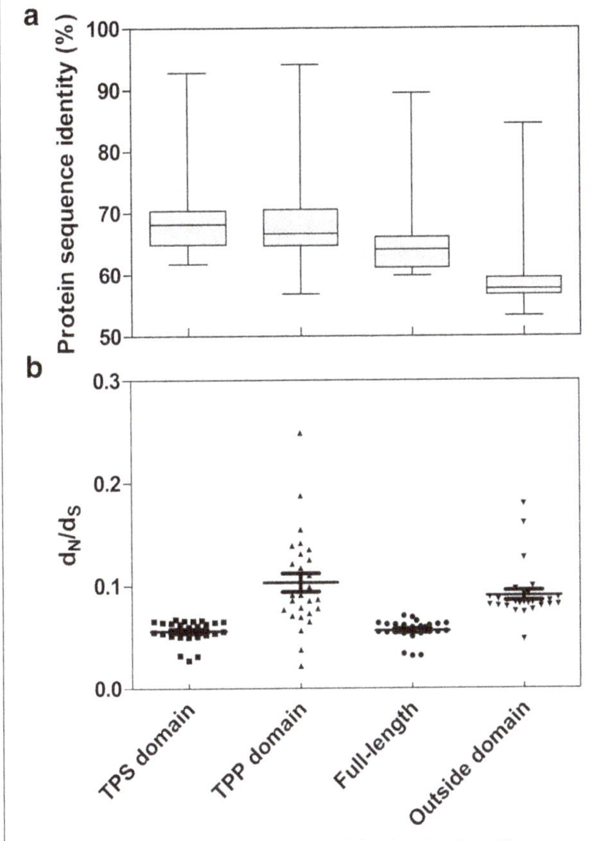

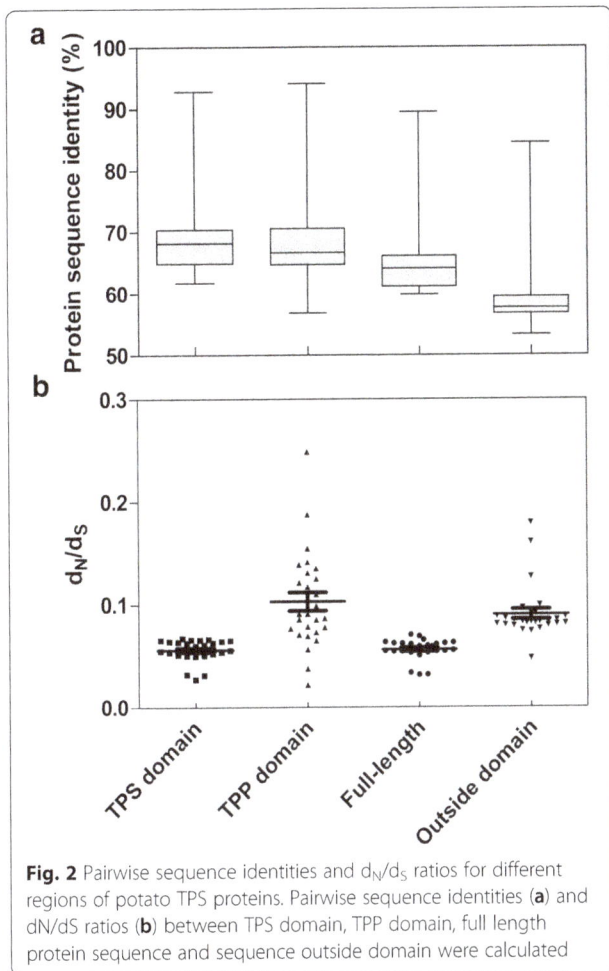

Fig. 2 Pairwise sequence identities and d_N/d_S ratios for different regions of potato TPS proteins. Pairwise sequence identities (**a**) and dN/dS ratios (**b**) between TPS domain, TPP domain, full length protein sequence and sequence outside domain were calculated

In contrast to the result of protein sequence identity, the d_N/d_S ratios in TPP domains were much higher than in TPS domains as well as regions outside TPS domains (Fig. 2b). This observation revealed that the sequence of TPP domains evolved faster than the TPS domain, which might be caused by relaxed purifying or positive selection in the TPP domain. The positive selective effect on residues of TPP domains might ultimately lead to changes in protein function.

Gene structures and protein domains of StTPSs

We then analyzed the exon/intron boundaries of *StTPS* genes, as this can provide additional evidence for the evolution of multiple gene families [37]. We observed that except *StTPS1* and *StTPS4*, most genes harbored two introns in the CDS region (Fig. 3). *StTPS* genes identified on the terminal node of the phylogenetic tree were more variable as compared with previous observations on *TPS* gene structure [33]. Moreover, in spite of the high similarity in CDS length (2571-2796 bp) among eight *StTPS* genes, their total gene length is more variable (3189-8406 bp).

The motif distribution in eight *StTPS* genes was investigated using the MEME program. MEME software identified a total of 20 conserved motifs in *StTPS* as well as their distribution (Fig. 4, Additional file 2: Figure S2). With the exception of three *StTPS* genes, including *StTPS2, StTPS3* and *StTPS4*, all motifs were found distributed diffusely among the other five genes. According to Fig. 4, TPS domains are composed of 12 motifs including motif 1, 2, 3, 4, 6, 9, 10, 13, 14, 15, 17 and 20, while TPP domains are composed of 2 motifs including 8 and 19. These features are consistent with those observed in other plants [38].

The promoter region regulates expression of genes in response to environmental stimuli. Determining promoter

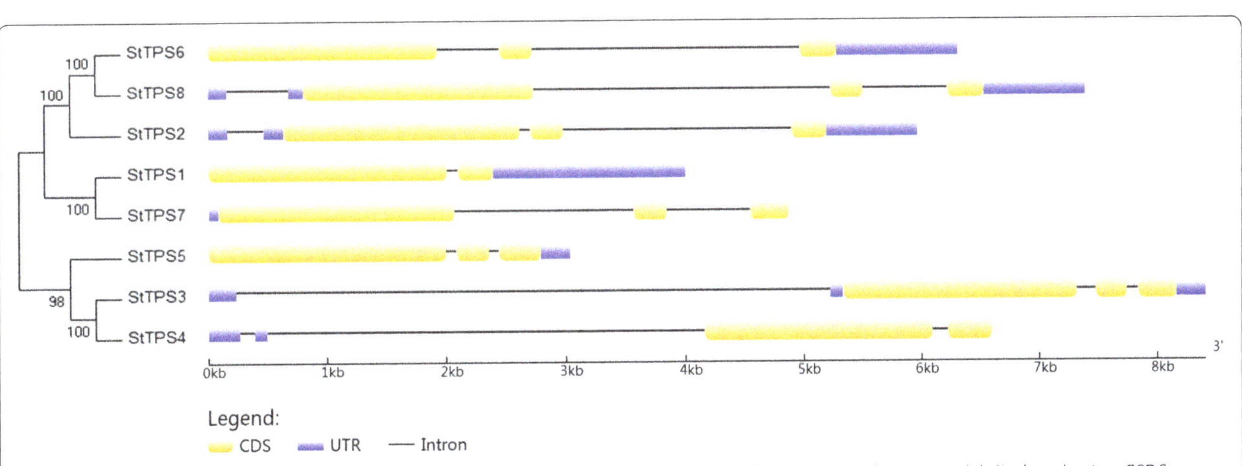

Fig. 3 Exon-intron structures of the identified *StTPS* genes. The graphic representation of the optimized gene model displayed using GSDS. Genes were grouped by an unrooted phylogenetic tree resulting from the full-length amino acid alignment of all the StTPS proteins as shown on the left side of the figure

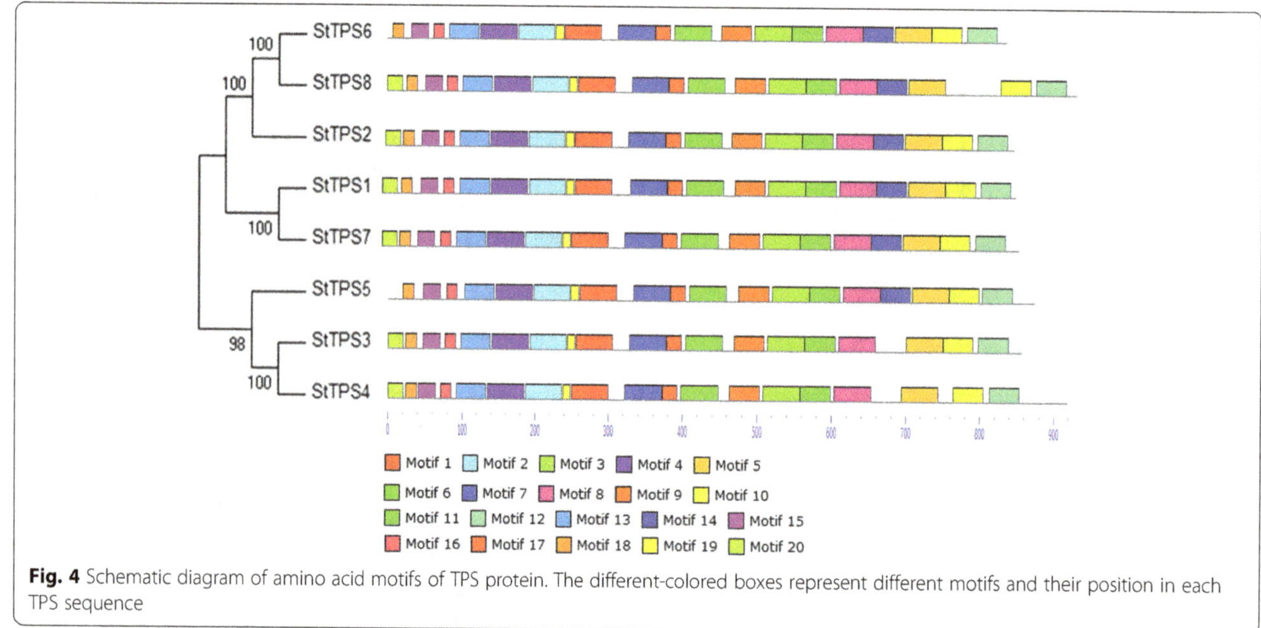

Fig. 4 Schematic diagram of amino acid motifs of TPS protein. The different-colored boxes represent different motifs and their position in each TPS sequence

region features, especially the *cis*-acting elements, is key to understanding the systems that regulate gene expression [39]. For instance, ABA-responsive elements (ABREs) regulate gene response to ABA, drought or salt signals [40, 41]. To identify *cis*-regulatory elements in *StTPS* genes, the 1.5 kb upstream region of the eight genes were extracted from the potato genome and analyzed using PlantCare server (Fig. 5). Several regulatory elements predicted in *StTPS* promoters were associated with phytohormones, abiotic stress and developmental processes. Further, we also identified a biotic stress response element (As-2-box) in *StTPS1*. These predicted cis-regulatory elements were evenly distributed throughout the promoter regions of the *StTPS* genes (Fig. 5). The presence of hormone-responsive elements (abscisic acid, auxin, gibberellin, salicylic acid and Jasmonic acid) could be interpreted as an indication that these *TPS* genes might be involved in various signaling pathway of phytohormones. In particular, *StTPS2* contained the largest number of phytohormones-responsive elements, suggesting an important role in phytohormone response. *StTPS* genes were predicted to contain various abiotic stress-responsive elements, most of which were involved in plant response to environmental stimuli. For example, *StTPS3* were found induced during both anaerobic and dark condition, indicating that it might involve in plant submergence response. *StTPS1* and *StTPS2* contain regulatory elements responsive to low temperature. These conclusions were supported by several reports that *TPS* expression levels increased under drought [42], salt and temperature stresses [18, 43] in various plants.

Evolution analysis of *TPS* genes

An unrooted Neighbor-Joining tree was created for the characterization of the evolutionary relationships between *StTPSs* from potato and *TPSs* from tomato, pepper, tobacco and petunia (Fig. 6). Based on the phylogenetic tree (Fig. 6), these *TPS* genes could be classified into two main subfamilies (I and II), which is in agreement with previous work [44]. To determine the paralogous and orthologous relations among this family, the subfamily II *TPS* genes were further assigned to five groups (II-1, 2, 3, 4, and 5) with high bootstrap support. The number of potato, tomato, pepper, tobacco and petunia *TPS* genes in each of groups were I (0, 2, 2, 3, 2), II-1 (1, 1, 1, 2, 1) , II-2 (1, 1, 1, 2, 1) , II-3 (2, 1, 2, 2, 1) , II-4 (3, 3, 3, 5, 3) and II-5 (1, 1, 1, 3, 2) respectively. At least one gene of the five species was present in each group with the exception that no *StTPS* genes was in group I-1 (Fig. 6). The phylogenetic relationships among the five *Solanaceae* species suggested that genes in the same group may have similar function.

We were then interested to see if any amino acid substitutions in subgroups of TPSs have caused adaptive functional diversification. For this purpose, we evaluated the type I and type II functional divergence, between groups of the TPS family by posterior analysis [45] (Table 2). It was found that most type I coefficients (θ_I) of functional divergence were significantly greater than zero ($P<0.01$), while few of the type II coefficients (θ_{II}) were statistically greater than zero, implying that type I functional divergence was the dominant pattern for the evolution of TPS family in these plants. The results also showed that site-specific selective constraints on most

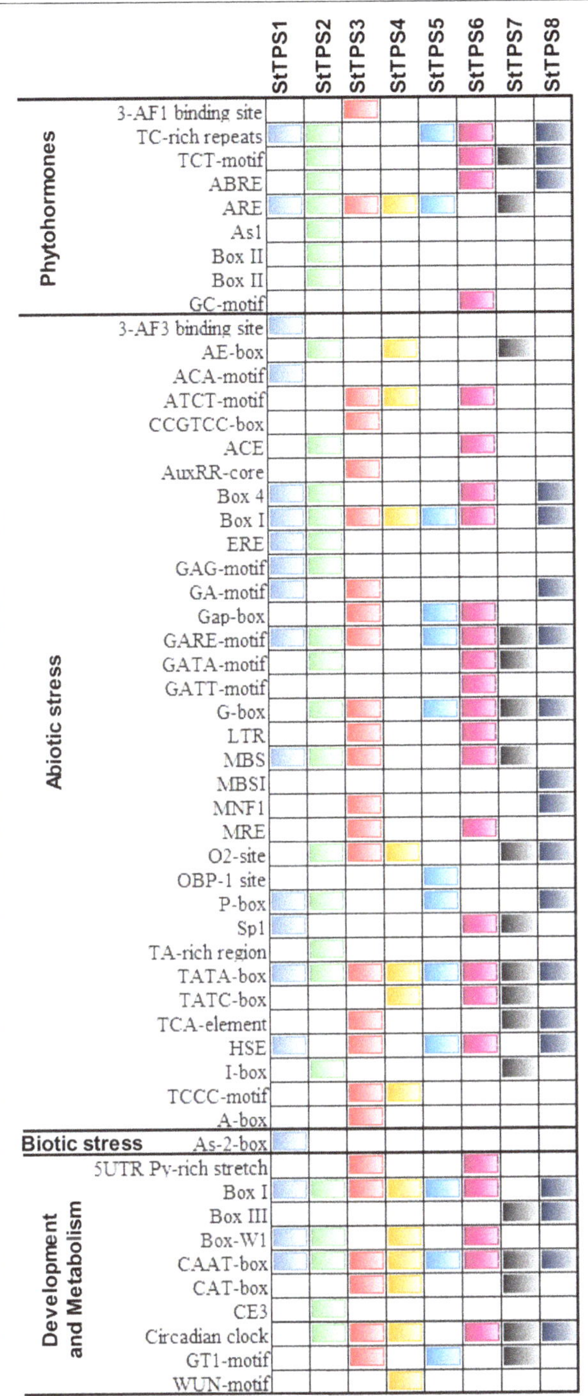

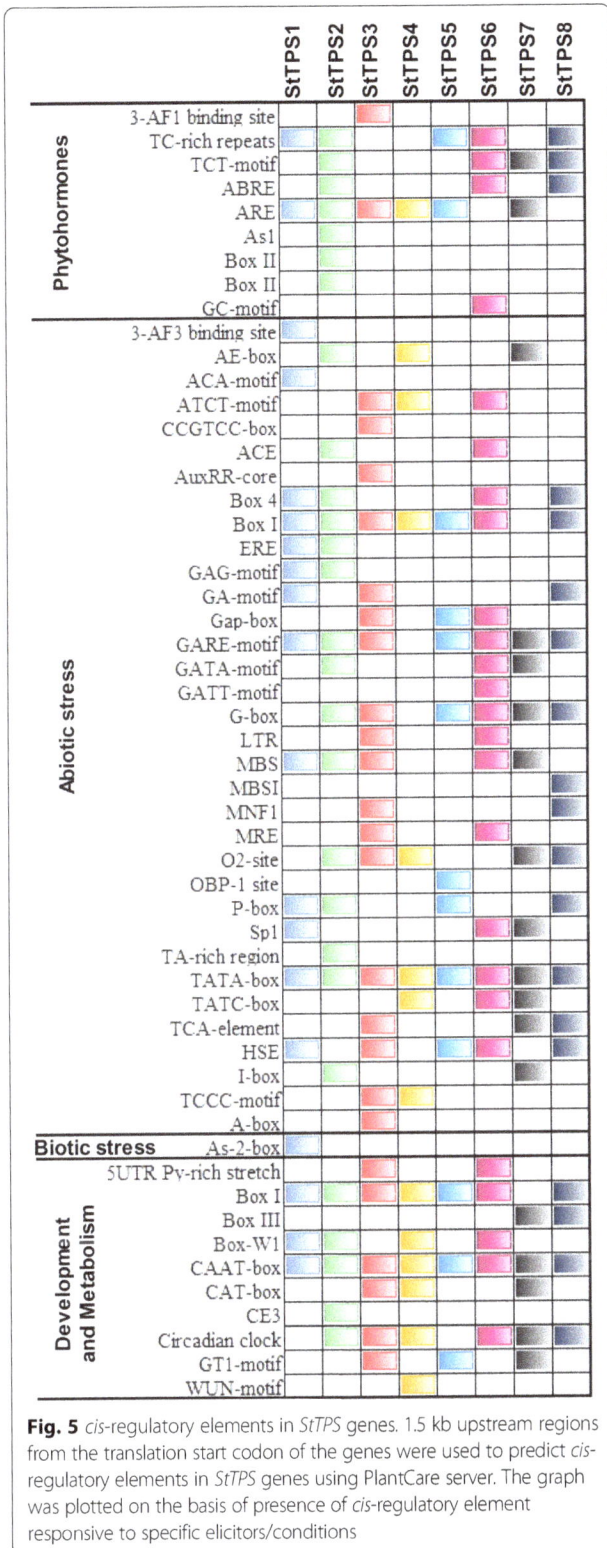

Fig. 5 *cis*-regulatory elements in *StTPS* genes. 1.5 kb upstream regions from the translation start codon of the genes were used to predict *cis*-regulatory elements in *StTPS* genes using PlantCare server. The graph was plotted on the basis of presence of *cis*-regulatory element responsive to specific elicitors/conditions

members of TPS family may contribute to a group-specific functional evolution after their diversification as the coefficients of all functional divergence (θ) values between these groups were less than 1. The group II-1/II-3 had the least θ_I value (0.001), revealed that the lowest evolutionary rate or site specific selective relaxation was between these two groups. By contrast, the theta value in group pair II-1/II-5 was the greatest (0.908), implying the largest divergence between them.

To gain more information on the critical amino acid residues responsible for the functional divergence, all pairs of groups with functional divergence were used for posterior analysis. A cut off value ($Q_k \geq 0.95$), as is frequently used in previous cited work [33], was used to identify type I functional divergence-related residues between groups. Most of the group pairs had at least one site in which the posterior probability was higher than 0.8. Among them, five pairs of groups had at least one site with posterior probability higher than 0.95 (Fig. 7). Similar to a previous report [33], the number and distribution of predicted sites for functional divergence within each pair are highly distinct. For example, only one critical amino acid site was predicted in the group II-2/I-1 pairs, while approximately 26 and 14 were predicted in the group II-3/I-1 and II-1/I-1 pairs, respectively. In total, 35 amino acid residues (656, 659, 679, 688, 700, 701, 705, 709, 767, 768, 769, 771, 775, 807, 809, 818, 821, 826, 840, 843, 865, 867, 869, 872, 877, 885, 935, 937, 964, 978, 979, 1069, 1073, 1096, 1103) in all comparisons were identified as being most important for the functional divergence (Fig. 7). It should be noted that all these amino acids were localized in the C-terminal region of TPSs.

Positive selection may be the most common factor that determines the retention of new genes after the duplication events, as many duplicated genes have been lost from the genome. Positive selection helps to accelerate the fixation of advantageous amino acids mutations which enable plants to adapt to its environment. By using the ML methods and codon substitution models, the selective pressure between the six groups of *TPS* genes were evaluated via the likelihood ratio tests [46, 47].

The ω of all *TPS* genes was estimated as 0.143 using one-ratio model (M0) (Table 3), which suggested that, on average, the *TPS* genes of five *Solanaceae* plants are under strong purifying selection. We then detected positive selection acting on particular group using a branch model in which each clade had its own ω (Table 3). Although the LRT statistic suggested that the ω of groups II-1 and II-3 were significantly different from other groups, the ω estimates for groups II-1 and II-3 still showed that they appear to have undergone purifying selection (Table 3).

As positive selection is unlikely to affect all sites over prolonged time, we thus estimated the evolutionary forces acting on individual codon site, using site-specific likelihood models of codon substitution [48, 49]. We use three pairs of models, forming

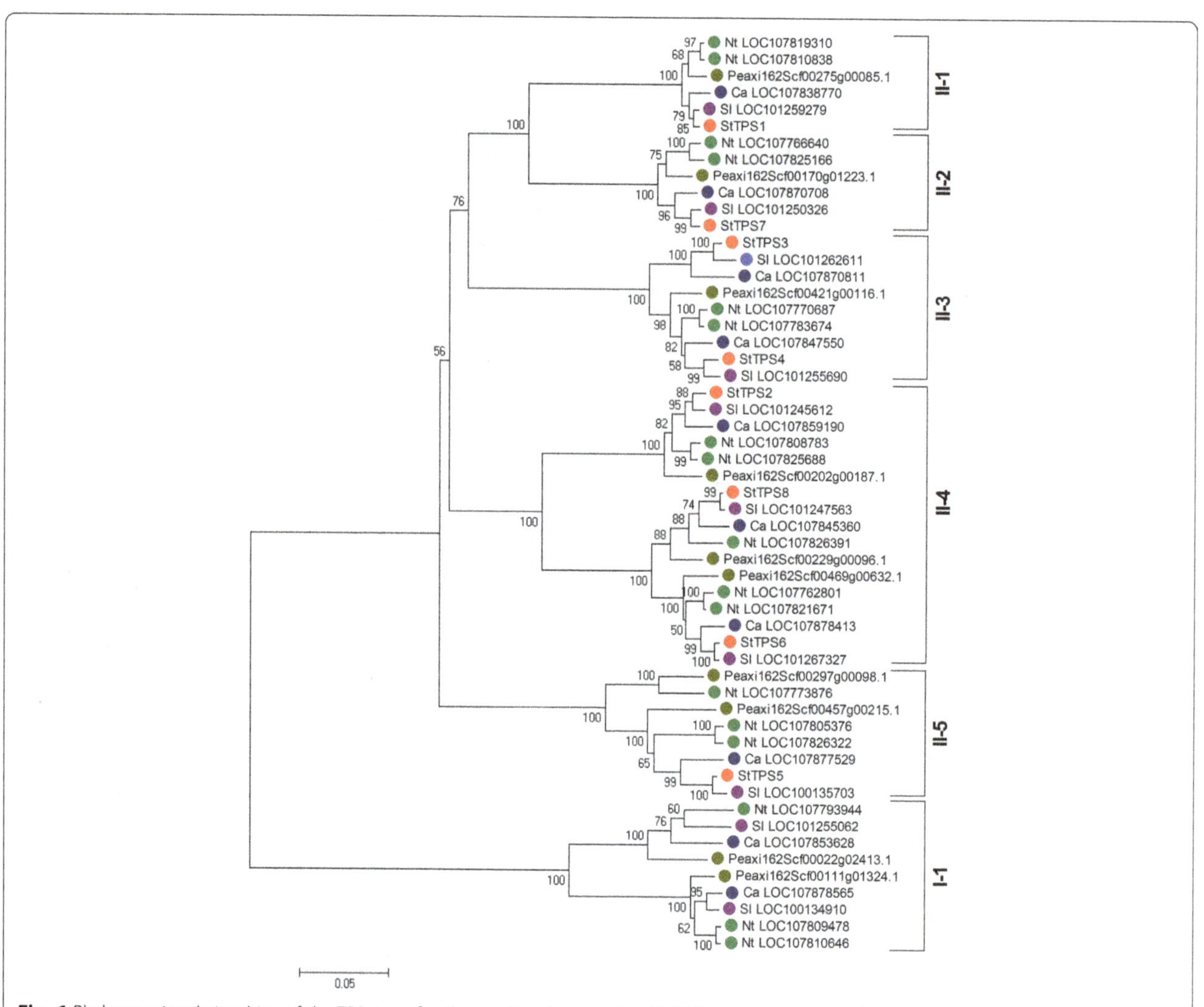

Fig. 6 Phylogenetic relationships of the TPS gene family members from potato (StTPS), tomato, pepper, tobacco, and petunia. The phylogenetic tree was constructed with MEGA 7.0 program by the neighbor-joining method

three LRTs: M1 (neutral) and M2 (selection), M0 (one ratio) and M3 (discrete), and M7 (beta) and M8 (beta & ω) [48, 49]. The results in Table 4 showed that model M2 is not significantly better than M1, although it suggested that 17.8% sites were nearly neutral with ω=1. The model M3 with K=5 suggested that 1.1% sites were under positive selection, and M3 model was significantly better than the one-ratio model. Model M8, which with an additional ω ratio estimated from the data, is significantly better than M7, also suggested that 5.3% sites were under positive selection. Model M3 and M8 identified 1 and 3 amino acid sites under positive selection at the 95% cutoff.

To enhance the power in detecting positive selection, the branch-site model [50] was also applied to evaluate the selection on a few amino acids of TPS genes at specific groups (Table 5). LRT test showed that model A fit the data significantly better than the site-specific model M1 ($p<0.01$) in groups I-1, II-3, 4 and 5. Model A suggested positive selection on 46.1%, 4.6%, 8.5% and 7.2% sites of *TPS* genes in groups I-1, II-3, 4 and 5 respectively. At the posterior probabilities (p) >95% level, there were 15 and 1 sites identified which were likely to be under positive selection along the groups I-1 and II-3 respectively (Fig. 8). Referring to first sequence of StTPS3, these positively selected sites were 293H, 255L, 360Q, 364S, 376V, 402R, 429D, 495E, 771S, 792P, 845W, 860Y, 985Q, 1095D and 1097S in group I-1 and 504 F in group II-3. It is interesting to note that half of the positively selected sites in group I-1 and the only site identified in II-3 appeared in TPS domains, which suggests that these positively selected sites might cause adaptive changes after gene duplications that separated into different groups.

Table 2 Functional divergence between groups of the *TPS* gene family in plant

	Type I				Type II	
	$\Theta_I \pm SE^a$	LRT	*P*	$Q_k \geq 0.95^b$	$\Theta_{II} \pm SE$	*P*
II-1/II-2	0.68±0.228	8.914	0.003	0	0.137±0.033	0
II-1/II-3	0.001±0.022	0	0.964	0	0.17±0.04	0
II-1/II-4	0.246±0.196	1.564	0.211	0	0.074±0.049	0.136
II-1/II-5	0.908±0.229	15.691	0	10	0.113±0.047	0.015
II-1/I-1	0.896±0.2	20.122	0	14	0.577±0.038	0
II-2/II-3	0.574±0.187	9.46	0.002	0	0.147±0.042	0
II-2/II-4	0.516±0.161	10.256	0.001	0	0.089±0.051	0.079
II-2/II-5	0.386±0.166	5.401	0.02	0	0.08±0.048	0.098
II-2/I-1	0.784±0.15	27.432	0	1	0.566±0.039	0
II-3/II-4	0.153±0.112	1.876	0.171	0	0.041±0.053	0.432
II-3/II-5	0.437±0.142	9.488	0.002	0	0.042±0.05	0.398
II-3/I-1	0.857±0.132	42.513	0	18	0.613±0.039	0
II-4/II-5	0.366±0.102	12.788	0	0	0.05±0.058	0.386
II-4/I-1	0.639±0.1	40.46	0	4	0.586±0.044	0
II-5/I-1	0.726±0.127	32.685	0	0	0.595±0.042	0

[a]The coefficient of functional divergence between the two subgroups and its standard error
[b]The number of critical amino acid residues with posterior probability (Q_k) >0.95

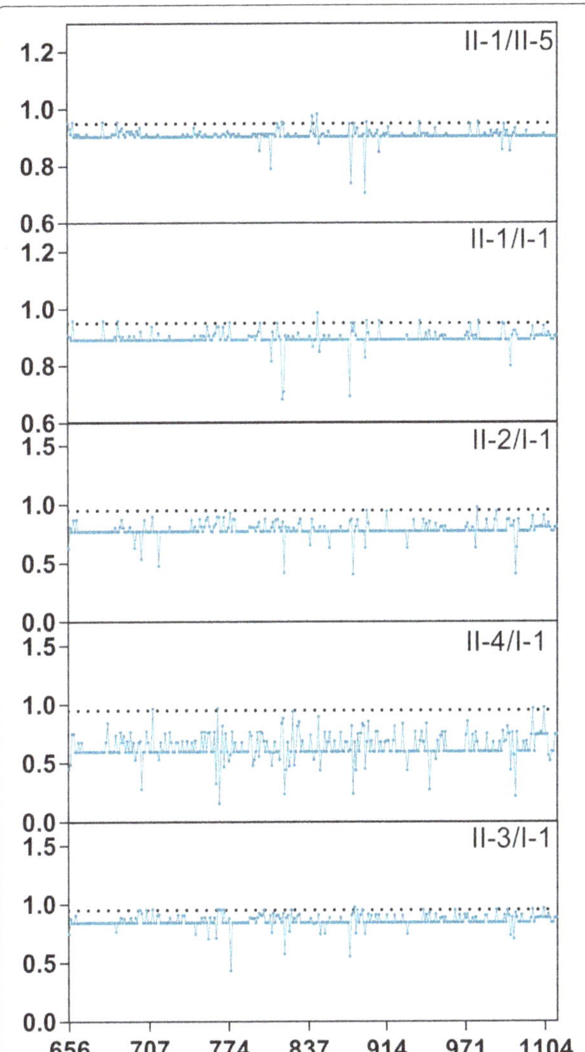

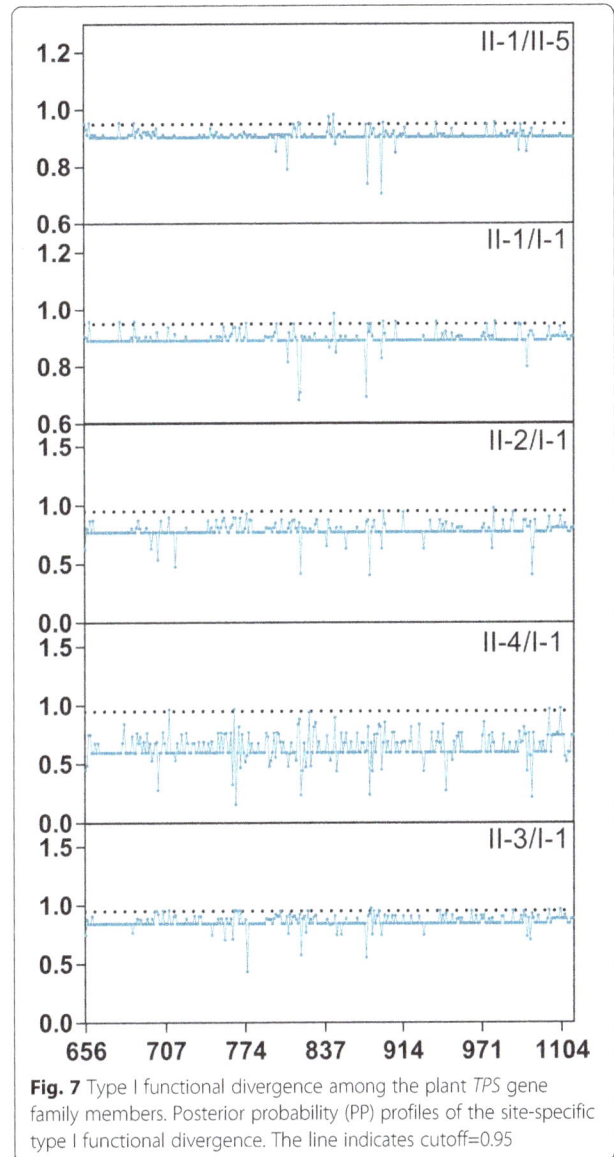

Fig. 7 Type I functional divergence among the plant *TPS* gene family members. Posterior probability (PP) profiles of the site-specific type I functional divergence. The line indicates cutoff=0.95

Expression pattern of *TPS* gene family in potato

TPS genes are known to be important in plants response to environmental stresses. In this study, we took advantage of available transcriptome data of potato, to analyze the complete set of *StTPS* genes in various tissues and under different phytohormones and abiotic stresses [51]. Transcripts of all *StTPS* gene family members were detected in all tested tissues of potato, although their abundance varied considerably.

Much work has been done in transgenic plants indicating that expressed *TPS* genes usually conferred higher tolerance to abiotic stresses [20, 52, 53]. In accordance with this, we found that *StTPS* genes showed differential expression patterns under various abiotic stresses (Fig. 9a). Under salt treatment, most *StTPS* genes were induced, whereas *StTPS2* and *StTPS8* were slightly downregulated. Under osmotic treatment (mannitol), only *StTPS1* and *StTPS5* exhibited increased expression. In contrast to salt stress, heat stress caused a large decline in transcriptional levels of most *StTPS* genes (*StTPS5* in particular), whereas *StTPS6* and *StTPS7* exhibited obvious increases in response to heat stress. *StTPS4* did not show obvious trends after heat treatment. Genes from the same group frequently showed similar expression pattern in various tissues. Based on the FPKM of different genes, the total transcript abundance of *StTPS* genes were highest in response to salt stress (Fig. 9b). Previous studies showed that *TPS* genes in maize were also upregulated in response to both salt and temperature stresses [43]. Enhanced *TPS* genes expression was observed for some "Resurrection plants", in response to extreme water deficit, where up to 99% of their water has been removed. Thus, it is not surprising that *StTPS* genes were induced in potato upon water deficit caused by salt, mannitol and heat stresses.

Phytohormones play crucial roles in coordinating regulatory networks and the signal transduction pathways associated with external stimuli. In potato, we found that under various phytohormones treatments including abscisic acid (ABA), 6-benzylaminopurine (BAP), gibberellic acid (GA_3), and indole-3-acetic acid (IAA), almost all the potato *TPS* genes were differentially downregulated except

Table 3 Parameter estimates and likelihood ratio tests for the branch model

Model	p^a	LnL[b]	Estimates of Parameters	2ΔI[c]	df	p	Positively selected sites
M0 (one ratio model)	1	-47935.395	ω=0.143	-	-	-	None
Branch-specific model (Model 2: two ratios)							
Estimate ω for I-1	2	-47935.222	ω_0=0.143, ω_{I1}= 0.083	Model 2 Vs M0: 0.346	1	0.556	-
Estimate ω for II-1	2	-47930.020	ω_0= 0.146, ω_{II1}= 0.083	Model 2 Vs M0: 10.750	1	0.001	-
Estimate ω for II-2	2	-47924.924	ω_0= 0.147, ω_{II2}= 0.065	Model 2 Vs M0: 20.943	1	0.091	-
Estimate ω for II-3	2	-47935.383	ω_0= 0.143, ω_{II3}= 0.147	Model 2 Vs M0: 0.024	1	0.000	-
Estimate ω for II-4	2	-47934.740	ω_0= 0.143, ω_{II4}= 0.199	Model 2 Vs M0: 1.310	1	0.252	-
Estimate ω for II-5	2	-47935.385	ω_0= 0.143, ω_{II5}= 0.148	Model 2 Vs M0: 0.020	1	0.888	-

[a]The number of free parameters for the ω ratios
[b]Likelihood of the model
[c]$2(l_1-l_0)$

StTPS2, StTPS3 and *StTPS5* which were slightly induced under GA treatment (Fig. 10a). The total transcript abundance of *StTPS* genes were extremely low in BAP treatment seedlings. BAP might be a key negative regulator of TPS abundance (Fig. 10b).

The expression data of *TPS* genes under various biotic stresses including leaves challenged with *Phytophthora infestans*, leaves wounded to mimic herbivory, and the elicitors acibenzolar-smethyl (BTH) and DL-ß-amino-n-butyric acid (BABA) were analyzed. *BABA* and BTH are well accepted inducers of resistance against pathogen infection. Under BTH and BABA treatment, five *TPS* genes including *StTPS1, StTPS2, StTPS3, StTPS4,* and *StTPS5* were differently induced. BTH and BABA exhibited differing effects on these five genes (Fig. 11a). For example, BTH could induce the expression of *StTPS1*, while BABA downregulated its expression level. As for the other four genes, BABA could induce expression of them when BTH downregulated. Upon *Phytoph*thora *infestans* infection, all the *StTPS* genes showed slightly decreased expression. Overall, either *Phytophthora infestans* infection or elicitors treatment showed less effect on *StTPS* genes. However, wounding leaves which mimicked herbivory caused obvious changes on expression of *StTPS* genes, especially on *StTPS2* and *StTPS7* genes. Wounding induced expression of *StTPS2* and *StTPS3*. Under all of these biotic stresses, *StTPS7* and *StTPS8* were always downregulated. However, the total transcript abundance of *StTPS* genes in wound treatment were obviously higher than other treatments (Fig. 11b).

Global gene expression analysis in various tissues revealed that *StTPS* genes were abundant in floral (stamens, sepals and petals) and root (average FPKM>40, four- fold higher than that in leaves) (Fig. 12). Moreover, *StTPS1* showed remarkably higher expression levels in almost every tissue, with average FPKM of 56 in different tissues, almost 28-fold higher than that of *StTPS7*, which has the lowest transcript level. Several studies

Table 4 Parameter estimates and likelihood ratio tests for the site models

Model	p^a	LnL[b]	Estimates of Parameters	2ΔI[c]	df	p	Positively selected sites
M0 (one ratio model)	1	-47935.395	ω=0.143	-	-	-	None
Site-specific models							
M1: Neutral (k=2)	1	-47271.650	p_0=0.774, (p_1=0.226)	-	-	-	Not allowed
M2 : Selction (k=3)	3	-47271.650	p_0=0.774, p_1=0.048, (p_2=0.178), ω_2=1.000	M2 vs M1: 0.000	1	1.000	None
M3: discrete (K=5)	5	-46728.169	p_0=0.244, p_1=0.389, p_2=0.263, p_3=0.094, p_4=0.011, ω_0=0.010, ω_1=0.085, ω_2=0.317, ω_3=0.799, ω_4=6.007	M3 vs M0: 2414.452	4	0.000	1 (p>0.95)
M7: beta	2	-46761.572	p=0.513, q=1.910	-	-	-	Not allowed
M8: beta & w>1	4	-46732.126	p_0=0.947, p=0.589, q=2.892, (p_1=0.053) ω=1.293	M8 vs M7: 58.891	2	0.000	6 (p>0.95), 3 (p>0.95)

[a]The number of free parameters for the ω ratios
[b]Likelihood of the model
[c]$2(l_1-l_0)$

Table 5 Parameter estimates and likelihood ratio tests for the branch-site models

Model	p^a	LnL^b	Estimates of Parameters	$2\Delta l^c$	df	p	Positively selected sites
M1: Neutral (k=2)	1	-47271.650	p_0=0.774, (p_1=0.226)				
Branch-site models							
Model A (I-1)	3	-47231.883	p_0= 0.416, p_1= 0.123, (p_{2a}+p_{2b}=0.461), ω_2=1.291	Model A vs M1: 79.535	2	0.000	Site for foreground lineage: 15 (at p>0.95)
Model A (II-1)	3	-47271.650	p_0=0.774, p_1= 0.226, (p_{2a}+p_{2b} =0.000), ω_2=2.639	Model A vs M1: 0.000	2	1.000	
Model A (II-2)	3	-47271.650	p_0= 0.774, p_1= 0.226, (p_{2a}+p_{2b}=0.000), ω_2=1.000	Model A vs M1: 0.000	2	1.000	
Model A (II-3)	3	-47253.663	p_0=0.740, p_1= 0.214, (p_{2a}+p_{2b} =0.046), ω_2=28.900	Model A vs M1: 35.974	2	0.000	Site for foreground lineage: 1 (at p>0.95)
Model A (II-4)	3	-47264.889	p_0=0.707, p_1= 0.207, (p_{2a}+p_{2b} =0.085), ω_2=1.506	Model A vs M1: 13.523	2	0.001	
Model A (II-5)	3	-47257.978	p_0=0.718,p_1= 0.210, (p_{2a}+p_{2b} =0.072), ω_2=3.022	Model A vs M1: 27.345	2	0.000	

[a]The number of free parameters for the ω ratios
[b]Likelihood of the model
[c]$2(l_1\text{-}l_0)$

using mutant plants have revealed the importance of trehalose metabolism in the control of plant development [7, 28, 54]. Moreover, there was some evidence showing that *AtTPS1* gene plays important roles in the control of stress response, cell and embryonic development, glucose sensing, and starch synthesis [7, 54, 55]. Beyond these established roles of *TPS* genes in plants, recent intriguing evidence has implicated these genes as important modulators of plant development and inflorescence architecture. Although less expressed, *StTPS7* was also found preferentially in floral tissues, indicating the role of *StTPS* genes in floral growth and development. Besides floral tissues, most *StTPS* genes show a slightly higher level of accumulation in root, shoot and callus. In different parts or growing stages of tuber, the general expression of *StTPS* genes are low except *StTPS1*, which showed relatively high expression levels in every part of the tuber.

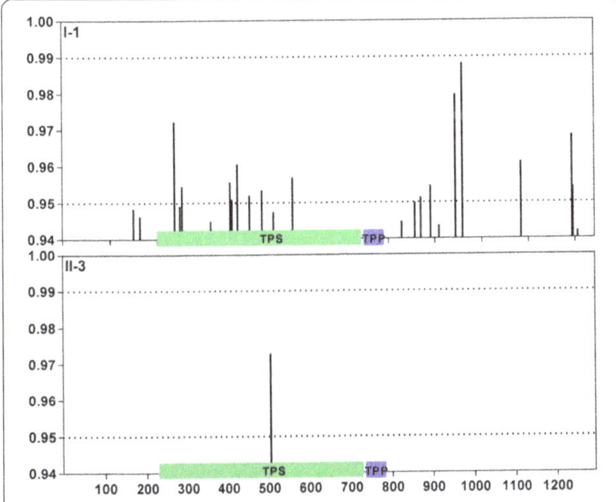

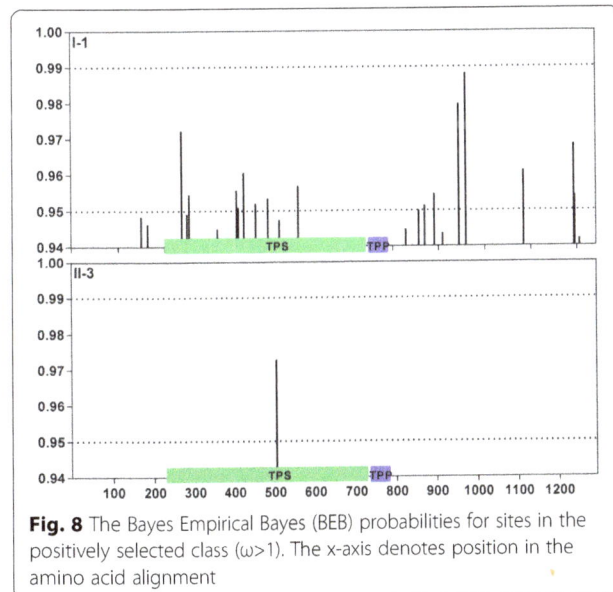

Fig. 8 The Bayes Empirical Bayes (BEB) probabilities for sites in the positively selected class (ω>1). The x-axis denotes position in the amino acid alignment

Validation of *StTPSs* differential expression

In *silico* analysis revealed that some *StTPS* genes are obviously regulated by different environmental stimuli. The differential expression of genes (fold changes >2) were chosen for qRT-PCR validation (Fig. 13). As expected, qRT-PCR results of genes under various treatment were similar in magnitude to those obtained by deep sequencing. qRT-PCR results suggested that two genes including *StTPS1* and *StTPS7* were frequently regulated by various treatments, which indicating they might be the

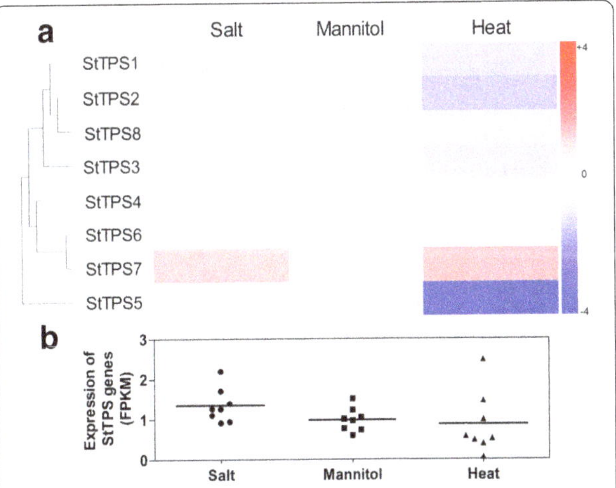

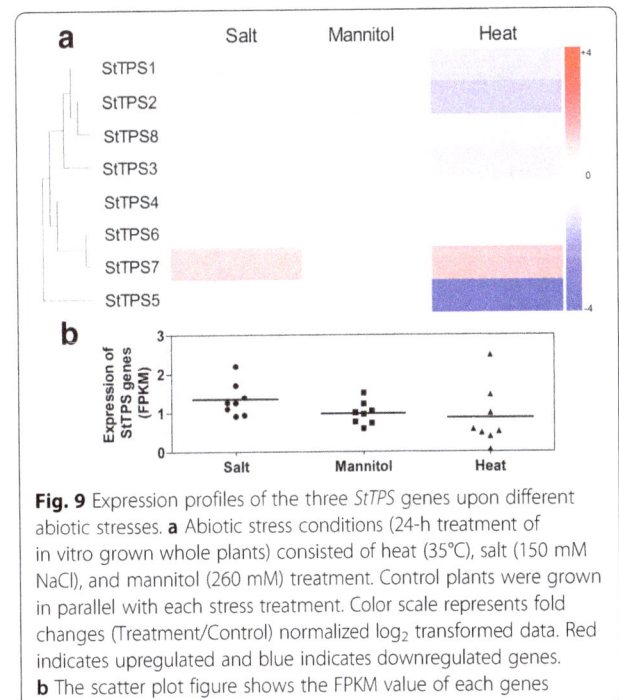

Fig. 9 Expression profiles of the three *StTPS* genes upon different abiotic stresses. **a** Abiotic stress conditions (24-h treatment of in vitro grown whole plants) consisted of heat (35°C), salt (150 mM NaCl), and mannitol (260 mM) treatment. Control plants were grown in parallel with each stress treatment. Color scale represents fold changes (Treatment/Control) normalized $\log_2$ transformed data. Red indicates upregulated and blue indicates downregulated genes. **b** The scatter plot figure shows the FPKM value of each genes

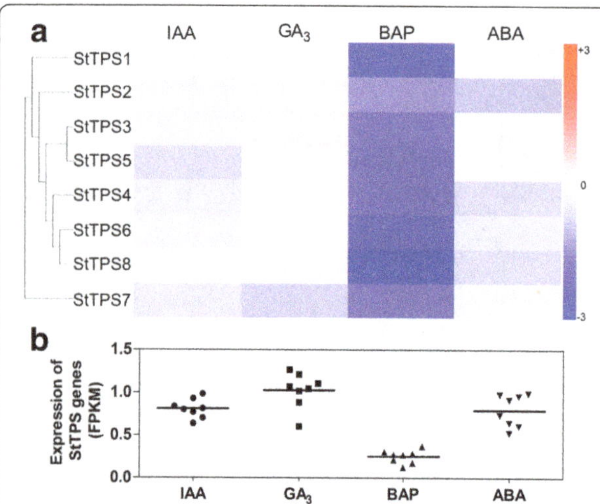

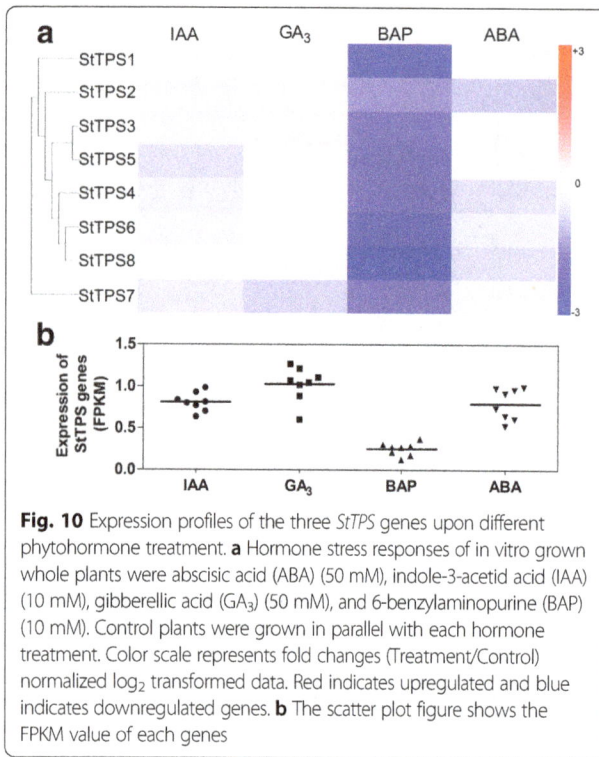

Fig. 10 Expression profiles of the three *StTPS* genes upon different phytohormone treatment. **a** Hormone stress responses of in vitro grown whole plants were abscisic acid (ABA) (50 mM), indole-3-acetid acid (IAA) (10 mM), gibberellic acid (GA_3) (50 mM), and 6-benzylaminopurine (BAP) (10 mM). Control plants were grown in parallel with each hormone treatment. Color scale represents fold changes (Treatment/Control) normalized $\log_2$ transformed data. Red indicates upregulated and blue indicates downregulated genes. **b** The scatter plot figure shows the FPKM value of each genes

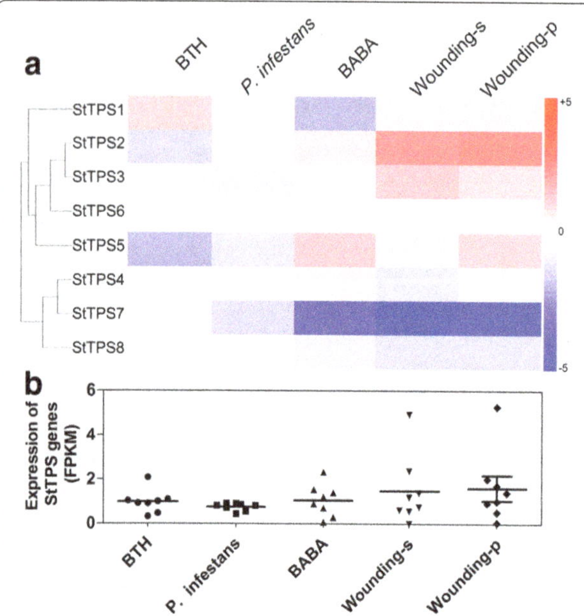

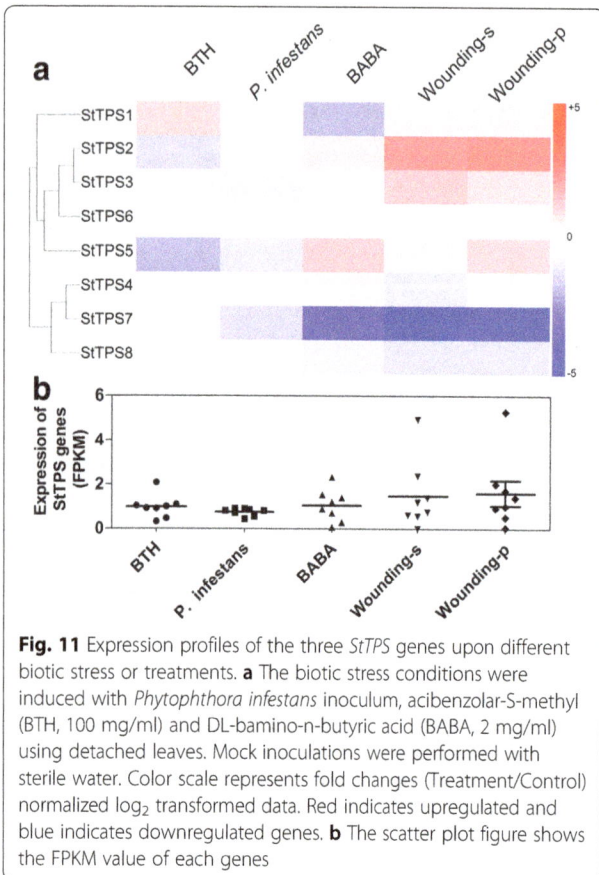

Fig. 11 Expression profiles of the three *StTPS* genes upon different biotic stress or treatments. **a** The biotic stress conditions were induced with *Phytophthora infestans* inoculum, acibenzolar-S-methyl (BTH, 100 mg/ml) and DL-bamino-n-butyric acid (BABA, 2 mg/ml) using detached leaves. Mock inoculations were performed with sterile water. Color scale represents fold changes (Treatment/Control) normalized $\log_2$ transformed data. Red indicates upregulated and blue indicates downregulated genes. **b** The scatter plot figure shows the FPKM value of each genes

primary TPSs involved in potato response to environment stimuli.

Conclusions

In summary, we identified eight *StTPS* genes from potato and characterized their conserved protein motif, gene structure, chromosomal distribution, *cis*-acting elements in promoter regions and molecular evolution. Collectively, this has led to greater functional characterization of potato *TPS* genes. Moreover, analyses of their expression profiles based on available transcriptome data and qRT-PCR validation of various potato tissues under biotic and abiotic stress treatments provides functional information of *StTPSs*. Our results provide important clues for future research on the function of *StTPS* gene family and StTPS-mediated signal transduction pathways, thereby advancing our knowledge of the molecular basis of genetic enhancements to potato.

Methods

Identification and classification of *TPS* genes

To extensively identify potato *TPS* genes, Hidden Markov models (HMMs) of the 'typical' TPS and TPP domain were used to search the latest version of the potato genome (v4.04) from Spud DB [51] and the genomes of four other *Solanaceae* species including tomato (*Solanum lycopersicum*, v3.1, id35173), pepper (*Capsicum annuum*, v2, id22828), tobacco (*Nicotiana tabacum*, TN90), petunia (*Petunia_axillaris*, v0.1, id24480) via HMMER v. 3.1 [32] with an E-value cut-off of <1e-10. When several variants of one gene were obtained, only the longest one was retained. All the candidate sequences were further confirmed to have complete PFAM TPS and TPP domains. Pseudogenes which only covered less than 50% of the PFAM domain models were eliminated from final *TPS* genes [56]. The basic physical and chemical properties of the protein sequences were analyzed using ProParam online tool in ExPASy. Subcellular localization predictor (http://cello.life.nctu.edu.tw/) was used to predict subcellular localizations. The chromosomal locations of *TPS* genes were drawn based on the potato genome database deposited [51]. MCScanX was used to analyze the syntenic relationships within genomes of melon, watermelon and cucumber respectively [57]. The diagrams were visualized using Circos software (v0.67).

Gene structure and conserved motifs analyses

Gene structures of *TPSs* were analyzed on the Gene Structure Display Server 2.0 (GSDS; http://gsds.cbi.pku.edu.cn/). Motifs in the candidate potato TPS protein sequences [58] were predicted using program MEME (http://meme-suite.org) with the default parameters.

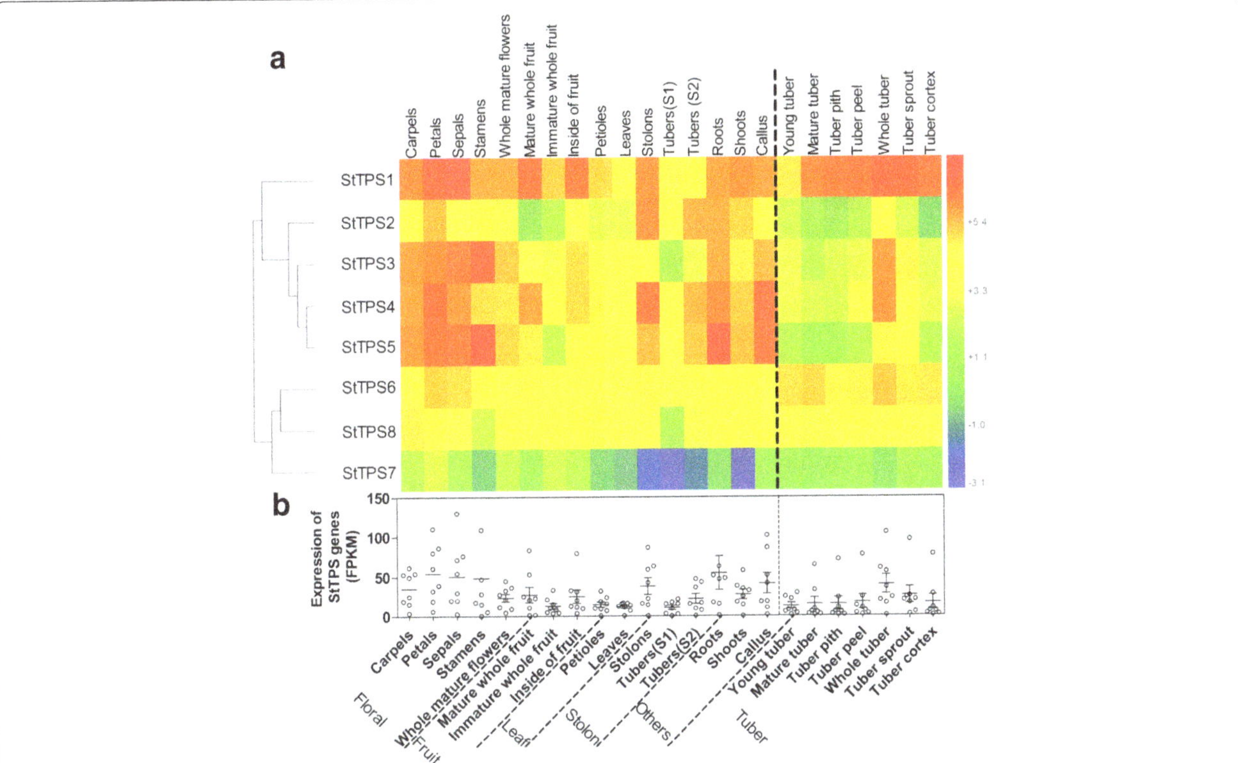

Fig. 12 Expression profiles of the three *StTPS* genes upon different biotic stress or treatment. **a** The developmental tissues represent vegetative (leaves, petioles, stolons, tubers sampled twice) and reproductive organs (Floral: carpels, petals, sepals, stamens, whole flowers; Fruit: mesocarp/endocarp, whole immature berries, whole mature berries) from greenhouse-grown plants. Shoots and roots from in vitro-grown plants were also included in the developmental series. Callus (10–11 weeks old) derived from leaves and stems were used to assess transcription in an undifferentiated tissue. Color scale represents FPKM normalized log2 transformed data. Red indicates high expression level and blue indicates low expression level. **b** The scatter plot figure shows the FPKM value of each genes

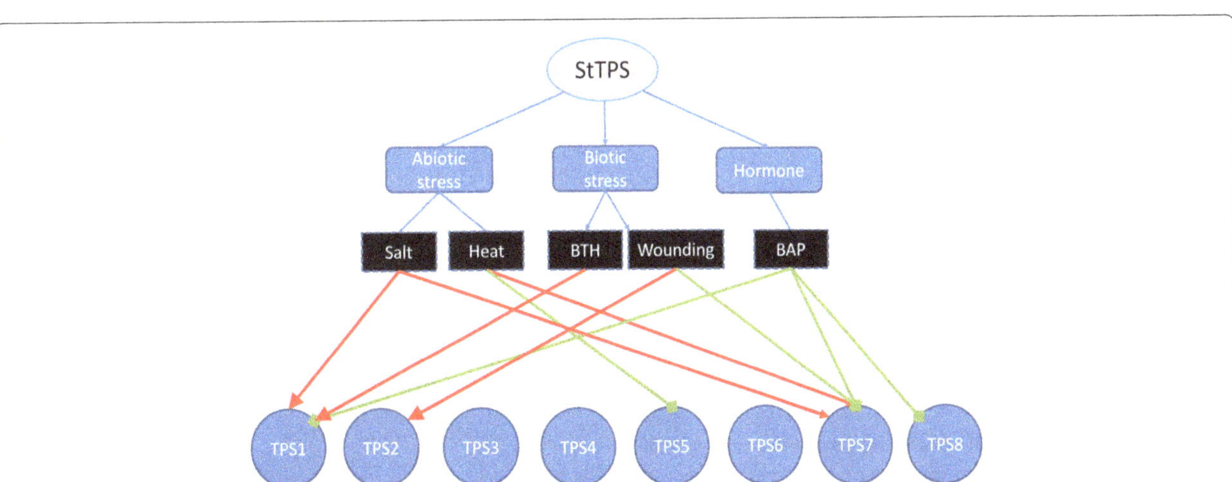

Fig. 13 Validation of selected *StTPS* genes during exogenous stimuli. Abiotic stress conditions (24-h treatment of in vitro grown whole plants) consisted of heat (35°C), salt (150 mM NaCl) treatment. Control plants were grown in parallel with each stress treatment. The biotic stress condition (24 hr) was induced with acibenzolar-S-methyl (BTH, 100 mg/ml). Wounded leaves were included to mimic herbivory. Mock inoculations were performed with sterile water. The red line indicates the gene was upregulated, ca. two-fold greater than control. The green line represents the gene was down regulated, being less than 50% of control. The relative transcript abundance was normalized using potato actin gene

Sequence alignment and phylogenetic analysis

Sequence alignment among *TPS* genes from potato or different species were performed using the ClustalX 1.83 software with full-length CDS sequences of *TPS* genes [59] and the alignment results were displayed with DNAMAN v5.2. MEGA v7 was used to construct the phylogenetic tree using the Neighbor-Joining method [60]. The percentage of replicate trees in which the associated taxa clustered together in the bootstrap test (1000 replicates) [61].

Functional divergence analyses

To estimate the level of functional divergence in the TPS subgroups, coefficients of Type-I and Type-II functional divergence were calculated using DIVERGE (version 2.0)X Gu [62].

Selection assessment and testing

The values of nonsynonymous substitutions (d_N), synonymous substitutions (d_S) and d_N/d_S ratio (or ω) were calculated via the program PAML version 4 [63], using branch-specific (model B), site-specific (neutral, selection, discrete, beta, beta & w>1), and branch-site models as implemented in PAML [50, 64]. Likelihood ratio test (LRT) were used to compare the fit of model pairs. The sites under positive selection were identified using Bayes methods [65].

In silico expression analysis of *StTPSs*

Transcriptome gene expression data were extracted from NCBI sequence Read Archive (SRA029323) and Spud DB [51, 66–68] to analyze the expression profiles of potato in organs and under different treatments as described in figure legends. The reads were mapped to *S. tuberosum* Group Phureja DM1-3 super scaffolds using Tophat (v1.4.1). The FPKM values were calculated by Cufflinks (v1.3.0) using v3.4 representative model.

Plant growth and treatments

Potato plants (*S. tuberosum* L. cultivar Shepody) were cultivated in a growth chamber in soil at 25 °C under a photoperiod of 16 h light/8 h dark. After growing for 30 d, the seedlings were used for treatments. Abiotic stress conditions (24-h treatment of in vitro grown whole plants) consisted of heat (35°C), salt (150 mM NaCl) treatment. Control plants were grown in parallel with each stress treatment. The biotic stress condition (24 h) was induced with acibenzolar-S-methyl (BTH, 100 mg/ml). Wounded leaves were included to mimic herbivory. Mock inoculations were performed with sterile water. After various treatments, the seedlings were sampled, then immediately frozen in liquid nitrogen, and stored at -80 °C until further analysis.

Real-time quantitative PCR

Total RNA was extracted from each sample, which were first homogenized with mortar and pestle in liquid nitrogen, using TRIzol reagent (Invitrogen, USA) according to the instructions supplied by the manufacturer. About 4 μg of total RNA was reverse-transcribed using an oligo(dT) primer and SuperScript Reverse Transcriptase (Invitrogen, USA). Real-time quantitative PCR was conducted using SYBR green (TaKaRa Biotechnology) on Mastercycler® ep *realplex* real-time PCR system (Eppendorf, Hamburg, Germany). Gene-specific primers for each *StTPS* gene were designed using Primer Premier 5.0 and optimized using oligo 7. The relative abundance of *Actin 1* was used as the internal standard. Primer pairs included: *StTPS1* (F-5'TGATGTAGTTGCCGATG C3' R-5' GATTGCCCTGGTTGTTGT3'); *StTPS2* (F- 5' AAATACCGTGTTTCTCGT 3' R- 5' CCTTTACTG ACTCCCTGA 3'); *StTPS5* (F- 5' AGGAAGGGATAC GCTCAG 3' R- 5' CCAAATGCCAAAGTCAGG 3'); *StTPS7* (F- 5' GAACGGAGAAGCTGGATG 3' R- 5' CTCTGCCTCGGAGACAAT 3'); *StTPS8* (F- 5' CTCA AGGCTTTGCTCTGT 3' R- 5' GATGCCTACTGTCC-TACCAT 3'); *Actin* (F- 5' CACCCTGTTCTGCTCACT 3' R- 5' CAGCCTGAATAGCAACATAC 3'). Real-time PCR reactions were performed in a total reaction volume of 25 μL using the following conditions: 94 °C /2 min; 40 cycles (94 °C /15 s, 58 °C /15 s, 72 °C /15 s). All reactions were performed in triplicate. Independent experiments were repeated three times. Relative gene expression was analyzed using the $2^{-\Delta\Delta c(t)}$ method [65].

Abbreviations

ABA: Abscisic acid; ABREs: ABA-responsive elements; BABA: DL-ß-amino-n-butyric acid; BAP: 6-benzylaminopurine; BTH: Acibenzolar-smethyl; d_N: Non-synonymous substitution rate; d_S: Synonymous substitution rate; GA_3: Gibberellic acid; IAA: Indole-3-acetic acid; T-6-P: Trehalose-6-P; TPP: T-6-P phosphatase; TPS: Trehalose-6-phosphate synthase

Acknowledgements

Not applicable.

Funding

This work was supported by National Natural Science Foundation of China (31501795, 31501610), the fundamental research funds for the central universities (KJQN201659), the China Postdoctoral Science Foundation funded project (2014M560432, 2015T80563), and the Natural Science Foundation of Jiangsu Province in China (BK20140695). The funding bodies had no role in study design, data collection and analysis, and in writing the manuscript.

Authors' contributions

JQJ and XYC conceived the study; JQJ WB YL and LX did data analysis and drafted the manuscript; NSM, WYJ and XYC revised the manuscript. All authors read and approved the final manuscript.

Competing interests
The authors declare that they have no competing interests.

Author details
[1]College of Horticulture, Nanjing Agricultural University, Nanjing 210095, China. [2]Horticulture Section, School of Integrative Plant Science, Cornell University, 134A Plant Science Bldg, Ithaca, NY 14853, USA. [3]Institute of Plant Protection, Jiangsu Academy of Agricultural Sciences, Nanjing 210095, China.

References
1. Chary SN, Hicks GR, Choi YG, Carter D, Raikhel NV. Trehalose-6-phosphate synthase/phosphatase regulates cell shape and plant architecture in Arabidopsis. Plant Physiol. 2008;146(1):97–107.
2. Lunn JE, Delorge I, Figueroa CM, Dijck PV, Stitt M. Trehalose metabolism in plants. Plant J. 2014;79(4):544–67.
3. Elbein AD, Pan YT, Pastuszak I, Carroll D. New insights on trehalose: a multifunctional molecule. Glycobiology. 2003;13(4):17R–27R.
4. Cai ZJ, Peng GX, Cao YQ, Liu YC, Jin K, Xia YX. Trehalose-6-phosphate synthase 1 from Metarhizium anisopliae: clone, expression and properties of the recombinant. J Biosci Bioeng. 2009;107(5):499–505.
5. López MF, Männer P, Willmann A, Hampp R, Nehls U. Increased trehalose biosynthesis in Hartig net hyphae of ectomycorrhizas. New Phytol. 2007; 174(2):389–98.
6. Gibson RP, Tarling CA, Roberts S, Withers SG, Davies GJ. The donor subsite of trehalose-6-phosphate synthase - Binary complexes with UDP-glucose and UDP-2-deoxy-2-fluoro-glucose at 2 angstrom resolution. J Biol Chem. 2004;279(3):1950–5.
7. van Dijken AJ, Schluepmann H, Smeekens SC. Arabidopsis trehalose-6-phosphate synthase 1 is essential for normal vegetative growth and transition to flowering. Plant Physiol. 2004;135(2):969–77.
8. Cabib E, Leloir LF. The biosynthesis of trehalose phosphate. J Biol Chem. 1958;231(1):259–75.
9. Márquez-Escalante JA, Figueroa-Soto CG, Valenzuela-Soto EM. Isolation and partial characterization of trehalose 6-phosphate synthase aggregates from Selaginella lepidophylla plants. Biochimie. 2006;88(10):1505–10.
10. Valenzuela-Soto EM, Márquez-Escalante JA, Iturriaga G, Figueroa-Soto CG. Trehalose 6-phosphate synthase from Selaginella lepidophylla : purification and properties. Biochem Biophys Res Commun. 2004;313(2):314–9.
11. Londesborough J, Vuorio OE. Purification of trehalose synthase from baker's yeast. Eur J Biochem. 1993;216(3):841–8.
12. Pan YT, Carroll JD, Elbein AD. Trehalose-phosphate synthase of Mycobacterium tuberculosis. Cloning, expression and properties of the recombinant enzyme. Eur J Biochem. 2002;269(24):6091–100.
13. Pan YT, Koroth EV, Jourdian WJ, Edmondson R, Carroll JD, Pastuszak I, Elbein AD. Trehalose synthase of Mycobacterium smegmatis: purification, cloning, expression, and properties of the enzyme. Eur J Biochem. 2004;271(21): 4259–69.
14. Deng YY, Wang XL, Guo H, Duan DLA. trehalose-6-phosphate synthase gene from *Saccharina japonica* (Laminariales, Phaeophyceae). Mol Biol Rep. 2014;41(1):529–36.
15. Thevelein JM. The RAS-adenylate cyclase pathway and cell cycle control in Saccharomyces cerevisiae. Antonie Van Leeuwenhoek. 1992;62(1-2):109–30.
16. Stiller I, Dulai S, Kodrak M, Tarnai R, Szabo L, Toldi O, Banfalvi Z. Effects of drought on water content and photosynthetic parameters in potato plants expressing the trehalose-6-phosphate synthase gene of Saccharomyces cerevisiae. Planta. 2008;227(2):299–308.
17. Romero C, Bellés JM, Vayá JL, Serrano R, Culiáñez-Macià FA. Expression of the yeast trehalose-6-phosphate synthase gene in transgenic tobacco plants: pleiotropic phenotypes include drought tolerance. Planta. 1997; 201(3):293–7.
18. Mu M, XK L, Wang JJ, Wang DL, Yin ZJ, Wang S, Fan WL, Ye WW. Genome-wide Identification and analysis of the stress-resistance function of the TPS (Trehalose-6-Phosphate Synthase) gene family in cotton. BMC Genet. 2016;17
19. Zentella R, Iturriaga GA. *Selaginella lepidophylla* trehalose-6-phosphate synthase complements growth and stress-tolerance defects in a yeast tps1 mutant. Plant Physiol. 1999;119(4):1473–82.
20. Garg AK, Kim JK, Owens TG, Ranwala AP, Yang DC, Kochian LV, Trehalose WRJ. accumulation in rice plants confers high tolerance levels to different abiotic stresses. Proc Natl Acad Sci U S A. 2002;99(25):15898–903.
21. In-Cheol Jang S-JO, Seo J-S, Choi W-B, Song SI, Kim CH, Kim YS, Seo H-S, Do Choi Y, Nahm BH, Kim J-K. Expression of a bifunctional fusion of the Escherichia coli genes for trehalose-6-phosphate synthase and trehalose-6-phosphate phosphatase in transgenic rice plants increases trehalose accumulation and abiotic stress tolerance without stunting growth. Plant Physiol. 2003;131(2):516–24.
22. Zang B, Li H, Li W, Deng XW, Wang X. Analysis of trehalose-6-phosphatesynthase (TPS) gene family suggests the formation of TPS complexes in rice. Plant Mol Biol. Plant Mol Bio. 2011;76(6):507–22.
23. Avonce N, Mendoza-Vargas A, Morett E, Iturriaga G. Insights on the evolution of trehalose biosynthesis. BMC Evol Biol. 2006;6(1):1–15.
24. Lunn JE. Gene families and evolution of trehalose metabolism in plants. Functional Plant Biology. 2007;34(6):550–63.
25. Leyman B, Dijck PV, Thevelein JM. An unexpected plethora of trehalose biosynthesis genes in Arabidopsis thaliana. Trends Plant Sci. 2001;6(11): 510–3.
26. Vandesteene M, Ramon K, Patrick D, Filip R. A single active trehalose-6-p synthase (TPS) and a family of putative regulatory TPS-Like proteins in Arabidopsis. Molecular Plant. 2010;3(2):406–19.
27. RAMON M, ID SMET, Vandesteene L, Naudts M, LEYMAN B, Dijck PV, ROLLAND F, Beeckman T, Thevelein JM. Extensive expression regulation and lack of heterologous enzymatic activity of the Class II trehalose metabolism proteins from Arabidopsis thaliana. Plant Cell Environ. 2009;32(8):1015–32.
28. Avonce N, Leyman B, Mascorrogallardo JO, Dijck PV, Thevelein JM, Iturriaga G. The *Arabidopsis* trehalose-6-P synthase AtTPS1 gene is a regulator of glucose, abscisic acid, and stress signaling. Plant Physiol. 2004;136(3):3649–59.
29. Wang GL, Zhao G, Feng YB, Xuan JS, Sun JW, Guo BT, Jiang GY, Weng ML, Yao JT, Wang B, et al. Cloning and comparative studies of seaweed trehalose-6-phosphate synthase genes. Mar Drugs. 2010;8(7):2065–79.
30. Zhang Y, Primavesi LF, Jhurreea D, Andralojc PJ, Mitchell RA, Powers SJ, Schluepmann H, Delatte T, Wingler A, Paul MJ. Inhibition of SNF1-related protein kinase1 activity and regulation of metabolic pathways by trehalose-6-phosphate. Chin Sci Bull. 2011;56(10):1055–62.
31. Cai Z, Peng G, Cao Y, Liu Y, Kai J, Xia Y. Trehalose-6-phosphate synthase 1 from Metarhizium anisopliae: clone, expression and properties of the recombinant. J Biosci Bioeng. 2009;107(5):499–505.
32. Eddy SRA. new generation of homology search tools based on probabilistic inference. Genome Informatics International Conference on. Genome Informatics. 2009:205–11.
33. Yang HL, Liu YJ, Wang CL, Zeng QY. Molecular evolution of trehalose-6-phosphate synthase (TPS) gene family in Populus, Arabidopsis and rice. PLoS One. 2012;7(8):e42438.
34. Consortium PGS, Xu X, Pan S, Cheng S, Zhang B, Mu D, Ni P, Zhang G, Yang S, Li R. Genome sequence and analysis of the tuber crop potato. Nature. 2011;475(7355):189.
35. Song J, Gao ZH, Huo XM, Sun HL, YS X, Shi T, Ni ZJ. Genome-wide identification of the auxin response factor (ARF) gene family and expression analysis of its role associated with pistil development in Japanese apricot (Prunus mume Sieb. et Zucc). Acta Physiologiae Plantarum. 2015;37(8)
36. Wang X, Shi X, Hao B, Ge S, Luo J. Duplication and DNA segmental loss in the rice genome: implications for diploidization. New Phytol. 2005;165(3):937–46.
37. Zhang Y, Mao L, Wang H, Brocker C, Yin X, Vasiliou V, Fei Z, Wang X. Genome-wide identification and analysis of grape aldehyde dehydrogenase (ALDH) gene superfamily. PLoS One. 2012;7(2):e32153.
38. Mu M, X-K L, Wang J-J, Wang D-L, Yin Z-J, Wang S, Fan W-L, Ye W-W. Genome-wide Identification and analysis of the stress-resistance function of the TPS (Trehalose-6-Phosphate Synthase) gene family in cotton. BMC Genet. 2016;17(1):1–11.
39. Doi K, Hosaka A, Nagata T, Satoh K, Suzuki K, Mauleon R, Mendoza MJ, Bruskiewich R, Kikuchi S. The development of a novel data mining tool to find ciselements in rice gene promoter regions. BMC Plant Biol. 2008;8:20.

40. Cao JM, Jiang M, Li P, Chu ZQ. Genome-wide identification and evolutionary analyses of the PP2C gene family with their expression profiling in response to multiple stresses in Brachypodium distachyon. BMC Genomics. 2016;17:175.
41. Li W, Liang W. Transcriptional regulation of Arabidopsis MIR168a and argonaute1 homeostasis in abscisic acid and abiotic stress responses. Plant Physiol. 2012;158(3):1279–92.
42. Kosmas SA, Argyrokastritis A, Loukas MG, Eliopoulos E, Tsakas S, Kaltsikes PJ. Isolation and characterization of drought-related trehalose 6-phosphate-synthase gene from cultivated cotton (*Gossypium hirsutum* L.). Planta. 2006; 223(2):329–39.
43. Jiang W, FL F, Zhang SZ, Wu L, Li WC. Cloning and Characterization of Functional Trehalose-6-Phosphate Synthase Gene in Maize. Journal of Plant Biology. 2010;53(2):134–41.
44. Zang B, Li H, Li W, Deng XW, Wang X. Analysis of trehalose-6-phosphatesynthase (TPS) gene family suggests the formation of TPS complexes in rice. Plant Mol Biol. Plant Mol Biol. 2011;76(6):507–22.
45. Gu X, Velden KVDIVERGE. phylogeny-based analysis for functional-structural divergence of a protein family. Bioinformatics. 2002;18(3):500–1.
46. Yang Z, Bielawski JP. Statistical methods for detecting molecular adaptation. Trends Ecol Evol. 2000;15(12):496–502.
47. Yang HL, Liu YJ, Wang CL, Zeng QY. Molecular evolution of trehalose-6-phosphate synthase (TPS) gene family in Populus, Arabidopsis and Rice. PLoS One. 2012;7(8):e42438.
48. Nielsen R, Yang Z. Likelihood models for detecting positively selected amino acid sites and applications to the HIV-1 envelope gene. Genetics. 1998;148(3):929.
49. Yang Z, Nielsen R, Goldman N, Pedersen AM. Codon-substitution models for heterogeneous selection pressure at amino acid sites. Genetics. 2000;155(1):431.
50. Yang Z, Nielsen R. Codon-Substitution Models for Detecting Molecular Adaptation at Individual Sites Along Specific Lineages. Mol Biol Evol. 2002; 19(6):908.
51. Hardigan MA, Crisovan E, Hamiltion JP, Kim J, Laimbeer P, Leisner CP, Manrique-Carpintero NC, Newton L, Pham GM, Vaillancourt B. Genome reduction uncovers a large dispensable genome and adaptive role for copy number variation in asexually propagated Solanum tuberosum. Plant Cell. 2016;41(1):81–8.
52. Ge LF, Chao DY, Shi M, Zhu MZ, Gao JP, Lin HX. Overexpression of the trehalose-6-phosphate phosphatase gene OsTPP1 confers stress tolerance in rice and results in the activation of stress responsive genes. Planta. 2008; 228(1):191–201.
53. Miranda JA, Avonce N, Suárez R, Thevelein JM, Dijck PV, Iturriaga GA. bifunctional TPS-TPP enzyme from yeast confers tolerance to multiple and extreme abiotic-stress conditions in transgenic Arabidopsis. Planta. 2007; 226(6):1411–21.
54. Gilday A, Li Y, Graham IA. Delayed embryo development in the ARABIDOPSIS TREHALOSE-6-PHOSPHATE SYNTHASE 1 mutant is associated with altered cell wall structure, decreased cell division and starch accumulation. Plant J. 2006;46(1):69–84.
55. Gilday A, Feil R, Lunn JE, Graham IA. AtTPS1-mediated trehalose 6-phosphate synthesis is essential for embryogenic and vegetative growth and responsiveness to ABA in germinating seeds and stomatal guard cells. Plant J. 2010;64:1–13.
56. Lehti-Shiu MD, Shiu SH. Diversity, classification and function of the plant protein kinase superfamily. Philos Trans R Soc Lond. 2012;367(1602):2619–39.
57. Wang Y, Tang H, Debarry JD, Tan X, Li J, Wang X, Lee TH, Jin H, Marler B, Guo H, et al. MCScanX: a toolkit for detection and evolutionary analysis of gene synteny and collinearity. Nucleic Acids Res. 2012;40(7):e49.
58. Bailey TL, Williams N, Misleh C, Li WWMEME. discovering and analyzing DNA and protein sequence motifs. Nucleic Acids Res. 2006;34(2):369–73.
59. Chenna R, Sugawara H, Koike T, Lopez R, Gibsom TJ, Higgins DG, Thompson JD. Multiple sequence alignment with the Clustal series of programs. Nucleic Acids Res. 2003;31(13):3497–500.
60. Saitou N, Nei M. The neighbor-joining method: a new method for reconstructing phylogenetic trees. Molbiolevol. 1987;4(6):406–25.
61. Jones DT, Taylor WR, Thornton JM. The rapid generation of mutation data matrices from protein sequences. Computer Applications in the Biosciences Cabios. 1992;8(3):275–82.
62. Maximum-likelihood GX. approach for gene family evolution under functionaldivergence. Mol Biol Evol. 2001;18(4):453–64.
63. Zhao Y, Fu L, Li R, Wang LN, Yang Y, Liu NN, Zhang CM, Wang Y, Liu P, PAML TBB. 4: phylogenetic analysis by maximum likelihood. Mol Biol Evol. 2007;24(24):1586–91.
64. Yang Z, Nielsen R. Synonymous and nonsynonymous rate variation in nuclear genes of mammals. J Mol Evol. 1998;46(4):409–18.
65. Yang Z, Wong WS, Nielsen R. Bayes empirical bayes inference of amino acid sites under positive selection. Mol Biol Evol. 2005;22(4):1107–18.
66. Xu X, Pan S, Cheng S, Zhang B, Mu D, Ni P, Zhang G, Yang S, Li R, Wang J. Genome Sequence and Analysis of the Tuber Crop Potato. Nature. 2011; 475(7355):189–95.
67. Massa AN, Childs KL, Buell CR. Abiotic and Biotic Stress Responses in Solanum tuberosum Group Phureja DM1-3 516 R44 as Measured through Whole Transcriptome Sequencing. Plant Genome. 2013;3:1–10.
68. Massa AN, Childs KL, Lin H, Bryan GJ, Giuliano G, Buell CR. The transcriptome of the reference potato genome *Solanum tuberosum* Group Phureja Clone DM1-3 516R44. PLoS ONE. 2011;6(10):e26801.

6

RNA-seq and Tn-seq reveal fitness determinants of vancomycin-resistant *Enterococcus faecium* during growth in human serum

Xinglin Zhang[1,2], Vincent de Maat[2], Ana M. Guzmán Prieto[2], Tomasz K. Prajsnar[3], Jumamurat R. Bayjanov[2], Mark de Been[2], Malbert R. C. Rogers[2], Marc J. M. Bonten[2], Stéphane Mesnage[3], Rob J. L. Willems[2] and Willem van Schaik[2,4*]

Abstract

Background: The Gram-positive bacterium *Enterococcus faecium* is a commensal of the human gastrointestinal tract and a frequent cause of bloodstream infections in hospitalized patients. The mechanisms by which *E. faecium* can survive and grow in blood during an infection have not yet been characterized. Here, we identify genes that contribute to growth of *E. faecium* in human serum through transcriptome profiling (RNA-seq) and a high-throughput transposon mutant library sequencing approach (Tn-seq).

Results: We first sequenced the genome of *E. faecium* E745, a vancomycin-resistant clinical isolate, using a combination of short- and long read sequencing, revealing a 2,765,010 nt chromosome and 6 plasmids, with sizes ranging between 9.3 kbp and 223.7 kbp. We then compared the transcriptome of *E. faecium* E745 during exponential growth in rich medium and in human serum by RNA-seq. This analysis revealed that 27.8% of genes on the *E. faecium* E745 genome were differentially expressed in these two conditions. A gene cluster with a role in purine biosynthesis was among the most upregulated genes in *E. faecium* E745 upon growth in serum. The *E. faecium* E745 transposon mutant library was then used to identify genes that were specifically required for growth of *E. faecium* in serum. Genes involved in de novo nucleotide biosynthesis (including *pyrK_2, pyrF, purD, purH*) and a gene encoding a phosphotransferase system subunit (*manY_2*) were thus identified to be contributing to *E. faecium* growth in human serum. Transposon mutants in *pyrK_2, pyrF, purD, purH* and *manY_2* were isolated from the library and their impaired growth in human serum was confirmed. In addition, the *pyrK_2* and *manY_2* mutants were tested for their virulence in an intravenous zebrafish infection model and exhibited significantly attenuated virulence compared to *E. faecium* E745.

Conclusions: Genes involved in carbohydrate metabolism and nucleotide biosynthesis of *E. faecium* are essential for growth in human serum and contribute to the pathogenesis of this organism. These genes may serve as targets for the development of novel anti-infectives for the treatment of *E. faecium* bloodstream infections.

Keywords: *Enterococcus faecium*, Transcriptome, Transposon mutant library screening, Nucleotide biosynthesis, Carbohydrate metabolism, Virulence, Zebrafish

* Correspondence: w.vanschaik@bham.ac.uk
[2]Department of Medical Microbiology, University Medical Center Utrecht, 3584CX Utrecht, the Netherlands
[4]Institute of Microbiology and Infection, College of Medical and Dental Sciences, The University of Birmingham, Birmingham B15 2TT, United Kingdom
Full list of author information is available at the end of the article

Background

Enterococci are commensals of the gastrointestinal tract of humans and animals, but some enterococcal species, particularly *E. faecium* and *E. faecalis,* are also common causes of hospital-acquired infections in immunocompromised patients [1]. While *E. faecalis* has been recognized as an important nosocomial pathogen for over a century, *E. faecium* has emerged as a prominent cause of hospital-acquired infections over the last two decades [2]. Since the 1980s, *E. faecium* acquired resistance to multiple antibiotics, including β-lactams, aminoglycosides and finally, to the glycopeptide vancomycin [3]. Nosocomial infections are almost exclusively caused by a specific sub-population of *E. faecium,* termed clade A-1, which has emerged from a background of human commensal and animal *E. faecium* strains [4]. Strains in clade A-1 carry genetic elements that are absent from animal or human commensal isolates and which contribute to gut colonization or pathogenicity [5–9]. Clade A-1 *E. faecium* strains are rarely found in healthy individuals but can colonize the gut of immunosuppressed, hospitalized patients to high-levels. These strains can then cause infections by direct translocation from the gut into the bloodstream [10–12]. In addition, due to faecal contamination of the skin in hospitalized patients, the use of intravenous catheters is another risk factor for the introduction of *E. faecium* into the bloodstream [3, 13, 14]. Currently, *E. faecium* causes approximately 40% of enterococcal bacteremias. Due to the accumulation of antibiotic resistance determinants in clade A-1 strains, *E. faecium* infections are more difficult to treat than infections caused by *E. faecalis* or other enterococci [15–17]. To cause bloodstream infections, *E. faecium* needs to be able to survive and multiply in blood, but the mechanisms by which it can do so, have not yet been studied. To thrive in the bloodstream, an opportunistic pathogen has to evade host immune mechanisms and to adjust its metabolism to an environment that is relatively poor in nutrients [18].

To identify genes that are conditionally essential in bacteria, high-throughput screening methods for transposon mutant libraries have been developed and optimized for many different bacterial species [19, 20]. To perform high-throughput functional genomics in ampicillin-resistant, vancomycin-susceptible clinical *E. faecium* strains, we previously developed a microarray-based transposon mutagenesis screening method which was used to identify genes involved in the development of endocarditis [7], resistance to ampicillin [21], bile [22] and disinfectants [23]. However, microarray-based methods for transposon mutant library screening are limited in their accuracy and can only be used in strains for which the microarray was designed. To address these limitations, several methods, including Tn-seq [24] and TraDIS [25], which are based on high-throughput sequencing of the junctions of the transposon insertion sites and genomic DNA, have been developed [26].

In this study, we set-up Tn-seq in the clinical *E. faecium* isolate E745 to identify genes that contribute to survival and growth in human serum. In addition, we determined the transcriptional response of *E. faecium* E745 in that same environment. Finally, we substantiated the role of two *E. faecium* genes that contribute to growth in serum and in virulence, in a zebrafish model of infection. Collectively, our findings show that metabolic adaptations are key to *E. faecium* growth in serum and contribute to virulence.

Results

The complete genome sequence of *E. faecium* E745

In this study, we implemented RNA-seq and Tn-seq analyses in *E. faecium* strain E745, an ampicillin- and vancomycin-resistant clinical isolate. *E. faecium* E745 was isolated from a rectal swab of a hospitalized patient as part of routine surveillance during an outbreak of VRE in the nephrology ward of a Dutch hospital in 2000 [27, 28]. To allow the application of RNA-seq and Tn-seq in *E. faecium* E745, we first determined the complete genome sequence of this strain through a combination of short-read Illumina sequencing and long-read sequencing on the RSII Pacific Biosciences and Oxford NanoPore's MinION systems. This resulted in a closed chromosomal sequence of 2,765,010 nt and 6 complete plasmids sequences, with sizes ranging between 9.3 kbp and 223.7 kbp (Additional file 1). Taken together, the chromosome and plasmids have 3095 predicted coding sequences. Phylogenetic analysis of the core genome of E745 and a set of 72 genomes representing global *E. faecium* diversity [4], showed that *E. faecium* E745 is a clade A-1 strain (Fig. 1). The E745 chromosome contains a pathogenicity island with the *esp* gene, which encodes a 207-kDa surface protein that is involved in biofilm formation and infection [6, 29, 30]. The vancomycin resistance genes of *E. faecium* E745 are of the *vanA* type [31] and are carried on the 32.4-kbp plasmid pE745-2. Additional antibiotic resistance genes in the *E. faecium* E745 genome are the trimethoprim resistance gene *dfrG* [32], which is located on plasmid pE745-6, and the chromosomally encoded macrolide resistance gene *msrC* [33].

Transcriptome of *E. faecium* E745 during growth in rich medium and in human serum

After confirming that serum can support the growth of *E. faecium* E745, though at a lower growth rate than the rich medium BHI (Additional file 2), the transcriptional profile of E745 was determined by RNA-seq during exponential growth in BHI and in heat-inactivated human

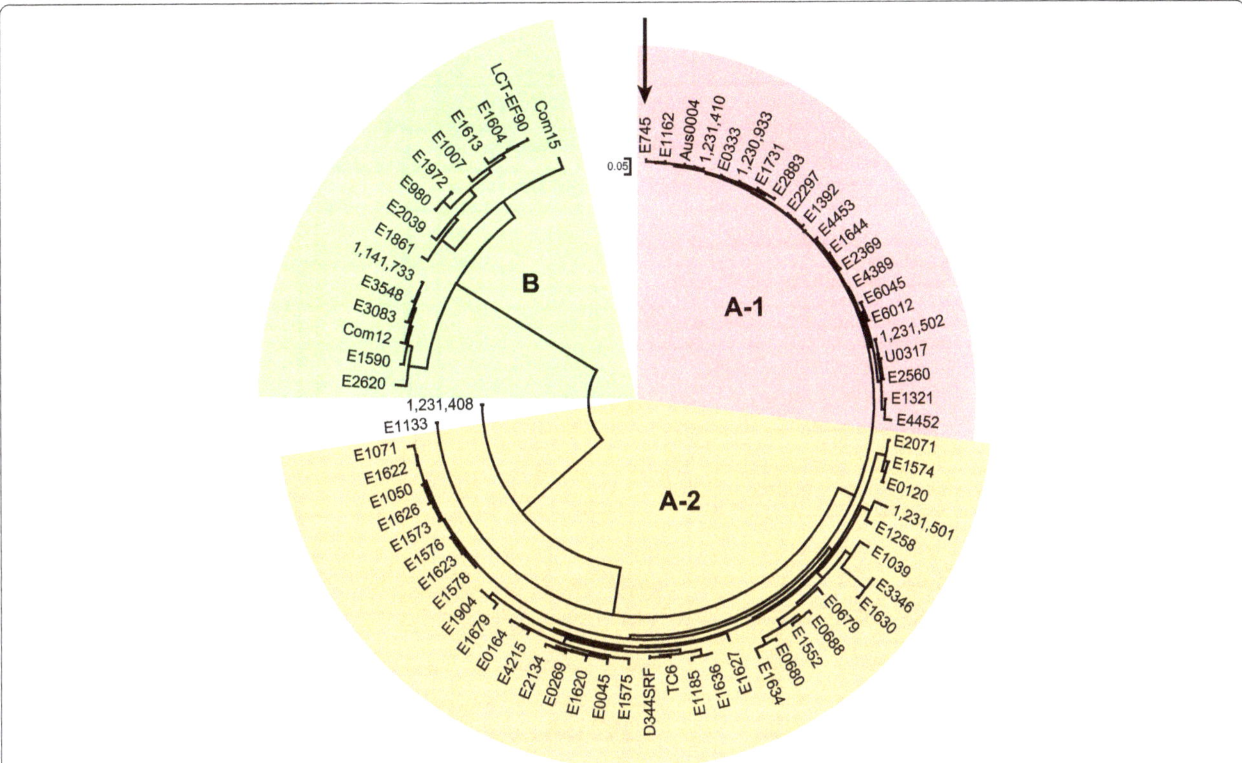

Fig. 1 Maximum likelihood phylogenetic tree of *E. faecium*. The phylogenetic tree was based on a core genome alignment of 1,545,750 positions that was generated by ParSNP [61]. The tree includes the *E. faecium* E745 genome sequence generated in this study and the 72 *E. faecium* whole genome sequences described in Lebreton et al. [4]. The tree was visualized and mid-point rooted using MEGA 7.0.26 [62]. The different *E. faecium* clades are indicated. The position of *E. faecium* E745 in the phylogenetic tree is highlighted by an arrow

serum. A total of 99.9 million (15.6–17.6 million per sample) 100 bp paired-end reads were successfully aligned to the genome, allowing the quantification of rare transcripts (Fig. 2). A total of 3217 transcription units were identified, including 651 predicted multi-gene operons, of which the largest contains 22 genes (Fig. 2a and Additional file 3).

A comparative analysis of the E745 transcriptome during growth in BHI and in human serum, showed that 860 genes exhibited significantly ($q < 0.001$ and a fold change in expression of >2 or <0.5 between cultures grown in BHI versus heat-inactivated serum) different expression between these conditions (Additional file 4). Among the genes with the highest difference in expression between growth in serum and in rich medium, we identified a gene cluster with a role in purine biosynthesis (Fig. 2b). In addition, we found a 58.4 kbp prophage-like gene cluster that exhibited higher expression in E745 during growth in serum (Fig. 2c). Analysis of this phage sequence against genome sequences deposited at NCBI Genbank revealed that an essentially identical element (with 99% nucleotide identity) could be identified in *E. faecium* ATCC 700221 but not in other genome sequences of *E. faecium* or other bacteria.

To confirm the RNA-seq analysis, we independently determined expression levels of eight genes during growth in serum versus growth in BHI by qPCR (Additional file 5). RNA-seq and qPCR data were highly concordant ($r^2 = 0.98$).

E. faecium E745 genes required for growth in human serum

A *mariner*-based transposon mutant library was generated in *E. faecium* E745 and Tn-seq [24] was performed on ten replicate transposon mutant libraries (after overnight growth in BHI at 37 °C), resulting in an average of 15 million Tn-seq reads for each library. To analyze the Tn-seq data, we divided the E745 genome in 25-nt windows. Of a total of 110,601 25-nt windows, 49,984 (45%) contained one or more sequence reads. No positional bias was observed in the transposon insertion sites in the chromosome and plasmids of *E. faecium* E745 (Additional file 6).

In order to identify genes that contribute to growth of *E. faecium* E745 in human serum, we performed Tn-seq on cultures of the *E. faecium* E745 transposon mutant libraries upon growth in rich medium (BHI) and in human serum. The human serum was either used natively,

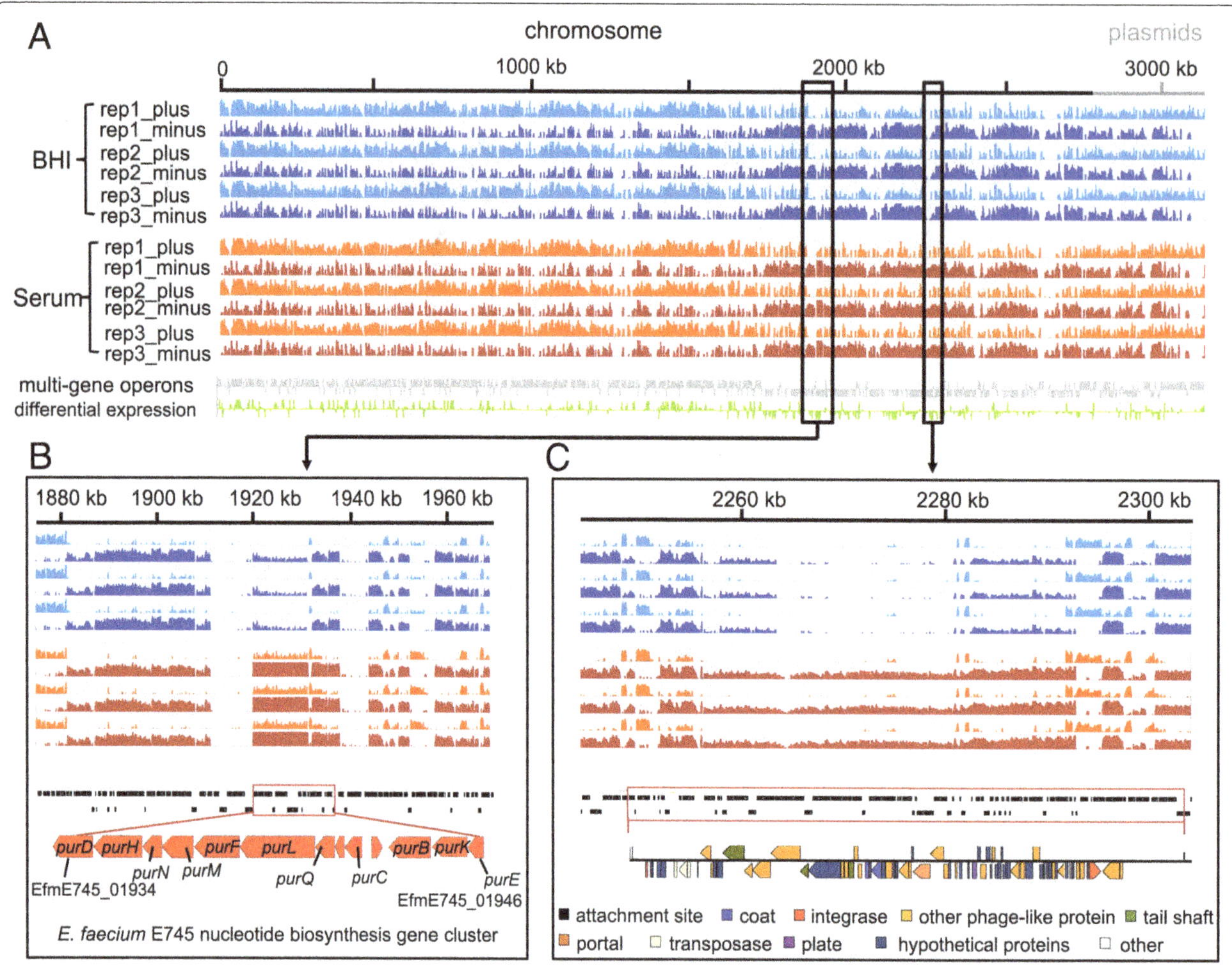

Fig. 2 Transcriptome analysis of *E. faecium* E745. Coverage plots of RNA-seq data aligning to chromosome and plasmid DNA are shown in panel **a**. The y-axis of each track indicates reads coverage and is represented on a log scale, ranging from 0 to 10,000. The x-axis represents the genomic location. Light blue (BHI) or orange (serum) tracks correspond to sequencing reads aligned to the plus strand of the replicon, and dark blue (BHI) or dark red (serum) tracks correspond to sequencing reads aligned to the minus strand of the replicon. The grey track corresponds to multi-gene operons. The green track corresponds to differentially expressed genes (BHI vs serum), with the height of the green bars indicative of differential expression. In panels **b** and **c**, two serum-induced regions are shown, i.e. a gene cluster involved in nucleotide biosynthesis (panel **b**) and a prophage (panel **c**). The RNA-seq experiments were performed using three biological replicates

or was heat-treated to inactivate the complement system [34]. Minor differences were observed among conditionally essential genes between the experiments performed in native human serum or heat-inactivated human serum (Additional file 7) and the following results correspond to the experiments obtained with heat-inactivated serum. This condition was chosen because it may be a more reproducible in vitro environment, particularly since the interaction between the complement system and Gram-positive bacteria remains to be fully elucidated [35, 36].

We identified 37 genes that significantly contributed to growth of E745 in human serum (Fig. 3 and Additional file 8): twenty-nine genes were located on the chromosome and eight genes were present on plasmids (six genes on pE745–5, two genes on plasmid pE745–6). The relatively large number of genes identified indicates that growth of *E. faecium* in human serum is a multifactorial process. The genes that conferred the most pronounced effect on growth of *E. faecium* in serum included genes that are involved in carbohydrate uptake (*manZ_3, manY_2, ptsL*), a putative transcriptional regulator (*algB*) and genes involved in the biosynthesis of purine and pyrimidine nucleotides (*guaB, purA, pyrF, pyrK_2, purD, purH, purL, purQ, purC*) (Fig. 3). Notably, the *purD, purH,* and *purL* genes were found to exhibit higher expression upon growth in human serum in the RNA-seq analysis (Fig. 2). Nine genes were identified as negatively contributing to growth in serum, i.e. the transposon mutants in these genes were significantly enriched upon growth in serum. The effects of these mutations were relatively limited (Additional file 8), compared to the major effects observed in the transposon mutants discussed above,

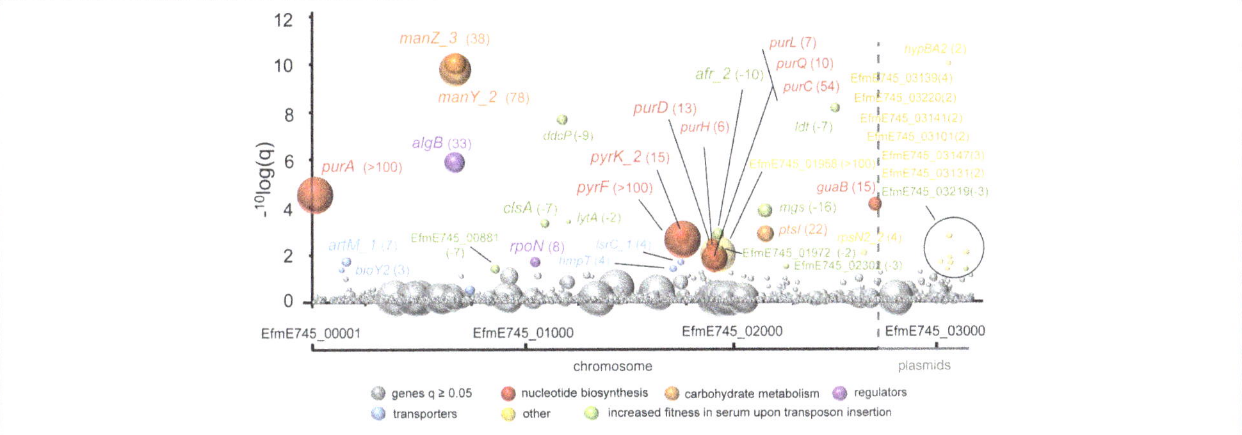

Fig. 3 Tn-seq analysis to identify *E. faecium* genes required for growth in human serum. Bubbles represent genes, and bubble size corresponds to the fold-changes (for visual reasons, a 100-fold change in transposon mutant abundance is set as a maximum) derived from the read-count ratio of libraries grown in BHI to libraries grown in human serum. On the x-axis genes are shown in order of their genomic location and the chromosome and plasmids are indicated. The outcome of statistical analysis of the Tn-seq data is indicated on the y-axis. Genes with a significant change ($q < 0.05$) in fitness in serum versus BHI are grouped by function and are labelled with different colors, and the name or locus tag and the change in abundance between the control condition and growth in serum is indicated next to the bubbles in parentheses. Negative values indicate that mutants in these genes outgrow other mutants in serum, suggesting that these mutants, compared to the wild-type strain *E. faecium* E745, have a higher fitness in serum

but it is notable that five (*clsA_1, ddcP,* ldt_{fm}, *mgs*, and *lytA_2*) of these genes have predicted roles in cell wall and cytoplasmic membrane biosynthesis.

We developed a PCR-based method (Additional file 9) to selectively isolate five transposon mutants (in the purine metabolism genes *purD* and *purH*, the pyrimidine metabolism genes *pyrF* and *pyrK_2* and the phosphotransferase system (PTS) gene *manY_2* from the transposon library. Growth in rich medium of these transposon insertion mutants was equal to the parental strain. However, all mutants were significantly impaired in their growth in human serum (Fig. 4a), confirming the results of the Tn-seq experiments.

E. faecium pyrK_2 and *manY_2* contribute to intravenous infection of zebrafish

Finally, we investigated whether the transposon insertion mutants in the *manY_2* and *pyrK_2* genes were attenuated in vivo (Fig. 4b). The mutants in these genes were selected because they represent the mutants in nucleotide and carbohydrate metabolism genes that were previously shown to contribute to the growth of *E. faecium* in human serum. As a model for intravenous infection, we used a recently described model in which *E. faecium* was injected into the circulation of zebrafish embryos to mimic systemic infections [37]. We showed that both the *manY_2* and the *pyrK_2* mutant were significantly less virulent than the parental strain. At 92 h post infection, survival of zebrafish embryos infected with the WT strain was 53%, as compared to 88% and 83% for zebrafish embryos that were infected with the transposon insertion mutants in *manY_2* and *pyrK_2*, respectively.

Discussion

E. faecium can contaminate the skin and from there colonize indwelling devices such as intravenous catheters, or it can translocate from the gastrointestinal tract in immunosuppressed patients, leading to the development of bacteremia and endocarditis. *E. faecium* infections are often difficult to treat, due to the multi-drug resistant character of the strains causing nosocomial infections [3, 4]. However, the bloodstream poses challenges for the proliferation and survival of *E. faecium*, including a scarcity of nutrients.

In the present study, we sequenced the complete genome of a vancomycin-resistant *E. faecium* strain, and identified *E. faecium* genes that were essential for growth in human serum. A total of 37 genes, including genes with roles in carbohydrate uptake and nucleotide biosynthesis, were found to be required for fitness of *E. faecium* E745 in serum. Previously, fitness determinants for growth in human serum have been identified through large-scale screening of mutant libraries in both a Gram-negative (*Escherichia coli*) and a Gram-positive (*Streptococcus pyogenes*) pathogen [38, 39]. Notably, these studies have also identified the ability for de novo synthesis of purines and pyrimidines as a crucial factor for growth in serum. In addition, in diverse pathogenic bacteria (including *Burkholderia cepacia, Pasteurella multocida, Acinetobacter baumannii, Salmonella enterica* serovar Typhimurium, *Bacillus anthracis*, and *Streptococcus pneumoniae*), nucleotide biosynthesis contributes importantly to virulence [40–45]. The ability to synthesize nucleotides de novo thus appears to be an essential trait for the success of a pathogen that spreads through the bloodstream [38]. The

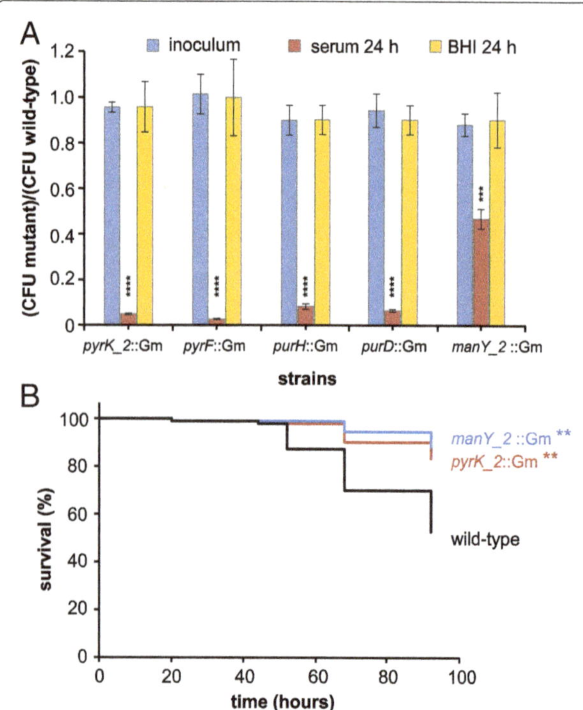

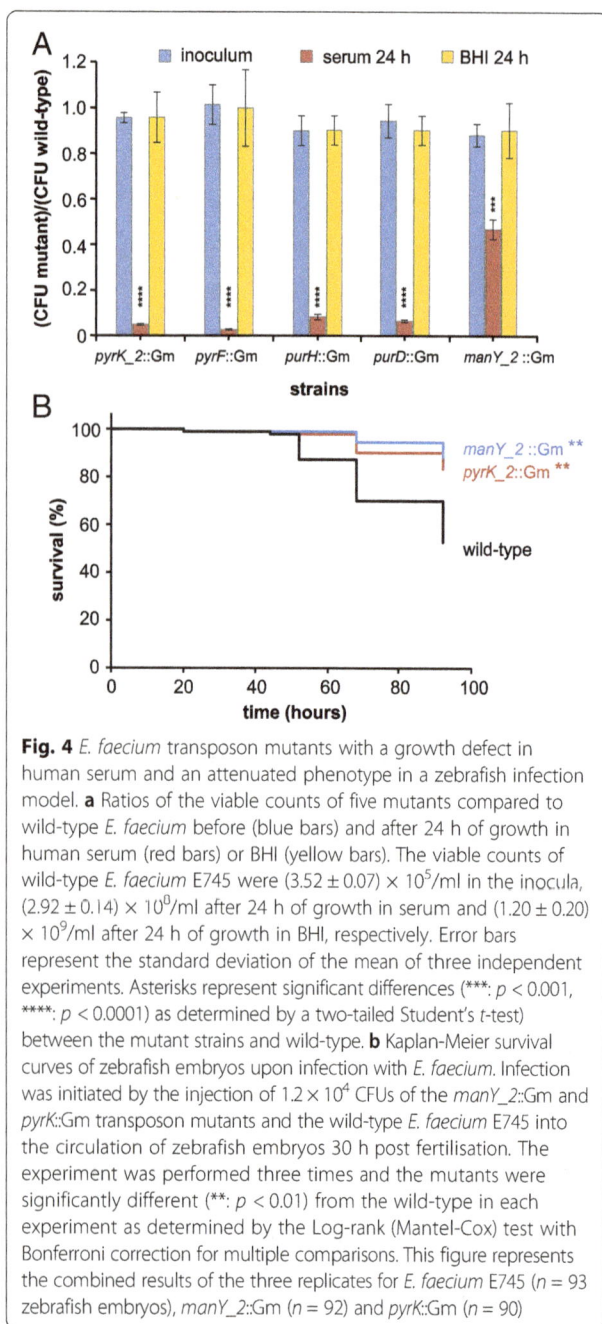

Fig. 4 *E. faecium* transposon mutants with a growth defect in human serum and an attenuated phenotype in a zebrafish infection model. **a** Ratios of the viable counts of five mutants compared to wild-type *E. faecium* before (blue bars) and after 24 h of growth in human serum (red bars) or BHI (yellow bars). The viable counts of wild-type *E. faecium* E745 were (3.52 ± 0.07) × 10^5/ml in the inocula, (2.92 ± 0.14) × 10^9/ml after 24 h of growth in serum and (1.20 ± 0.20) × 10^9/ml after 24 h of growth in BHI, respectively. Error bars represent the standard deviation of the mean of three independent experiments. Asterisks represent significant differences (***: $p < 0.001$, ****: $p < 0.0001$) as determined by a two-tailed Student's *t*-test) between the mutant strains and wild-type. **b** Kaplan-Meier survival curves of zebrafish embryos upon infection with *E. faecium*. Infection was initiated by the injection of 1.2 × 10^4 CFUs of the *manY_2*::Gm and *pyrK*::Gm transposon mutants and the wild-type *E. faecium* E745 into the circulation of zebrafish embryos 30 h post fertilisation. The experiment was performed three times and the mutants were significantly different (**: $p < 0.01$) from the wild-type in each experiment as determined by the Log-rank (Mantel-Cox) test with Bonferroni correction for multiple comparisons. This figure represents the combined results of the three replicates for *E. faecium* E745 ($n = 93$ zebrafish embryos), *manY_2*::Gm ($n = 92$) and *pyrK*::Gm ($n = 90$)

data presented here indicate that de novo biosynthesis of nucleotides is also required for *E. faecium* growth in serum and virulence. The nucleotide biosynthesis pathway of *E. faecium* may be a promising target for the development of novel antimicrobials for the treatment of *E. faecium* bloodstream infections. Indeed, compounds that target guanine riboswitches, thereby inhibiting nucleotide biosynthesis, have already shown their efficacy in a *Staphylococcus aureus* infection model [46].

Three genes, *ptsL*, *manY_2* and *manZ_3*, encoding subunits of *E. faecium* PTSs were found to contribute to growth in serum in our Tn-seq experiments. The *ptsL* gene is predicted to encode an enzyme that confers a phosphate group from phosphoenolpyruvate to Enzyme I of PTS, while *manY_2* and *manZ_3* are predicted to form the IIA and IIBC components of a permease system that is homologous (64% and 69% amino acid identity, respectively) to the PtnAB PTS permease of *Lactococcus lactis* MG1363. PtnAB is one of the glucose uptake systems of *L. lactis* [47] and the *E. faecium* homolog may have a similar function, which could explain its essential role during growth in serum, as glucose is the only carbohydrate that occurs in the free state in appreciable amounts in serum [18].

It is notable that among the nine genes that exhibited increased fitness upon inactivation by transposon insertion, five genes are predicted to have a role in cell wall or cytoplasmic membrane biosynthesis. The protein encoded by *ddcP* was previously characterized as a low-molecular-weight penicillin-binding protein with D-alanyl-D-alanine carboxypeptidase activity [21], while ldt_{fm} acts as a peptidoglycan L,D transpeptidase [48]. The predicted α-monoglucosyldiacylglycerol synthase gene *mgs* is orthologous (73% amino acid identity) to *bgsB* in *E. faecalis*, which is required for the biosynthesis of membrane glycolipids [49]. The *clsA_1* gene is predicted to be responsible for the synthesis of cardiolipin (bisphosphatidylglycerol) and its inactivation may modulate the physical properties of the cytoplasmic membrane [50]. Finally, *lytA_2* is predicted to encode an autolysin, which may be involved in the turnover of peptidoglycan in the cell wall [51]. The transposon mutants in these genes were not further characterized in this study, but our findings suggest that non-essential pathways of cell wall or cytoplasmic membrane remodelling may confer subtle fitness defects to *E. faecium* when growing in a nutrient-poor environment, like serum.

Our RNA-seq-based transcriptional profiling of *E. faecium* E745 during mid-exponential growth in serum showed pervasive changes in gene expression compared to exponential growth in rich medium. The large number of differentially expressed genes may not all reflect the different growth conditions (serum and BHI) per se, but could also be influenced by the difference in growth rate during mid-exponential growth in serum and BHI (Additional file 2). The purine metabolism genes *purL*, *purH*, *purD*, which were found to be required for growth in serum in our Tn-seq experiments, were among those that were significantly upregulated during growth in serum compared to growth in rich medium. Notably, a single prophage was expressed at higher levels during growth in serum than in rich medium. The abundance of prophage elements in the genome of *E. faecium* has been noted before [4, 52]. Interestingly, in the related bacterium *Enterococcus faecalis* prophages encode platelet-binding proteins [53] and may have a role in intestinal colonization [54]. The contribution of *E. faecium* prophages to traits that are important for colonization and infection may provide important insights into the success of *E. faecium* as a nosocomial pathogen.

Conclusions

Our data indicate that nucleotide biosynthesis and carbohydrate metabolism are critical metabolic pathways for the proliferation and survival of *E. faecium* in the bloodstream. The proteins encoded by the genes required for growth in human serum that were identified in this study, could serve as candidates for the development of novel anti-infectives for the treatment of bloodstream-infections by multi-drug resistant *E. faecium*.

Methods

Bacterial strains, plasmids, growth conditions, and oligonucleotides

The vancomycin-resistant *E. faecium* strain E745 was used throughout this study. This strain was isolated from a rectal swab of a hospitalized patient, during routine surveillance of a VRE outbreak in a Dutch hospital [27, 28]. Unless otherwise mentioned, *E. faecium* was grown in brain heart infusion broth (BHI; Oxoid) at 37 °C. The *E. coli* strains DH5α (Invitrogen) was grown in Luria-Bertani (LB) medium. When necessary, antibiotics were used at the following concentrations: chloramphenicol 4 μg ml^{-1} for *E. faecium* and 10 μg ml^{-1} for *E. coli*, and gentamicin 300 μg ml^{-1} for *E. faecium* and 25 μg ml^{-1} for *E. coli*. All antibiotics were obtained from Sigma-Aldrich. Growth was determined by measuring the optical density at 660 nm (OD_{660}). The sequences of all oligonucleotides used in this study are listed in Additional file 10.

Genome sequencing, assembly and bioinformatic analysis

E. faecium E745 was sequenced using a combination of Illumina HiSeq 100 bp paired-end sequencing, long-read sequencing using the Pacific Biosciences RS II SMRT technology and the MinION system with R7 flowcell chemistry (Oxford Nanopore Technologies). Corrected PacBio reads were assembled using the Celera assembler (version 8.1) [55] and assembled contigs were then corrected by aligning Illumina reads using BWA (version 0.7.9a), with default parameters for index creation and the BWA-MEM algorithm with the *-M* option for the alignment [56]. This approach resulted in 15 contigs, including one contig covering the entire 2.77 Mbp chromosome. After discarding contigs with low-coverage, the remaining contigs constituted 5 circular plasmid sequences and 5 non-overlapping contigs. These 5 contigs were aligned against the NCBI Genbank database and all were found to be part of the *E. faecium* plasmid pMG1 [57]. Based on this alignment the presumed order of contigs was determined and confirmed by gap-spanning PCRs and sequencing of the products. A single gap between two contigs, could not be closed by PCR. Thus, we assembled Illumina reads together with MinION 2D reads using the SPAdes assembler (version 3.0) [58], which produced a contig that closed the gap, resulting in a complete assembly of this plasmid. Sequence coverage of chromosomal and plasmid sequences was determined with SAMtools (version 0.1.18) using short read alignments to the assembly, which were generated using BWA (version 0.7.9a). SAMtools was also used to identify possible base-calling and assembly errors, by aligning short reads to assembled contigs. A base was corrected using the consensus of aligned reads [59]. The corrected sequences were annotated using Prokka (version 1.10) [60]. A maximum likelihood phylogenetic tree based on the core genome of *E. faecium* E745 and an additional 72 *E. faecium* strains representing the global diversity of the species [4], was generated using ParSNP [61] with settings -c (forcing inclusion of all genome sequences) and -x (enabling recombination detection and filtering). The resulting phylogenetic tree was visualized using MEGA 7.0.26 [62]. Antibiotic resistance genes in the assembled genome sequence of *E. faecium* E745 were identified using ResFinder [63]. The annotated genome of *E. faecium* E745 is available from NCBI Genbank database under accession numbers CP014529 – CP014535.

RNA-seq

Approximately 3×10^7 cfu of *E. faecium* E745 were inoculated into 14 ml of BHI broth and heat-inactivated serum, and grown at 37 °C until exponential phase. Cultures were centrifuged at room temperature (15 s; 21.380 *g*), and pellets were flash frozen in liquid N_2 prior to RNA extraction, which was performed as described previously [21]. The ScriptSeq Complete Kit (Bacteria) (Epicentre Biotechnologies, WI) was used for rRNA removal and strand-specific library construction. Briefly, rRNA was removed from 2.5 μg of total RNA. To generate strand specific RNA-seq data, approximately 100 ng of rRNA-depleted RNA was fragmented and reverse transcribed using random primers containing a 5′ tagging sequence, followed by 3′ end tagging with a terminal-tagging oligo to yield di-tagged, single-stranded cDNA. Following magnetic-bead based purification, the di-tagged cDNA was amplified by PCR (15 cycles) using ScriptSeq Index PCR Primers (Epicentre Biotechnologies, WI). Amplified RNA-seq libraries were purified using AMPure XP System (Beckman Coulter) and sequenced by a 100 bp paired end reads sequencing run using the Illumina HiSeq 2500 platform (University of Edinburgh, United Kingdom). Data analysis was performed using Rockhopper [64] using the default settings for strand specific analysis.

Validation of RNA-seq results by quantitative real-time RT-PCR (qRT-PCR).

Total RNA isolated as described previously was used to confirm the transcriptome analysis by qRT-PCR. cDNA was synthesized as described above and qRT-PCR on these cDNAs was performed using the Maxima SYBR Green/ROX qPCR Master Mix (Thermo Scientific, Breda, The Netherlands) and a StepOnePlus instrument (Life Technologies). The expression of *tufA* was used as a housekeeping control. Ct values were calculated using the StepOne analysis software v2.2. Transcript levels, relative to *tufA*, of the assayed genes were calculated using REST 2009 V2.0.13 (Qiagen, Venlo, the Netherlands). This experiment was performed with three biological replicates.

Generation of *mariner* transposon mutant library in *E. faecium*

To create a transposon mutant library in *E. faecium* E745 suitable for Tn-seq, the *mariner* transposon cassette (carrying a gentamicin resistance gene) in the transposon delivery plasmid pZXL5 [21] was adapted as follows. The transposon from pZXL5 was amplified by PCR using the set of primers: pZXL5_MmeI_SacII_Fw and pZXL5_MmeI_SacII_Rv. These primers introduced MmeI restriction sites in the inverted repeats on both sides of the transposon. The modified transposon delivery vector, termed pGPA1, was generated by the digestion of pZXL5 with SacII, followed by the insertion of the SacII-digested *mariner* transposon that contained MmeI restriction sites at its extreme ends. pGPA1 was electroporated into *E. faecium* E745 and the transposon mutant library was generated by selecting for gentamicin-resistant transposon mutants as described previously [21].

Tn-seq analysis of conditionally essential genes in *E. faecium* E745

The transposon mutant library created in E745 was prepared for Tn-seq analysis, similar to previously described procedures [65]. To identify genes that are essential for the viability of *E. faecium* in BHI, we used ten experimental replicates of the mutant library. Aliquots (20 µl) of the transposon mutant library, containing approximately 10^7 cfu, were used to inoculate 20 ml BHI broth and grown overnight at 37 °C. Subsequently, 1 ml aliquots of the cultures were spun down (15 s, 21.380 *g*) and used for the extraction of genomic DNA (Wizard genomic DNA purification kit, Promega Benelux). 2 µg of the extracted DNA was digested for 4 h at 37 °C using 10 U MmeI (New England Biolabs) and immediately dephosphorylated with 1 U of calf intestine alkaline phosphatase (Invitrogen) during 30 min at 50 °C. DNA was isolated using phenol-chloroform extraction and subsequently precipitated using ethanol. The DNA pellets were then dissolved in 20 µl water. The samples were barcoded and prepared for Tn-seq sequencing as described previously [65]. The sequence reads from all ten experimental replicates were mapped to the genome, and the mapped read-counts were then tallied for the analysis of the essentiality of the genes in the *E. faecium* E745 genome (further described below).

To identify genes that are required for growth in human serum, 20 µl aliquots of the frozen mutant library in E745 were inoculated in BHI broth and grown overnight as described above. Subsequently, bacterial cells were washed with physiological saline solution. Approximately 3×10^7 cfu were inoculated into 14 ml BHI broth, and approximately 3×10^6 cfu were inoculated into 14 ml human serum obtained from Sigma (Cat. No. H4522; Sterile filtered type-AB human serum) or heat-inactivated human serum (the same, after incubation for 30 min at 56 °C). The different inoculum-sized were used in order for a similar number of divisions to occur during the experiment. Cells were incubated at 37 °C for 24 h without shaking and then further processed for Tn-seq [65]. This experiment was performed in triplicate.

Tn-seq samples were sequenced (50 nt, single-end) on one lane of a Illumina Hiseq 2500 (Baseclear, Leiden, the Netherlands and Sequencing facility University Medical Center, Utrecht, The Netherlands), generating an average of 15 million high quality reads per sample.

Tn-seq data analysis

Raw Illumina sequence reads from Illumina sequencing were split, based on their barcode, using the Galaxy platform [66], and 16-nucleotide fragments of each read that corresponded to E745 sequences, were mapped to the E745 genome using Bowtie 2 [67]. The results of the alignment were sorted and counted by IGV [68] using a 25-nucleotide window size and then summed over the gene. Read mapping to the final 10% of a gene were discarded as these insertions may not inactivate gene function. Read counts per gene were then normalized to the total number of reads that mapped to the genome in each replicate, by calculating the normalized read-count RPKM (Reads Per Kilobase per Million input reads) via the following formula: RPKM = (number of reads mapped to a gene × 10^6) / (total mapped input reads in the sample x gene length in kbp). Statistical analysis of the RPKM-values between the experimental conditions was performed using Cyber-T [69]. Genes were determined to be significantly contributing to growth in human serum when the Benjamini-Hochberg corrected *P*-value was <0.05 and the difference in abundance of the transposon mutant during growth in BHI and serum was >2.

Isolation of mutants from the transposon mutant library pool

To recover a targeted transposon mutant from the complete mutant pool, a PCR-based screening strategy was developed (Additional data file 9). 40 μl of the transposon mutant library was inoculated into 40 ml of BHI broth with gentamicin and grown overnight at 37 °C with shaking (200 rpm). The overnight culture, containing approximately 10^9 cfu/ml, was then diluted to approximately 20 cfu/ml in 500 ml of BHI with gentamicin and kept on ice. Subsequently, 200 μl aliquots were transferred to wells of sterile 96 wells plates (n = 12, Corning Inc.). After overnight incubation at 37 °C without shaking, aliquots (15 μl) of each one of the 96 wells, were further pooled into a single new 96 well plate, as described in Additional file 9.

PCRs were performed on the final plate in which the transposon mutants were pooled, to check for the presence of the Tn-mutants of interest, using the primer ftp_tn_both_ends_MmeI, which is complementary to the repeats flanking the transposon sequence, in combination with a gene-specific primer. When a PCR was found to be positive in one of the wells of this plate, the location of the Tn-mutant was tracked backwards to the wells containing approximately 4 independent transposon mutants, by performing PCRs mapping the presence of the transposon mutant in each step. Cells from the final positive well were plated onto BHI with gentamicin and colony PCR was performed to identify the desired transposon mutant.

Growth of *E. faecium* E745 and individual mutants in human serum

Wild-type E745 and the mutant strains were grown overnight at 37 °C in BHI broth. Subsequently, bacterial cells were washed with physiological saline and approximately 3×10^5 cfu were inoculated into 1.4 ml BHI broth or heat-inactivated serum. Cells were grown in 1.5 ml tubes (Eppendorf) in triplicate for each condition and incubated at 37 °C for 24 h without shaking. Bacterial growth was determined by assessing viable counts, for which the cultures were serially diluted using physiological saline solution and plated onto BHI agar followed by overnight incubation at 37 °C.

Intravenous infection of zebrafish embryos

London wild-type (LWT) inbred zebrafish embryos, provided by the aquarium staff of The Bateson Center (University of Sheffield), were used for infection experiments. The parental E745 strain and its *pyrK_2* and *manY_2* transposon mutants were grown in BHI broth until they reached an optical density at 600 nm of approximately 0.5 and were then harvested by centrifugation (5500 *g*, 10 min). Bacteria were microinjected into the circulation of dechorionated zebrafish embryos at 30 h post fertilization, as previously described [70]. Briefly, anesthetized embryos were embedded in 3% (*w*/*v*) methylcellulose and injected individually with approximately 1.2×10^4 cfu using microcapillary pipettes. For each strain, 29 to 32 infected embryos were observed for survival up to 90 h post infection (hpi). This experiment was performed in triplicate.

Additional files

Additional file 1: Genome sequence information for *E. faecium* E745. (XLSX 10 kb)

Additional file 2: Growth of *E. faecium* E745 in BHI and serum. (PDF 962 kb)

Additional file 3: Operons identified by RNA-seq in *E. faecium* E745. (XLSX 40 kb)

Additional file 4: *E. faecium* E745 genes that exhibited significant (q < 0.001 and fold-change > 2) differential expression in human serum, as determined by RNA-seq. (XLSX 70 kb)

Additional file 5: qRT-PCR validation of RNA-seq experiments. Correlation of RNA-seq and qRT-PCR expression ratios for the seven genes with various expression levels and genomic locations. The gene expression ratios obtained from both qRT-PCR and RNA-seq were normalized by a housekeeping control gene Efm745_00056 (*tufA*). The experiment was performed with three biological replicates. (PDF 121 kb)

Additional file 6: Characterization of the *E. faecium* E745 transposon mutant library, showing the number of reads that were mapped to the *E. faecium* E745 chromosome and plasmids. The height of each peak represents the read abundance at a specific insertion site. On the y-axis, the number of mapped reads is shown on a log scale. (PDF 1427 kb)

Additional file 7: Tn-seq data: comparison of heat-inactivated and native serum (XLSX 139 kb)

Additional file 8: *E. faecium* E745 genes that significantly (q < 0.05 and fold-change <–2 or >2) contribute to growth in human serum, as identified by Tn-seq. (XLSX 13 kb)

Additional file 9: Isolation of mutants from the transposon mutant library pool. (A) Schematic representation of the PCR reaction designed to find a particular Tn-mutant within the transposon mutant library. This PCR uses a combination of a gene-specific primer (blue arrow) and a transposon specific primer primer (yellow arrow). Positive PCR products, indicated by the green check marks, should occur when the transposon (depicted as a yellow triangle) is inserted in the gene of interest (depicted in blue). If the transposons inserted in adjacent genes or intergenic regions, no PCR product can be amplified (red crosses). (B) Schematic workflow to isolate Tn-mutants from the mutant library. The transposon mutant library is split into 12 plates (A1 - A12) of 96 wells, with each 200 μl well containing an average of 4 mutants. Plates were incubated overnight (Step 2). Plate A1 was then pooled into the first column of a new 96 well plate, denominated plate B1 (Step 3) and the same was done for plates A2 to A12. Subsequently, plate B1 was pooled again into the first column of a third plate, denominated C1 (Step 4). PCR using the gene-specific primer and the transposon specific primer was performed on the 8 wells of plate C1 (Step 5). A positive PCR was suggestive of the presence of a particular transposon-mutant (depicted as a red dot). The presence of the transposon mutant was then confirmed by PCR in plate B1 (step 6) and the corresponding plate A (step 7). Once a transposon-mutant was located to a particular well in plate A, the well was plated on BHI plates containing gentamicin, and the colonies were screened for the presence of the transposon mutant by PCR (step 8). (PDF 277 kb)

Additional file 10: Oligonucleotides used in this study. (XLSX 10 kb)

Abbreviations

BHI: Brain Heart Infusion broth; cDNA: complementary DNA; cfu: colony forming units; hpi: hours post-infection; kbp: kilo-base pair; LB: Luria-Bertani; LWT: London wild-type; NCBI: National Center for Biotechnology Information; nt: nucleotides; PTS: phosphotransferase system; qPCR: quantitative polymerase chain reaction; RNA-seq: RNA sequencing; RPKM: reads per kilobase per million input reads; rRNA: ribosomal RNA; Tn-seq: transposon sequencing; TraDIS: transposon directed insertion sequencing; U: unit(s)

Acknowledgements

Not applicable.

Funding

This work was supported by the European Union Seventh Framework Programme (FP7-HEALTH-2011-single-stage) "Evolution and Transfer of Antibiotic Resistance" (EvoTAR) under grant agreement number 282004 and by a grant from the Netherlands Organization for Scientific Research (VIDI: 917.13.357) to W.v.S. The funders had no role in study design, data collection and analysis, decision to publish, or preparation of the manuscript.

Authors' contributions

XZ, VdM, AMGP, and TKP performed experiments. JRB, MdB and MRCR contributed bioinformatic analyses. XZ, MJMB, SM, RW and WvS designed the study. XZ, AMGP, JRB and WvS drafted the manuscript. All authors read and approved the final manuscript.

Competing interests

The authors declare that they have no competing interests

Author details

[1]College of Biosystems Engineering and Food Science, Zhejiang University, Hangzhou 310058, China. [2]Department of Medical Microbiology, University Medical Center Utrecht, 3584CX Utrecht, the Netherlands. [3]Krebs Institute, University of Sheffield, Sheffield S10 2TN, United Kingdom. [4]Institute of Microbiology and Infection, College of Medical and Dental Sciences, The University of Birmingham, Birmingham B15 2TT, United Kingdom.

References

1. Dupont H, Friggeri A, Touzeau J, Airapetian N, Tinturier F, Lobjoie E, et al. Enterococci increase the morbidity and mortality associated with severe intra-abdominal infections in elderly patients hospitalized in the intensive care unit. J Antimicrob Chemother. 2011;66:2379–85.
2. Guzman Prieto AM, van Schaik W, Rogers MRC, Coque TM, Baquero F, Corander J, et al. Global emergence and dissemination of enterococci as nosocomial pathogens: attack of the clones? Front Microbiol. 2016;7:788.
3. Arias CA, Murray BE. The rise of the *Enterococcus*: beyond vancomycin resistance. Nat Rev Microbiol. 2012;10:266–78.
4. Lebreton F, van Schaik W, Manson McGuire A, Godfrey P, Griggs A, Mazumdar V, et al. Emergence of epidemic multidrug-resistant *Enterococcus faecium* from animal and commensal strains. MBio. 2013;4:e00534-13.
5. Zhang X, Top J, de Been M, Bierschenk D, Rogers M, Leendertse M, et al. Identification of a genetic determinant in clinical *Enterococcus faecium* strains that contributes to intestinal colonization during antibiotic treatment. J Infect Dis. 2013;207:1780–6.
6. Heikens E, Singh KV, Jacques-Palaz KD, Van Luit-Asbroek M, Oostdijk EAN, Bonten MJM, et al. Contribution of the enterococcal surface protein Esp to pathogenesis of *Enterococcus faecium* endocarditis. Microbes Infect. 2011;13: 1185–90.
7. Paganelli FL, Huebner J, Singh KV, Zhang X, van Schaik W, Wobser D, et al. Genome-wide screening identifies phosphotransferase system permease BepA to be involved in *Enterococcus faecium* endocarditis and biofilm formation. J Infect Dis. 2016;214:189–95.
8. Sillanpää J, Prakash VP, Nallapareddy SR, Murray BE. Distribution of genes encoding MSCRAMMs and pili in clinical and natural populations of *Enterococcus faecium*. J Clin Microbiol. 2009;47:896–901.
9. Montealegre MC, Singh KV, Somarajan SR, Yadav P, Chang C, Spencer R, et al. Role of the Emp pilus subunits of *Enterococcus faecium* in biofilm formation, adherence to host extracellular matrix components, and experimental infection. Infect Immun. 2016;84:1491–500.
10. Kamboj M, Blair R, Bell N, Sun J, Eagan J, Sepkowitz K. What is the source of bloodstream infection due to vancomycin-resistant enterococci in persons with mucosal barrier injury? Infect Control Hosp Epidemiol. 2014;35:99–101.
11. de Regt MJA, van Schaik W, van Luit-Asbroek M, Dekker HAT, van Duijkeren E, Koning CJM, et al. Hospital and community ampicillin-resistant *Enterococcus faecium* are evolutionarily closely linked but have diversified through niche adaptation. PLoS One. 2012;7:e30319.
12. Ubeda C, Taur Y, Jenq RR, Equinda MJ, Son T, Samstein M, et al. Vancomycin-resistant *Enterococcus* domination of intestinal microbiota is enabled by antibiotic treatment in mice and precedes bloodstream invasion in humans. J Clin Invest. 2010;120:4332–41.
13. Bouza E, Kestler M, Beca T, Mariscal G, Rodríguez-Créixems M, Bermejo J, et al. The NOVA score: a proposal to reduce the need for transesophageal echocardiography in patients with enterococcal bacteremia. Clin Infect Dis. 2015;60:528–35.
14. Arias CA, Murray BE. Emergence and management of drug-resistant enterococcal infections. Expert Rev Anti-Infect Ther. 2008;6:637–55.
15. Coombs GW, Pearson JC, Daly DA, Le TT, Robinson JO, Gottlieb T, et al. Australian Enterococcal sepsis outcome Programme annual report, 2013. Commun Dis Intell Q Rep. 2014;38:E320–6.
16. de Kraker MEA, Jarlier V, Monen JCM, Heuer OE, van de Sande N, Grundmann H. The changing epidemiology of bacteraemias in Europe: trends from the European antimicrobial resistance surveillance system. Clin Microbiol Infect. 2013;19:860–8.
17. Hidron AI, Edwards JR, Patel J, Horan TC, Sievert DM, Pollock DA, et al. NHSN annual update: antimicrobial-resistant pathogens associated with healthcare-associated infections: annual summary of data reported to the National Healthcare Safety Network at the Centers for Disease Control and Prevention, 2006-2007. Infect Control Hosp Epidemiol. 2008;29:996–1011.
18. Krebs HA. Chemical composition of blood plasma and serum. Annu Rev Biochem. 1950;19:409–30.
19. van Opijnen T, Camilli A. Transposon insertion sequencing: a new tool for systems-level analysis of microorganisms. Nat Rev Microbiol. 2013;11:435–42.
20. Barquist L, Boinett CJ, Cain AK. Approaches to querying bacterial genomes with transposon-insertion sequencing. RNA Biol. 2013;10:1161–9.
21. Zhang X, Paganelli FL, Bierschenk D, Kuipers A, Bonten MJM, Willems RJL, et al. Genome-wide identification of ampicillin resistance determinants in *Enterococcus faecium*. PLoS Genet. 2012;8:e1002804.
22. Zhang X, Bierschenk D, Top J, Anastasiou I, Bonten MJ, Willems RJ, et al. Functional genomic analysis of bile salt resistance in *Enterococcus faecium*. BMC Genomics. 2013;14:299.
23. Prieto AMG, Wijngaarden J, Braat JC, Rogers MRC, Majoor E, Brouwer EC, et al. The two component system ChtRS contributes to chlorhexidine tolerance in *Enterococcus faecium*. Antimicrob Agents Chemother. 2017;61: e02122-16.

24. van Opijnen T, Bodi KL, Camilli A. Tn-seq: high-throughput parallel sequencing for fitness and genetic interaction studies in microorganisms. Nat Methods. 2009;6:767–72.
25. Langridge GC, Phan M-D, Turner DJ, Perkins TT, Parts L, Haase J, et al. Simultaneous assay of every *Salmonella Typhi* gene using one million transposon mutants. Genome Res. 2009;19:2308–16.
26. van Opijnen T, Lazinski DW, Camilli A. Genome-wide fitness and genetic interactions determined by Tn-seq, a high-throughput massively parallel sequencing method for microorganisms. Curr Protoc Microbiol. 2015;36:1E.3. 1–1E.3.24.
27. Leavis HL, Willems RJL, van Wamel WJB, Schuren FH, Caspers MPM, Bonten MJM. Insertion sequence-driven diversification creates a globally dispersed emerging multiresistant subspecies of *E faecium*. PLoS Pathog. 2007;3:e7.
28. Mascini EM, Troelstra A, Beitsma M, Blok HEM, Jalink KP, Hopmans TEM, et al. Genotyping and preemptive isolation to control an outbreak of vancomycin-resistant *Enterococcus faecium*. Clin Infect Dis. 2006;42:739–46.
29. van Schaik W, Top J, Riley DR, Boekhorst J, Vrijenhoek JEP, Schapendonk CME, et al. Pyrosequencing-based comparative genome analysis of the nosocomial pathogen *Enterococcus faecium* and identification of a large transferable pathogenicity island. BMC Genomics. 2010;11:239.
30. Heikens E, Bonten MJ, Willems RJ. Enterococcal surface protein Esp is important for biofilm formation of enterococcus faecium E1162. J Bacteriol. 2007;189:8233–40.
31. Courvalin P. Vancomycin resistance in Gram-positive cocci. Clin Infect Dis. 2006;42:S25–34.
32. Sekiguchi J, Tharavichitkul P, Miyoshi-Akiyama T, Chupia V, Fujino T, Araake M, et al. Cloning and characterization of a novel trimethoprim-resistant dihydrofolate reductase from a nosocomial isolate of *Staphylococcus aureus* CM.S2 (IMCJ1454). Antimicrob Agents Chemother. 2005;49:3948–51.
33. Singh KV, Malathum K, Murray BE. Disruption of an *Enterococcus faecium* species-specific gene, a homologue of acquired macrolide resistance genes of staphylococci, is associated with an increase in macrolide susceptibility. Antimicrob Agents Chemother. 2001;45:263–6.
34. Joisel F, Leroux-Nicollet I, Lebreton JP, Fontaine MA. Hemolytic assay for clinical investigation of human C2. J Immunol Methods. 1983;59:229–35.
35. Berends ETM, Dekkers JF, Nijland R, Kuipers A, Soppe JA, van Strijp JAG, et al. Distinct localization of the complement C5b-9 complex on gram-positive bacteria. Cell Microbiol. 2013;15:1955–68.
36. Pence MA, Rooijakkers SHM, Cogen AL, Cole JN, Hollands A, Gallo RL, et al. Streptococcal inhibitor of complement promotes innate immune resistance phenotypes of invasive M1T1 group a *Streptococcus*. J Innate Immun. 2010; 2:587–95.
37. Prajsnar TK, Renshaw SA, Ogryzko NV, Foster SJ, Serror P, Mesnage S. Zebrafish as a novel vertebrate model to dissect enterococcal pathogenesis. Infect Immun. 2013;81:4271–9.
38. Samant S, Lee H, Ghassemi M, Chen J, Cook JL, Mankin AS, et al. Nucleotide biosynthesis is critical for growth of bacteria in human blood. PLoS Pathog. 2008;4:e37.
39. Le Breton Y, Mistry P, Valdes KM, Quigley J, Kumar N, Tettelin H, et al. Genome-wide identification of genes required for fitness of group a *Streptococcus* in human blood. Infect Immun. 2013;81:862–75.
40. Jenkins A, Cote C, Twenhafel N, Merkel T, Bozue J, Welkos S. Role of purine biosynthesis in *Bacillus anthracis* pathogenesis and virulence. Infect Immun. 2011;79:153–66.
41. Polissi A, Pontiggia A, Feger G, Altieri M, Mottl H, Ferrari L, et al. Large-scale identification of virulence genes from *Streptococcus pneumoniae*. Infect Immun. 1998;66:5620–9.
42. Wang N, Ozer EA, Mandel MJ, Hauser AR. Genome-wide identification of *Acinetobacter baumannii* genes necessary for persistence in the lung. MBio. 2014;5:e01163–14.
43. Fuller TE, Kennedy MJ, Lowery DE. Identification of *Pasteurella multocida* virulence genes in a septicemic mouse model using signature-tagged mutagenesis. Microb Pathog. 2000;29:25–38.
44. Schwager S, Agnoli K, Kothe M, Feldmann F, Givskov M, Carlier A, et al. Identification of *Burkholderia cenocepacia* strain H111 virulence factors using nonmammalian infection hosts. Infect Immun. 2013;81:143–53.
45. Chaudhuri RR, Peters SE, Pleasance SJ, Northen H, Willers C, Paterson GK, et al. Comprehensive identification of *Salmonella enterica* serovar Typhimurium genes required for infection of BALB/c mice. PLoS Pathog. 2009;5:e1000529.
46. Mulhbacher J, Brouillette E, Allard M, Fortier L-C, Malouin F, Lafontaine DA. Novel riboswitch ligand analogs as selective inhibitors of guanine-related metabolic pathways. PLoS Pathog. 2010;6:e1000865.
47. Castro R, Neves AR, Fonseca LL, Pool WA, Kok J, Kuipers OP, et al. Characterization of the individual glucose uptake systems of *Lactococcus lactis*: mannose-PTS, cellobiose-PTS and the novel GlcU permease. Mol Microbiol. 2009;71:795–806.
48. Mainardi J-L, Fourgeaud M, Hugonnet J-E, Dubost L, Brouard J-P, Ouazzani J, et al. A novel peptidoglycan cross-linking enzyme for a beta-lactam-resistant transpeptidation pathway. J Biol Chem. 2005;280:38146–52.
49. Theilacker C, Sava I, Sanchez-Carballo P, Bao Y, Kropec A, Grohmann E, et al. Deletion of the glycosyltransferase *bgsB* of *Enterococcus faecalis* leads to a complete loss of glycolipids from the cell membrane and to impaired biofilm formation. BMC Microbiol. 2011;11:67.
50. Davlieva M, Zhang W, Arias CA, Shamoo Y. Biochemical characterization of cardiolipin synthase mutations associated with daptomycin resistance in enterococci. Antimicrob Agents Chemother. 2013;57:289–96.
51. Vollmer W, Joris B, Charlier P, Foster S. Bacterial peptidoglycan (murein) hydrolases. FEMS Microbiol Rev. 2008;32:259–86.
52. Mikalsen T, Pedersen T, Willems R, Coque TM, Werner G, Sadowy E, et al. Investigating the mobilome in clinically important lineages of *Enterococcus faecium* and *Enterococcus faecalis*. BMC Genomics. 2015;16:282.
53. Matos RC, Lapaque N, Rigottier-Gois L, Debarbieux L, Meylheuc T, Gonzalez-Zorn B, et al. *Enterococcus faecalis* prophage dynamics and contributions to pathogenic traits. Hughes D. PLoS Genet. 2013;9:e1003539.
54. Duerkop BA, Clements CV, Rollins D, Rodrigues JLM, Hooper LVA. Composite bacteriophage alters colonization by an intestinal commensal bacterium. Proc Natl Acad Sci. 2012;109:17621–6.
55. Goldberg SMD, Johnson J, Busam D, Feldblyum T, Ferriera S, Friedman R, et al. A Sanger/pyrosequencing hybrid approach for the generation of high-quality draft assemblies of marine microbial genomes. Proc Natl Acad Sci U S A. 2006;103:11240–5.
56. Li H, Durbin R. Fast and accurate short read alignment with burrows-wheeler transform. Bioinformatics. 2009;25:1754–60.
57. Tanimoto K, Ike Y. Complete nucleotide sequencing and analysis of the 65-kb highly conjugative *Enterococcus faecium* plasmid pMG1: identification of the transfer-related region and the minimum region required for replication. FEMS Microbiol Lett. 2008;288:186–95.
58. Bankevich A, Nurk S, Antipov D, Gurevich AA, Dvorkin M, Kulikov AS, et al. SPAdes: a new genome assembly algorithm and its applications to single-cell sequencing. J Comput Biol. 2012;19:455–77.
59. Li H, Handsaker B, Wysoker A, Fennell T, Ruan J, Homer N, et al. The sequence alignment/map format and SAMtools. Bioinformatics. 2009;25:2078–9.
60. Seemann T. Prokka: rapid prokaryotic genome annotation. Bioinformatics. 2014;30:2068–9.
61. Treangen TJ, Ondov BD, Koren S, Phillippy AM. The Harvest suite for rapid core-genome alignment and visualization of thousands of intraspecific microbial genomes. Genome Biol. 2014;15:524.
62. Kumar S, Stecher G, Tamura K. MEGA7: Molecular Evolutionary Genetics Analysis version 7.0 for bigger datasets. Mol Biol Evol. 2016;33:1870–4.
63. Zankari E, Hasman H, Cosentino S, Vestergaard M, Rasmussen S, Lund O, et al. Identification of acquired antimicrobial resistance genes. J Antimicrob Chemother. 2012;67:2640–4.
64. McClure R, Balasubramanian D, Sun Y, Bobrovskyy M, Sumby P, Genco CA, et al. Computational analysis of bacterial RNA-Seq data. Nucleic Acids Res. 2013;41:e140.
65. Burghout P, Zomer A, van der Gaast-de Jongh CE, Janssen-Megens EM, Françoijs K-J, Stunnenberg HG, et al. *Streptococcus pneumoniae* folate biosynthesis responds to environmental CO_2 levels. J Bacteriol. 2013;195:1573–82.

66. Goecks J, Nekrutenko A, Taylor J. Galaxy team. Galaxy: a comprehensive approach for supporting accessible, reproducible, and transparent computational research in the life sciences. Genome Biol. 2010;11:R86.
67. Langmead B, Salzberg SL. Fast gapped-read alignment with bowtie 2. Nat Methods. 2012;9:357–9.
68. Robinson JT, Thorvaldsdóttir H, Winckler W, Guttman M, Lander ES, Getz G, et al. Integrative genomics viewer. Nat Biotechnol. 2011;29:24–6.
69. Baldi P, Long ADA. Bayesian framework for the analysis of microarray expression data: regularized t -test and statistical inferences of gene changes. Bioinformatics. 2001;17:509–19.
70. Prajsnar TK, Cunliffe VT, Foster SJ, Renshaw SAA. Novel vertebrate model of *Staphylococcus aureus* infection reveals phagocyte-dependent resistance of zebrafish to non-host specialized pathogens. Cell Microbiol. 2008;10:2312–25.

7

A compendium of transcription factor and Transcriptionally active protein coding gene families in cowpea (*Vigna unguiculata* L.)

Vikram A. Misra[1], Yu Wang[1,2] and Michael P. Timko[1*]

Abstract

Background: Cowpea (*Vigna unguiculata* (L.) Walp.) is the most important food and forage legume in the semi-arid tropics of sub-Saharan Africa where approximately 80% of worldwide production takes place primarily on low-input, subsistence farm sites. Among the major goals of cowpea breeding and improvement programs are the rapid manipulation of agronomic traits for seed size and quality and improved resistance to abiotic and biotic stresses to enhance productivity. Knowing the suite of transcription factors (TFs) and transcriptionally active proteins (TAPs) that control various critical plant cellular processes would contribute tremendously to these improvement aims.

Results: We used a computational approach that employed three different predictive pipelines to data mine the cowpea genome and identified over 4400 genes representing 136 different TF and TAP families. We compare the information content of cowpea to two evolutionarily close species common bean (*Phaseolus vulgaris*), and soybean (*Glycine max*) to gauge the relative informational content. Our data indicate that correcting for genome size cowpea has fewer TF and TAP genes than common bean (4408 / 5291) and soybean (4408/ 11,065). Members of the GROWTH-REGULATING FACTOR (GRF) and Auxin/indole-3-acetic acid (Aux/IAA) gene families appear to be over-represented in the genome relative to common bean and soybean, whereas members of the MADS (Minichromosome maintenance deficient 1 (MCM1), AGAMOUS, DEFICIENS, and serum response factor (SRF)) and C2C2-YABBY appear to be under-represented. Analysis of the AP2-EREBP APETALA2-Ethylene Responsive Element Binding Protein (AP2-EREBP), NAC (NAM (no apical meristem), ATAF1, 2 (Arabidopsis transcription activation factor), CUC (cup-shaped cotyledon)), and WRKY families, known to be important in defense signaling, revealed changes and phylogenetic rearrangements relative to common bean and soybean that suggest these groups may have evolved different functions.

Conclusions: The availability of detailed information on the coding capacity of the cowpea genome and in particular the various TF and TAP gene families will facilitate future comparative analysis and development of strategies for controlling growth, differentiation, and abiotic and biotic stress resistances of cowpea.

Keywords: Cowpea, Common bean, Phylogenetic analysis, Soybean, Transcription factor

* Correspondence: mpt9g@virginia.edu
[1]Department of Biology, University of Virginia, Gilmer Hall 044, Charlottesville, VA 22904, USA
Full list of author information is available at the end of the article

Background

Cowpea (*Vigna unguiculata L. Walp*) is an important grain legume in the sub-tropics and the most important food and forage legume in sub-Saharan Africa [1, 2]. Estimates by the Food and Agriculture Organization (FAO) of the United Nations [3] indicate that 5.59 million metric tons of cowpea was produced worldwide, the majority of which (81%) is produced by low-input subsistence farmers in Western Africa [3–5], followed by Eastern (8.68%) and Central Africa (4.37%) [3]. In these regions cowpea grains are an important source of protein and carbohydrates [5, 6]. In addition to the fruit, leaves are also eaten [7], and cowpea stems are an effective fodder for livestock [8]. Moreover, cowpea can be used to restore nitrogen to soils [9], making it an effective companion crop to cereals [10, 11]. Furthermore, cowpea can withstand dry conditions and low quality soils relatively well [12].

Like all plants, cowpea faces a myriad of challenges from abiotic and biotic factors that constrain its growth and productivity [5, 6, 13–16]. Among the most significant stresses are attacks from root parasitic angiosperms [5], drought and increased soil salinity [17], which lead to significant or even total losses of yield [18–20]. Despite the social and economic importance of the crop, until recently genomic scale information was not available for cowpea putting its improvement at a disadvantage relative to other legumes such as soybean, common bean, and chickpea [21–25]. Initial attempts to capture genomic scale information [26] using reduced representation cloning and sequencing provided information on about 70% of the estimated 620 Megabase (Mb) genome, including information of transcription factors and resistance related genes. Recently, a draft genome sequence assembly providing 65× coverage has been reported by Muñoz-Amatriaín et al. (2017) [27] making possible a much more robust analysis of the informational content of this species.

Prior studies on the genomic contents of plants have tackled uncovering the content and complexity of transcription factors (TFs) and transcriptionally active proteins (TAPs, syn. Transcription associated proteins) present in the genome. Beginning with the first genome-scale analyses of TFs and TAPs in *Arabidopsis thaliana* almost two decades ago (Riechmann et al., 2000) [28], genome scale studies of these important regulatory molecules have appeared for a wide variety of plant species including rice (Gao et al., 2006) [29], poplar (Zhu et al., 2007) [30], soybean (Schmutz et al., 2010) [31] and other legumes (Richardt et al., 2007; Udvardi et al., 2007 [32, 33]) and tobacco (Rushton et al., 2008) [34].

Comparative information on TF and TAP content in genomes can be found in various databases based upon different discovery pipelines (e.g., PlantTAPDB (Plant Transcription Associated Protein Database) (Richardt et al., 2007) [32], PlnTFDB (Plant Transcription Factor Database) (Riano-Pachon et al., 2007; Perez-Rodriguez et al., 2010) [35, 36], PlantTFDB (Plant Transcription Factor Database) [37], GreenPhylDB (Rouard et al., 2011, 2014; http://www.greenphyl.org/cgi-bin/index.cgi) [38, 39], and PlantTFcat pipeline [40]). Some of these databases identify only TFs, while others include transcription regulators (TRs) and TAPs, and chromatin remodelers (CRs). In some cases, the pipelines do not include the range of all known TFs and TAPs. The most comprehensive data to date on the TF and TAP content of legumes can be found in the iTAK Plant Transcription Factor & Protein Kinase Identifier and Classifier database [41, 42], which contains TF and TAP content of 74 plant species, including the legumes *Medicago truncatula*, *Lotus japonicus*, chickpea, soybean, pigeon pea and common bean, but not cowpea. At present the only information on TFs for cowpea exists on the *Vigna unguiculata* Gene Expression Atlas (VuGEA) database [43], but this information relies solely on transcriptomic data.

Therefore, to address the lack of information available for this species, we have utilized the recently published draft genome assembly for cowpea [27] and applied existing and novel computation pipelines to create a comprehensive dataset of TFs and TAPs for cowpea. We also compared the genomic content of cowpea with two of its evolutionarily close relative legumes, common bean (*Phaseolus vulgaris*) and soybean (*Glycine max*). We also highlight information on several selected TF families in cowpea involved in stress responses and compare the content and phylogenetic organization of these families in cowpea to their counterparts in common bean.

Results

Identification of TFs and TAPs

A draft genome assembly has been generated that provides 67X coverage of the estimated 620 Mb cowpea genome [27]. The assembly includes 39 Gb of Illumina GAII (Genome Analyzer II) paired-end sequences (70–130 base), and ~250,000 gene-space sequences (GSSs) (average length of 609 nucleotides). About 97% of all previously reported cowpea expressed sequence tags (ESTs) can be found in the assembly by Basic Local Alignment and Search Tool Nucleotide (BLASTN) and a large proportion of the assembly is composed of scaffold sequences containing two or more overlapping contigs [27]. To identify genes encoding TFs and TAPs present in the draft cowpea genome assembly, we used three different identification pipelines: the PlantTFcat pipeline [40], the iTAK pipeline [41] and a novel pipeline developed for this study that uses a strict set of rules for gene family membership as defined by Lang et al. (2010) [44]. The latter pipeline is capable of identifying 111 TF and TAP families and is based on 223 rules (134 "mandatory" and 89 "forbidden") focused on the presence/absence of specific domains in certain families. In this case, 124 domain hidden Markov models (HMMs) and 108 domains obtained from the Pfam protein family database were used to identify sequences as TFs and TAPs. This pipeline originally used the TavernaPBS software [45], which was made using the Taverna software [46] to identify and characterize TFs and TAPs on a PBS (Portable Batch System) cluster (Additional file 1). We adapted this pipeline into a Bash shell script that we developed to run on a SLURM (Simple Linux Utility for Resource Management) cluster.

Using these three approaches, we identified a total of 5460 TF- and TAP-encoding domains from 4416 sequences falling into 136 families. Multiple TF-encoding regions came from the same sequence in part due to the translation of TF-encoding transcript sequences to protein, which may have yielded different TFs on different reading frames. When sequences with multiple open reading frames (ORFs) are taken into account, these 5460 TFs come from 4416 sequences, which represent 7.26% of the 60,838 transcript genes (56,626 of which directly code for protein) in the 620 Mb of the cowpea genome. We also applied these same rules and pipelines to identify the TFs and TAPs in two evolutionarily close relatives, common bean and soybean. The number of TFs and TAPs found were 6468 and 13,419 sequences, respectively, for common bean and soybean. At 4416 TF-encoding sequences, cowpea has ~ 31.7% fewer TF/TAPs than common bean, which, like cowpea, is diploid, and only 32.9% of the content of soybean, a tetraploid legume (Additional file 2a). When only the genes and not the gene models are taken into account, the 4416 TF sequences together come from 4408 TF genes (Additional file 2b), and common bean and soybean have 5291 and 11,065 TF genes, respectively, making the number of cowpea TF/TAP genes 16.7% smaller than in common bean, and only 39.8% as many as in soybean (Additional file 2b). Thus, the number of TFs and TAPs in cowpea is relatively small for a diploid legume.

In order to assess the quality of our identification and classification approach we compared our results with selected publications in which detailed analyses had been carried out for TF/TAP families of other plant species. The combination of the three pipelines used in this study suggests that the TFs and TAPs in cowpea is relatively small in number of TFs and TAPs to other diploid plants, and that in terms of percentage of protein-coding genes, cowpea has a small proportion of TFs and TAPs. Compared to data from other databases, though, cowpea (and related legumes) has a relatively large number of TFs, although this could be due to differences between the pipelines used in other databases and the combination of three pipelines used in this study. The cowpea TF and TAP repertoires resulting from analyses using the PlantTFcat, iTAK and the TavernaPBS pipelines individually are shown in Additional file 3. According to data on the iTAK Plant Transcription Factor & Protein Kinase Identifier and Classifier Database [41, 42], the adzuki bean genome (*Vigna angularis*) [47], has 2755 total TF-encoding and TAP-encoding genes (2260 TFs and 495 TRs) across 92 families [42]; since adzuki bean has 26,857 genes, TFs and TAPs account for approximately 10.3% of protein-coding genes in adzuki bean. The common bean genome from Schmutz et al. (2014) [48], which consists of 27,197 genes, is found on iTAK to have 2779 TFs and TAPs (2314 TFs and 465 TRs), or 10.2% of genes, across 89 families [42]. *Medicago truncatula* genome v4.0v1 [49], which has 50,894 genes, is found on iTAK to have 3670 TFs and TRs (2948 TFs and 722 TRs), or 7.2% of genes, across 89 families [42].

Others who studied the repertoire of cowpea TFs, TRs and CRs did not find data as fully comprehensive as the data found in this study. PlantTFDB v4.0 [37] contains only 488 TFs from 48 families in cowpea [37]. The VuGEA database [43], which used the PlantTFcat pipeline to identify TFs, TRs and CRs in the cowpea transcriptome, found 2485 TFs, TRs and CRs out of 24,866 cowpea unigenes (10% of unigenes). Thus, this study represents the most comprehensive study yet on cowpea TFs and TAPs.

Here, it must be noted that PlantTFcat, iTAK and the TavernaPBS pipeline used in this study were used to analyze the cowpea protein assembly, as well as the cowpea transcripts assembly. For comparison purposes, all three pipelines were also used to analyze the raw cowpea assembly. According to the PlantTFcat pipeline, the largest TF families in the raw cowpea genome assembly are Zinc-finger, CCHC-type (CCHC(Zn)) (2548), C2H2 (702) and WD-40-like (449) (Additional file 3). According to the iTAK pipeline, the largest three families in the raw genome are Myeloblastosis-related (MYB-related) (249), WRKY (175) andAP2/ERF-ERF (syn. ERF, ethylene response factor) (173). According to the TavernaPBS pipeline, MYB-related (396) and WRKY (302) are the second and third largest families behind B3 (known in the TavernaPBS pipeline as ABI3/VP1 (Abscisic Acid Insensitive 3 / Viviparous1)) (409). Here, it must be noted that the raw cowpea genome statistics on TF families is significantly different from those for the cowpea protein and transcript assemblies, with some families that are under-represented in the raw cowpea genome as opposed to the protein and transcript assemblies (e.g., ARF (Auxin Response Factor)) (Additional file 2), and other families being represented in the raw cowpea genome and not found in protein and transcript assemblies (e.g., Rel (Relish)). The presence of TF families in the raw cowpea genome and not the protein and transcript families could be an effect of the MAKER [50] and AUGUSTUS [51, 52] methods of annotation used for the cowpea genome v0.03 (see Materials and Methods). Due to the un-curated nature of the raw cowpea genome assembly, the TF families only found in the raw cowpea genome are not included in the TF statistics in Additional file 2.

In this study, several families were found that were not present in VuGEA, due to the use of iTAK and the TavernaPBS pipeline in this study. Such families include the NF-X1 (nuclear factor X-box binding 1), NF-YC (nuclear factor Y subunit C) (syn. CCAAT-HAP5 (CCAAT motif, heme-associated protein 5), SOH1 (suppressor of hyper-recombination 1), Rel, RF-X (regulatory factor X) and zn-clus (Zn(2)-Cys(6) binuclear cluster domain). In the cowpea protein and transcript assemblies, the NF-X1 (1 member), CCAAT-HAP5 (16), and SOH1 (1) are represented. In the raw cowpea genome assembly, members of four families were found that were not found in the cowpea protein or transcript assemblies. These families were JmjC-ARID (Jumonji C-terminal, AT-Rich Interaction Domain) (1), Rel (40), RF-X (40) and zn-clus (1).

It must be noted that in VuGEA, 3 cowpea transcripts were reported as belonging to the family ABTB (Ankyrin Broad Complex, tramtrack and bric a brac) (a sub-family of TRAF (Tumor necrosis factor receptor-associated factor)), and 2 others as members of CW-Zn-B3_VAL (CW-like zinc finger, B3, VP1/ABI3-Like) [53]. None of the pipelines in this study have found members of either family in cowpea, but our pipelines have found that 5 ABTB family members and 7 members of the CW-Zn-B3_VAL family exist in common bean. Selected common bean ABTB and CW-Zn-B3_VAL nucleotide sequences were used as queries against the raw cowpea v0.03 assembly in a FASTA search (FASTA version 36.3.8e [54]; E-value 10e-3). The FASTA search yielded 36 raw cowpea sequences that were homologous to ABTB and 28 raw sequences homologous to CW-Zn-B3_VAL (E-value < 10e-3).

When the cowpea protein and transcript assemblies are taken into account, the largest TF/TAP family in cowpea is C2H2 (511, 11.57% of cowpea repertoire), followed by Polycomb Group Fertilization-Independent Endosperm (PcG_FIE) (i.e., WD-40) (462, 10.46%), MYB-HB-like (MYB Homeobox like) (311, 7.04%), and basic helix-loop-helix (bHLH) (214, 4.85%) (Fig. 1a, Additional file 2a). Similar to cowpea, the three largest TF families in common bean are C2H2 (974, 15.06% of bean TF repertoire), PcG_FIE (489, 7.56%), and MYB-HB-like (468, 7.24%). This was also true of soybean, with C2H2 (2144, 15.98% of soy TF repertoire), followed by PcG_FIE (1208, 9.00%), and MYB-HB-like (1153, 8.59%) being the three largest families. It is worth noting, though, that of these families, cowpea is under-represented in C2H2 and MYB-HB-like, and over-represented in PcG_FIE.

When only the genes and not the gene models are taken into account, the largest cowpea TF/TAP family is C2H2 (416, 9.44%), PcG_FIE (307, 6.96%), MYB-HB-like (240, 5.44%), bHLH and CCHC_Zn (155 each, 3.52% each) (Fig. 1b, Additional file 2b). For both common bean and soybean, C2H2 is the largest TF family (699 genes in common bean, 1253 genes in soybean). However, in common bean and soybean, unlike in cowpea, MYB-HB-like is the second largest (322 in common bean and 654 in soybean), and not the third largest family (which is PcG_FIE; 315 in common bean and 632 in soybean) (Additional file 2b).

Based on our prediction, Dicer is not represented in cowpea, but is represented in common bean and soybean. According to that same pipeline, CCAAT-Down regulator of transcription 1 (CCAAT-Dr1), Runt and transcriptional enhancer activator (TEA) are represented in cowpea (with each having one sequence), but not in common bean.

In determining whether certain cowpea TF families were over- or under-represented in relation to common bean and soybean, we compared cowpea TF families to common bean and soybean with respect to percentage of the respective TF repertoires. In order to determine whether or not the over- or under-representation was affected by gene number, we used two other criteria to compare cowpea TF repertoires to those of common bean and soybean: 1) raw numbers of TFs, and 2) percent of total protein-encoding transcript sequences. Over- or under-representation of a certain TF family in cowpea is depicted as a ratio; for example, the number of members of a TF family in cowpea divided by the number of members of that same TF family in common bean or soybean. If the ratio was between 0.9 and 1.1 for a certain family, that family was determined to be neither over- nor under-represented in cowpea. If the ratio exceeded 1.1, that family was determined to be over-represented in cowpea. If such ratios were smaller than 0.9, that family was under-represented in cowpea. The results for the comparisons between cowpea and common bean and soybean TF families based on these ratios are shown in Additional files 2 and 4.

Here it is important to note that common bean has 36,995 protein-coding transcripts and a genome size of 587 Mb [34], and that soybean has 88,647 protein-coding transcripts and a genome size of approximately 978.5 Mb [55]. For the statistics represented in Additional files 2 and 4, because both protein and transcripts are studied, and because in cowpea the protein coding sequences (56,626) are a subset of the set of transcripts (60,838), the percentage of total genes is represented by percentage of protein-coding transcripts.

When each cowpea TF family was compared to its counterparts in common bean and soybean with respect

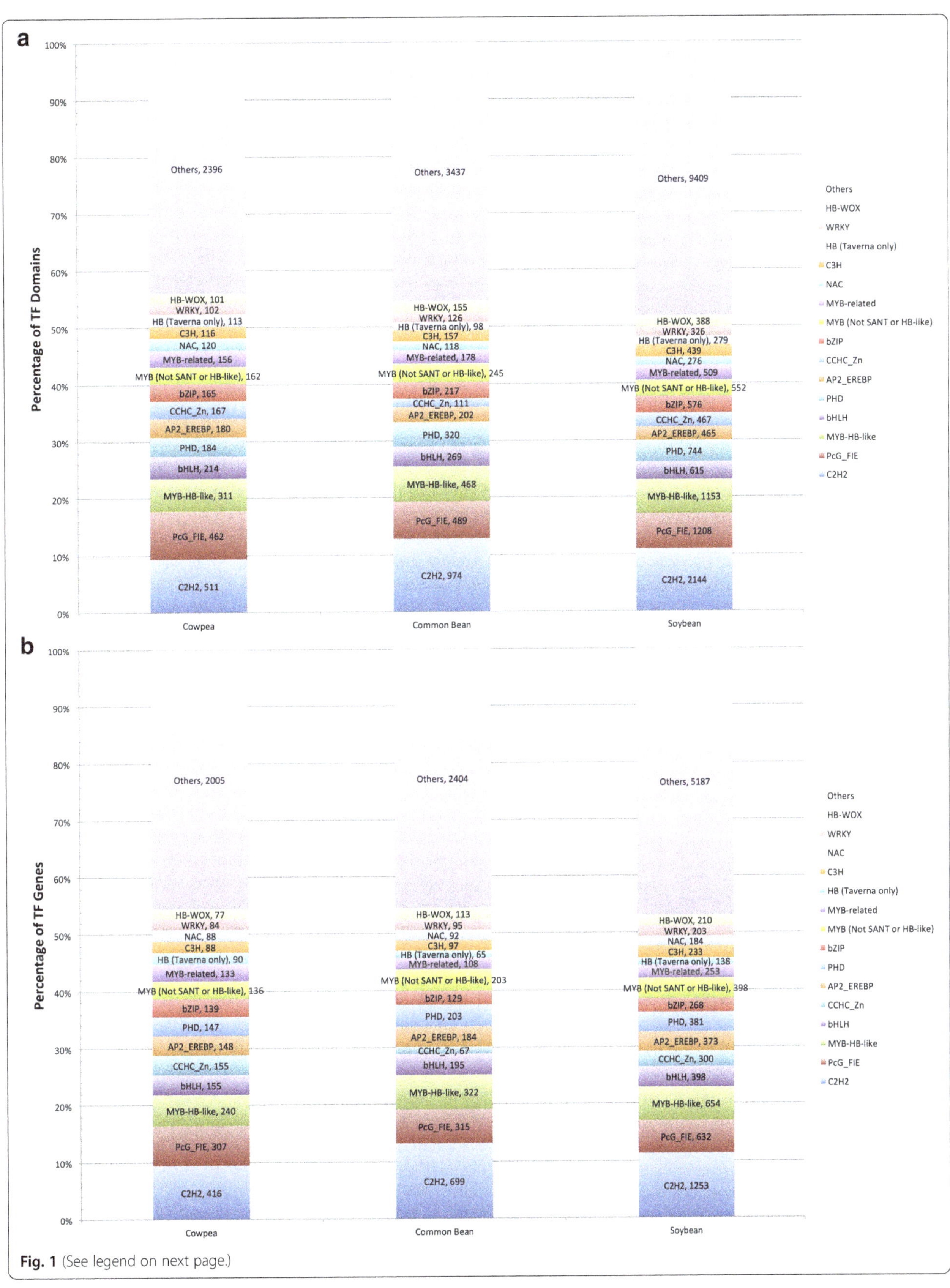

Fig. 1 (See legend on next page.)

(See figure on previous page.)
Fig. 1 The largest cowpea TF and TAP families, and their sizes in common bean and soybean. Domains from all gene models are counted in (**a**) whereas genes (not counting gene models) are counted in (**b**). The largest fifteen families in cowpea account for approximately 56% of the cowpea TF, TR and TAP domains (and 54.5% of genes), whereas in soybean, these same families account for almost 52% of domains (and 53.1% of genes). Moreover, the over-representation of families such as PcG_FIE and the under-representation of families such as C2H2 in cowpea compared to common bean and soybean is shown

to percentage of their respective TF/TAP repertoires, common bean and soybean showed significant difference from cowpea (Additional files 2 and 4a).

When cowpea sequences (with all gene models accounted for) were compared to the diploid common bean on the basis of percentage of TF repertoire, cowpea was found to have 51 families under-represented, 31 families similarly proportioned, and 50 TF families over-represented compared to common bean. Also on the basis of percentage of TF repertoire, when only genes and not gene models are taken into account, results are similar to when gene models are taken into account. Cowpea, when compared to common bean, is under-represented in 51 families, similarly represented in 26 families, and over-represented in 55 families.

These results suggest that despite both cowpea and common bean being diploid legumes, that the cowpea TF repertoire is significantly different from that of common bean.

When cowpea sequences (with all gene models) were compared to the tetraploid soybean on the basis of percentage of TF repertoire, cowpea was found to have 29 TF families over-represented, 20 families similarly proportioned, and 87 TF families under-represented in terms of percentage of TF repertoire. While this result may be expected due to cowpea being diploid and soybean being tetraploid, it is important to note the presence of a significant proportion of cowpea TF families that were over-represented compared to both common bean and soybean. Of the 50 cowpea TF families that were over-represented compared to common bean, 21 were over-represented and 13 were similarly proportioned compared to their counterparts in soybean. When only genes are taken into account, compared to soybean, cowpea is under-represented in 48 families, similarly represented in 33 families, and over-represented in 55 families. Of the 55 TF families that were over-represented in common bean, 42 were over-represented and 4 were similarly proportioned compared to their counterparts in soybean. This presence of these families shows that the cowpea TF repertoire is unique in composition (Additional file 2).

Moreover, when each cowpea TF family was compared to its counterparts in common bean and soybean with respect to raw number of TFs, all but five of the cowpea TF families were found to be under-represented compared to soybean, while 15 cowpea TF families were over-represented, 19 families were similarly proportioned, and 98 were under-represented compared to common bean (Additional file 4b). This is expected since cowpea and common bean are diploid and soybean is tetraploid.

When the raw number of TFs and TAPs in cowpea and soybean were compared in such a way that each number of TFs in soybean was halved to provide a comparison that accounted for the tetraploidy of soybean and the diploidy of cowpea, cowpea was found to have 8 TF families that were over-represented compared to soybean, 7 families similarly proportioned to their counterparts in soybean, and the remaining 121 cowpea TF families were under-represented compared to soybean (Additional file 2). Here it is important to note that the number of soybean TF and TAP sequences divided in half is 6710, which is similar to the number of TFs in common bean (6468), while cowpea has 4416 TF and TAP sequences. Thus a comparison of cowpea to common bean and soybean can most accurately be based on percentage of their respective TF repertoires.

When each cowpea TF family was compared to its counterparts in common bean and soybean with respect to percentage of total protein genes, cowpea was, compared to common bean and soybean, under-represented in almost all TF families (Additional file 4c). This is consistent with expectations because in cowpea, TFs comprise 7.26% of all protein coding genes, which is a small percentage compared to common bean (17.5%) and soybean (15.1%) (Additional file 2).

These results suggest that the significant differences between cowpea TF families and their counterparts in fellow legumes common bean and soybean in terms of percentage of TF repertoire are not significantly affected by differences in gene number or genome size. For this reason, in all statistical analyses in which only the genes and not gene models are counted, any comparison between cowpea, common bean and soybean are done in terms of percentage of TF repertoire. Cowpea TF families whose size and complexity differ with respect to common bean and soybean have been identified (Additional file 2). For example, when all gene models are counted, cowpea TF families that were over-represented in cowpea compared to common bean are PcG_FIE (462), followed by CCHC(Zn) (167) and NAC (120). The largest cowpea TF families that were similarly proportioned to their counterparts in common bean are bHLH (214), AP2-EREBP (180) and basic leucine zipper domain (bZIP) (165). The C2H2 (511), MYB-HB-like (311) and Plant Homeodomain (PHD) (184) families are the largest cowpea TF families that are under-represented with respect to common bean. Similarly, in a comparison of TF families between cowpea and soybean based on percentage of respective TF repertoires, PcG_FIE (462), followed by AP2-EREBP (180) and NAC (120) were the largest cowpea TF families that were over-represented compared to soybean. Of the cowpea TF families that are similarly proportioned to their counterparts in soybean, bHLH (214), CCHC(Zn) (167) and MYB-related (156) are the largest. C2H2 (511), MYB-HB-like (311) and PHD (184) are the largest cowpea TF families to be under-represented compared to soybean (Additional file 2a).

When only genes and not gene models are taken into account, the largest cowpea families over-represented compared to common bean are PcG_FIE (307), CCHC_Zn (155) and bZIP (139). The largest cowpea families similarly proportioned to their counterparts in common bean are bHLH (155), AP2_EREBP (148), and C3H (88). The largest cowpea families under-represented compared to common bean are C2H2 (416), MYB-HB-like (240) and PHD (147). The largest cowpea families over-represented compared to soybean are PcG_FIE (307), CCHC_Zn (155), and bZIP (139). The largest cowpea families similarly proportioned to their counterparts in soybean are MYB-HB-like (240), bHLH (155) and AP2-EREBP (148). The largest cowpea families under-represented compared to soybean are C2H2 (416), MYB (136) and CCAAT_HAP3 (62) (Additional file 2b).

Interestingly, according to the TavernaPBS pipeline, some families, such as CCAAT-HAP3 (CCAAT motif, heme-associated protein 3) (syn. NF-YB (nuclear factor Y subunit B)), are fewer in number than expected (Additional file 2), especially given that other organisms were found to have several CCAAT-HAP3 sequences, like the fifteen CCAAT-HAP3 sequences found in tobacco [34]. This could be due to an artifact in the TavernaPBS pipeline used in this study to identify and classify TFs: when the pipeline found a sequence with required domains for two or more TF families, the pipeline would sometimes not assign a TF family to the sequence. For example, a sequence with similarity to an NF-YB domain, a required domain of the CCAAT-HAP3 TF family, may have also been found to have similarity to an NF-YC domain, a domain required for CCAAT-HAP5 [30]. In this situation, the pipeline did not classify the sequence into any particular TF family. One possible future approach to improving the TavernaPBS pipeline used in this study is to assign a sequence to a family based on the TF domain to which it has the strongest similarity.

In this study, we compensated for the artifacts in the TavernaPBS pipeline by incorporating statistics from a search for cowpea TFs using the PlantTFcat [40] and iTAK pipelines [41]. These pipelines yielded statistics for CCAAT-HAP3 and CCAAT-HAP5 that were closer to expected for a diploid legume (Additional files 2 and 3).

Phylogenetic analysis of TF families

The AP2-EREBP, NAC and WRKY families were chosen for deeper analysis due to their involvement with defense response in plants [25, 56, 57] and because each family has easily recognizable conserved domains [56, 58–60]. After a multiple sequence alignment using MAFFT (Multiple Alignment using Fast Fourier Transform) L-ins-i version 7.245 [61], a phylogenetic analysis using maximum likelihood in RAxML (Randomized Axelerated Maximum Likelihood) [62] with 100 bootstrap replicates was performed on each of the three families. In these phylogenetic analyses, amino acid sequences, and therefore all gene models, are used.

AP2-EREBP family

The AP2-EREBP superfamily first described by Ohme-Takagi & Shinshi (1995) [58] in tobacco controls multiple processes in plants from development [63] to defense against abiotic and biotic stresses [64, 65]. The AP2-EREBP superfamily consists of three families: ERF, Related to ABI3/VP1 (RAV) and AP2 [66, 67]. The ERF family has distinguishing conserved motifs such as an N-terminal AEIRD motif and a WLG [66]. The RAV family has an AP2 and a B3 domain [67]. The AP2 family usually has two AP2 domains [66]. The first (N-terminal) domain, known as the R1 repeat, usually has a YEAH or WESHI at the 5′ end and a YDRAA or LAALKY at the 3′ end, whereas the second (C-terminal) domain, known as the R2 repeat, has a WQAR or WEAR at the 5′ end and a NAVT or YDIAAI at the 3′ end [66, 68]. The AP2 family also has a conserved YLG instead of the WLG and AEIRD found in ERFs [66, 68].

The *Arabidopsis* and rice sequences were chosen in order to facilitate the classification of AP2-EREBP sequences into clades. The phylogenetic tree for the AP2-EREBP family (shown in Figs. 2 and 3) include 122 AP2-EREBP cowpea sequences along with the 22 *Arabidopsis* and 15 rice sequences representative of each clade of the AP2-EREBP superfamily. When cowpea ERFs were grouped according to the grouping scheme in Dietz et al. (2010) [69], namely groups Dehydration-responsive element (DRE)-binding (DREB) A1-A6 and ERF B1-B6, the Dietz et al. (2010) [69] and Sharoni et al. (2011) [70] classification was found not to be entirely consistent with the classification in Nakano et al. (2006) [64]. For example, group II in Nakano et al. (2006) [64] is supposed to contain only members of the DREB A-4 clade. Instead, group II sequences contained DREB-A4 and DREB-A5 sequences (Figs. 2 and 3). Moreover, group V in Nakano et al. (2006) [64] contained members of ERF B-2 and ERF B-6 from Dietz et al. (2010) [69]

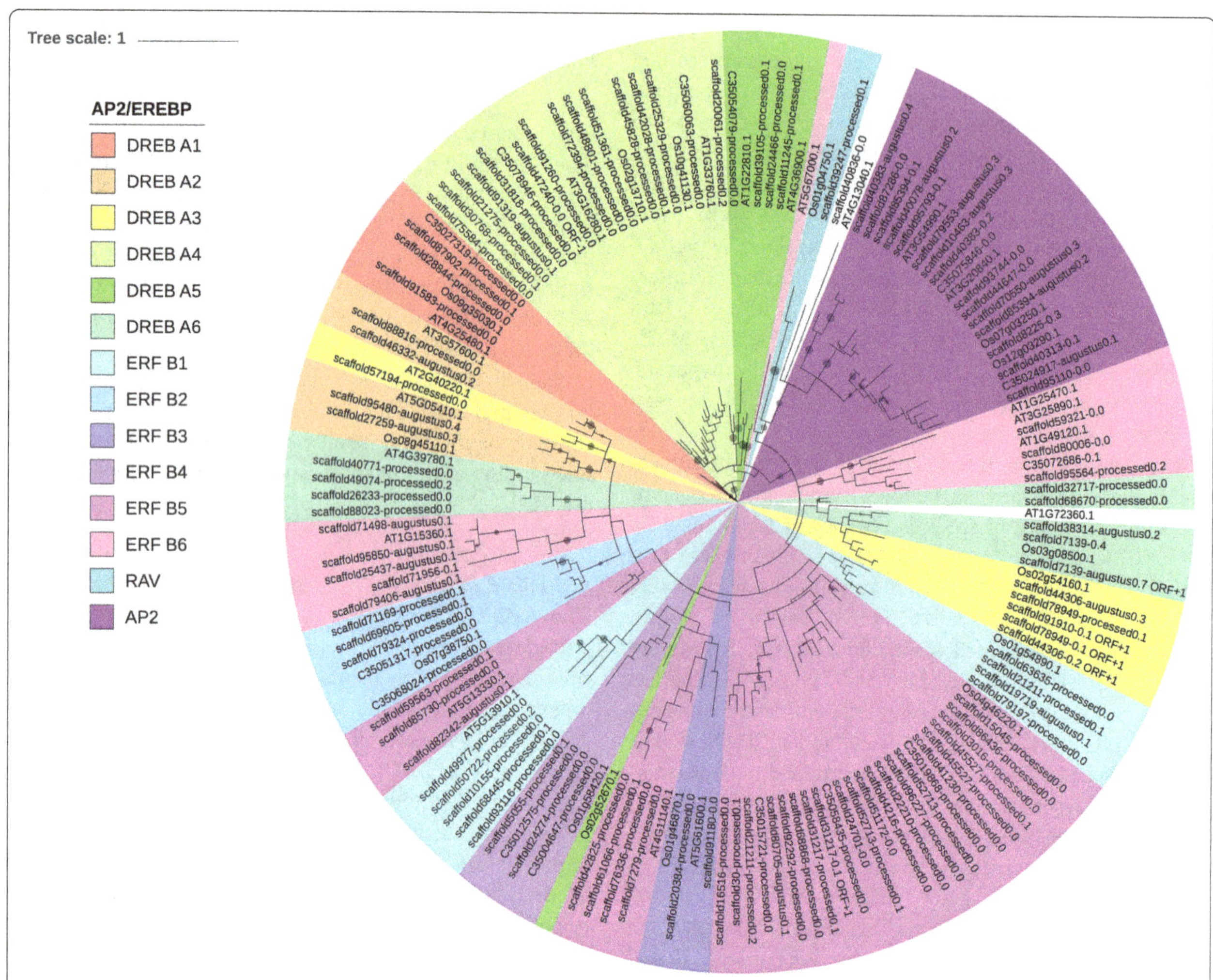

Fig. 2 Cowpea AP2-EREBP genes according to the DREB A1-A6 and ERF B1-B6 grouping. This grouping follows the classification schemes of Dietz et al. (2010) [69] and Sharoni et al. (2011) [70]. The trees were generated using RAxML [62] with 100 bootstrap values with the optimal amino acid substitution model automatically chosen in RAxML (i.e., the PROTGAMMAAUTO option). The circles on the branches are bootstrap support values from 50 to 100, with the largest circles representing the greatest bootstrap support

and Sharoni et al. (2011) [70] (Figs. 2 and 3). According to Nakano et al. (2006) [64], group V ERFs should only contain sequences from the ERF B-6 clade. Here it must be noted that Nakano et al. (2006) [64] divided *Arabidopsis* AP2-EREBP TFs into 12 clades and AP2-EREBP members in rice into 15 clades, and that the clades in this study are the 12 clades found in *Arabidopsis*. 11 of these clades are held in common between monocots and dicots [64].

NAC family

Methods of classifying NAC TFs vary greatly with one classification scheme by Ooka et al. (2003) [71] separating the family into two broad categories (I and II) with category I consisting of 15 groups named after individual members (e.g., *Arabidopsis* transcription activation factor 2 (ATAF2), Senescence Upregulated 5 (Senu5)) and category II containing 3 groups (e.g., ONAC003 (*Oryza sativa* NAC 003)). Rushton et al. (2008) [34] simplified the organization defining six clades (1 through 6) in most species and 3 clades unique to Solanaceae. Zhu et al. (2012) [72] classify NACs into ten numbered groups (I-X), some of which contain several subgroups. According to the classification scheme different species will have NAC families with different groups, depending on whether the plant is a monocot, dicot, moss or lycophyte, with dicots usually having Groups Ia-c, II, IIIa-c, IVa-d, Va(1), Va(2), Vb, VIa, VIc, VII, and VIII. All of these groups, with the exception of groups IVb and VIII, are found in the cowpea NAC family (Fig. 4).

Some more subtle differences exist between the cowpea NAC family organization found here and that reported by Zhu et al. (2012) [72] for dicots. First, Group II in cowpea does not sit on the same location within the NAC tree (i.e., between Groups I and III) as reported in Zhu et al. (2012) [72]. Second, Group VIa in cowpea

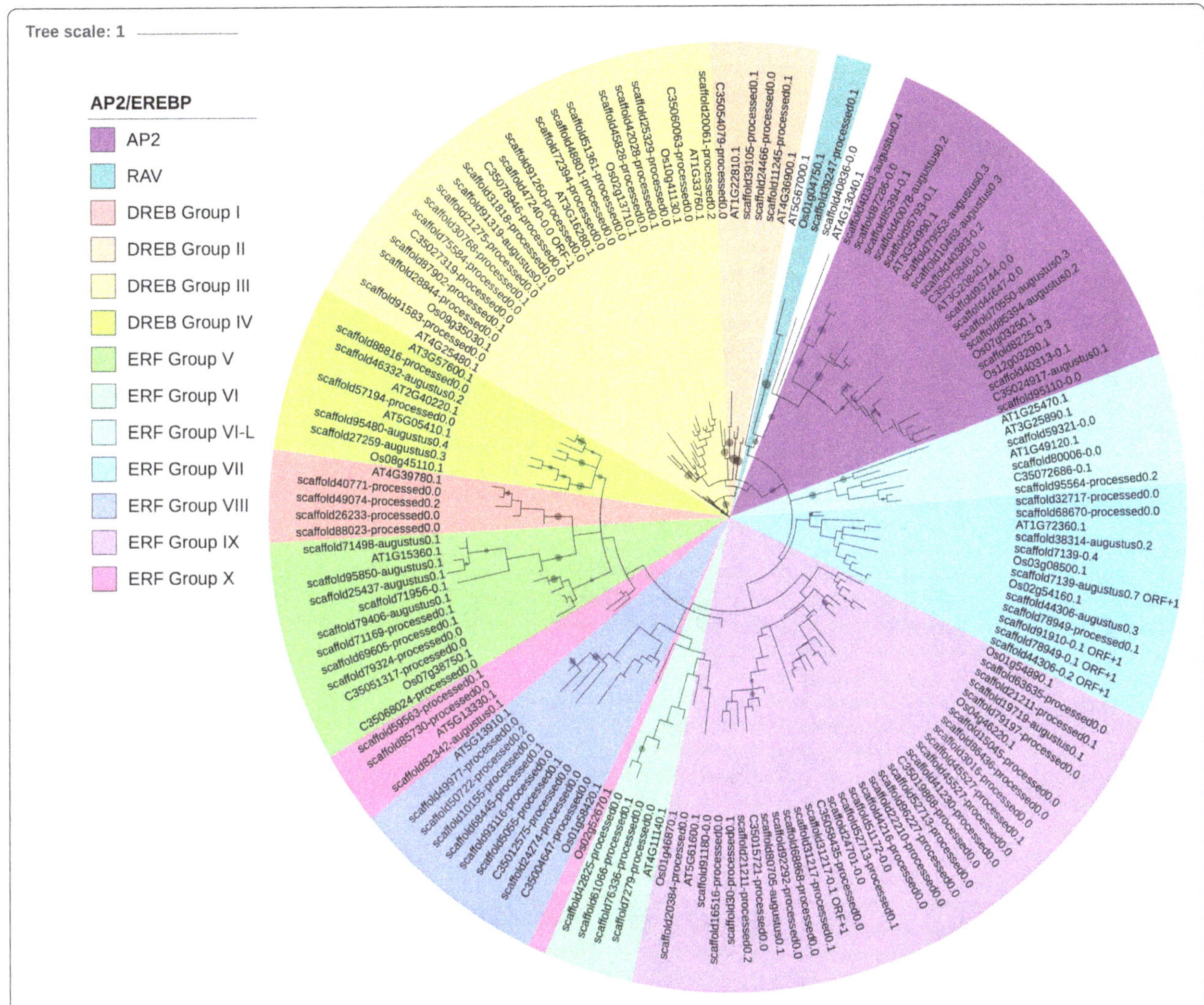

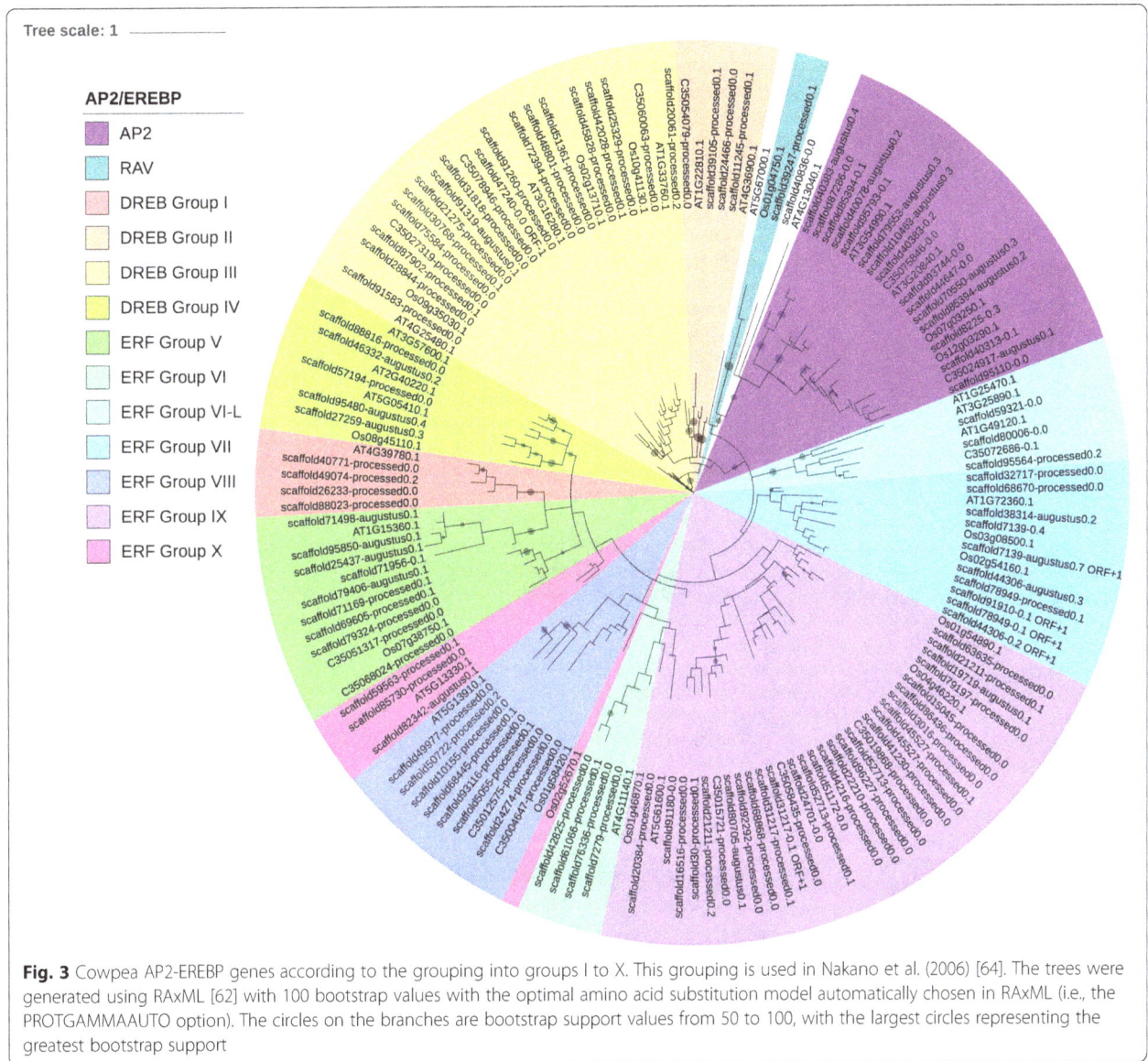

Fig. 3 Cowpea AP2-EREBP genes according to the grouping into groups I to X. This grouping is used in Nakano et al. (2006) [64]. The trees were generated using RAxML [62] with 100 bootstrap values with the optimal amino acid substitution model automatically chosen in RAxML (i.e., the PROTGAMMAAUTO option). The circles on the branches are bootstrap support values from 50 to 100, with the largest circles representing the greatest bootstrap support

groups next to IIIa, which is consistent with the Bayesian tree in Zhu et al. (2012) [72], but not the maximum likelihood tree from that same paper.

The WRKY family

The WRKY TF family is characterized by a conserved N-terminal WRKYGQK motif and a motif resembling a zinc finger [59, 73]. Variations in the conserved parts of the WRKY domain allow separation into three major groups (Groups I-III) [59]. Group I WRKY TFs have two WRKY domains, with their C-terminal domains being functionally distinct from the N-terminal domains [74]. Group II is the most variable group in terms of amino acid sequence, with five subgroups designated IIa –IIe [59, 60, 74]. Group III WRKYs differ in zinc finger structure from group-I and -II WRKY TFs; a group III WRKY zinc finger has a C_2-HC structure, as opposed to the C_2-H_2 in the other two WRKY groups [59, 60].

The relationship among the three groups of WRKY TFs can also be seen in the phylogenetic clustering of family members as depicted in Fig. 5. Notable about the tree are the following. Subgroups IId and IIe cluster together, similar to what was reported initially by Timko et al. (2008) [26] for the family. However, unlike the early phylogeny [26], subgroup IIb does not split into two distinct clades. The most likely explanation is that the earlier study had incomplete sequence data and the split was an artifact in the ClustalW alignment caused by truncated WRKY domains. This split did not happen in this study, probably because the alignment used in this study was based on MAFFT L-ins-i [61]. In

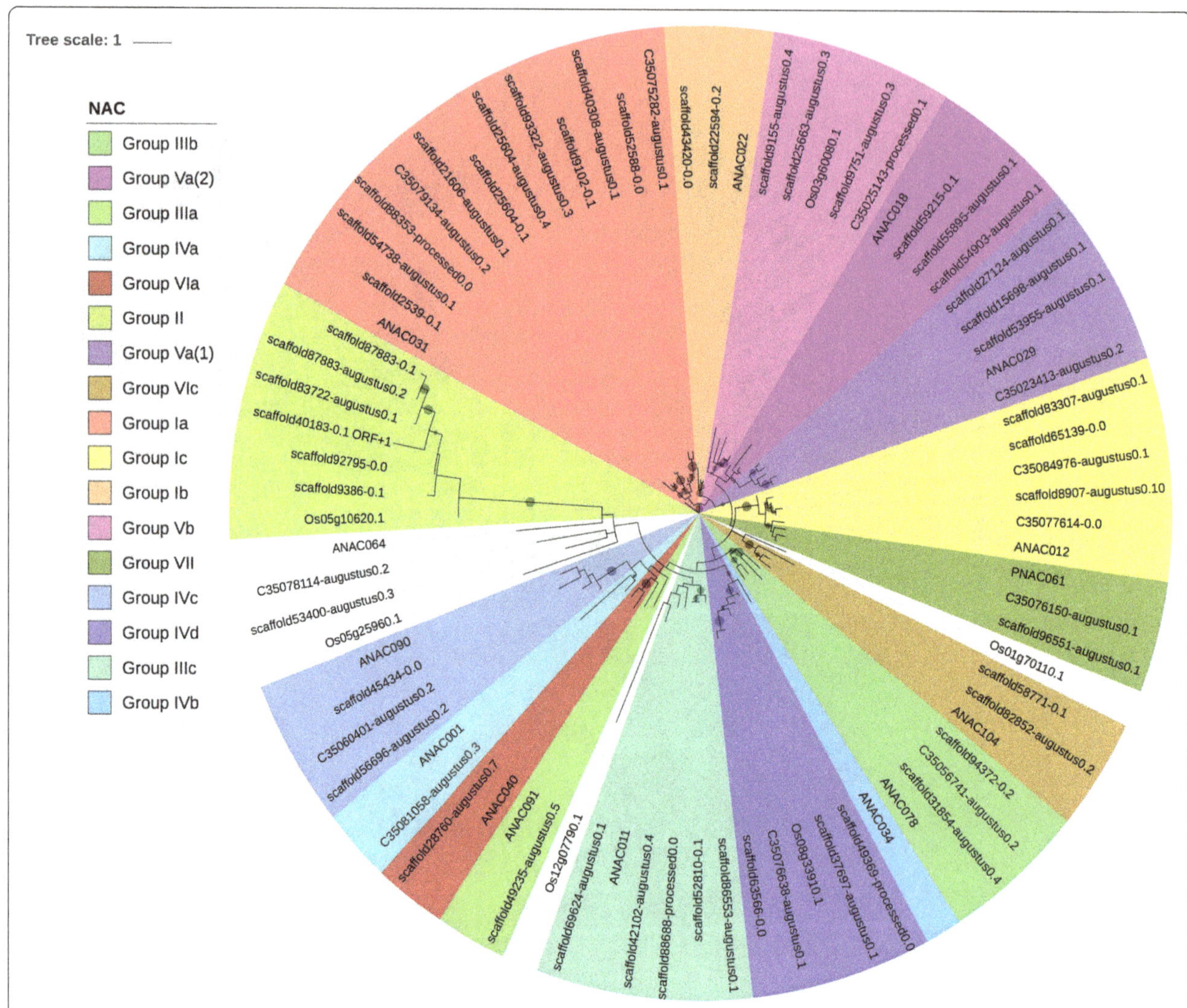

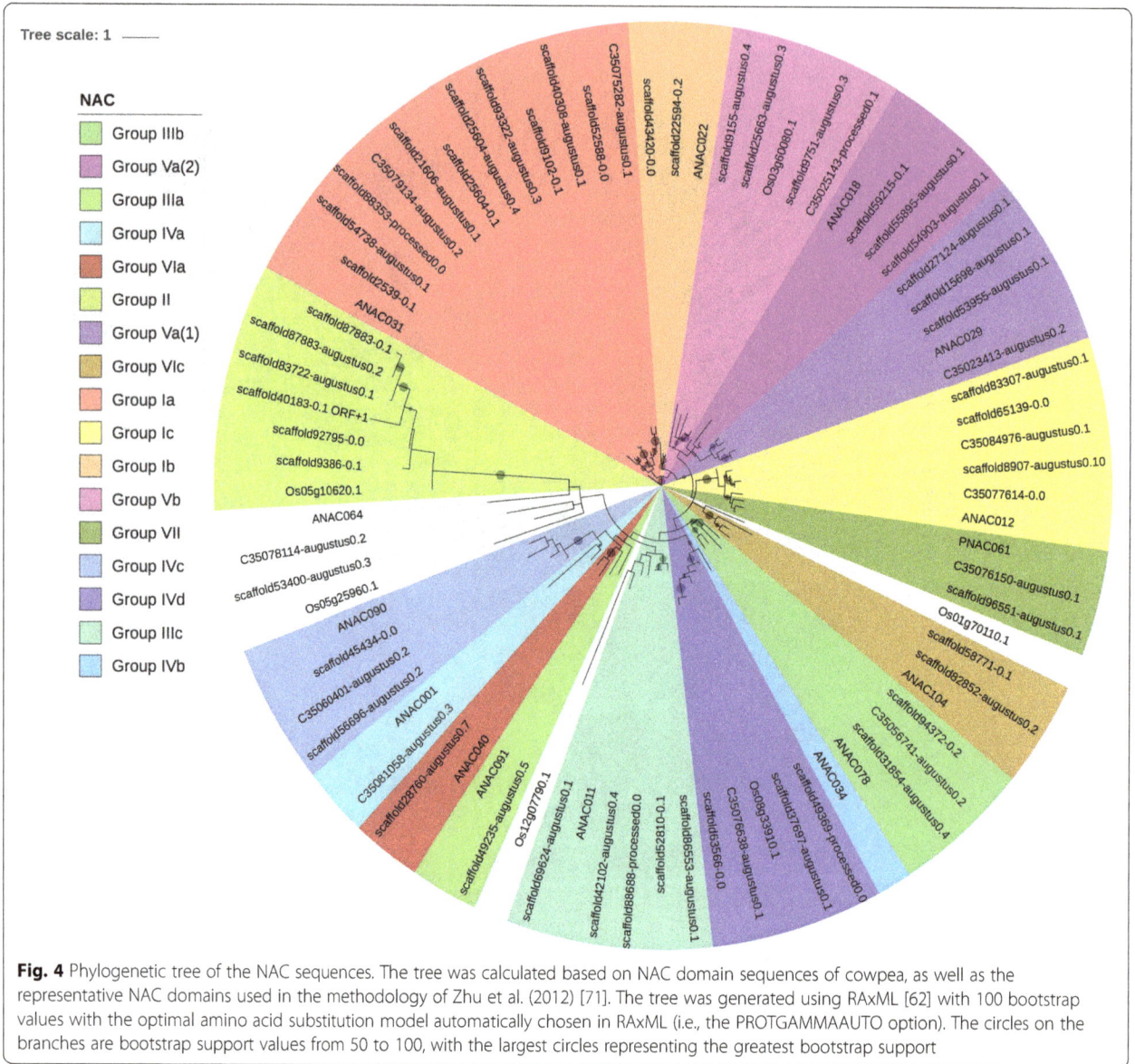

Fig. 4 Phylogenetic tree of the NAC sequences. The tree was calculated based on NAC domain sequences of cowpea, as well as the representative NAC domains used in the methodology of Zhu et al. (2012) [71]. The tree was generated using RAxML [62] with 100 bootstrap values with the optimal amino acid substitution model automatically chosen in RAxML (i.e., the PROTGAMMAAUTO option). The circles on the branches are bootstrap support values from 50 to 100, with the largest circles representing the greatest bootstrap support

addition, the sequences in this study were scaffold sequences, which were likely to contain two or more GSR or EST sequences, and hence would be more likely to have a complete WRKY domain.

Comparison of cowpea and common bean TF families

Several families were chosen for phylogenetic comparison between cowpea and common bean: the NAC family (Figs. 4 and 6), selected due to its roles in many cellular processes, including development and stress response, two families in which the number of cowpea genes present are under-represented compared to common bean (tify and B3), two families in which the sizes of the gene families are similar in the two species (BES/BZR (Brassinosteroid insensitive 1 (BRI1)-ethyl methanesulfonate (EMS)-suppressor / Brassinazole Resistant) and CCAAT-HAP5), two families in which the sizes of the family are larger (over-represented) relative to common bean (mitochondrial transcription termination factor (mTERF) and TUBBY (TUB)), and two families in which the size of the gene family in cowpea is significantly larger (strongly over-represented) relative to common bean (GRF and Aux/IAA) (see Figs. 7, 8, 9, 10 and 11, Additional files 5, 6 and 7). Here it is important to note that when only genes and not gene models are counted, BES, B3 and CCAAT-HAP5 are under-represented in cowpea compared to common bean, and that mTERF, tify, TUBBY, GRF, NAC and Aux/IAA are over-represented compared to common bean (Additional file 2b). Overall, it was found that cowpea and common bean were more likely to differ in families

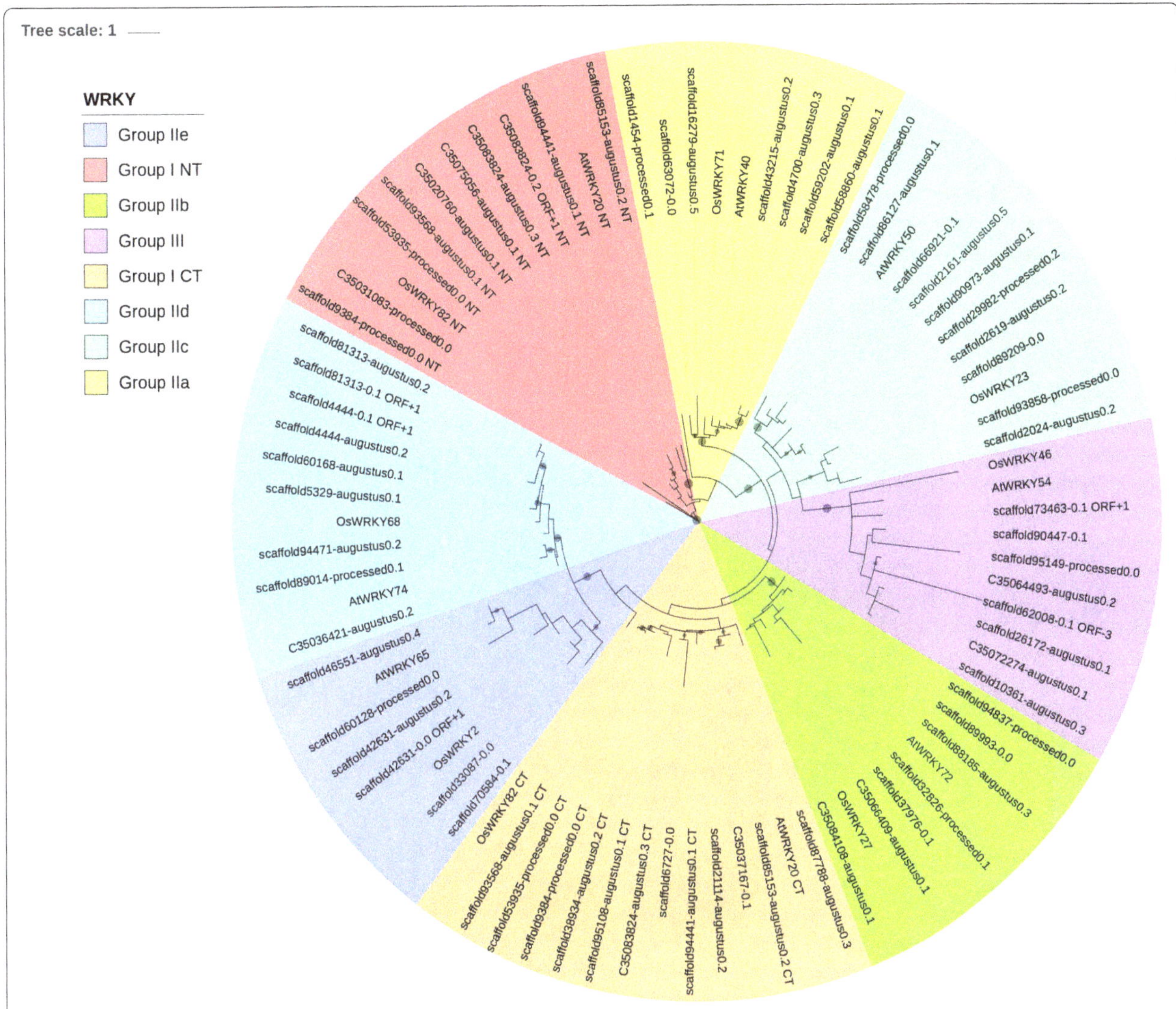

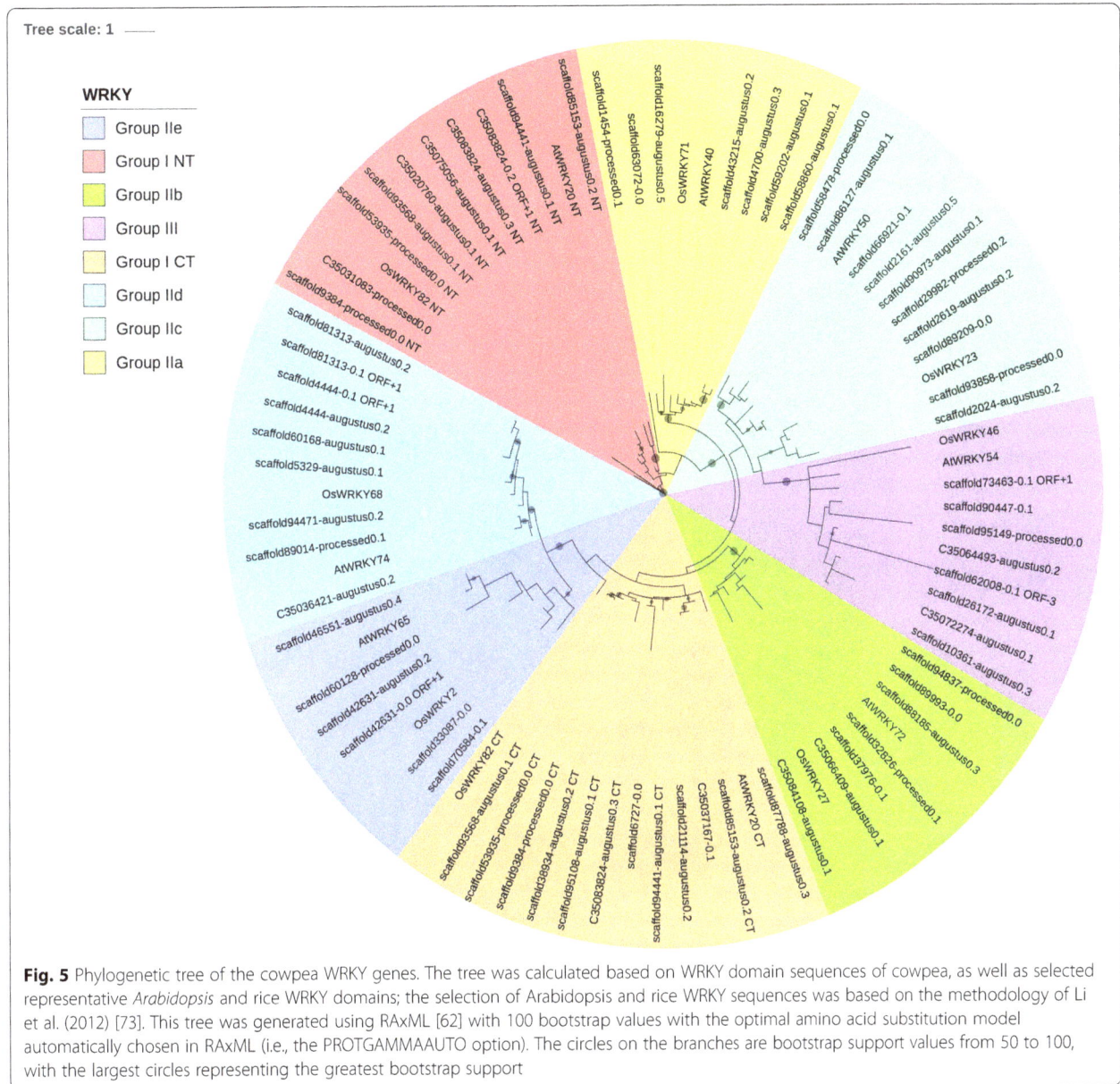

Fig. 5 Phylogenetic tree of the cowpea WRKY genes. The tree was calculated based on WRKY domain sequences of cowpea, as well as selected representative *Arabidopsis* and rice WRKY domains; the selection of Arabidopsis and rice WRKY sequences was based on the methodology of Li et al. (2012) [73]. This tree was generated using RAxML [62] with 100 bootstrap values with the optimal amino acid substitution model automatically chosen in RAxML (i.e., the PROTGAMMAAUTO option). The circles on the branches are bootstrap support values from 50 to 100, with the largest circles representing the greatest bootstrap support

that are known for having roles in development and stress response.

Figure 6 shows the phylogenetic tree of the NAC family of common bean. In comparison to the NAC family of cowpea (see Fig. 4), three key differences can be found. First, there are five members of group Ic in cowpea, while there are nine members of that same clade in common bean. Second, there are six members of Group II in cowpea as opposed to eight in common bean. Finally, Group IVa has one member in cowpea and four members in common bean (Fig. 6). As stated in the Discussion below, the differences in group Ic and II may be due to possible differences in how cowpea and common bean use NAC TFs in processes such as development and stress response. The differences between cowpea and common bean group IVa seem to indicate a difference in cell cycle control between cowpea and common bean [75].

Figures 7, 8, 9, 10, 11 shows the phylogenetic trees of B3, Aux/IAA, GRF, TUB and tify families. In the phylogenic trees for each family, there are groups of sequences that are unique to cowpea, and in the case of TUB and tify, there are also groups that are unique to common bean. These phylogenetic differences occur in families that have roles in growth and development, as well as in stress response, and thus reflect possible differences in growth, developmental and stress response mechanisms between the two legumes. For example, the B3 family, being a part of the auxin, gibberellic acid (GA) and abscisic acid (ABA) pathways [76], is

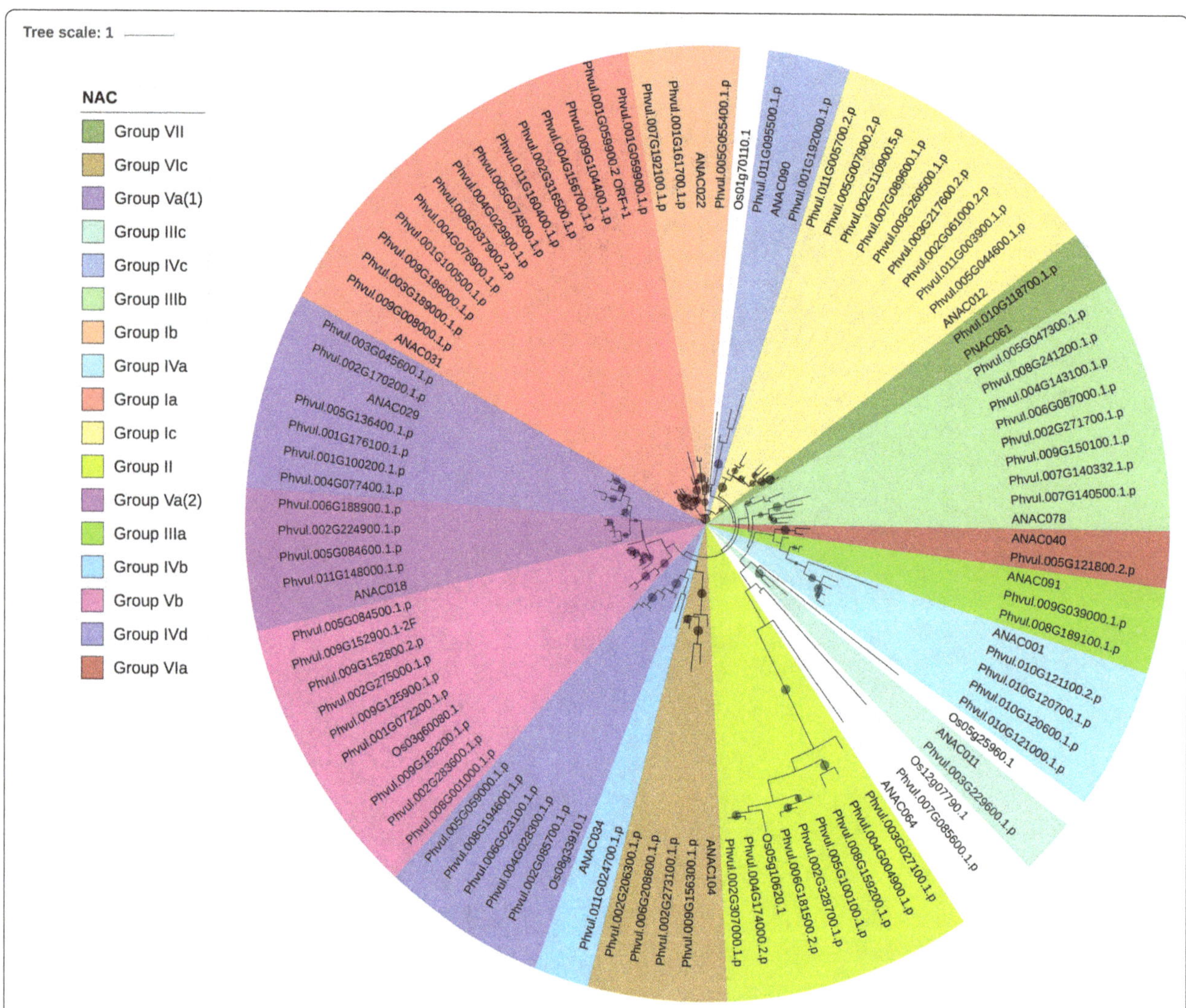

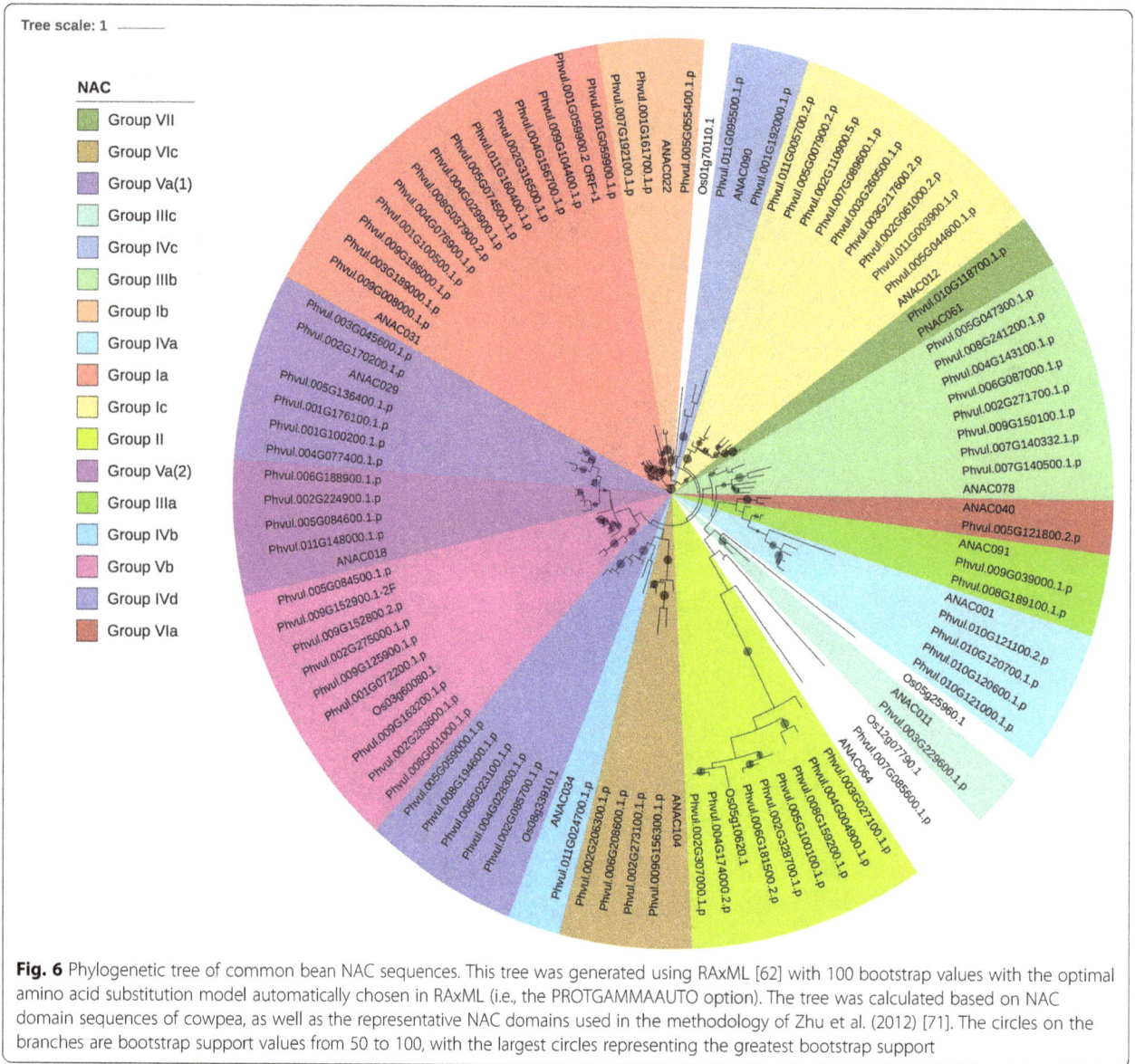

Fig. 6 Phylogenetic tree of common bean NAC sequences. This tree was generated using RAxML [62] with 100 bootstrap values with the optimal amino acid substitution model automatically chosen in RAxML (i.e., the PROTGAMMAAUTO option). The tree was calculated based on NAC domain sequences of cowpea, as well as the representative NAC domains used in the methodology of Zhu et al. (2012) [71]. The circles on the branches are bootstrap support values from 50 to 100, with the largest circles representing the greatest bootstrap support

important in seed development [75; 76] and dehydration response [77].

Additional files 5, 6 and 7 show the phylogenetic trees of the BES/BZR, CCAAT-HAP5, and mTERF families for cowpea and common bean. Cowpea and common bean are similar in phylogenetic organization for these families. The functions of the BES/BZR, CCAAT-HAP5 and mTERF families significantly differ from each other. The BES/BZR family is important in brassinosteroid signaling, which regulates stem cell quiescence in root stem cells [78, 79]. CCAAT-HAP5 TFs, also known as NF-Y subunit C (NF-YC), regulate light-mediated development [80]. The mTERF family has roles in regulating organelle gene expression for a variety of processes [81]. The phylogenetic similarities between cowpea and common bean in these families suggest that a diverse array of cowpea TF families is similarly organized to their counterparts in common bean.

Expression analyses

To find the stage of development and location of each cowpea TF and TAP sequence in this study, a BLASTN search was performed against the VuGEA database [41] to find for each cowpea TF and TAP in this study, the cowpea transcript with strongest homology (Additional file 8). In the B3, Aux/IAA, GRF, TUB and tify families (Figs. 7, 8, 9, 10 and 11), where cowpea and common bean had different phylogenetic organizations, cowpea TFs were somewhat more likely to express in roots and in pods; in pods, expression showed a slight tendency to be stronger 16 days after planting. In these families, the cowpea TFs that showed stronger expression patterns in roots and pods

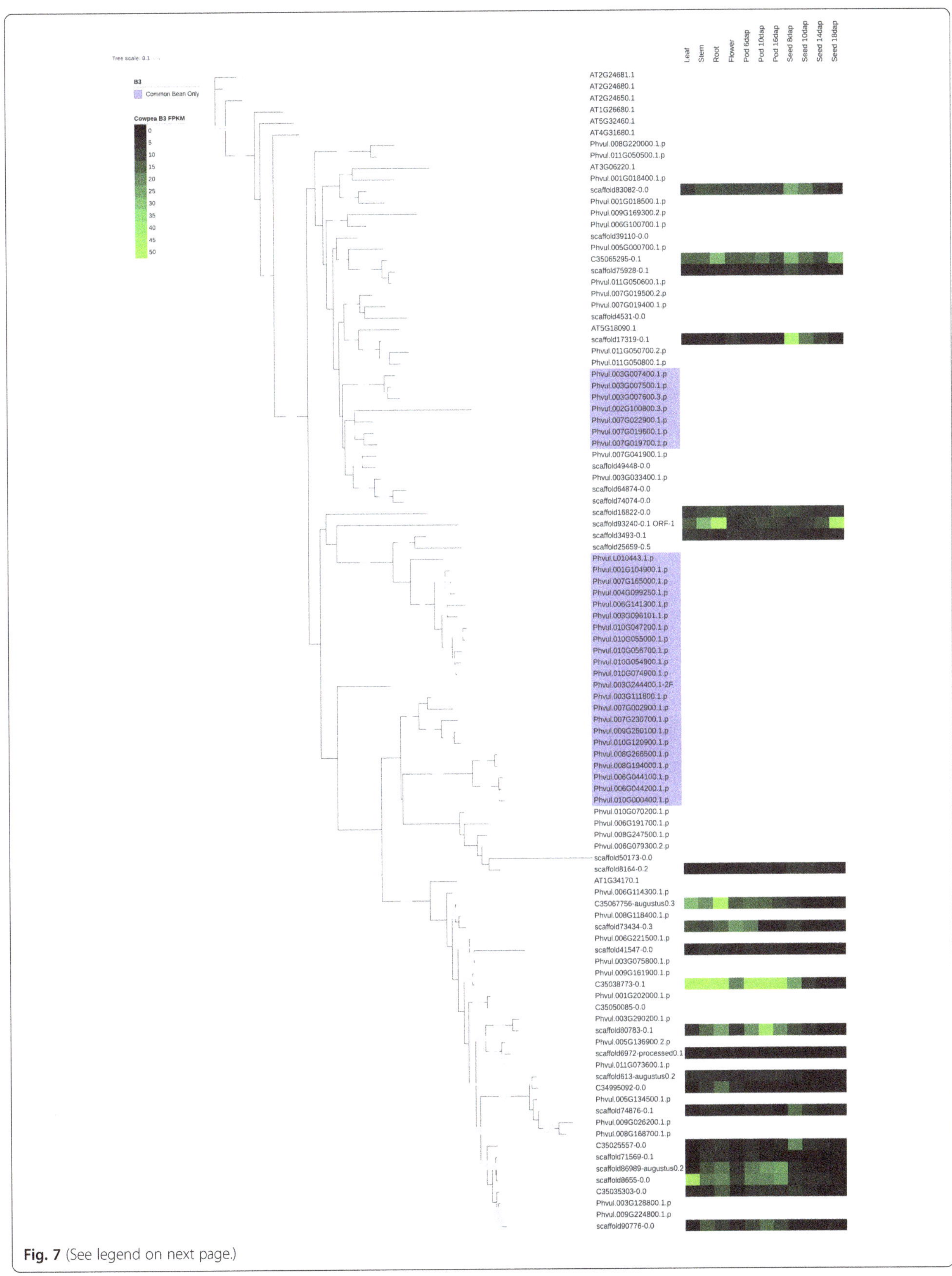

Fig. 7 (See legend on next page.)

(See figure on previous page.)

Fig. 7 Cowpea and common bean B3 sequences, with predictive heatmaps for cowpea TFs based on FPKM expression values from cowpea transcriptome data on VuGEA [43]. This tree was generated using RAxML [62] with 100 bootstrap values with the optimal amino acid substitution model automatically chosen in RAxML (i.e., the PROTGAMMAAUTO option). Sequences starting with "C3" or "scaffold" are cowpea sequences, while sequences starting with "Phvul" (*Phaseolus vulgaris*) are from common bean. The circles on the branches are bootstrap support values from 50 to 100, with the largest circles representing the greatest bootstrap support

were more likely to be in groups unique to cowpea. In families where cowpea and common bean show similar organization, by contrast (Additional files 5, 6 and 7), cowpea TFs and TAPs show their strongest expression in varying tissues and phases of development. While these results seem to suggest differences in nutrient gathering, root development and pod development between cowpea and common bean, further study is clearly needed to examine these potential differences.

Discussion

The availability of a draft genome for cowpea providing 67X coverage [27] has permitted a detailed characterization of the TF and TAP gene families in the genome. While

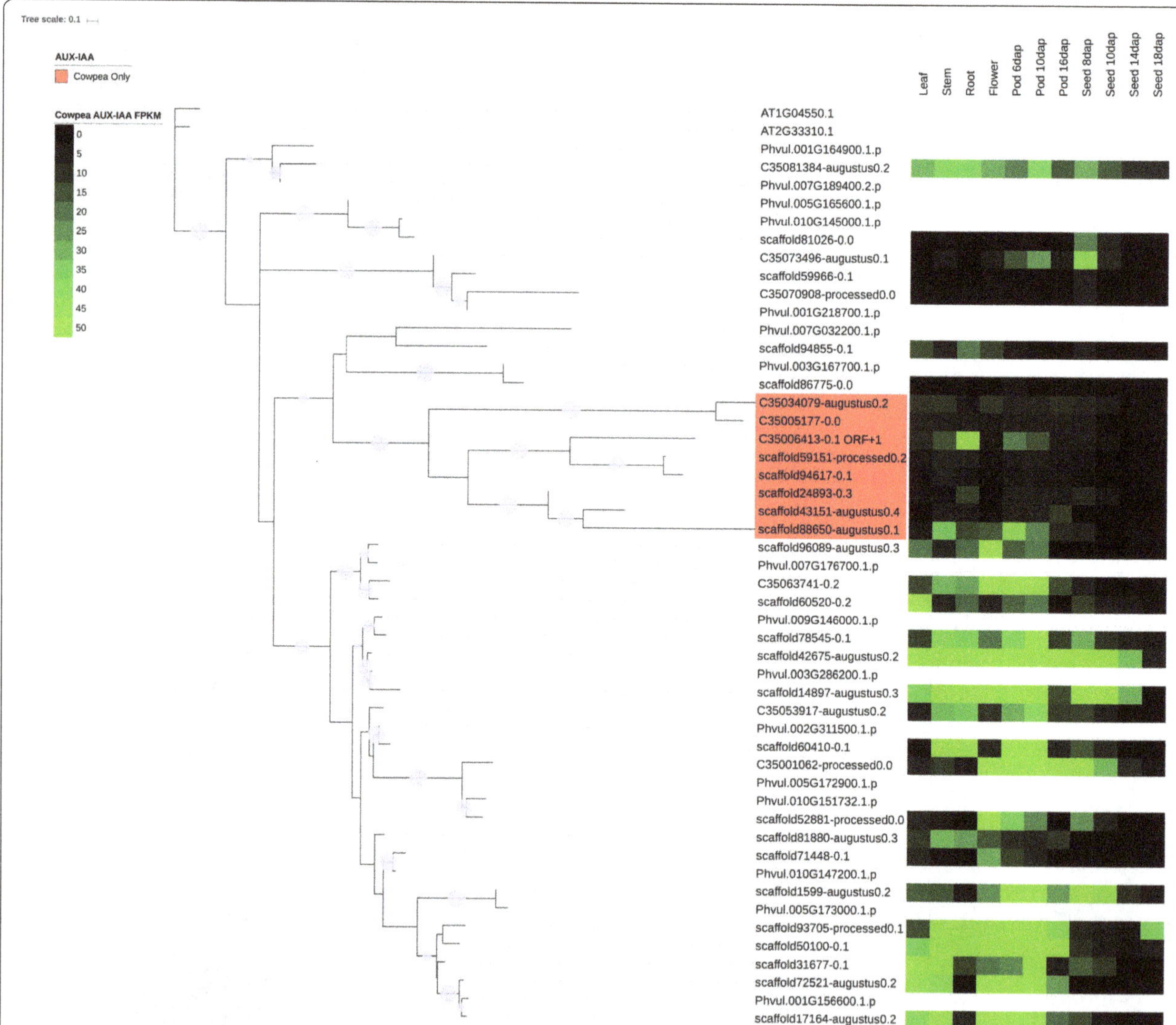

Fig. 8 Cowpea and common bean AUX_IAA sequences, with predictive heatmaps for cowpea TFs based on FPKM expression values from cowpea transcriptome data on VuGEA [43]. This tree was generated using RAxML [62] with 100 bootstrap values with the optimal amino acid substitution model automatically chosen in RAxML (i.e., the PROTGAMMAAUTO option). Sequences starting with "C3" or "scaffold" are cowpea sequences, while sequences starting with "Phvul" are from common bean. The circles on the branches are bootstrap support values from 50 to 100, with the largest circles representing the greatest bootstrap support

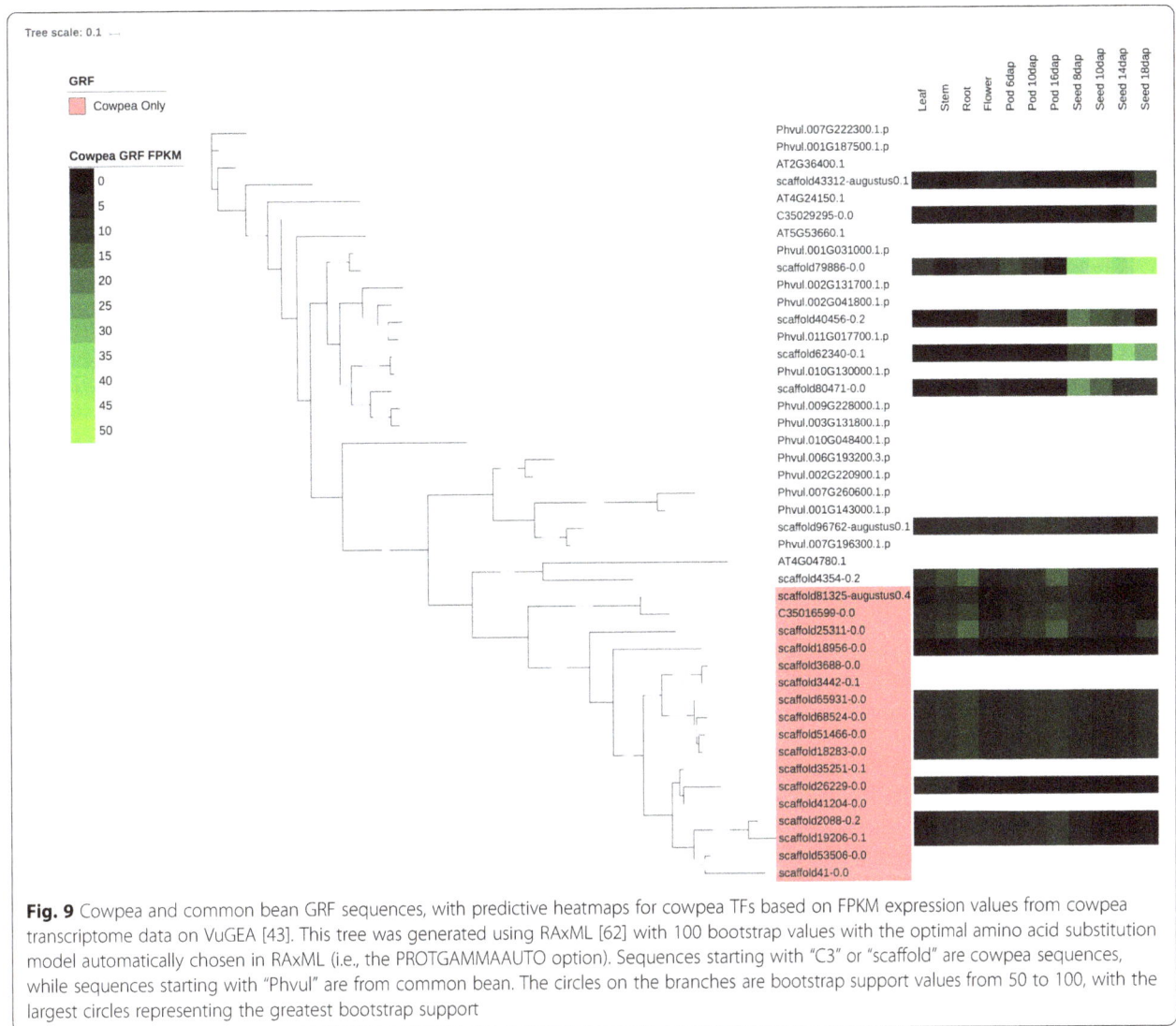

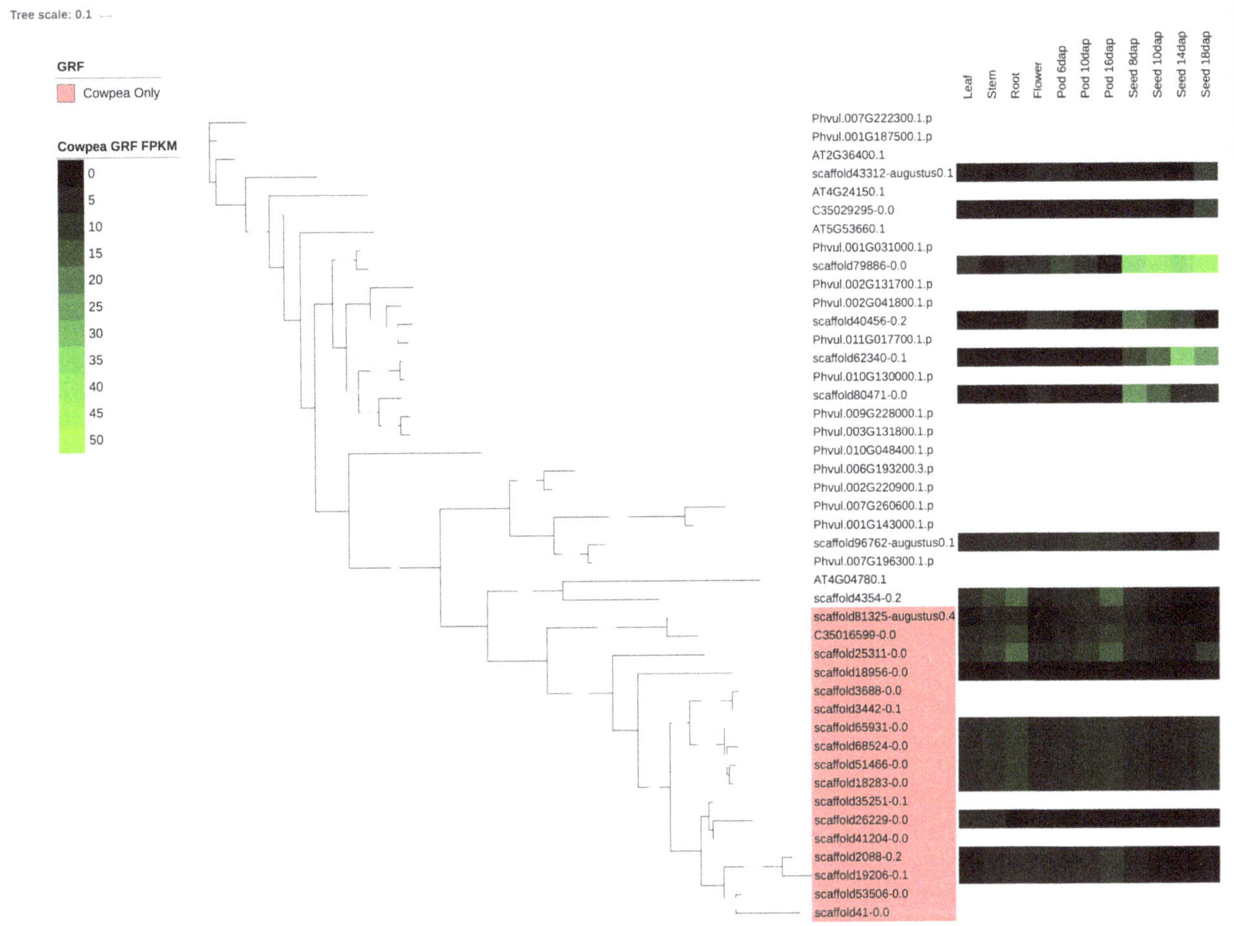

Fig. 9 Cowpea and common bean GRF sequences, with predictive heatmaps for cowpea TFs based on FPKM expression values from cowpea transcriptome data on VuGEA [43]. This tree was generated using RAxML [62] with 100 bootstrap values with the optimal amino acid substitution model automatically chosen in RAxML (i.e., the PROTGAMMAAUTO option). Sequences starting with "C3" or "scaffold" are cowpea sequences, while sequences starting with "Phvul" are from common bean. The circles on the branches are bootstrap support values from 50 to 100, with the largest circles representing the greatest bootstrap support

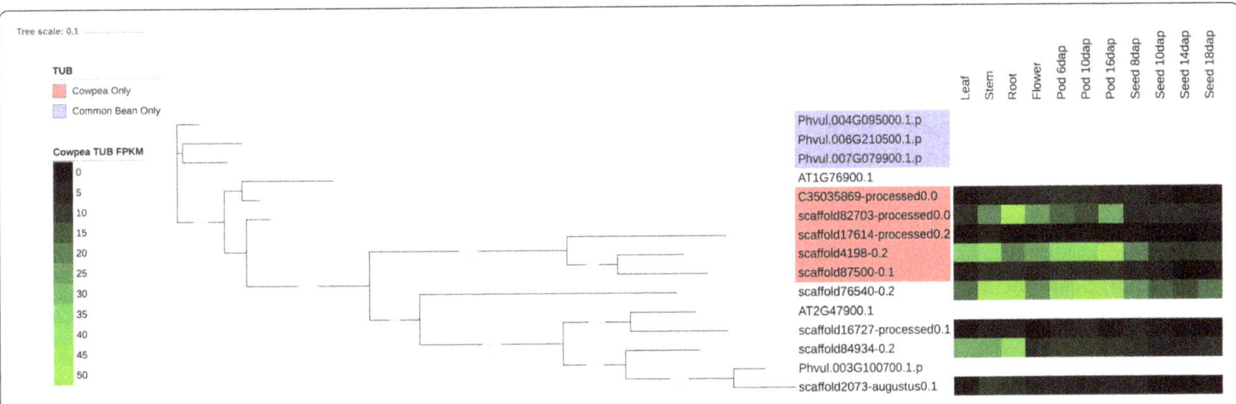

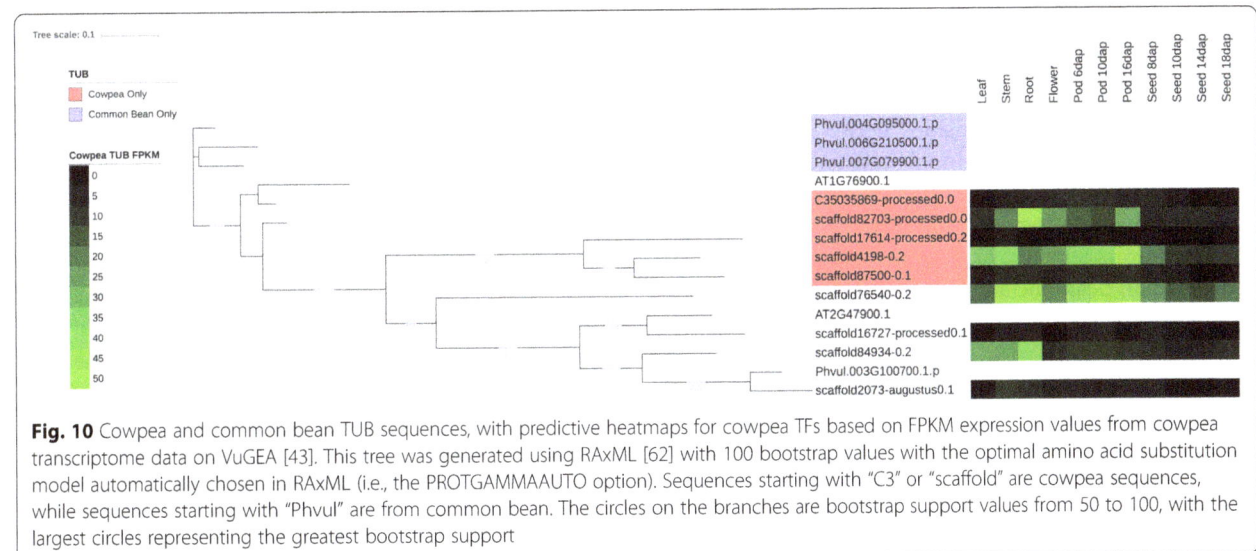

Fig. 10 Cowpea and common bean TUB sequences, with predictive heatmaps for cowpea TFs based on FPKM expression values from cowpea transcriptome data on VuGEA [43]. This tree was generated using RAxML [62] with 100 bootstrap values with the optimal amino acid substitution model automatically chosen in RAxML (i.e., the PROTGAMMAAUTO option). Sequences starting with "C3" or "scaffold" are cowpea sequences, while sequences starting with "Phvul" are from common bean. The circles on the branches are bootstrap support values from 50 to 100, with the largest circles representing the greatest bootstrap support

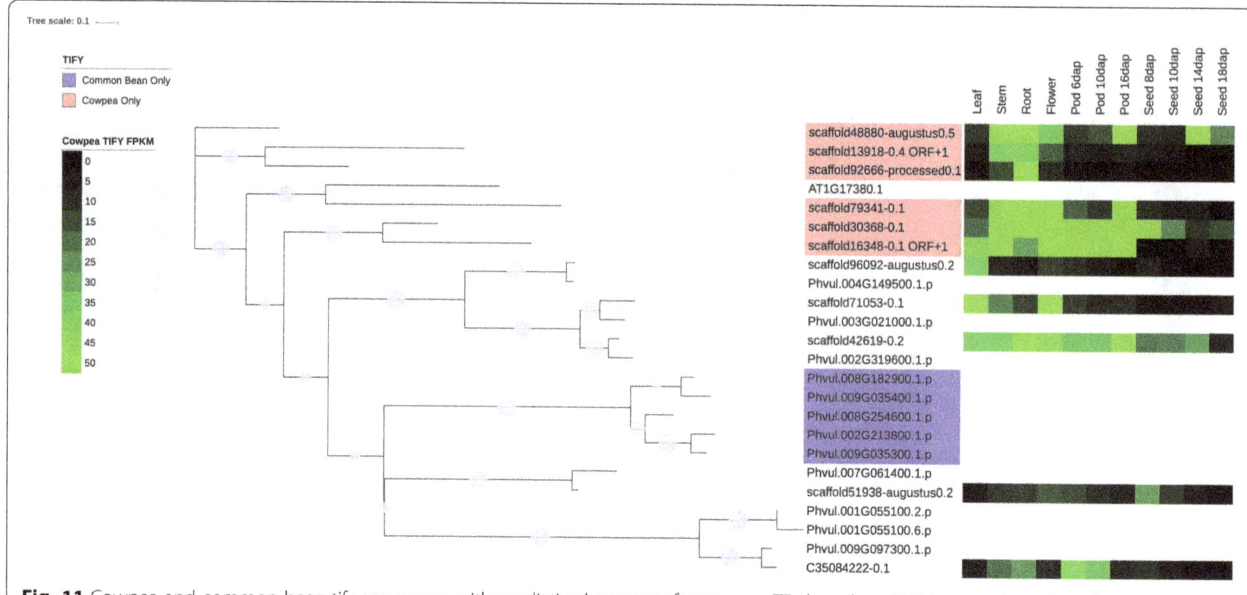

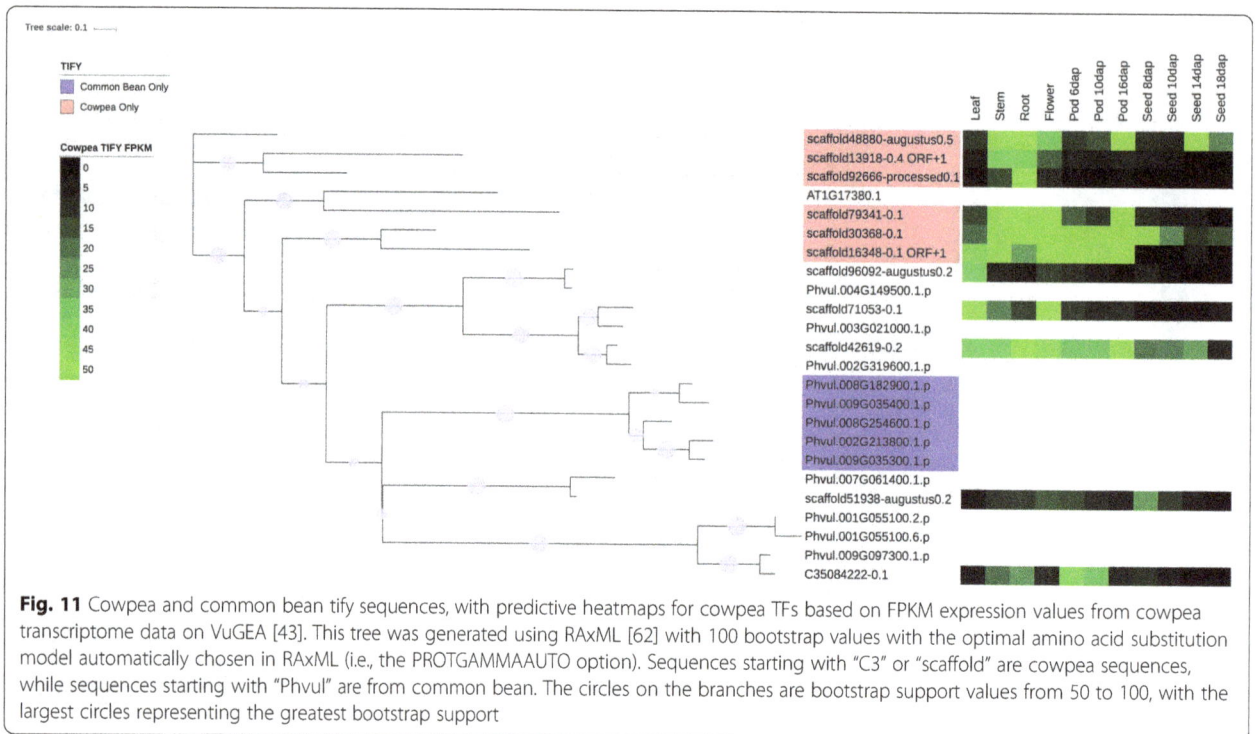

Fig. 11 Cowpea and common bean tify sequences, with predictive heatmaps for cowpea TFs based on FPKM expression values from cowpea transcriptome data on VuGEA [43]. This tree was generated using RAxML [62] with 100 bootstrap values with the optimal amino acid substitution model automatically chosen in RAxML (i.e., the PROTGAMMAAUTO option). Sequences starting with "C3" or "scaffold" are cowpea sequences, while sequences starting with "Phvul" are from common bean. The circles on the branches are bootstrap support values from 50 to 100, with the largest circles representing the greatest bootstrap support

other studies [37, 43] have attempted to identify the repertoire of these regulatory factors in cowpea, prior studies were compromised either by incomplete datasets or by methods that yielded an incomplete repertoire. Our results show that almost all of the known TF and TAP families recognized in other higher plants are represented in the cowpea genome. Cowpea contains a complement of TFs and TAPs that is two thirds as many as its close relative common bean and about one third as many as soybean. Interestingly, when the TF and TAP families in cowpea, common bean and soybean are compared as percentages of their respective total complement, the TFs and TAPs in cowpea are distributed in a manner that is significantly different from common bean and soybean.

This contrasts with expectations given the close relationships between cowpea and other legumes, as well as the strong syntenic relationships between cowpea and soybean [82], between common bean and soybean [83, 84], and between *Vigna* and *Phaseolus* species [85]. However, the discovery by Vascoconcelos et al. (2015) [86] of chromosomal rearrangements between cowpea and common bean, including translocations and duplications, is worth noting.

In the present study, 4416 sequences (from 4408 genes) encoding 5460 TF and TAP encoding domains were found in cowpea, a significant contrast to the 2485 TFs and TAPs found in an analysis of the cowpea transcriptome [44]. This could be due to differences between the annotation in cowpea genome assembly v0.03 [27] and that of the cowpea transcriptome studied in Yao et al. (2016) [44]. Moreover, this study incorporates three separate TF classification pipelines, PlantTFcat [40], iTAK [41] and the TavernaPBS pipeline used in this study [45]; as stated in the Results section, the individual results for each of these pipelines are presented in Additional file 3.

Phylogenetic analysis of selected TF families

AP2-EREBP

The AP2-EREBP family has been investigated in many plants, from *Arabidopsis* and rice [64, 69, 70], tobacco [34], maize [87], and soybean [88], to tomato [89] and cotton [90]. Some of these studies placed AP2, ERF and RAV into the same phylogenetic tree [69, 87], while some others exclusively investigate the ERF family [63; 88], or separate the ERF family from the AP2 and RAV families [90]. Two different classification schemes have emerged, one in which the ERF family is separated into twelve groups (i.e., DREB A1-A6 and ERF B1-B6) [69], and the other in which the ERF family is separated into groups I-X, VI-L and Xb-L [64]. Sharoni et al. (2011) [70] separates the AP2-EREBP into two trees, ERF and AP2/RAV. In this study, the AP2, ERF and RAV families were placed in the same tree (Figs. 2 and 3).

When the twelve-group DREB/ERF classification [69, 70] was employed we found that many of the groups in cowpea are polyphyletic. Dietz et al. (2010) [69] reported some polyphyletic groups in the ERF family of *Arabidopsis*, but not to the extent observed in cowpea. These polyphyletic groups were not found in the ERF family in soybean [88] or rice [70].

When the Nakano et al. (2006) [64] classification scheme was applied, the cowpea ERF family separates more completely into distinct groups, with the exception that group Xb-L appears to be absent in cowpea, and that one group X sequence, Os02g52670.1, does not group with the rest of group X (Fig. 3). The phylogenetic trees presented in Figs. 2 and 3 are similar to that reported by Timko et al. (2008) [26] based upon cowpea GSS data with Groups I-IV forming the DREB clade, and Groups VI, VIII and IX forming the ERF clade. However, Timko et al. (2008) [26] found that Group V ERFs separated between the DREB clade and the ERF clade. In addition, Group IX ERFs separated into two separate clades, with the group VII ERFs between them [26]. In contrast, in this study we observed that the Group V ERFs were in the ERF clade and Groups IX and X are not split by Group VII, but by groups VI and VIII. These differences reflect both the completeness and quality of the sequence data sets and difference in the consistency and accuracy of the alignment tools and phylogenetic methods applied here.

The grouping of the cowpea ERFs into the groups characterized by Nakano et al. (2006) [64] was neater than the grouping into DREB A1-A6 and ERF B1-B6 in Dietz et al. (2010) [69] and Sharoni et al. (2011) [70]. For example, a Group X ERF in rice according to Nakano et al. (2006) [64] was classified as a DREB A5 in Sharoni et al. (2011) [70]; according to Nakano et al. (2006) [64], Group X ERFs represent ERF group B-3 and B-4. Group X in this study was associated with the ERF B-4 clade, consistent with Nakano et al. (2006) [64]. In another example, two DREB sequences from Sharoni et al. (2011) [70] were classified as group VII in Nakano et al. (2006) [64], when the group-VII ERFs in Nakano et al. (2006) [64] were classified as ERF B-2. This could explain why group VII cowpea ERFs grouped with the DREBs in this study (Figs. 2 and 3), unlike in Timko et al. (2008) [26]. The details of the contrast between the classification schemes of Nakano et al. (2006) [64] and Sharoni et al. (2011) [70] are outlined in Additional file 9.

NAC family

While methods of classifying NAC TFs into groups vary greatly throughout the literature, regardless of which clustering algorithm one employs, it appears that some legume NAC families are organized differently from other dicots. This difference is evident in the organization of the NAC family in soybean [91] and for rice and *Arabidopsis* [71]. For example, clades Senu5 and Tobacco Elicitor-Responsive NAC protein-like (TERN) (which are VIa and IVc, respectively, in Zhu et al. (2012) [72]), group much farther apart in rice and *Arabidopsis* [71].

When one compares the phylogenetic organization of the cowpea NAC TFs to that observed in other plants using the relative position of each group (closest to furthest from the root of their respective trees) cowpea, not surprisingly, is somewhat closer in organization to the NAC family in Zhu et al. (2012) [72] than to the soybean NAC family in Pinheiro et al. (2009) [90]. Moreover, the cowpea NAC TF family (Fig. 4) differs from common bean in abundance of group Ic, II and IVa NACs (Fig. 6). Our results suggest that even among legumes, the cowpea NAC family may be relatively unique in organization.

Since the NAC family is functionally diverse with regulatory functions ranging from defense against biotic and abiotic stresses, to hormone signaling, and control of reproduction [56] it may not be surprising that plants such as cowpea highly adapted for growth in dry savannah regions would have diversified its NAC family. Thus, the value of our analysis lies in the ability to find those members of this gene family that may confer beneficial growth and development properties onto cowpea that could be exploited for improvement of close relatives like soybean.

WRKY family

Both similarities and differences are observed in the WRKY family of cowpea relative to those of other plants. For example, the cowpea WRKY TFs in this study have groups IId and IIe group together as in a Rushton et al. (2010) [60] tree containing soybean, *Arabidopsis thaliana*, rice and poplar. However, in Rushton et al. (2010) [60], Group I N-terminal (NT) and Group I C-terminal (CT) group together, whereas these groups are distant in the cowpea WRKY TFs from this study (Fig. 5). In Li et al. (2012) [74], a phylogenetic tree showing WRKY sequences from *Arabidopsis thaliana*, rice and castor bean shows that like the cowpea WRKY tree in Fig. 5, the WRKYs in Li et al. (2012) [74] are such that Groups I NT and I CT are farther apart. However, groups IIa and IIc in this study group between I and III, whereas in Li et al. (2012) [74], group IIc groups with IIa, b, d and e. When the cowpea WRKY TFs from this study and the WRKYs from Li et al. (2012) [74] are compared from closest to furthest from the root of their respective trees, the following orders are found for this study: Group I NT - > IIa - > IIc - > III - > IIb - > I CT - > IId - > IIe, compared to that of Li et al. (2012) [74]: Group III - > I NT - > IIb - > IIa - > IId - > IIe - > IIc - > I CT. These findings suggest WRKY TF family organization in cowpea is unique.

Phylogenetic comparison between cowpea and common bean

Overall, TF families that regulate a wide variety of processes in cowpea and common bean are similar in their phylogenetic organization. This is to be expected since cowpea and common bean are close together in the Millettoid (i.e., Phaseoloid) clade of legumes [92]. The BES/BZR, CCAAT-HAP5, and mTERF families in cowpea were all similar in phylogenetic organization to common bean. Thus if a TF family in cowpea is similar in phylogenetic organization but larger in size than in common bean, it may be possible that there are instances of gene duplication (without whole-genome duplication) in that TF family in cowpea; likewise a TF family that is larger in common bean but phylogenetically similar to cowpea may show instances of gene duplication for that TF family in common bean. Instances of such processes have been known to occur throughout plant evolution [30; 93], and they may lead to the presence of sequences with new functions [93].

Among the families analyzed, the B3, Aux/IAA, GRF, TUB, tify and NAC families in cowpea showed some differences in phylogenetic organization from their counterparts in common bean (Figs. 3, 5 and 6). It is possible that different forms of duplication on a small scale are involved in these families. These phylogenetic differences occur in families that have roles in growth and development, as well as in stress response, and thus reflect possible differences in growth, developmental and stress response mechanisms between the two legumes.

The B3, GRF, aux/IAA, TUB and tify families

The B3, GRF, Aux/IAA, TUB and tify families have roles in stress response, or in growth and development. As stated earlier, the B3 family has roles in the ABA and GA pathways [76] and in seed development [76, 94] and response to dehydration [77]. The GRF family is associated with a wide array of growth and developmental processes including leaf growth and floral organ development [95, 96], but the Aux/IAA is associated with tolerance to stress [97, 98]. However, it is important to note that Aux/IAA TFs can also be involved in repressing stomatal development in seedlings [99] and root development [100]. The presence of a sequence that is predicted to express most strongly in roots in a group of the Aux/IAA family that is unique to cowpea (Fig. 8) suggests that cowpea root development evolved differently from common bean. The TUB family may have roles in response to abscisic acid (ABA), and thus may be important in stress response [101]. Moreover, experimental studies by Yulong et al. (2016) [101] and Lai et al. (2004) [102] suggest that TUB transcription factors have roles in germinating seedlings. The tify family may have roles in the jasmonic acid (JA) stress response pathway [103]. Since clades exist in these families (with the exception of B3) that are unique to cowpea (Figs. 8, 9, 10 and 11), it is possible that cowpea is evolving to have genes with new functions in growth, development and stress response. Common bean, having expansions in clades of the B3, TUB and tify families (Figs. 7, 10 and 11), may also be evolving to have new genes in these processes, albeit in different ways than in cowpea.

NAC

As stated in the Results section, a comparison between the cowpea NACs in Fig. 4 and the common bean NACs in Fig. 6 shows that unlike in cowpea, common bean has a larger clade II and Ic. A search of Os05g10620, a group II NAC, on the Rice Functional Genomic Express Database (RiceGE) database at the Salk Institute Genomic Analysis Laboratory (SIGnAL) [104] shows that the Gene Ontology (GO) annotation of Os05g10620 has "development" as a GO term. Given that Zhu et al. (2012) [72] note that group II is noted to be part of the same monophyletic lineage, the under-representation of group II in cowpea compared to common bean suggests differences in regulation of developmental processes between cowpea and common bean.

Group Ic, also called the Secondary wall-associated NAC Domain (SND) clade [72] has roles in secondary wall development [96]. The greater abundance of group Ic NAC genes in common bean (Fig. 6) further suggests differences from cowpea in regulating developmental processes.

Group IVa NAC sequences are involved in cell cycle control [75], and the over-representation of this group in common bean (Fig. 6) compared to cowpea (Fig. 4) suggests gene duplication and thus the possible presence of genes with new functions [93] in cell cycle control in common bean.

The availability of whole genomic data for cowpea, a non-model legume with significant importance in the developing world, is a significant step forward in orphan legume research. Placing value on this genome sequence by characterization of the various gene families of TFs and TAPs, their organization and phylogenetic relationships, will facilitate future comparative analysis and development of strategies for controlling growth, differentiation, and abiotic and biotic stress resistances.

Overall, our analysis revealed that cowpea, like many of its diploid relatives within the Leguminoseae show gene contents similar to other diploid dicotyledonous plants, and that cowpea, despite having fewer TFs and TAPs in number, has genes coding for almost all of the TF and TAP families in other plants. In many cases the phylogenetic organization of the TF and TAP families in cowpea mimicked their counterparts in common bean, whether or not the number of members of certain TF families in cowpea were significantly different in size compared to their counterparts in common bean. However, aspects of the TF and TAP families of cowpea are unique in composition and organization when compared with its evolutionarily close relatives, common bean and soybean. The functional relevance of these variations can now be explored in greater detail, particularly with regards to growth and developmental processes as well as response to stress.

Conclusions

We used a computational approach employing three different predictive pipelines to data mine the recently release cowpea genome and identified over 4400 genes representing 136 different TF and TAP families. The availability of detailed information on the coding capacity of the cowpea genome and in particular the various TF and TAP gene families will facilitate future comparative analysis and development of strategies for controlling growth, differentiation, and abiotic and biotic stress resistances of cowpea. By comparison to other closely related legumes we also provide a starting point for additional comparative evolutionary studies.

Methods

Genomic data sources and analysis

The analysis presented here used a genomic assembly from cowpea genotype IT97K-499-35 that was an updated version (version 0.03) [27] of the draft assembly (version 0.02) previously described by Pottorff et al. (2012) [105]. The updated version used for this work included one long-insert paired end library (5 kb), which improved the scaffold lengths [27]. Cowpea genome v0.03 is available for BLAST searches and sequence retrieval [106]. All Perl, Bioperl and Bash scripts used in this study are custom scripts that we developed (Additional file 1).

For exon detection and translation of the cowpea assembly v0.03 [27] to protein two approaches were used, one in which the low-complexity regions were masked via AUGUSTUS [51, 52] before using the MAKER annotation pipeline [50], and one that was simply annotated via MAKER and then processed through AUGUSTUS (Stephen M. Turner, personal communication, October 29, 2013).

The soybean genome [31] version Wm82.a2.v1 and the *Phaseolus vulgaris* genome version 2.1 [48, 107] were downloaded from Phytozome v11 [108, 109].

Since two approaches were used for exon detection and translation to proteins as mentioned above, two different versions of the set of protein and nucleotide sequences in the cowpea genome Version 0.03 were produced. These versions were combined into a non-redundant set of sequences using a custom BioPerl script that we designed to make sure that only sequences with unique amino acid compositions were included in the set (Additional file 1).

In addition, to prepare all cowpea, common bean and soybean nucleotide sequences for searches using the TavernaPBS [45] TF classification pipeline, which required amino acid sequences, we used a Bioperl script that we developed to perform a six-frame translation of every nucleotide sequence.

The characterization of TFs and TAPs in cowpea involved identifying TFs and TAPs at the amino acid and the transcript level. In a separate analysis, TFs and TAPs were identified in the raw cowpea genome assembly (the assembly produced prior to exon detection and translation); this was done in order for a comparison with protein and transcript assemblies.

Classification of transcription factors and transcriptionally active proteins

Transcription factors (TFs) and transcriptionally active proteins (TAPs) were identified by using three approaches: the PlantTFcat pipeline [40], the iTAK pipeline [41], and a TavernaPBS [45] workflow (see Additional file 1) that uses HMMER 3 [110] for Hidden Markov models and PFAM [111] for finding protein domains. TF and TAP classification was done using the rules for membership in each of 136 TF and TAP families, which include 111 different TF and TAP families described by Lang et al. (2010) [44] with only minor adjustments, as well as 43 additional TF and TAP families described in PlantTFcat and iTAK, eighteen of which were synonymous with the 111 TF and TAP families in Lang et al. (2010) [44].

Since both proteins and transcripts were investigated, and three pipelines were used, reducing the sequences to a non-redundant set needed to be done using a Perl script and a Bash script that we developed. Here, the Perl script was used to sort sequences into families with the aid of a text file that not only listed all the TF/TAP families, but took into account the synonyms that different databases used for the same family. The Bash script was developed to follow the following procedure: first, if the protein was present, it was accepted into the set. If not, then translated transcripts from PlantTFcat were accepted. If the PlantTFcat transcript for a certain open reading frame (ORF) (e.g., +2) was present and that same transcript with the same ORF was present in a set of sequences found by iTAK, the latter transcript was eliminated. Otherwise, the iTAK transcript was accepted into the sequence set. Then, if the iTAK transcript for a certain ORF (e.g., +2) was present and that same transcript with the same ORF was present in a set of sequences found by the TavernaPBS pipeline, the latter transcript was eliminated. Otherwise, the transcript was accepted. PlantTFcat was given higher preference than iTAK due to the comprehensive nature of the InterPro database [112] that PlantTFcat uses [40]. iTAK was given higher preference than the TavernaPBS pipeline because the TF and TR classification scheme in iTAK was developed as a consensus between PlantTFDB [37] and PlnTFDB [35, 36] and was developed more recently than the TavernaPBS pipeline based on Lang et al. (2010) [44].

When none of these pipelines identified TFs of a certain family in cowpea, FASTA version 36.3.8e [54] was used to search the raw cowpea genome for the closest homologs of that family, with an E-value of 1e-3. Common bean members of the TF families not found in cowpea were used as queries in these searches.

MEME (Multiple Em for Motif Elicitation) version 4.11.2 [113] was used to identify non-overlapping motifs in each sequence and to aid in curating each TF and TAP family. In motif discovery, an E-value threshold of 0.001, a motif width range of 8–200 amino acids, the "any repeat" (anr) mode active, and maximum number of motifs of 100 was used. We designed two custom scripts that use MEME motif data to curate the sequences, adding each motif in the order that they appear in the sequence: a BioPerl script to parse the MEME data into a table containing sequence name, motif start and end coordinates, and motif sequence; and the other script, a Perl script to use that table to create a curated sequence consisting of the MEME motifs for that sequence in order of start coordinate.

We applied the identical approach to analyzing the TF and TAP contents of common bean and soybean using the soybean and common bean genomes. When comparing common bean sequences for comparison with cowpea sequences, the redundant sequences within the common bean sequence set for a family were removed using the same custom BioPerl script that we designed to remove redundant sequences in the cowpea genome.

Multiple sequence alignment of TF sequences

Multiple sequence alignment was done using MAFFT version 7.245 [61]. Any duplicate sequence or sequence that covered less than 30% of the alignment was eliminated, and columns that consisted of at least 90% gaps were removed; this was done using trimAl v1.2 [114]. After this, the TF family being analyzed had sequences with 95% similarity or more removed using a custom BioPerl script that we designed, and was re-aligned. This cycle was repeated until the number of sequences and number of alignment columns stabilized.

Phylogenetic analysis

Phylogenetic trees were generated using RAxML version 8.2.9 [62] with 100 bootstrap replicates. The trees seen in this study were generated using Interactive Tree of Life (iTOL) version 3.3.3 [115]. The trees and their associated alignment files can be found on TreeBASE under the accession 21,817 (http://purl.org/phylo/treebase/phylows/study/TB2:S21817?x-access-code=bc32e308174e24a9e59381 01a324a744&format=html) [116].

Analysis of AP2-EREBP, NAC and WRKY families

For the AP2-EREBP family, selected sequences from each clade of the *Arabidopsis* AP2-EREBP superfamily analyzed in Dietz et al. (2010) [69] and from each clade of the rice AP2-EREBP superfamily analyzed in Sharoni et al. (2011) [70] were used to determine the clades to which cowpea AP2-EREBP sequences belonged, and the AP2-EREBP phylogenetic organization scheme from Dietz et al. (2010) [69] and Sharoni et al. (2011) [70] were compared to that from Nakano et al. (2006) [64]. To analyze the NAC family, sequences from each clade of the NAC family analyzed by Zhu et al. (2012) [72] were used to classify cowpea NAC sequences into clades. To analyze the WRKY family, sequences from each clade of the WRKY family analyzed by Li et al. (2012) [74] were used to classify the cowpea WRKY sequences into clades.

Expression analysis

All cowpea TF sequences were searched via BLASTN (E-value = 1e-5) against VuGEA [43] to find for each cowpea TF in the cowpea genome, its cowpea transcript with strongest homology in VuGEA, as well as the FPKM expression for that transcript. Upon downloading the FPKM data, we developed custom Perl scripts to link each cowpea TF and TAP to the FPKM expression of the transcript to which it has the greatest similarity. We developed and used a custom Bioperl script to parse the BLASTN search result files from VuGEA, and we designed and used custom Perl scripts to construct heatmaps of the FPKM data, as well as a table of the data showing the cowpea gene, gene model, transcript from VuGEA with strongest homology, and corresponding FPKM values. This table is available in Additional file 8.

Additional files

Additional file 1: The TavernaPBS workflow used to classify TFs and TAPs in cowpea, common bean and soybean. This workflow was adapted to work in a SLURM environment. Figure adapted from an image of the workflow used in TavernaPBS [45]). This image was generated using Taverna 2.5.0 [46]. The Bioperl scripts and bash scripts and their purposes are also given. (DOC 484 kb)

Additional file 2: Numbers of cowpea TFs, TAPs and TRs compared to common bean and soybean. The comparisons are made with respect to percentage of TF repertoires, raw number and percentage of total protein coding sequences. (XLS 133 kb)

Additional file 3: Cowpea TFs, TAPs and TRs in the A) PlantTFcat, B) iTAK and C) TavernaPBS pipelines. For each pipeline, the numbers shown are from the protein, transcript and raw cowpea assemblies. For A), the cowpea TFs and TRs in VuGEA, which used the PlantTFcat pipeline [40], are presented for comparison. (XLS 76 kb)

Additional file 4: Comparisons of cowpea TF families to their counterparts in common bean and soybean. These comparisons are made with respect to: a) percentage of TF repertoires, b) raw number of TFs, and c) percentage of total protein-coding genes. (JPEG 3544 kb)

Additional file 5: Cowpea and common bean BES_BZR sequences, with predictive heatmaps based on FPKM expression values from cowpea transcriptome data on VuGEA [43]. This tree was generated using RAxML [62] with 100 bootstrap values with the optimal amino acid substitution model automatically chosen in RAxML (i.e., the PROTGAMMAAUTO option). Sequences starting with "C3" or "scaffold" are cowpea sequences, while sequences starting with "Phvul" are from common bean. The circles on the branches are bootstrap support values from 50 to 100, with the largest circles representing the greatest bootstrap support. (JPEG 484 kb)

Additional file 6: Cowpea and common bean CCAAT-HAP5 sequences, with predictive heatmaps based on FPKM expression values from cowpea transcriptome data on VuGEA [43]. This tree was generated using RAxML [62] with 100 bootstrap values with the optimal amino acid substitution model automatically chosen in RAxML (i.e., the PROTGAMMAAUTO option). Sequences starting with "C3" or "scaffold" are cowpea sequences, while sequences starting with "Phvul" are from common bean. The circles on the branches are bootstrap support values from 50 to 100, with the largest circles representing the greatest bootstrap support. (JPEG 954 kb)

Additional file 7: Cowpea and common bean mTERF sequences, with predictive heatmaps based on FPKM expression values from cowpea transcriptome data on VuGEA [43]. This tree was generated using RAxML [62] with 100 bootstrap values with the optimal amino acid substitution model automatically chosen in RAxML (i.e., the PROTGAMMAAUTO option). Sequences starting with "C3" or "scaffold" are cowpea sequences, while sequences starting with "Phvul" are from common bean. The circles on the branches are bootstrap support values from 50 to 100, with the largest circles representing the greatest bootstrap support. (JPEG 2395 kb)

Additional file 8: Cowpea TFs and TAPs with corresponding transcripts in VuGEA, FPKM values, GO annotations, and gene families. This data shows a prediction of where and when cowpea TFs and TAPs express most strongly. (XLS 1840 kb)

Additional file 9: Two different groupings of *Arabidopsis thaliana* and rice (*Oryza sativa*) ERF sequences. The two schemes are the I-X grouping used in Nakano et al. (2006) [64], and DREB A1-A6 / ERF B1-B6 grouping from Dietz et al. (2010) [69] and Sharoni et al. (2011) [70]. (XLS 34 kb)

Abbreviations

AP2/ERF-ERF: ERF - Ethylene Response Factor; ABA: Abscisic acid; ABI3/VP1: Abscisic Acid Insensitive 3 - Viviparous1; ABTB: Ankyrin Broad Complex, tramtrack and bric a brac; Anr: Any repeat; AP2: APETALA2; AP2-EREBP: APETALA2-Ethylene Responsive Element Binding Protein; ARF: Auxin Response Factor; ATAF2: Arabidopsis transcription activation factor 2; BES/BZR: Brassinosteroid insensitive 1 (BRI1)-ethyl methanesulfonate (EMS)-suppressor / Brassinazole resistant bHLHBasic helix-loop-helix; BLASTN: Basic Local Alignment and Search Tool Nucleotide; bZIP: Basic leucine zipper domain; CCAAT-Dr1: Down regulator of transcription 1; CCAAT-HAP3: CCAAT motif, Heme-associated protein 3; CCAAT-HAP5: CCAAT motif, Heme-associated protein 5; CCHC(Zn): Zinc finger, CCHC-type (CysCysHisCys); CRs: Chromatin Remodelers; CT: C-terminal; CW-Zn-B3_VAL: CW-like zinc finger, B3, VP1/ABI3-Like (CysTrp); DREB: Dehydration-responsive element (DRE)-binding; ERF: ERF - Ethylene Response Factor; ESTs: Expressed sequence tags; FAO: Food and Agriculture Organization; GA: Gibberellic acid; GAII: Genome Analyzer II; GO: Gene Ontology; GRF: Growth-Regulating Factor; GSR: Gene space reads (or use GSS for gene space sequences); HMMs: Hidden Markov Models; iTOL: Interactive Tree of Life; JA: Jasmonic acid; JmjC-ARID: Jumonji C-terminal - AT-rich interaction domain; MADS: Minichromosome maintenance deficient (MCM1), AGAMOUS, DEFICIENS, and serum response factor (SRF); MAFFT: Multiple Alignment using Fast Fourier Transform; Mb: Megabase; MEME: Multiple Em for Motif Elicitation; mTERF: Mitochondrial transcription termination factor; MYB: Myeloblastosis; MYB-HB-like: MYB Homeobox like; MYB-related: Myeloblastosis-related; NF-X1: Nuclear factor, X-box binding 1; NF-YB: Nuclear factor Y subunit B; NF-YC: NF-Y subunit C; NF-YC: Nuclear factor Y subunit C; NT: N-terminal; ONAC: *Oryza sativa* NAC; ORF: Open reading frame; PBS: Portable Batch System; PcG_FIE: Polycomb Group Fertilization-Independent Endosperm; PHD: Plant Homeodomain; Phvul: *Phaseolus vulgaris*; PlantTAPDB: Plant Transcription Associated Protein Database; PlantTFDB: Plant Transcription Factor Database; PlnTFDB: Plant Transcription Factor Database; RAV: Related to ABI3/VP1; RAxML: Randomized Axelerated Maximum Likelihood; RF-X: Regulatory factor X; RiceGE: Rice Functional Genomic Express Database; Senu5: Senescence Upregulated 5; SIGnAL: Salk Institute Genomic Analysis Laboratory; SLURM: Simple Linux Utility for Resource Management; SND: Secondary Wall-Associated NAC Domain; SOH1: Suppressor of hyper-recombination 1 (hpr1); TAPs: Transcriptionally Active Proteins / Transcription Associated Proteins; TEA: Transcriptional enhancer activator; TERN: Tobacco Elicitor-Responsive NAC protein-like; TFs: Transcription Factors; TRAF: Tumor necrosis factor receptor-associated factor; TRs: Transcriptional Regulators; TUB: TUBBY; VuGEA: *Vigna unguiculata* Gene Expression Atlas; zn-clus: Zn(2)-Cys(6) binuclear cluster domain

Acknowledgements

We wish to thank Tim Close, Phil Roberts and their colleagues for making the draft version 0.03 of the cowpea genome available prior to publication and for helpful comments on the work. We also thank Mark Lawson and Aaron Mackey for their assistance in developing the computational pipeline used for data mining and the various members of the Timko laboratory, past and present, who contributed comments and suggestions on this work.

Funding

This work was supported in part by grants from the National Science Foundation (DBI-0701748 and IBN-0322420) and Kirkhouse Trust to MPT.

Authors' contributions

MPT conceived of the project and was responsible for directing all of the research activities. VAM carried out the computational analysis of the TFs with the assistance of YW. All authors have assisted in the writing of the manuscript and have read and approved the final submitted version of the manuscript.

Competing interests

The authors declare that they have no competing interests, and no financial relationships that would comprise a conflict of interest.

Author details

[1]Department of Biology, University of Virginia, Gilmer Hall 044, Charlottesville, VA 22904, USA. [2]Center for Quantitative Sciences, Vanderbilt University, Nashville, TN 37232-6848, USA.

References

1. Ehlers J, Hall A. Cowpea (*Vigna unguiculata* L. Walp.). Field Crop Res. 1997; 53(1):187–204.
2. Timko MP, Singh B. Cowpea, a multifunctional legume, Genomics of tropical crop plants; 2008. p. 227–58.
3. The Food and Agriculture Organization of the United Nations. FAOSTAT. 2014. Retrieved from faostat3.fao.org on February 10, 2017.
4. Langyintuo A, Lowenberg-DeBoer J, Faye M, Lambert D, Ibro G, Moussa B, et al. Cowpea supply and demand in west and Central Africa. Field Crop Res. 2003;82(2):215–31.
5. Timko MP, Ehlers JD, Roberts PA. Cowpea. In: Genome Mapping and Molecular Breeding in Plants, Volume 3 Pulses, Sugar and Tuber Crops. Springer-Verlag: Berlin Heidelberg; 2007. p. 49–67.
6. Singh B. Cowpea [*Vigna unguiculata* (L.) Walp]. In: Genetic resources, chromosome engineering, and crop improvement: grain legumes, vol. 1; 2005. p. 117–62.
7. Nielsen S, Ohler T, Mitchell C. Cowpea leaves for human consumption: production, utilization, and nutrient composition. Adv Cowpea Res. 1997;326
8. Singh B, Ajeigbe H, Tarawali SA, Fernandez-Rivera S, Abubakar M. Improving the production and utilization of cowpea as food and fodder. Field Crop Res. 2003;84(1):169–77.
9. Elowad HO, Hall AE. Influences of early and late nitrogen fertilization on yield and nitrogen fixation of cowpea under well-watered and dry field conditions. Field Crop Res. 1987;15(3):229–44.
10. Oseni TO. Evaluation of sorghum-cowpea intercrop productivity in savanna agro-ecology using competition indices. J Agric Sci. 2010;2(3):229.
11. Dahmardeh M, Ghanbari A, Syasar B, Ramrodi M. Intercropping maize (Zea Mays L.) and cow pea (*Vigna unguiculata* L.) as a whole-crop forage: effects of planting ratio and harvest time on forage yield and quality. J Food Agric Environ. 2009;7(2):505–9.
12. Hall AE. Breeding for adaptation to drought and heat in cowpea. Eur J Agron. 2004;21(4):447–54.
13. Boukar O, Fatokun CA, Huynh B-L, Roberts PA, Close TJ. Genomic tools in cowpea breeding programs: status and perspectives. Front Plant Sci. 2016;7: 757.
14. Roberts P, Matthews W, Ehlers J. New resistance to virulent root-knot nematodes linked to the Rk locus of cowpea. Crop Sci. 1996;36(4):889–94.
15. Roberts P, Ehlers J, Hall A, Matthews W. Characterization of new resistance to root-knot nematodes in cowpea. In: Singh BB, Mohan Raj DR, Dashiell KE, Jackai LEN, editors. Advances in cowpea research. Nigeria: IITA Ibadan; 1997. p. 207–14.
16. Das S, Ehlers J, Close T, Roberts P. Transcriptional profiling of root-knot nematode induced feeding sites in cowpea (Vigna Unguiculata L. Walp.) using a soybean genome array. BMC Genomics. 2010;11(1):480.
17. Wests D, Francois L. Effects of salinity on germination, growth and yield of cowpea. Irrig Sci. 1982;3(3):169–75.
18. Alonge SO, Lagoke STO, Ajakaiye CO. Cowpea reactions to Striga Gesnerioides I. Effect on growth. Crop Prot. 2005;24(6):565–73.
19. Alonge SO, Lagoke STO, Ajakaiye CO. Cowpea reactions to Striga Gesnerioides II. Effect on grain yield and nutrient composition. Crop Prot. 2005;24(6):575–80.
20. Cardwell KF, Lane JA. Effect of soils, cropping system and host phenotype on incidence and severity of Striga Gesnerioides on cowpea in West Africa. Agric Ecosyst Environ. 1995;53(3):253–62.
21. MacQuarrie KL, Fong AP, Morse RH, Tapscott SJ. Genome-wide transcription factor binding: beyond direct target regulation. Trends Genet. 2011;27(4): 141–8.
22. Shore P, Sharrocks AD. The MADS-box family of transcription factors. Eur J Biochem. 1995;229(1):1–13.
23. Kubo M, Udagawa M, Nishikubo N, Horiguchi G, Yamaguchi M, Ito J, Mimura T, Fukuda H, Demura T. Transcription switches for protoxylem and metaxylem vessel formation. Genes Dev. 2005;19(16):1855–60.
24. Zhao C, Avci U, Grant EH, Haigler CH, Beers EP. XND1, a member of the NAC domain family in Arabidopsis Thaliana, negatively regulates lignocellulose synthesis and programmed cell death in xylem. Plant J. 2008; 53(3):425–36.
25. Singh K, Foley RC, Onate-Sanchez L. Transcription factors in plant defense and stress responses. Curr Opin Plant Biol. 2002;5(5):430–6.
26. Timko MP, Rushton PJ, Laudeman TW, Bokowiec MT, Chipumuro E, Cheung F, Town CD, Chen X. Sequencing and analysis of the gene-rich space of cowpea. BMC Genomics. 2008;9:103.
27. Muñoz-Amatriaín M, Mirebrahim H, Xu P, Wanamaker SI, Luo M, Alhakami H, Alpert M, Atokple I, Batieno BJ, Boukar O. Genome resources for climate-resilient cowpea, an essential crop for food security. Plant J. 2017;89(5): 1042–54.
28. Riechmann JL, Ratcliffe OJ. A genomic perspective on plant transcription factors. Curr Opin Plant Biol. 2000;3(5):423–34.
29. Gao G, Zhong Y, Guo A, Zhu Q, Tang W, Zheng W, Gu X, Wei L, Luo J. DRTF: a database of rice transcription factors. Bioinformatics. 2006;22(10):1286–7.
30. Zhu Q-H, Guo A-Y, Gao G, Zhong Y-F, Xu M, Huang M, Luo J. DPTF: a database of poplar transcription factors. Bioinformatics. 2007;23(10):1307–8.
31. Schmutz J, Cannon SB, Schlueter J, Ma J, Mitros T, Nelson W, Hyten DL, Song Q, Thelen JJ, Cheng J, et al. Genome sequence of the palaeopolyploid soybean. Nature. 2010;463(7278):178–83.
32. Richardt S, Lang D, Reski R, Frank W, Rensing SA. PlanTAPDB, a phylogeny-based resource of plant transcription-associated proteins. Plant Physiol. 2007;143(4):1452–66.
33. Udvardi MK, Kakar K, Wandrey M, Montanari O, Murray J, Andriankaja A, Zhang JY, Benedito V, Hofer JM, Chueng F, et al. Legume transcription factors: global regulators of plant development and response to the environment. Plant Physiol. 2007;144(2):538–49.
34. Rushton PJ, Bokowiec MT, Han S, Zhang H, Brannock JF, Chen X, Laudeman TW, Timko MP. Tobacco transcription factors: novel insights into transcriptional regulation in the Solanaceae. Plant Physiol. 2008; 147(1):280–95.
35. Riano-Pachon DM, Ruzicic S, Dreyer I, Mueller-Roeber B. PlnTFDB: an integrative plant transcription factor database. BMC Bioinformatics. 2007; 8:42.
36. Perez-Rodriguez P, Riano-Pachon DM, Correa LG, Rensing SA, Kersten B, Mueller-Roeber B. PlnTFDB: updated content and new features of the plant transcription factor database. Nucleic Acids Res. 2010;38(Database issue): D822–7.
37. Jin J, Tian F, Yang D-C, Meng Y-Q, Kong L, Luo J, Gao G. PlantTFDB 4.0: toward a central hub for transcription factors and regulatory interactions in plants. Nucleic Acids Res. 2017;45(D1):D1040–5.
38. Rouard M, Guignon V, Aluome C, Laporte M-A, Droc G, Walde C, Zmasek CM, Périn C, Conte MG. GreenPhylDB v2. 0: comparative and functional genomics in plants. Nucleic Acids Res. 2011;39(suppl 1):D1095–102.
39. Rouard M, Guignon V, Dufayard JF, Conte M, Briois S, Cenci A, et al. (2014) GreenPhyl v4. 2014. http://www.greenphyl.org/cgi-bin/index.cgi. Accessed 10 Jan 2016.
40. Dai X, Sinharoy S, Udvardi M, Zhao PX. PlantTFcat: an online plant transcription factor and transcriptional regulator categorization and analysis tool. BMC Bioinformatics. 2013;14(1):321.
41. Zheng Y, Jiao C, Sun H, Rosli HG, Pombo MA, Zhang P, Banf M, Dai X, Martin GB, Giovannoni JJ. iTAK: a program for genome-wide prediction and classification of plant transcription factors, transcriptional regulators, and protein kinases. Mol Plant. 2016;9(12):1667–70.
42. iTAK - Plant transcription factor & protein Kinase identifier and classifier. Fei bioinformatics lab. 2016. http://bioinfo.bti.cornell.edu/cgi-bin/itak/index.cgi. Accessed 10 Sep 2016.
43. Yao S, Jiang C, Huang Z, Torres-Jerez I, Chang J, Zhang H, Udvardi M, Liu R, Verdier J. The Vigna Unguiculata gene expression atlas (VuGEA) from de novo assembly and quantification of RNA-seq data provides insights into seed maturation mechanisms. Plant J. 2016;88(2):318–27.
44. Lang D, Weiche B, Timmerhaus G, Richardt S, Riano-Pachon DM, Correa LG, Reski R, Mueller-Roeber B, Rensing SA. Genome-wide phylogenetic comparative analysis of plant transcriptional regulation: a timeline of loss, gain, expansion, and correlation with complexity. Genome Biol Evol. 2010;2:488–503.
45. Lawson M, Shuber P. TavernaPBS. 2010. https://sourceforge.net/projects/tavernapbs/. Accessed 8 Sep 2016.

46. Wolstencroft K, Haines R, Fellows D, Williams A, Withers D, Owen S, Soiland-Reyes S, Dunlop I, Nenadic A, Fisher P. The Taverna workflow suite: designing and executing workflows of web services on the desktop, web or in the cloud. Nucleic Acids Res. 2013;41(W1):W557–61.
47. Kang YJ, Satyawan D, Shim S, Lee T, Lee J, Hwang WJ, Kim SK, Lestari P, Laosatit K, Kim KH. Draft genome sequence of adzuki bean, Vigna Angularis. Sci Rep. 2015;5
48. Schmutz J, McClean PE, Mamidi S, Wu GA, Cannon SB, Grimwood J, Jenkins J, Shu S, Song Q, Chavarro C. A reference genome for common bean and genome-wide analysis of dual domestications. Nat Genet. 2014;46(7):707–13.
49. Tang H, Krishnakumar V, Bidwell S, Rosen B, Chan A, Zhou S, Gentzbittel L, Childs KL, Yandell M, Gundlach H. An improved genome release (version Mt4. 0) for the model legume Medicago Truncatula. BMC Genomics. 2014; 15(1):312.
50. Cantarel BL, Korf I, Robb SM, Parra G, Ross E, Moore B, Holt C, Sanchez Alvarado A, Yandell M. MAKER: an easy-to-use annotation pipeline designed for emerging model organism genomes. Genome Res. 2008;18(1):188–96.
51. Stanke M, Morgenstern B. AUGUSTUS: a web server for gene prediction in eukaryotes that allows user-defined constraints. Nucleic Acids Res. 2005; 33(suppl 2):W465–7.
52. Stanke M, Steinkamp R, Waack S, Morgenstern B. AUGUSTUS: a web server for gene finding in eukaryotes. Nucleic Acids Res. 2004;32(suppl 2):W309–12.
53. *Vigna unguiculata* Gene Expression Atlas. Noble Research Institute, Ardmore. 2016. https://vugea.noble.org/list_tf.php. Accessed on 3 Dec 2016.
54. FASTA version 36.3.8e. The Pearson Lab at the University of Virginia, Charlottesville. 2016. http://faculty.virginia.edu/wrpearson/fasta/fasta36/. Accessed 27 Nov 2016.
55. Phytozome. *Glycine max* Wm82.a2.v1 (soybean). US Department of Energy: Office of Science; 2016. https://phytozome.jgi.doe.gov/pz/portal.html#!info?alias=Org_Gmax Accessed 1 Sep 2016
56. Olsen AN, Ernst HA, Leggio LL, Skriver K. NAC transcription factors: structurally distinct, functionally diverse. Trends Plant Sci. 2005;10(2):79–87.
57. Van Verk MC, Gatz C, Linthorst HJM. Transcriptional regulation of plant defense responses. In: LC VL, editor. Plant innate immunity, vol. 51. London: Academic Press Ltd-Elsevier Science Ltd; 2009. p. 397–438.
58. Ohme-Takagi M, Shinshi H. Ethylene-inducible DNA binding proteins that interact with an ethylene-responsive element. Plant Cell Online. 1995;7(2): 173–82.
59. Eulgem T, Rushton PJ, Robatzek S, Somssich IE. The WRKY superfamily of plant transcription factors. Trends Plant Sci. 2000;5(5):199–206.
60. Rushton PJ, Somssich IE, Ringler P, Shen QJ. WRKY transcription factors. Trends Plant Sci. 2010;15(5):247–58.
61. Katoh K, Standley DM. MAFFT multiple sequence alignment software version 7: improvements in performance and usability. Mol Biol Evol. 2013; 30(4):772–80.
62. Stamatakis A. RAxML version 8: a tool for phylogenetic analysis and post-analysis of large phylogenies. Bioinformatics. 2014;30(9):1312–3.
63. Byzova MV, Franken J, Aarts MG, de Almeida-Engler J, Engler G, Mariani C, Van Lookeren Campagne MM, Angenent GC. Arabidopsis STERILE APETALA, a multifunctional gene regulating inflorescence, flower, and ovule development. Genes Dev. 1999;13(8):1002–14.
64. Nakano T, Suzuki K, Fujimura T, Shinshi H. Genome-wide analysis of the ERF gene family in Arabidopsis and Rice. Plant Physiol. 2006;140(2):411–32.
65. Xu ZS, Chen M, Li LC, Ma YZ. Functions and application of the AP2/ERF transcription factor family in crop improvement. J Integr Plant Biol. 2011; 53(7):570–85.
66. Riechmann JL, Meyerowitz EM. The AP2/EREBP family of plant transcription factors. Biol Chem. 1998;379(6):633–46.
67. Kagaya Y, Ohmiya K, Hattori T. RAV1, a novel DNA-binding protein, binds to bipartite recognition sequence through two distinct DNA-binding domains uniquely found in higher plants. Nucleic Acids Res. 1999;27(2):470–8.
68. Rushton, P. J. TOBFAC: Family AP2. 2008. http://compsysbio.achs.virginia.edu/tobfac/browse_family.pl?family=AP2. Accessed on 22 Jun 2013.
69. Dietz KJ, Vogel MO, Viehhauser A. AP2/EREBP transcription factors are part of gene regulatory networks and integrate metabolic, hormonal and environmental signals in stress acclimation and retrograde signalling. Protoplasma. 2010;245(1–4):3–14.
70. Sharoni AM, Nuruzzaman M, Satoh K, Shimizu T, Kondoh H, Sasaya T, Choi IR, Omura T, Kikuchi S. Gene structures, classification and expression models of the AP2/EREBP transcription factor family in rice. Plant Cell Physiol. 2011; 52(2):344–60.
71. Ooka H, Satoh K, Doi K, Nagata T, Otomo Y, Murakami K, Matsubara K, Osato N, Kawai J, Carninci P, et al. Comprehensive analysis of NAC family genes in Oryza Sativa and Arabidopsis Thaliana. DNA Res. 2003;10(6):239–47.
72. Zhu T, Nevo E, Sun D, Peng J. Phylogenetic analyses unravel the evolutionary history of NAC proteins in plants. Evolution. 2012;66(6): 1833–48.
73. Rushton PJ, Macdonald H, Huttly AK, Lazarus CM, Hooley R. Members of a new family of DNA-binding proteins bind to a conserved cis-element in the promoters of alpha-Amy2 genes. Plant Mol Biol. 1995;29(4):691–702.
74. Li H-L, Zhang L-B, Guo D, Li C-Z, Peng S-Q. Identification and expression profiles of the WRKY transcription factor family in Ricinus Communis. Gene. 2012;503(2):248–53.
75. Kim Y-S, Kim S-G, Park J-E, Park H-Y, Lim M-H, Chua N-H, Park C-M. A membrane-bound NAC transcription factor regulates cell division in Arabidopsis. Plant Cell. 2006;18(11):3132–44.
76. Carbonero P, Iglesias-Fernández R, Vicente-Carbajosa J. The AFL subfamily of B3 transcription factors: evolution and function in angiosperm seeds. J Exp Bot. 2016;68(4):871–80.
77. Bedi S, Sengupta S, Ray A, Chaudhuri RN. ABI3 mediates dehydration stress recovery response in Arabidopsis Thaliana by regulating expression of downstream genes. Plant Sci. 2016;250:125–40.
78. Vilarrasa-Blasi J, González-García M-P, Frigola D, Fàbregas N, Alexiou KG, López-Bigas N, Rivas S, Jauneau A, Lohmann JU, Benfey PN. Regulation of plant stem cell quiescence by a brassinosteroid signaling module. Dev Cell. 2014;30(1):36–47.
79. Salazar-Henao JE, Lehner R, Betegón-Putze I, Vilarrasa-Blasi J, Caño-Delgado AI. BES1 regulates the localization of the brassinosteroid receptor BRL3 within the provascular tissue of the Arabidopsis primary root. J Exp Bot. 2016;67(17):4951–61.
80. Myers ZA, Kumimoto RW, Siriwardana CL, Gayler KK, Risinger JR, Pezzetta D, Holt BF III. NUCLEAR FACTOR Y, subunit C (NF-YC) transcription factors are positive regulators of photomorphogenesis in Arabidopsis Thaliana. PLoS Genet. 2016;12(9):e1006333.
81. Robles P, Micol JL, Quesada V. Unveiling plant mTERF functions. Mol Plant. 2012;5(2):294–6.
82. Muchero W, Diop NN, Bhat PR, Fenton RD, Wanamaker S, Pottorff M, Hearne S, Cisse N, Fatokun C, Ehlers JD, et al. A consensus genetic map of cowpea [Vigna Unguiculata (L) Walp.] and synteny based on EST-derived SNPs. Proc Natl Acad Sci U S A. 2009;106(43):18159–64.
83. McClean P, Mamidi S, McConnell M, Chikara S, Lee R. Synteny mapping between common bean and soybean reveals extensive blocks of shared loci. BMC Genomics. 2010;11(1):184.
84. Galeano C, Fernandez A, Gomez M, Blair M. Single strand conformation polymorphism based SNP and Indel markers for genetic mapping and synteny analysis of common bean (Phaseolus Vulgaris L.). BMC Genomics. 2009;10(1):629.
85. Boutin S, Young N, Olson T, Yu Z-H, Vallejos C, Shoemaker R. Genome conservation among three legume genera detected with DNA markers. Genome. 1995;38(5):928–37.
86. Vasconcelos EV, de Andrade Fonsêca AF, Pedrosa-Harand A, de Andrade Bortoleti KC, Benko-Iseppon AM, Da Costa AF, Brasileiro-Vidal AC. Intra-and interchromosomal rearrangements between cowpea [Vigna Unguiculata (L.) Walp.] and common bean (Phaseolus Vulgaris L.) revealed by BAC-FISH. Chromosom Res. 2015;23(2):253.
87. Zhuang J, Deng D-X, Yao Q-H, Zhang J, Xiong F, Chen J-M, Xiong A-S. Discovery, phylogeny and expression patterns of AP2-like genes in maize. Plant Growth Regul. 2010;62(1):51–8.
88. Zhang G, Chen M, Chen X, Xu Z, Guan S, Li L-C, Li A, Guo J, Mao L, Ma Y. Phylogeny, gene structures, and expression patterns of the ERF gene family in soybean (Glycine max L.). J Exp Bot. 2008;59(15):4095–107.
89. Mosa K, El-din EH, Ismail A, El-Feky F, El-Refy A. Molecular characterization of two AP2/ERF transcription factor genes from Egyptian tomato cultivar (Edkawy). Plant Sci Today. 2017;4(1):12–20.
90. Liu C, Zhang T. Expansion and stress responses of the AP2/EREBP superfamily in cotton. BMC Genomics. 2017;18(1):118.
91. Pinheiro GL, Marques CS, Costa MDBL, Reis PAB, Alves MS, Carvalho CM, Fietto LG, Fontes EPB. Complete inventory of soybean NAC transcription factors: sequence conservation and expression analysis uncover their distinct roles in stress response. Gene. 2009;444(1–2): 10–23.

92. Lavin M, Herendeen PS, Wojciechowski MF. Evolutionary rates analysis of Leguminosae implicates a rapid diversification of lineages during the tertiary. Syst Biol. 2005;54(4):575–94.
93. Moore RC, Purugganan MD. The evolutionary dynamics of plant duplicate genes. Curr Opin Plant Biol. 2005;8(2):122–8.
94. Swaminathan K, Peterson K, Jack T. The plant B3 superfamily. Trends Plant Sci. 2008;13(12):647–55.
95. Omidbakhshfard MA, Proost S, Fujikura U, Mueller-Roeber B. Growth-regulating factors (GRFs): a small transcription factor family with important functions in plant biology. Mol Plant. 2015;8(7):998–1010.
96. Zhong R, Lee C, Zhou J, McCarthy RL, Ye Z-H. A battery of transcription factors involved in the regulation of secondary cell wall biosynthesis in Arabidopsis. Plant Cell. 2008;20(10):2763–82.
97. Shani E, Salehin M, Zhang Y, Sanchez SE, Doherty C, Wang R, Mangado CC, Song L, Tal I, Pisanty O. Plant stress tolerance requires auxin-sensitive Aux/IAA transcriptional repressors. Curr Biol. 2017;27(3):437–44.
98. Jung H, Lee D-K, Do Choi Y, Kim J-K. OsIAA6, a member of the rice aux/IAA gene family, is involved in drought tolerance and tiller outgrowth. Plant Sci. 2015;236:304–12.
99. Balcerowicz M, Ranjan A, Rupprecht L, Fiene G, Hoecker U. Auxin represses stomatal development in dark-grown seedlings via aux/IAA proteins. Development. 2014;141(16):3165–76.
100. Zhang Y, Paschold A, Marcon C, Liu S, Tai H, Nestler J, Yeh C-T, Opitz N, Lanz C, Schnable PS. The aux/IAA gene rum1 involved in seminal and lateral root formation controls vascular patterning in maize (Zea Mays L.) primary roots. J Exp Bot. 2014;65(17):4919–30.
101. Yulong C, Wei D, Baoming S, Yang Z, Qing M. Genome-wide identification and comparative analysis of the TUBBY-like protein gene family in maize. Genes Genomics. 2016;1(38):25–36.
102. Lai C-P, Lee C-L, Chen P-H, Wu S-H, Yang C-C, Shaw J-F. Molecular analyses of the Arabidopsis TUBBY-like protein gene family. Plant Physiol. 2004;134(4): 1586–97.
103. Saha G, Park J-I, Kayum MA, Nou I-S. A genome-wide analysis reveals stress and hormone responsive patterns of TIFY family genes in Brassica Rapa. Front Plant Sci. 2016;7
104. RiceGE: Rice Functional Genomic Express Database. Salk Institute Genomic Analysis Laboratory (SIGnAL). 2013. http://signal.salk.edu/cgi-bin/RiceGE Accessed 19 Jan 2017.
105. Pottorff M, Ehlers JD, Fatokun C, Roberts PA, Close TJ. Leaf morphology in cowpea [Vigna Unguiculata (L.) Walp]: QTL analysis, physical mapping and identifying a candidate gene using synteny with model legume species. BMC Genomics. 2012;13(1):234.
106. HarvEST Blast Search. The Close Lab at University of California, Riverside. 2013. http://138.23.178.42/blast/blast.html Accessed on 1 Jun 2016.
107. Phytozome. 2016. https://phytozome.jgi.doe.gov/pz/portal.html Accessed 10 May 2016.
108. Goodstein DM, Shu S, Howson R, Neupane R, Hayes RD, Fazo J, Mitros T, Dirks W, Hellsten U, Putnam N, et al. Phytozome: a comparative platform for green plant genomics. Nucleic Acids Res. 2012;40(Database issue):D1178–86.
109. *Phaseolus vulgaris* v2.1. DOE-JGI and USDA-NIFA. 2014. http://phytozome.jgi.doe.gov/ Accessed 1 Sep 2016.
110. Eddy SR. Accelerated profile HMM searches. PLoS Comput Biol. 2011;7(10): e1002195.
111. Finn RD, Coggill P, Eberhardt RY, Eddy SR, Mistry J, Mitchell AL, Potter SC, Punta M, Qureshi M, Sangrador-Vegas A. The Pfam protein families database: towards a more sustainable future. Nucleic Acids Res. 2016;44(D1): D279–85.
112. Finn RD, Attwood TK, Babbitt PC, Bateman A, Bork P, Bridge AJ, Chang H-Y, Dosztányi Z, El-Gebali S, Fraser M, et al. InterPro in 2017—beyond protein family and domain annotations. Nucleic Acids Res. 2017;45(D1):D190–9.
113. Bailey TL, Boden M, Buske FA, Frith M, Grant CE, Clementi L, Ren J, Li WW, Noble WS. MEME SUITE: tools for motif discovery and searching. Nucleic Acids Res. 2009;37(suppl_2):W202–8.
114. Capella-Gutiérrez S, Silla-Martínez JM. Gabaldón T: trimAl: a tool for automated alignment trimming in large-scale phylogenetic analyses. Bioinformatics. 2009;25(15):1972–3.
115. Letunic I, Bork P. Interactive tree of life (iTOL) v3: an online tool for the display and annotation of phylogenetic and other trees. Nucleic Acids Res. 2016;44(W1):W242–5.
116. Anwar N, Hunt E. Improved data retrieval from TreeBASE via taxonomic and linguistic data enrichment. BMC Evol Biol. 2009;9(1):93.

8

Transcriptomics analysis of the flowering regulatory genes involved in the herbicide resistance of Asia minor bluegrass (*Polypogon fugax*)

Fengyan Zhou[1*†], Yong Zhang[1†], Wei Tang[2], Mei Wang[1] and Tongchun Gao[1]

Abstract

Background: Asia minor bluegrass (*Polypogon fugax, P. fugax*), a weed that is both distributed across China and associated with winter crops, has evolved resistance to acetyl-CoA carboxylase (ACCase) herbicides, but the resistance mechanism remains unclear. The goal of this study was to analyze the transcriptome between resistant and sensitive populations of *P. fugax* at the flowering stage.

Results: Populations resistant and susceptible to clodinafop-propargyl showed distinct transcriptome profiles. A total of 206,041 unigenes were identified; 165,901 unique sequences were annotated using BLASTX alignment databases. Among them, 5904 unigenes were classified into 58 transcription factor families. Nine families were related to the regulation of plant growth and development and to stress responses. Twelve unigenes were differentially expressed between the clodinafop-propargyl-sensitive and clodinafop-propargyl-resistant populations at the early flowering stage; among those unigenes, three belonged to the ABI3VP1, BHLH, and GRAS families, while the remaining nine belonged to the MADS family. Compared with the clodinafop-propargyl-sensitive plants, the resistant plants exhibited different expression pattern of these 12 unigenes.

Conclusion: This study identified differentially expressed unigenes related to ACCase-resistant *P. fugax* and thus provides a genomic resource for understanding the molecular basis of early flowering.

Keywords: RNA sequencing, Transcriptomics, Herbicide resistance, *Polypogon fugax*, Clodinafop-propargyl, Flowering

Background

Common annual Asia minor bluegrass (*Polypogon fugax*) is a weed that is both distributed across China and associated with winter crops. This weed has evolved resistance to clodinafop-propargyl, an acetyl-CoA carboxylase (ACCase) herbicide [1]. The mechanisms of herbicide resistance have been intensively studied in the past twenty years and at least three have been identified: 1) target site change; 2) closure or translocation of herbicides; and 3) alteration in the rate of herbicide metabolism [2–4]. Nevertheless, these three mechanisms alone often fail to explain the development of herbicide resistance from an evolutionary and ecological perspective [5]. Herbicide resistance will increase the fitness of resistant individuals and hence their ability to produce the next generation. Identifying which biological characteristics play a major role on fitness and interactions with environmental factors is essential for predicting herbicide resistance.

It is generally believed that the initial occurrence of major resistance genes in weed populations is the main factor that influences the dynamic evolution of a resistance under herbicide selection [6]. In *Lolium rigidum*, carrying the common Leu-1781 in ACCase affects the fitness of resistant mutants, and the germination rate of the resistant biotype is lower than that of the sensitive biotype [7]. To palliate this lower germination rate, the mutant *Setaria viridis* produces more seeds than the

* Correspondence: zbszhoufy@163.com
†Equal contributors
[1]Institute of Plant Protection and Agro-Products Safety, Anhui Academy of Agricultural Sciences, Hefei 230001, China
Full list of author information is available at the end of the article

sensitive population [8]. In addition, herbicide-resistant *Setaria* flowers earlier than the susceptible population, with more tillers and panicle, and with lighter seeds [9]. These mechanisms contribute to a more important spread of the resistant populations compared with susceptible ones.

Stress-induced flowering has recently received increased attention. Early flowering and seed setting of resistant plants allow the resistant seeds better access to resources [9]. Photoperiodic flowering and vernalization have been well characterized, and the regulatory mechanisms are well known [10–12]. In addition, flowering physiologists had reported that plants tend to flower when grown under unsuitable conditions, indicating that stress is a flower-inducing factor [13–16]. Stress-induced flowering is now considered as the third category of flowering responses alongside regulated autonomous flowering and environment-induced flowering [17].

Nevertheless, the mechanisms underlying early flowering remain poorly understood, particularly among herbicide-resistant weeds. Stress-induced flowering changes the life cycle and might alter fitness. In addition, stress adaptation extends to the evolution of the flowering characteristics [17]. Transcription factors are proteins regulating gene expression and specific transcription factors selectively regulate the transcriptional expression of specific genes. Therefore, in the present study, we aimed to investigate the transcription factors that regulate plant flowering in order to elucidate the relationship between early flowering and selection pressure (herbicide application) of ACCase-resistant *P. fugax*, and to identify candidate genes responsible for early flowering in resistant plants.

Methods

Plants

The seeds of a putative resistant population of *P. fugax* (RP population) were collected from Qingsheng County (29° 54′ 1″ N, 103° 48′ 57″ E), Sichuan Province, China, where clodinafop-propargyl has been used for more than five years and has failed to control *P. fugax* growth. A sensitive population of *P. fugax* (susceptible plant (SP) population) was sampled from a non-cultivated area in Xichang City, Sichuan Province (27° 50′ 56″ N, 102° 15′ 53″ E). The plants were collected without permissions being sought for the nature of scientific research according to the law of the People's Republic of China.

Because gene expression can differ due to genetic background, genetically homogenized plant material was generated by controlled pairings to narrow the difference. In brief, F1 plants were transplanted to individual 1-L pots in a greenhouse that has an 18/15 °C day/night temperature and a 14-h photoperiod. At the four-tiller stage, the plants were subjected to vegetative propagation: all individual tillers of each plant were separated and transplanted to individual pots. Four clones were therefore obtained. At the three-leaf growth stage, ACCase herbicide was applied. The herbicide sensitivity of each F1 plant was assessed by spraying with clodinafop-propargyl (45 g. active ingredient (a.i.)/ha). Sensitive and resistant F1 plants were then crossed, yielding an F2 population. Visual phenotype rating of the F2 plants was carried out by clodinafop-propargyl selection. The F2 plants with contrasting phenotypes were selected as resistant and sensitive plants. And the F2 generation was used for transcriptome sequencing.

Sample collection

RPs and SPs ($n = 18$/group) at the seedling stage (3-leaf stage) were selected, and treated with clodinafop-propargyl (R: 45 g.a.i./ha; S: 0.2 g.a.i/ha) ($n = 9$ plants/group-); the remaining plants were treated with an equal volume of water. After 72 h, the aerial parts were collected ($n = 3$ plants/group) and immediately cryopreserved in liquid nitrogen. Afterward, samples were taken at the tillering (four tillers, $n = 3$ plants/group) and flowering (early flowering, $n = 3$ plants/group) stages in the same manner and stored in liquid nitrogen.

Genomic library construction and sequencing

The total RNA was extracted using TRIzol (Invitrogen Inc., Carlsbad, CA, USA) and DNase I (Takara, Otsu, Japan) in accordance with the manufacturer's instructions. Magnetic beads with oligo (dT) (Takara, Otsu, Japan) were used to isolate mRNA, and the mRNA was mixed together with fragmentation buffer by a Thermo-Mixer (Thermo Fisher Scientific, Waltham, MA, USA), breaking the mRNA into short fragments. cDNA was synthesized using the RevertAid First Strand cDNA Synthesis Kit (Fermentas, Thermo Fisher Scientific, Waltham, MA, USA) in accordance with the manufacturer's instructions. Short fragments were purified and resolved with EB buffer (10 mM Tris·Cl, pH 8.5) to repair their ends by the addition of a single adenine nucleotide. The short fragments were then connected with adaptors (BGI, Beijing, China). Suitable fragments were selected as templates for PCR amplification. The constructed sample library was quantified and characterized using an Agilent 2100 Bioanalyzer (Agilent Technologies, Santa Clara, CA, USA) and an ABI StepOnePlus Real-Time PCR system (Applied Biosystems, Foster City, CA, USA) and then sequenced using a HiSeq 4000 system (Illumina, Inc., San Diego, CA, USA).

Sequencing and de novo assembly

To obtain high-quality clean reads for de novo assembly, the raw reads were generated from transcriptome sequencing in accordance with the following steps. First, the adaptor sequences were removed. Then, reads with more than 5% of unknown nucleotides were removed and reads with more than 50% of low-quality bases (base quality ≤ 20) were discarded. The clean reads that remained were assembled into unigenes using the Trinity software with an optimized K-mer length of 25 for de novo assembly, as previously published [18]. The expression of unigenes was calculated via the reads per kb per million reads (RPKM, ≥ 0.5), which is a general method of quantifying gene expression from RNA sequencing data by normalizing for total read length and the number of sequencing reads [19].

Data analysis

Genes expressed at different levels in RPs and SPs (i.e., differentially expressed genes (DEGs)) were subjected to Gene Ontology (GO) functional analysis and Kyoto Encyclopedia of Genes and Genomes (KEGG) pathway analysis. We used getorf software (http://emboss.bioinformatics.nl/cgi-bin/emboss/getorf) to identify the open reading frame (ORF) of each unigene. Then, the ORFs were aligned to the transcription factor (TF) domains using hmmsearch software (http://hmmer.org/) [20].

The false discovery rate (FDR) statistical method was used in multiple hypothesis testing to correct the *P*-value. A smaller FDR and larger ratio indicate a greater difference in the expression level between two samples. In this analysis, we chose samples with an FDR ≤ 0.001 and a ratio greater than 2.

Differential gene expression analysis

Gene expression levels were estimated using RSEM for each sample, as previously described [21]. Clean reads were mapped back onto the assembled transcriptome, and read count for each gene was obtained from the mapping results and normalized as FPKM (fragments per kilobase of transcript per million mapped reads). Differential expression analysis of the two groups was performed using the DESeq 2. The resulting *P*-values were adjusted using the Benjamini and Hochberg's approach for controlling the false discovery rate. Genes with an adjusted P-value < 0.05 (fold change ≥ 2) found by DESeq were determined as being differentially expressed. Differentially expressed genes (DEGs) were analyzed by GO and KEGG enrichment analysis.

GO and KEGG enrichment analyses

All DEGs were mapped to terms in the GO database (http://www.geneontology.org//) and the number of genes corresponding to each GO term was calculated. We established a gene list and gene numbers for each GO term and then used a hypergeometric test to identify DEG GO terms whose enrichment was significantly greater than that in the genome background.

The KEGG pathway database contains networks of molecular interactions in cells and variants specific to particular organisms. We used pathway enrichment analysis to identify DEG metabolic pathways or signal transduction pathways whose enrichment was significantly greater than that in the whole genomic background. After multiple corrections, we selected pathways with an FDR value ≤ 0.001 to represent pathways significantly enriched in DEGs.

Real-time PCR

The total RNA was extracted as described above and cDNA was synthesized using an M-MLV Rtase Kit (Thermo Fisher Scientific, Waltham, MA, USA) in accordance with the manufacturer's instructions. The qRT-PCR mix (25 μl) contained 12.5 μl of SYBR Green Mix (Thermo Fisher Scientific, Waltham, MA, USA), 0.5 μl of each primer (10 μM), 2 μl of cDNA, and 9.5 μl of RNase-free water. The reaction was performed on an ABI 7300 real-time PCR system (Applied Biosystems, Foster City, CA, USA). The qRT-PCR program consisted of 95 °C for 10 min, 40 cycles of 95 °C for 15 s and 60 °C for 45 s, and finally 60 °C for 15 s. *EF1* was used as a reference gene for normalization. GraphPad Prism 5 software (GraphPad Software, Inc., La Jolla, CA, USA) was used for data analysis. Expression was calculated in accordance with the $2^{-\Delta\Delta Ct}$ method. Each experiment was repeated at least three times and consisted of three replicates. Primer sequences are listed in Additional file 1: Table S1.

Data access

The raw reads have been deposited in the NCBI Sequence Read Archive (SRA) database (BioProject PRJNA385696, SRP106591). The data are available at https://trace.ncbi.nlm.nih.gov/Traces/sra_sub/sub.cgi?acc=SRP106591&focus=SRP106591&from=list&action=show:STUDY.

Results

Polypogon fugax Transcriptome sequencing and data assembly

Compared with *P. fugax* SPs, RPs flowered 10-15 days earlier, and their inflorescence was morphologically altered, RPs produced 1.9 times more seed than did SPs Additional file 2: Figure S1). Therefore, the transcriptome of SPs and RPs were compared to explore the potential genes involved in this process.

Strand-specific RNA-Seq was applied to assess RNAs from three pairs of RPs and SPs at the seedling, tillering, and flowering stages with or without clodinafop-propargyl treatment (seedling stage) to comprehensively identify the unigenes associated with herbicide resistance.

The sequencing reads containing low-quality, adaptor-contaminated, or high contents of unknown base (N) reads were removed before downstream analyses. Afterward, 150-base single-end sequence raw reads were subjected to quality control using the Phred scaled quality score. Overall, 1.39 billion raw reads and 1.07 billion clean reads (average clean read ratio of 77.05%) were obtained; 92.8% of the clean reads had a quality score ≥ 30, and 97.7% of the clean reads were quality filtered and matched the Illumina's quality requirements. Read quality metrics after filtering are shown in Additional file 3: Table S2. De novo assembly of the 150-base reads yielded 206,041 unique sequences ranging from 300 to 3000 nt in length (Table 1) (including 14,166 unigenes with sequences of up to 3000 nt in length) Additional file 3: Table S3). The length distribution of the assembled contigs is shown in Additional file 4: Figure S2. Among the detected 206,041 unigenes, 165,901 unique sequences were annotated based on BLASTX alignment (E-value <0.00001) searches of seven databases: the NCBI non-redundant (NR), NCBI non-redundant nucleotide (NT), Swiss-Prot protein, KEGG, Cluster of Orthologous Groups of proteins (COG), InterPro protein, and GO databases (Additional file 3: Table S4). The 153,591 unique sequences were annotated by reference to the NR database, and then compared to those encoded in the genomes of all grass (*Poaceae*) species whose genome is fully sequenced, i.e., *Brachypodium distachyon*, *Hordeum vulgare*, *Aegilops tauschii*, and *Triticum urartu*. (Additional file 5: Figure S3).

Annotation of assembled unigenes

To further examine the integrity and effectiveness of the annotation process, the number of unigenes (that have NR matches) with a COG classification was calculated. A total of 74,434 unigenes were identified with a COG classification (Additional file 3: Table S5). Among the 25 COG categories, the cluster of "General function prediction only" had the highest number (21,107, 28.36%), followed by "Transcription" (15,595, 20.95%), and "Function unknown" (14,816, 19.90%). Categories of "Extracellular structures" (86, 0.001%) and "Nuclear structure" (11, 1.48 e^{-4}) had the fewest matching genes (Additional file 6: Figure S4).

GO and KEGG enrichment analyses were used to classify the functions of the predicted *P. fugax* unigenes. Based on homologous genes, 80,312 sequences (Additional file 3: Table S6) from all unigenes of 36 *P. fugax* libraries were categorized into 56 GO terms comprising three domains: biological process, cellular component, and molecular function (Fig. 1). Most were categorized in "cellular process", "metabolic process", "cell", and "cell part". A high percentage of genes were also assigned to "binding", "catalytic activity", "organelle", and "membrane" as well as "biological regulation", "development process", "transporter activity", and "reproductive process" (Fig. 1).

There were 121,526 unigenes that mapped onto KEGG pathways (Additional file 3: Table S7). A total of 53,337 annotated unigenes between NR, COG, KEGG, SwissProt, and InterPro databases were identified with Venn diagrams (Fig. 2).

Table 1 Summary of *Polypogon fugax* transcriptome sequencing, assembly, and annotation

	Reference transcriptome
Total clean reads	275,659,060
Assembled unigenes	206,041
Average read length	1337
N50 contig size	1851
N90 contig size	705
Annotation in NR	153,591
Annotation in NT	145,977
Annotation in KO	121,526
Annotation in SwissProt	111,290
Annotation in GO	80,312
Annotation in COG	74,434
Annotation in Interpro	93,000

NR, NCBI non-redundant protein sequences database
NT, NCBI nucleotide sequences database
KO, KEGG (Kyoto Encyclopedia of Genes and Genomes) Orthology database
GO, Gene Ontology
COG, Clusters of Orthologous Groups of Proteins

TF prediction of assembled unigenes

Next, we studied unigenes that encoded TFs. The list of unigenes that encode TFs is shown in Additional file 3: Table S8. We also performed TF family classification, and found that 5904 unigenes were classified into 58 TF families (Fig. 3). Among those families, the MYB family had the highest number (742, 12.57%), followed by the MYB-related family (604, 10.23%) and the AP2-EREBP family (453, 7.67%). Genes of the MADS-box family were also found (131, 0.02% %); these genes are associated with plant development and adversity responses.

qRT-PCR validation of *P. fugax* expression data

To verify the gene expression patterns, qRT-PCR analyses were performed on Unigene12462, CL1441.Contig20, CL21112.Contig3, CL4600.Contig9 and CL4600.Contig2, CL12188.Contig2, and *EF1* served as candidate reference genes for RT-PCR normalization.

The expression of CL12188.Contig2, Unigene12462, and CL1441.Contig20 was higher in the treated sensitive population when flowering (TFS) than in the untreated sensitive population when flowering (UFS). CL4600.Contig2 and CL21112.Contig3 were higher in the UFS than in the TFS. These results confirmed the reliability of the data (Fig. 4).

Functional analysis of DEGs

To screen the flowering regulatory genes related to resistance, we analyzed DEGs among the seedling,

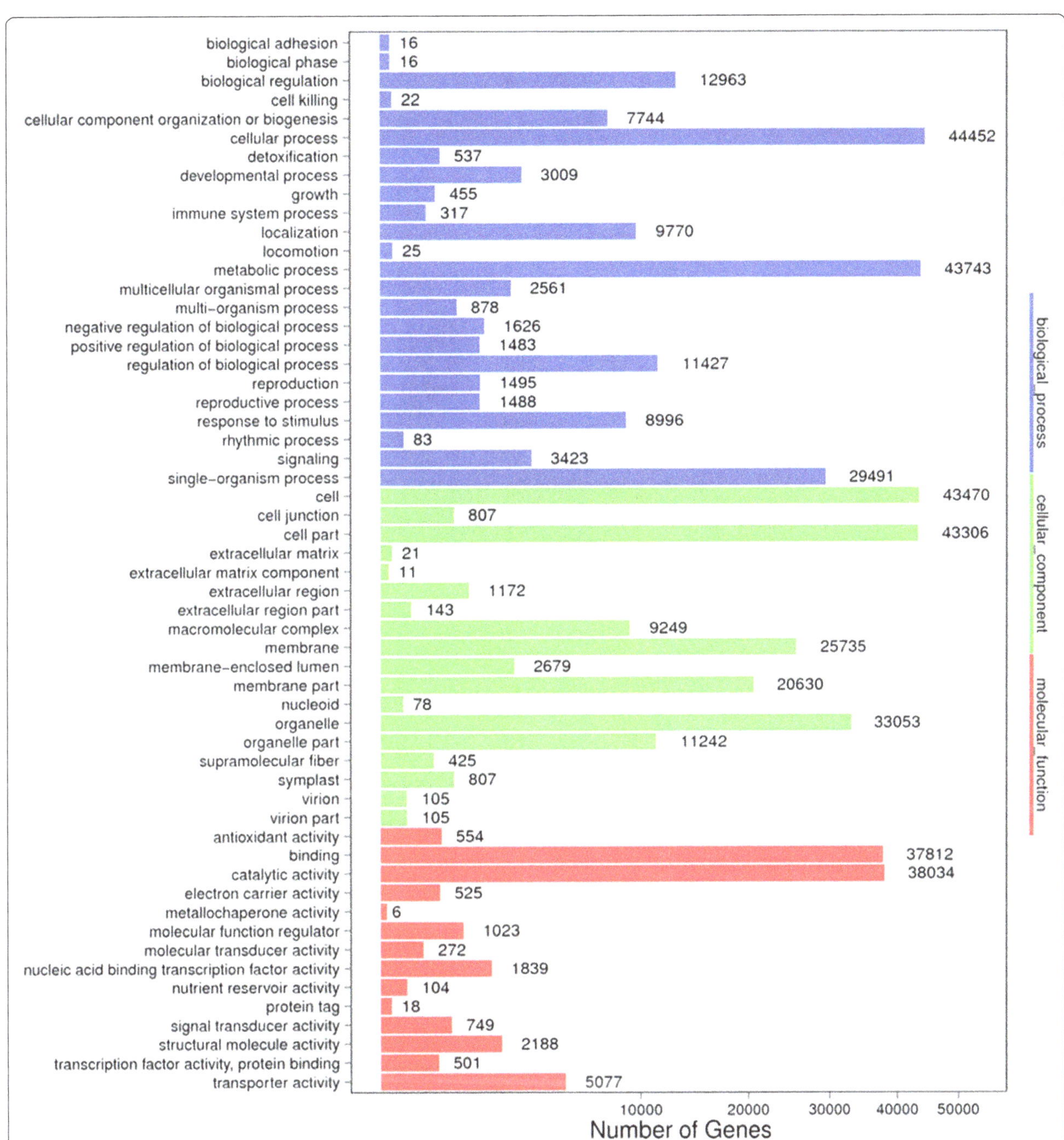

Fig. 1 Histogram of GO classification. Based on the homologous genes, 80,312 sequences from all unigenes of 36 *P. fugax* libraries were categorized into 56 GO terms comprising three domains: biological process, cellular component, and molecular function. The majority of the terms were categorized as "cellular process", "metabolic process", "cell", and "cell part". A high percentage of genes were also assigned to "binding", "catalytic activity", "organelle", and "membrane", and as well as "biological regulation", "development process", "transporter activity", and "reproductive process"

tillering, and flowering stages under different treatments. The genes at the seedling and tillering stages served as a background to identify the specific DEGs at the flowering stage. Cluster analysis was used to compare DEGs, and the parameter was the log2 ratio of gene expression in the difference comparison scheme. The Euclidean distance (calculation of gene distance) referred to genes differentially expressed among all groups. Inter- and inner-group comparisons were performed by the same methods (Fig. 5a-b). Fifty-eight TF families were ultimately predicted, and nine families were related to the regulation of plant growth and development and stress response. We analyzed the different expression values of the TF

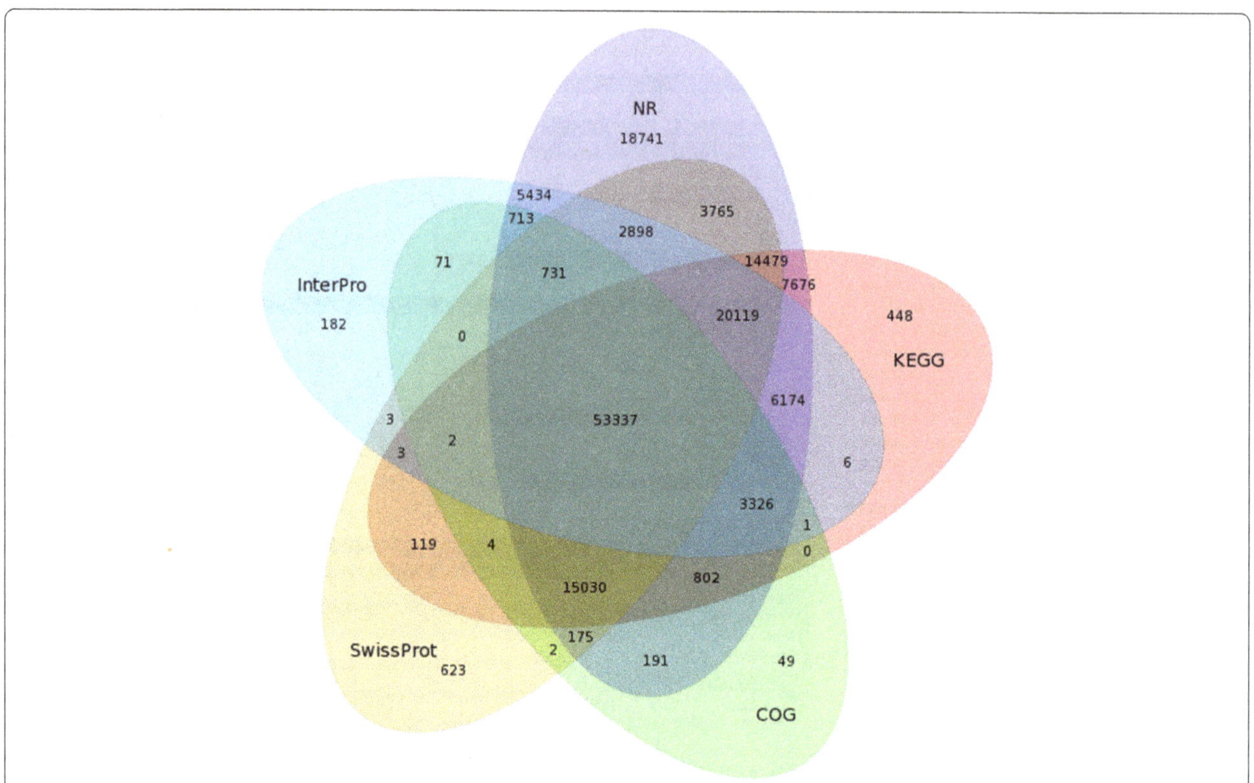

Fig. 2 Venn diagram of the NR, COG, KEGG, SwissProt, and InterPro databases. A total of 53,337 annotated unigenes between the NR, COG, KEGG, SwissProt, and InterPro databases were identified by Venn diagram

families (log2-fold change) of the different populations (resistant vs. sensitive) under the same treatment at the same time, including the MYB (Additional file 3: Table S9), MYB-related (Additional file 3: Table S10), MADS (Additional file 3: Table S11), NAC (Additional file 3: Table S12), mTERF (Additional file 3: Table S13), ABI3VP1 (Additional file 3: Table S14), AP2-EREBP (Additional file 3: Table S15), bHLH (Additional file 3: Table S16), and GRAS (Additional file 3: Table S17) families.

For the DEGs in the nine abovementioned TF families, the screening criteria were as follows: 1) the expression levels of the RPs at the three stages were all higher than those of the SPs regardless of herbicide application; 2) the gene expression of the RPs was higher than that of the SPs after spraying; 3) the DEGs were expressed only at the flowering stage and not at the seedling or tillering stage; and 4) the expression in the sprayed resistant population was higher than that in the sensitive population only at the flowering stage (Additional file 3: Tables S9-S17, sheet 1). Afterward, a total of 30 candidate genes were selected for screening resistance related flowering-regulated genes. qRT-PCR was then carried out to verify the expression of these 30 genes in four samples: the UFS, untreated resistant population at the flowering stage(UFR), TFS, and treated resistant population at the flowering stage(TFR). Twelve DEGs were related to the regulation of plant development, flowering, and stress response (Fig. 6a-b, Table 2); the remaining 18 genes were false positives (data not shown). The ABI3VP1, BHLH, and GRAS families each had one gene (CL18402.Contig2, CL6193.Contig3, and CL20691.Contig17), and the other nine genes belonged to the MADSbox family. The expression of four genes (CL4600.contig2, CL278.contig6, CL10951.contig2, and CL18402.contig2) was higher in the RPs than in the SPs under herbicide and water treatments. The expression of eight genes (CL15323.contig1, CL6626.contig8, CL10710.contig2, CL19935.contig11, CL7805.contig1, CL19935.contig11, CL6193.contig3, and CL20691.contig17) was slightly lower in the RPs than in the SPs under water treatment, but their expression levels increased rapidly after herbicide application and, consequently, were significantly higher than those of the SPs. Interspecific comparisons showed that the expression of 12 unigenes in the RPs was higher than that in the SPs under herbicide selective pressure, suggesting that these genes in the RPs likely promote reproductive growth (flowering and fruiting) under stress conditions: this phenomenon constitutes an unknown resistance mechanism (Fig. 6a).

Under all treatment conditions, expression levels of the 12 unigenes were significantly higher in the UFS

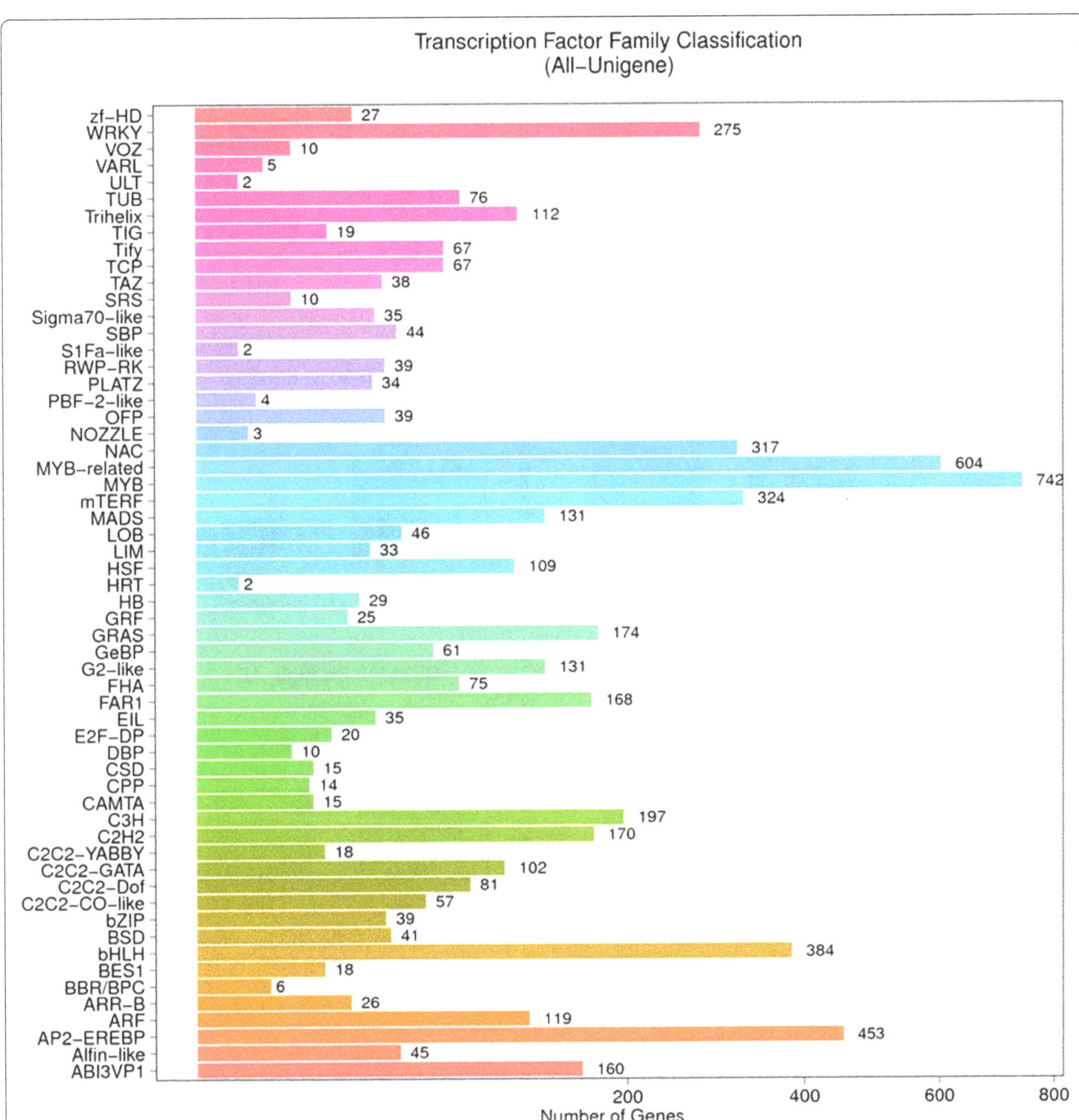

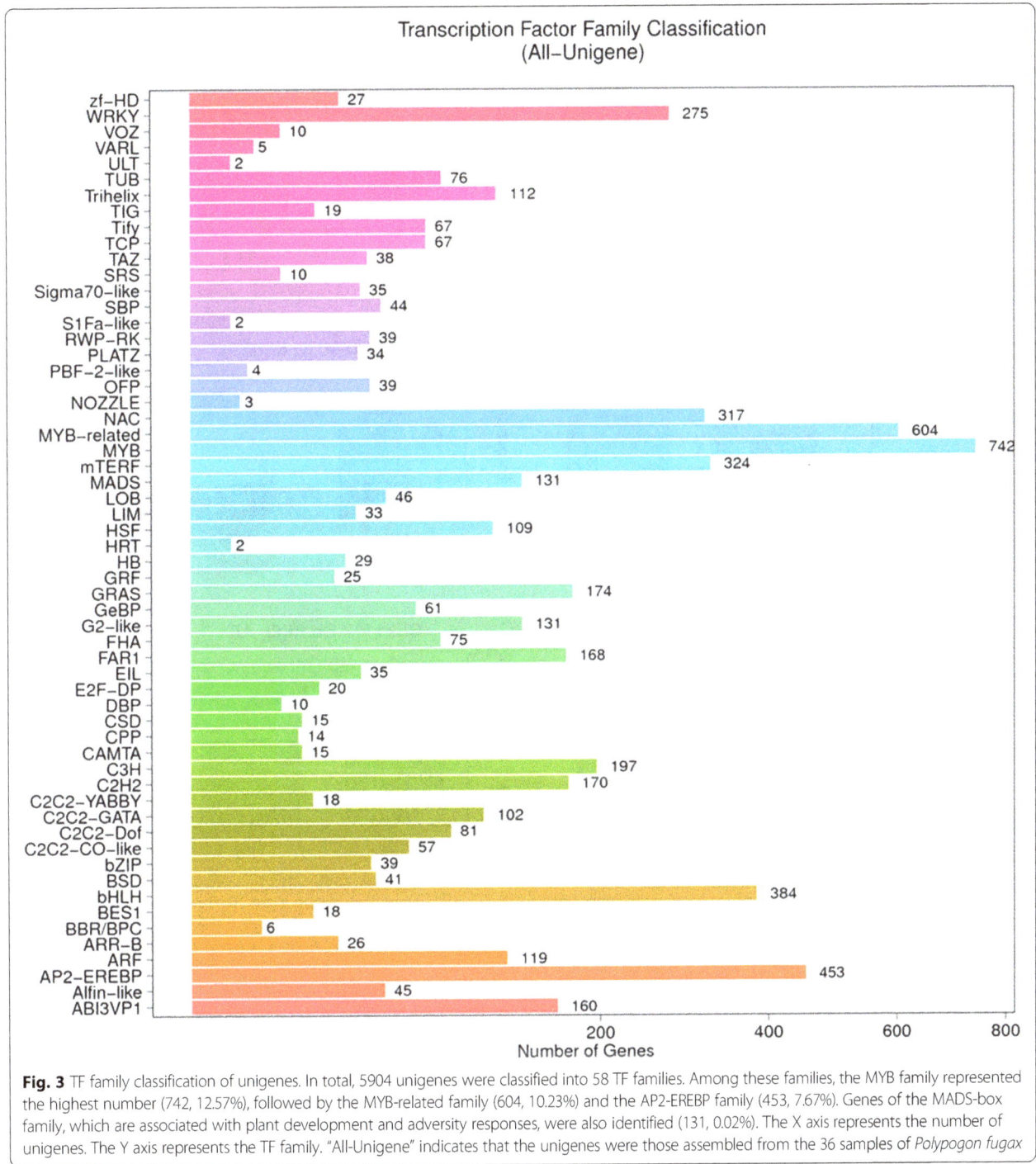

Fig. 3 TF family classification of unigenes. In total, 5904 unigenes were classified into 58 TF families. Among these families, the MYB family represented the highest number (742, 12.57%), followed by the MYB-related family (604, 10.23%) and the AP2-EREBP family (453, 7.67%). Genes of the MADS-box family, which are associated with plant development and adversity responses, were also identified (131, 0.02%). The X axis represents the number of unigenes. The Y axis represents the TF family. "All-Unigene" indicates that the unigenes were those assembled from the 36 samples of *Polypogon fugax*

than in the TFS, which means that the expression of these genes in SPs was inhibited by herbicide selection. In the RPs, the expression of four unigenes (CL6193.contig3, CL20691.contig17, CL18402.contig2, and CL19935. contig9) in the UFR did not differ from that in the TFR (Fig. 6a). The expression of four unigenes (CL4600.contig2, CL19935.contig11, CL15323.contig1, and CL6626.contig8) was slightly lower in the UFR than in the TFR (Fig. 6a), and the expression of four unigenes (CL10710.contig2, CL7805.contig1, CL278.contig6, and CL10951.contig2) was slightly higher in the UFR than in the TFR (Fig. 6a). The results of the intraspecific comparison showed that herbicide selection pressure did not significantly influence the expression of these genes in the RPs (Fig. 6a). The results indicate that these genes have genetically adapted in RPs, while herbicides did not affect the growth and development of those plants.

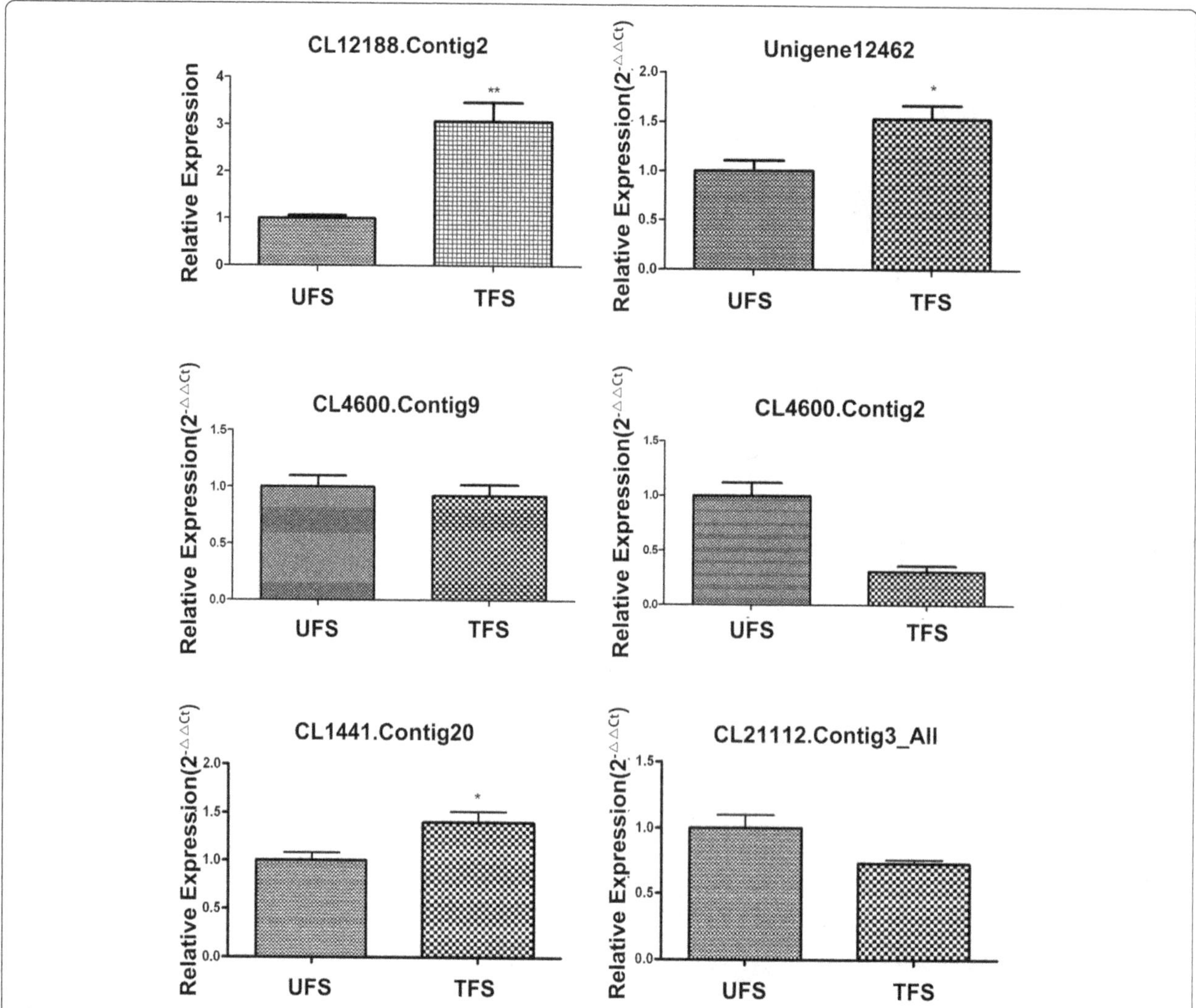

Fig. 4 qRT-PCR validation of RNA-Seq results. qRT-PCR analysis of six randomly selected genes was conducted to confirm the expression patterns indicated by sequencing. The expression of CL12188.Contig2, Unigene12462, and CL1441.Contig20 was higher in the TFS than in the UFS. CL4600.Contig2 and CL21112.Contig3 was higher in the UFS than in the TFS. These results confirmed the reliability of the data. The error bars represent the standard error of the mean. U: untreated; T: treated; F: flowering stage; S: sensitive population; R: resistant population. *significant difference p>0.05, **significant difference 0.01<P<0.05.

Discussion

In this study, a *P. fugax* population resistant to clodinafop-propargyl and a susceptible population were selected. The RPs flowered 10-15 days earlier than did the SPs, but the mechanisms remain unclear. The goal of this study was to establish a resource to study transcriptomic patterns in *P. fugax* resistant or sensitive to clodinafop-propargyl in the absence or presence of herbicide and at different stages (seeding, tillering, and flowering), using experimental conditions as similar as possible to field conditions. The results revealed DEGs related to clodinafop-propargyl resistance in *P. fugax*. The assembled, annotated transcriptomes provide a genomic resource for understanding the molecular basis of *P. fugax* herbicide resistance.

Because there is no genomic resource for *P. fugax*, the Illumina technology was selected for sequencing, as it is the technology of choice for de novo transcriptome deep sequencing and assembly when a reference genome is absent [22]. Three replicates of 12 samples were sequenced by an Illumina HiSeq 4000 in this study, generating approximately 160.52 Gb bases in total. After discarding improper sequences, 206,041 unigenes were obtained; the total length, average length, N50, and GC content of these unigenes were 275,659,060 bp, 1337 bp, 1851 bp, and 51.51%, respectively. The N50 size of the contigs in this study was 1851 bp, which is higher than that recently obtained for plant de novo transcriptome assemblies based on Illumina sequence reads [23, 24]. The average contig size was 1337 bp, which matched the

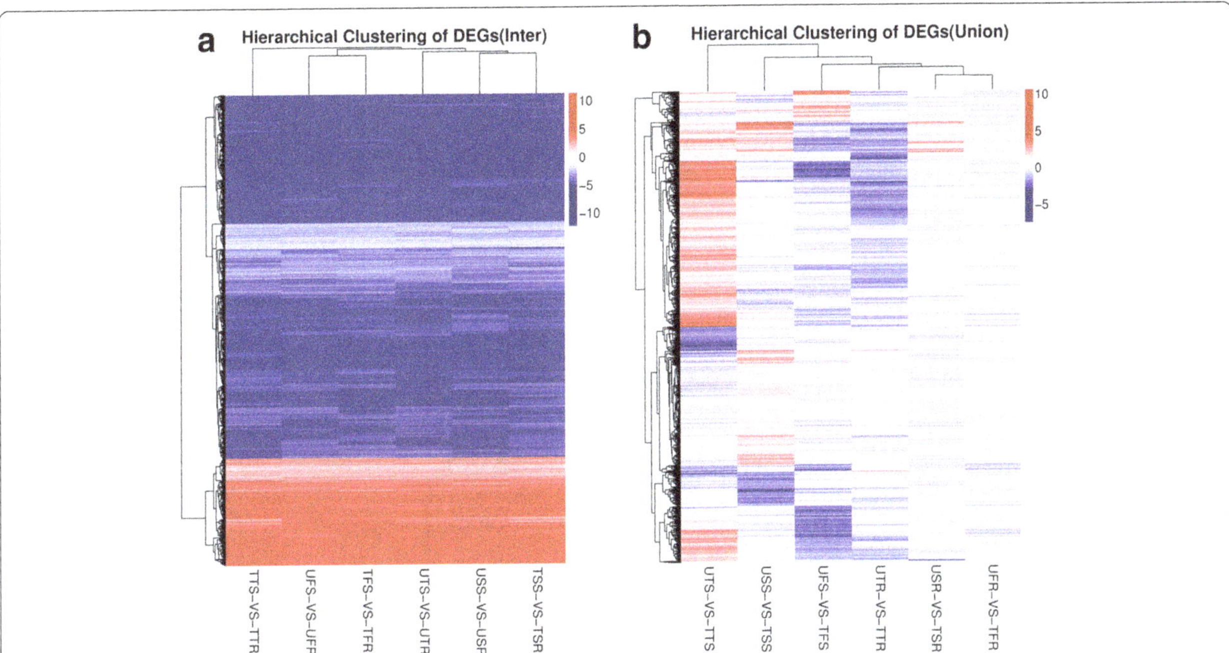

Fig. 5 Cluster analysis of DEGs. Cluster analyses of inter (**a**) and inner (**b**) group comparisons were analyzed by heat maps. Fifty-eight TF families were ultimately predicted, nine families were related to the regulation of plant growth and development and stress responses. We analyzed the different expression values of the TF families (log2-fold change) of the different populations (resistant vs. sensitive) under the same treatment at the same time, including the MYB family, MYB-related family, MADS family, NAC family, mTERF family, ABI3VP1 family, AP2-EREBP family, bHLH family, and GRAS family

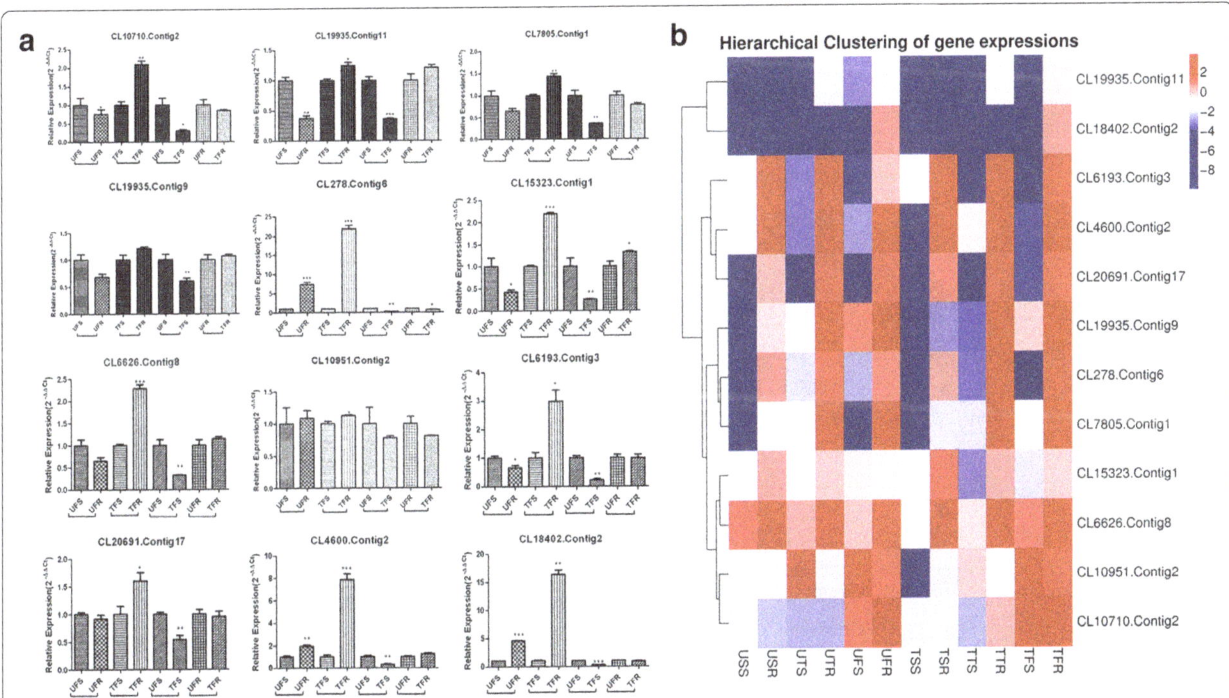

Fig. 6 a Comparison of the expression levels of 12 unigenes in differently treated samples and groups at the flowering stage. Three replicates were performed for each of the three biological replicates. The error bars represent the standard error of the mean. U: untreated; T: treated; F: flowering stage; S: sensitive population; R: resistant population. *significant difference p>0.05, **significant difference 0.01<P<0.05, ***significant difference P<0.01. **b** Heat map analysis of the 12 unigenes

Table 2 Differentially expressed genes (DEGs) as candidate transcription factors (TFs) between the susceptible plants (SPs) and resistant plants (RPs) of *Polypogon fugax* at the early flowering stage[a]

Unigene ID	Contig length	Gene family	Gene annotations	ID in database	Log_2Ratio (UFR/UFS)	P_{adj}	Log_2 ratio (TFR/TFS)	P_{adj}
CL10710.contig2	1396	MADS	MIKC-type MADS-box transcription factor WM29B [*Triticum aestivum*]	CAM59077.1	2.44949	7.16E-07	2.08657	0.00886
CL19935.contig11	1314	MADS	VRN1 [*Festuca arundinacea*]	ACN81330.1	1.41482	0.65251	3.95796	0.05880
CL7805.contig1	1423	MADS	MADS3 [*Lolium perenne*]	AAO45875.1	6.75310	5.83E-09	4.31204	5.94E-05
CL19935.contig9	1496	MADS	MADS-box protein 1 [*Lolium temulentum*]	AAD10625.1	2.71062	1.03E-07	3.93220	1.90E-06
CL278.contig6	814	MADS	predicted protein [*Hordeum vulgare subsp. vulgare*]	BAK01668.1	2.79704	0.034962	4.56666	0.00195
CL15323.contig1	985	MADS	MADS-box transcription factor 42 [*Brachypodium distachyon*]	AIG21850.1	1.60674	0.38702	1.88285	0.20711
CL6626.contig8	1089	MADS	unnamed protein product [Triticum aestivum]	CDM85805.1	1.49045	0.17262	1.01250	0.38150
CL10951.contig2	1176	MADS	MADS-box transcription factor WM21B[Triticum aestivum]	AM502888.1	−1.65321	0.02961	−1.30507	0.22610
CL4600.contig2	1233	MADS	VRT2 [Festuca arundinacea]	ADK55060.1	4.11115	0.00056	5.05851	0.00019
CL19114.Contig2	1056	MADS	MADS-box transcription factor 31 [*Aegilops tauschii*]	EMT17017.1	5.10271	7.61E-05	4.99548	0.00040
CL3687.Contig1	1187	MYB	hypothetical protein OsJ_36546 [*Oryza sativa* Japonica Group]	EAZ20907.1	6.04532	0.00030	5.34815	0.00197
Unigene19990	1015	MYB	Myb-related protein MYBAS2 [Aegilops tauschii]	EMT18692.1	4.06133	1.33E-07	4.23625	9.57E-05
Unigene7488	1353	MYB	hypothetical protein SETIT_ 017851 mg [*Setaria italica*]	KQL30198.1	5.44547	0.00105	4.25490	0.01766
Unigene58296	2044	BHLH	predicted protein [Hordeum vulgare subsp. vulgare]	BAJ85853.1	3.22304	0.03804	5.82932	0.00086
CL9764.Contig4	1172	BHLH	hypothetical protein F775_42459 [Aegilops tauschii]	EMT14257.1	3.79240	0.01288	4.66703	0.01675
CL20691.Contig14	2701	GRAS	predicted protein [Hordeum vulgare subsp. vulgare]	BAJ89210.1	3.77490	5.83E-05	5.10209	0.00015
CL3809.Contig22	2464	GRAS	predicted protein [Hordeum vulgare subsp. vulgare]	BAJ97549.1	3.44933	3.96E-05	3.64527	0.00743
CL18402.contig2	1556	ABI3VP1	B3 domain-containing protein Os01g0234100-like [Brachypodium distachyon]	XP_010240699.1	5.06971	0.00733	4.55668	0.02181
CL12188.Contig2	4196	ABI3VP1	hypothetical protein OsI_35443 [Oryza sativa Indica Group]	EEC67837.1	5.77226	7.86E-07	4.04356	5.17E-05
Unigene13022	1367	ABI3VP1	unnamed protein product [Triticum aestivum]	CDM85878.1	5.75417	0.00076	5.08982	0.00602
Unigene51942	1567	AP2-EREBP	hypothetical protein BRADI_ 1 g46690 [Brachypodium distachyon]	KQK19167.1	6.70426	1.01E-23	7.43238	3.86E-14
Unigene12462	2057	AP2-EREBP	DRF-like transcription factor DRFL2a [Triticum aestivum]	ABC74510.1	9.51535	9.86E-13	9.17289	3.86E-11
CL8782.Contig4	1181	AP2-EREBP	ethylene-responsive transcription factor-like protein At4g13040 [Brachypodium distachyon]	XP_003574993.2	−1.89557	0.13175	4.77569	0.01210
CL1441.Contig20	1766	NAC	predicted protein [Hordeum vulgare subsp. vulgare]	BAK08125.1	8.03481	2.09E-08	6.60687	4.59E-08
Unigene25784	1076	NAC	hypothetical protein MNEG_0811 [Monoraphidium neglectum]	KIZ07145.1	6.80800	6.75E-06	6.49068	1.84E-05

Table 2 Differentially expressed genes (DEGs) as candidate transcription factors (TFs) between the susceptible plants (SPs) and resistant plants (RPs) of *Polypogon fugax* at the early flowering stage[a] *(Continued)*

Unigene ID	Contig length	Gene family	Gene annotations	ID in database	Log_2Ratio (UFR/UFS)	P_{adj}	Log_2 ratio (TFR/TFS)	P_{adj}
CL21112.Contig3	1951	NAC	NAC transcription factor [Hordeum vulgare subsp. vulgare]	CBZ41156.1	4.96265	8.40E-08	7.80505	1.52E-07
Unigene31436	2102	NAC	predicted protein [Hordeum vulgare subsp. vulgare]	BAJ93083.1	6.13811	2.79E-07	5.17007	8.91E-08

[a]Limitations of all significantly different expressed genes between the susceptible plants (SP) and resistant plants (RP) of *Polypogon fugax* are based on $P_{adj} < 1$ and the absolute value of $\log_2$ ratio ≥ 1. The $\log_2$ ratio indicated the change of gene expression; a positive number means up-regulation and a negative one means down-regulatio

average length of gene coding sequences in grasses (1000 to 1300 bp) [25]. The use of functional annotation results revealed 152,966 coding DNA sequences (CDS), ESTScan (v3.0.2) was then used to predict the remaining unigenes, after which 11,716 additional CDS were obtained. Given that only 48 nucleotide sequences from *P. fugax* had been deposited in GenBank on May 22nd, 2017, our work tremendously increased the sequence data available for *P. fugax*.

The predicted peptide content of the *P. fugax* transcriptome was compared to that of five fully sequenced grass genomes. The five grass species belong to three major subfamilies of the *Poaceae*: *Pooideae* (*Brachypodium distachyon* and *H. vulgare*), *Panicoideae* (*Zea mays* and *Sorghum bicolor*), and *Ehrhartoideae* (*Oryza sativa*) [26]. The five grass genomes shared 58.64% of the identified protein families. These proportions were in agreement with a previous genome-wide study showing that genome peptide contents are largely shared among grass species, including peptide family representations [27]. In addition, good representation of the transcriptome of the aerial parts of *P. fugax* at different stages and confirmation of the relevance of the RNA-Seq-based expression data using qRT-PCR make the sequences a reliable resource for investigating the transcriptomic response to herbicide stress promoting early flowering in *P. fugax*.

Stress-induced flowering is a response to stress and is the ultimate stress adaptation, as plants can survive as a species if they flower and produce seeds under severe stress even when they cannot survive as individuals [17]. To validate the function of the candidate genes with respect to flowering in the resistant population, gene expression at the seedling and tillering stages served as a background. The present study used herbicide selection pressure when the RPs and SPs were about to flower. qRT-PCR was subsequently applied to analyze the candidate gene expression at the flowering stage to understand the relationship and mechanism of flowering and response to herbicide application in *P. fugax*.

The replication and diversity of MADS-box genes may be the factors affecting the morphological diversity of land plants and some angiosperms [28]. This gene family encodes conserved TFs and plays an important role in both vegetative and reproductive development. Among these genes, *APETALA1* (*AP1*) is one of the earliest and most intensively studied genes. In *Arabidopsis*, *AP1* deletion mutations delay flowering time and show a high frequency of floral meristem to inflorescence meristem transformation after flowering. On the other hand, constitutive expression of the *AP1* gene changes the floral bud meristem to a floral meristem, and ectopic expression of the *AP1* gene can significantly promote early flowering [29, 30]. The *AP1* gene in herbaceous [31] and woody [32] plants also plays an important role in the initiation and development of flowers, and constitutive expression of *AP1*-like genes also contributes to early flowering. In contrast to the *AP1* gene, overexpression of some MYB TFs (such as *EPR1* and *AtMYB44*) in *Arabidopsis* leads to delayed flowering time [33, 34].

Table 2 summarizes the differentially expressed TFs identified in the present study. CL4600.Contig2 was annotated to the *JOINTLESS*-like protein (ADK55060.1). *JOINTLESS* (*J*) is a MADS-box gene that belongs in the same clade as the *Arabidopsis* flowering time genes *SHORT VEGETATIVE PHASE* (*SVP*) and *AGAMOUS LIKE 24* (*AGL24*) [35]. Loss of *J* function causes premature termination of flower formation during inflorescence and reversion to a vegetative sympodial growth [36]. In addition, the formation of an inflorescence in tomato requires the interaction of *J* and a target of *SINGLE FLOWER TRUSS* (*SFT*) in the meristem [37].

CL10951.Contig2 was annotated to *SUPPRESSOR OF OVEREXPRESSION OF CONSTANS1* (*SOC1*), which plays an important role in the regulation of flowering by integrating multiple flowering signals in *Arabidopsis thaliana* [38]. In the photoperiod pathway, *SOC1* is regulated by *CONSTANS* (*CO*) through *FLOWERING LOCUS T* (*FT*), causing early flowering [39]. In addition, similar to *SOC1*, *SOC1*-like genes promote flowering when overexpressed, but some are also involved in floral development. Ectopic expression of *UNSHAVEN* (*UNS*), a *SOC1*-like gene of *Petunia* hybrids, leads to ectopic trichome formation on floral organs and the formation of petals into organs that exhibit leaf-like features and an early flowering phenotype [40, 41]. CL10951.Contig2 may be related to the promotion of flowering in *P. fugax* under herbicide stress [42].

AGAMOUS-like proteins (CL10710.Contig2, CL19935. Contig11, CL19935. Contig9, CL278.Contig6, and CL7805.Contig1) belong to the *AG* subfamily whose members are involved in the specification of floral reproductive organs and are also required for the normal development of carpels and fruits in *Arabidopsis* and *Gossypium hirsutum* [43] as well as for both drought stress responses [44] and regulation of post-germination growth [45]. *AGAMOUS*-like proteins under herbicide selection pressure were identified in this study, suggesting that these genes may promote early flowering and increased seed yield in resistant *P. fugax* plants.

The remaining three unigenes (CL18402.Contig2, CL6193.Contig3, and CL20691.Contig17) belong to the ABI3VP1, BHLH, and GRAS families, respectively. The genes of these three families play important roles in plant growth and development and stress responses [46–48], but their specific roles in *P. fugax* under herbicide stress need to be studied further.

The present study is not without limitations. Only two cultivars of *P. fugax* were studied, it is possible that cultivars from different regions could yield different results. In addition, transcriptome analysis is limited by available comparative data. It must be stressed that the present study does not provide any mechanistic data, but rather makes available transcriptomic data that could be used for the determination of those mechanisms. Additional studies are necessary to improve our knowledge and understanding of transcriptomics in plants.

Conclusions

In conclusion, the present study compared ACCase-resistant to ACCase-sensitive *P. fugax*, as RPs flower earlier and yield more seeds. The results revealed nine related resistance genes in the MADS-box family (which regulate flowering), and three genes involved in the regulation of growth and development. These data lay the foundation for the further exploration of the specific functions of these genes and for the study of transcriptomics in grasses.

Additional files

Additional file 1: Table S1. Primers used for qRT-PCR. (DOC 63 kb)

Additional file 2: Figure S1. Morphological characteristics in resistant (R) and susceptible (S) Asian minor bluegrass plants at different growth stages. (TIFF 1872 kb)

Additional file 3: Table S2. Summary of sequencing reads after filtering. **Table S3.** Quality metrics of unigenes. **Table S4.** Summary of functional annotation result. **Table S5.** All-unigenes with a COG classification. **Table S6.** All-unigenes categorized by GO terns. **Table S7.** Unigenes mapped onto KEGG pathways. **Table S8.** TF prediction results. **Table S9.** MYB family. **Table S10.** MYB-related family. **Table S11.** MADS-box family. **Table S12.** NAC family. **Table S13.** mTERF family. **Table S14.** ABI3VP1 family. **Table S15.** AP2-EREBP family. **Table S16.** bHLH family. **Table S17.** GRAS family. (ZIP 3364 kb)

Additional file 4: Figure S2. Length distribution of unigenes. (TIFF 5378 kb)

Additional file 5: Figure S3. Distribution of annotated species. (TIFF 3199 kb)

Additional file 6: Figure S4. Functional distribution of COG annotation. (TIFF 6509 kb)

Abbreviations

ACCase: acetyl-CoA carboxylase; COG: Cluster of Orthologous Groups of proteins; DEGs: differently expressed genes; FDR: false discovery rate; GO: Gene Ontology; KEGG: Kyoto Encyclopedia of Genes and Genomes; NR: non-redundant; NT: nucleotide; ORF: open reading frame; RPKM: reads per kilobase per million reads; SRA: Short Read Archive; TFs: transcription factors

Acknowledgements

Not applicable.

Funding

This research was supported by the National Natural Science Foundation of China Project (31501658), the National Key Research and Development Program of China (2016YFD0201305), and the Special Scientific Research Fund of Agricultural Public Welfare Profession of China (201403030).

Authors' contributions

FYZ and YZ conceived and coordinated the study: designed, performed and analyzed the experiments and wrote the paper. WT, MW and TCG carried out the data collection, analyzed the data, and revised the paper. All authors reviewed the results and approved the final version of the manuscript.

Competing interests

The authors declare that they have no competing interests.

Author details

[1]Institute of Plant Protection and Agro-Products Safety, Anhui Academy of Agricultural Sciences, Hefei 230001, China. [2]State Key Laboratory of Rice Biology, China National Rice Research Institute, Hangzhou 311400, China.

References

1. Tang W, Zhou F, Chen J, Zhou X. Resistance to ACCase-inhibiting herbicides in an Asia minor bluegrass (Polypogon Fugax) population in China. Pestic Biochem Physiol. 2014;108:16–20.
2. Xu HL, Zhu XD, Wang HC, Li J, Dong LY. Mechanism of resistance to fenoxaprop in Japanese foxtail (Alopecurus Japonicus) from China. Pestic Biochem Physiol. 2013;107:25–31.

3. Yu Q, Abdallah I, Han H, Owen M, Powles S. Distinct non-target site mechanisms endow resistance to glyphosate, ACCase and ALS-inhibiting herbicides in multiple herbicide-resistant Lolium Rigidum. Planta. 2009;230:713–23.
4. Shaner DL, Lindenmeyer RB, Ostlie MH. What have the mechanisms of resistance to glyphosate taught us? Pest Manag Sci. 2012;68:3–9.
5. Neve P. Challenges for herbicide resistance evolution and management: 50 years after Harper. Weed Res. 2007;47:365–9.
6. Preston C, Powles SB. Evolution of herbicide resistance in weeds: initial frequency of target site-based resistance to acetolactate synthase-inhibiting herbicides in Lolium Rigidum. Heredity (Edinb). 2002;88:8–13.
7. Vila-Aiub MM, Neve P, Steadmank J, Powles SB. Ecological fitness of a multiple-resistant Lolium Rigidum population: dynamics of seed germination and seedling emergence of resistant and susceptible phenotypes. J Appl Ecol. 2005;42:288–98.
8. Wang T, Darmency H. Comparison of growth and yield of foxtail millet (Setaria italica) resistant and susceptible to acetyl-coenzyme A carboxylase inhibiting herbicides. In: 10th Colloque International Biologiedes Mauvaises Herbes de Dijon. Paris: Association Francaise de Protection des Plantes; 1996. p. 203-10.
9. Wang T, Picard JC, Tian X, Darmency H. A herbicide-resistant ACCase 1781 Setaria mutant shows higher fitness than wild type. Heredity (Edinb). 2010;105:394–400.
10. Bernier G, Perilleux C. A physiological overview of the genetics of flowering time control. Plant Biotechnol J. 2005;3:3–16.
11. Turnbull C. Long-distance regulation of flowering time. J Exp Bot. 2011;62:4399–413.
12. Song YH, Shim JS, Kinmonth-Schultz HA, Imaizumi T. Photoperiodic flowering: time measurement mechanisms in leaves. Annu Rev Plant Biol. 2015;66:441–64.
13. Yaish MW, Colasanti J, Rothstein SJ. The role of epigenetic processes in controlling flowering time in plants exposed to stress. J Exp Bot. 2011;62:3727–35.
14. Pieterse AH. Is flowering in Lemnaceae stress-induced? A review Aquatic Botany. 2012;104:1–4.
15. Riboni M, Robustelli Test A, Galbiati M, Tonelli C, Conti L. Environmental stress and flowering time: the photoperiodic connection. Plant Signal Behav. 2014;9:e29036.
16. Kazan K, Lyons R. The link between flowering time and stress tolerance. J Exp Bot. 2016;67:47–60.
17. Takeno K. Stress-induced flowering: the third category of flowering response. J Exp Bot. 2016;67:4925–34.
18. Garber M, Grabherr MG, Guttman M, Trapnell C. Computational methods for transcriptome annotation and quantification using RNA-seq. Nat Methods. 2011;8:469–77.
19. Mortazavi A, Williams BA, McCue K, Schaeffer L, Wold B. Mapping and quantifying mammalian transcriptomes by RNA-Seq. Nat Methods. 2008;5:621–8.
20. Mistry J, Finn RD, Eddy SR, Bateman A, Punta M. Challenges in homology search: HMMER3 and convergent evolution of coiled-coil regions. Nucleic Acids Res. 2013;41:e121.
21. Li B, Dewey CN. RSEM: accurate transcript quantification from RNA-Seq data with or without a reference genome. BMC Bioinformatics. 2011;12:323.
22. Ward JA, Ponnala L, Weber CA. Strategies for transcriptome analysis in nonmodel plants. Am J Bot. 2012;99:267–76.
23. Gardin JA, Gouzy J, Carrere S, Delye C. ALOMYbase, a resource to investigate non-target-site-based resistance to herbicides inhibiting acetolactate-synthase (ALS) in the major grass weed Alopecurus Myosuroides (black-grass). BMC Genomics. 2015;16:590.
24. Chen J, Huang H, Wei S, Huang Z, Wang X, Zhang C. Investigating the mechanisms of glyphosate resistance in goosegrass (Eleusine Indica (L.) Gaertn.) by RNA sequencing technology. Plant J. 2017;89:407–15.
25. Carels N, Bernardi G. Two classes of genes in plants. Genetics. 2000;154:1819–25.
26. Bolot S, Abrouk M, Masood-Quraishi U, Stein N, Messing J, Feuillet C, et al. The 'inner circle' of the cereal genomes. Curr Opin Plant Biol. 2009;12:119–25.
27. Davidson RM, Gowda M, Moghe G, Lin H, Vaillancourt B, Shiu SH, et al. Comparative transcriptomics of three Poaceae species reveals patterns of gene expression evolution. Plant J. 2012;71:492–502.
28. Kaufmann K, Melzer R, Theissen G. MIKC-type MADS-domain proteins: structural modularity, protein interactions and network evolution in land plants. Gene. 2005;347:183–98.
29. Mandel MA, Yanofsky MF. A gene triggering flower formation in Arabidopsis. Nature. 1995;377:522–4.
30. Weigel D, Nilsson O. A developmental switch sufficient for flower initiation in diverse plants. Nature. 1995;377:495–500.
31. Chi YJ, Huang F, Liu HC, Yang SP, Yu DY. An APETALA1-like gene of soybean regulates flowering time and specifies floral organs. J Plant Physiol. 2011;168:2251–9.
32. Wang J, Zhang X, Yan G, Zhou Y, Zhang K. Over-expression of the PaAP1 gene from sweet cherry (Prunus Avium L.) causes early flowering in Arabidopsis Thaliana. J Plant Physiol. 2013;170:315–20.
33. Kuno N, Moller SG, Shinomura T, Xu X, Chua NH, Furuya M. The novel MYB protein EARLY-PHYTOCHROME-RESPONSIVE1 is a component of a slave circadian oscillator in Arabidopsis. Plant Cell. 2003;15:2476–88.
34. Jung CK, Seo JS, Han SW, Koo YJ, Kim CH, Song SI, Nahm BH, Choi YD, Cheong JJ. Overexpression of ATMYB44 enhances stomatal closure to confer abiotic stress tolerance in transgenic arabidopsis. Plant Physiol 2008; 146:623-35.
35. Mao L, Begum D, Chuang HW, Budiman MA, Szymkowiak EJ, Irish EE, et al. JOINTLESS is a MADS-box gene controlling tomato flower abscission zone development. Nature. 2000;406:910–3.
36. Quinet M, Dubois C, Goffin MC, Chao J, Dielen V, Batoko H, et al. Characterization of tomato (Solanum Lycopersicum L.) mutants affected in their flowering time and in the morphogenesis of their reproductive structure. J Exp Bot. 2006;57:1381–90.
37. Thouet J, Quinet M, Lutts S, Kinet JM, Perilleux C. Repression of floral meristem fate is crucial in shaping tomato inflorescence. PLoS One. 2012;7:e31096.
38. Moon J, Suh SS, Lee H, Choi KR, Hong CB, Paek NC, et al. The SOC1 MADS-box gene integrates vernalization and gibberellin signals for flowering in Arabidopsis. Plant J. 2003;35:613–23.
39. Yoo SK, Chung KS, Kim J, Lee JH, Hong SM, Yoo SJ, et al. Constans activates SUPPRESSOR OF OVEREXPRESSION OF CONSTANS 1 through FLOWERING LOCUS T to promote flowering in Arabidopsis. Plant Physiol. 2005;139:770–8.
40. Ferrario S, Busscher J, Franken J, Gerats T, Vandenbussche M, Angenent GC, et al. Ectopic expression of the petunia MADS box gene UNSHAVEN accelerates flowering and confers leaf-like characteristics to floral organs in a dominant-negative manner. Plant Cell. 2004;16:1490–505.
41. Zhong X, Dai X, Xv J, Wu H, Liu B, Li H. Cloning and expression analysis of GmGAL1, SOC1 homolog gene in soybean. Mol Biol Rep. 2012;39:6967–74.
42. Liu S, Ma T, Ma L, Lin X. Ectopic expression of PVSOC1, a homolog of SOC1 form Phyllostachys Violascens, promotes flowering in Arabidopsis and rice. Acta Physiol Plant. 2016;38:166–74.
43. Favaro R, Pinyopich A, Battaglia R, Kooiker M, Borghi L, Ditta G, et al. MADS-box protein complexes control carpel and ovule development in Arabidopsis. Plant Cell. 2003;15:2603–11.
44. Bechtold U, Penfold CA, Jenkins DJ, Legaie R, Moore JD, Lawson T, et al. Time-series Transcriptomics reveals that AGAMOUS-LIKE22 affects primary metabolism and developmental processes in drought-stressed Arabidopsis. Plant Cell. 2016;28:345–66.
45. Yu LH, Wu J, Zhang ZS, Miao ZQ, Zhao PX, Wang Z, et al. Arabidopsis MADS-box transcription factor AGL21 acts as environmental surveillance of seed germination by regulating ABI5 expression. Mol Plant. 2017;10:834–45.
46. Chen J, Tan RK, Guo XJ, Fu ZL, Wang Z, Zhang ZY, et al. Transcriptome analysis comparison of lipid biosynthesis in the leaves and developing seeds of Brassica Napus. PLoS One. 2015;10:e0126250.
47. Ranjan R, Khurana R, Malik N, Badoni S, Parida SK, Kapoor S, et al. bHLH142 regulates various metabolic pathway-related genes to affect pollen development and anther dehiscence in rice. Sci Rep. 2017;7:43397.
48. Song L, Tao L, Cui HP, Ling L, Guo CH. Genome-wide identification and expression analysis of the GRAS family proteins in Medicago truncatula. Acta Physiologiae Plantarum. 2017;39(4):93.

9

From genomes to genotypes: molecular epidemiological analysis of *Chlamydia gallinacea* reveals a high level of genetic diversity for this newly emerging chlamydial pathogen

Weina Guo[1,2†], Martina Jelocnik[3†], Jing Li[1], Konrad Sachse[4], Adam Polkinghorne[3], Yvonne Pannekoek[5], Bernhard Kaltenboeck[6], Jiansen Gong[7], Jinfeng You[1] and Chengming Wang[1,6*]

Abstract

Background: *Chlamydia* (*C.*) *gallinacea* is a recently identified bacterium that mainly infects domestic chickens. Demonstration of *C. gallinacea* in human atypical pneumonia suggests its zoonotic potential. Its prevalence in chickens exceeds that of *C. psittaci*, but genetic and genomic research on *C. gallinacea* is still at the beginning. In this study, we conducted whole-genome sequencing of *C. gallinacea* strain JX-1 isolated from an asymptomatic chicken, and comparative genomic analysis between *C. gallinacea* strains and related chlamydial species.

Results: The genome of *C. gallinacea* JX-1 was sequenced by single-molecule, real-time technology and is comprised of a 1,059,522-bp circular chromosome with an overall G + C content of 37.93% and sequence similarity of 99.4% to type strain 08-1274/3. In addition, a plasmid designated pJX-1, almost identical to p1274 of the type strain, except for two point mutations, was only found in field strains from chicken, but not in other hosts. In contrast to chlamydial species with notably variable polymorphic membrane protein (*pmp*) genes and plasticity zone (PZ), these regions were conserved in both *C. gallinacea* strains. There were 15 predicted *pmp* genes, but only B, A, E1, H, G1 and G2 were apparently intact in both strains. In comparison to chlamydial species where the PZ may be up to 50 kbp, *C. gallinacea* strains displayed gene content reduction in the PZ (14 kbp), with strain JX-1 having a premature STOP codon in the *cytotoxin* (*tox*) gene, while *tox* gene is intact in the type strain. In multilocus sequence typing (MLST), 15 *C. gallinacea* STs were identified among 25 strains based on cognate MLST allelic profiles of the concatenated sequences. The type strain and all Chinese strains belong to two distinct phylogenetic clades. Clade of the Chinese strains separated into 14 genetically distinct lineages, thus revealing considerable genetic diversity of *C. gallinacea* strains in China.

Conclusions: In this first detailed comparative genomic analysis of *C. gallinacea*, we have provided evidence for substantial genetic diversity among *C. gallinacea* strains. How these genetic polymorphisms affect *C. gallinacea* biology and pathogenicity should be addressed in future studies that focus on phylogenetics and host adaption of this enigmatic bacterial agent.

Keywords: *Chlamydia gallinacea*, Whole-genome sequence, Comparative genomics analysis, MLST, Phylogenetic analysis

* Correspondence: wangche@auburn.edu
[†]Equal contributors
[1]Jiangsu Co-Innovation Center for Prevention and Control of Important Animal Infectious Diseases and Zoonoses, Yangzhou University College of Veterinary Medicine, Yangzhou, Jiangsu 225009, People's Republic of China
[6]College of Veterinary Medicine, Auburn University, Auburn, AL, USA
Full list of author information is available at the end of the article

Background

The obligate intracellular bacteria in the genus *Chlamydia* are globally widespread and represent successful pathogens that infect a wide range of animals as well as humans. However, some of them are frequently overlooked as these infections typically remain latent and only rarely lead to overt clinical signs. For a long time, *Chlamydia (C.) psittaci*, an avian pathogen with well-documented zoonotic potential, was considered the only chlamydial species infecting domestic and wild birds. However, recent reports showed that *C. gallinacea* and *C. avium* are two emerging chlamydial agents that can also be involved in avian chlamydiosis [1]. To date, *C. avium* has been detected in pigeons and psittacine birds, while *C. gallinacea* has been mainly detected in chickens, ducks, guinea fowl, turkey, backyard poultry and cattle [2, 3]. Interestingly, the high prevalence of *C. gallinacea* in poultry flocks across Europe and China determined by PCRs surpassed that of *C. psittaci* [4, 5]. This organism is known to occasionally be in transmission with *C. psittaci* in the same flock and also can co-infect individuals [6, 7]. Beyond the potential role of this emerging pathogen in animal health, an earlier study of an outbreak of atypical pneumonia in a slaughterhouse, where workers were exposed to *C. gallinacea*-infected chickens, raised questions over its zoonotic potential as well [8].

Whole-genome sequencing and subsequent comparative genomic analysis has become standard in analysis of the biology, virulence factors, evolution and phylogenetic relationships of chlamydial organisms [9, 10]. While there is plentiful genomic data on the related chlamydial species, the only completely assembled genomic sequence of *C. gallinacea* currently available is that of the type strain 08-1274/3, which was isolated from a chicken in France [11]. So far, this limited genomic information for *C. gallinacea* has allowed only little insight into its developmental cycle and potential virulence factors. Likewise, intra-species genetic diversity and phylogenetic relationships have yet to be investigated. Little information available from partial multi-locus sequence analysis (MLSA) of five *C. gallinacea* strains revealed limited genetic diversity within the species [12]. However, this contrasts with the findings of our own genotyping studies targeting the *ompA* gene, which encodes the chlamydial major outer membrane protein (MOMP), where we found 13 diverse *ompA* genotypes of *C. gallinacea* in Chinese poultry [4].

In the present study, we describe the second whole-genome sequence (WGS) of *C. gallinacea*, which originates from the Chinese chicken isolate JX-1, and report findings from comparative genomic analysis between *C. gallinacea* strains and closely related species in the genus *Chlamydia*. To understand the epidemiology and genetic diversity of *C. gallinacea* infections in chickens, we conducted previously described *Chlamydiales* multi-locus sequence typing (MLST) [13] on 23 *C. gallinacea*-positive samples from nine farms located in nine provinces across China. This enabled us to provide a detailed description of genomic features and assess naturally occurring genetic diversity of this pathogen.

Methods

Description of *C. gallinacea* isolate JX-1 and clinical samples used in this study

C. gallinacea JX-1 strain, used for genome sequencing and plasmid characterization, was isolated from a cloacal swab of an asymptomatic chicken in the Jiangxi province of China [4]. In the present study, we also used DNA from 45 previously tested *C. gallinacea*-positive clinical swabs taken from oral and cloacal anatomical sites of chickens, pigeons, ducks and geese from various farms across China [4] (Additional file 1: Table S1). Ethics approval was not needed as the DNA used in this study was extracted from the chickens in a previous study [4].

Whole-genome sequencing and assembly

C. gallinacea strain JX-1 was propagated via yolk sac inoculation on a 7-day-old chicken embryo followed by yolk membrane harvesting, in order to perform genomic DNA extraction using the QIAgen® DNA Mini Kit (Qiagen, Valencia, CA, USA). The obtained total DNA was subjected to quality control, by running 1 μl of DNA on an agarose gel and quantification by Qubit. The genome of *C. gallinacea* JX-1 was sequenced by Single-Molecule, Real-Time (SMRT) technology at the Beijing Novogene Bioinformatics Technology Co., Ltd. (China). SMRT Analysis 2.3.0 was used to filter low-quality reads, following assembly into a single gap-free contig using filtered reads. Low-quality reads were filtered by the SMRT Analysis v2.3.0 software, and then the genome was subjected to de novo assembly by the SMRT portal software according to the valid sequencing data. The draft *C. gallinacea* JX-1 genome was automatically annotated using the NCBI Prokaryotic Genomes Annotation Pipeline (NCBI_PGAP), and the genome sequence was deposited in the NCBI database under GenBank accession number CP019792.

C. gallinacea plasmid screening

In order to assess whether the *C. gallinacea* JX-1 carries a plasmid, 16 paired primers were designed based on the plasmid sequence of type strain 08-1274/3 (Additional file 2: Table S2) to amplify the complete plasmid sequence. PCR conditions and reaction mixes are described in the section below. Each amplified fragment was purified using the QIAquick PCR Purification Kit (Qiagen), and sent for Sanger sequencing to GenScript,

Jiangsu, Nanjing, China. The chromatograms of the sequenced plasmid fragments were mapped against the p1274 sequence, and the complete *C. gallinacea* JX-1 plasmid (pJX-1) was extracted and annotated using RAST [14] and deposited in GenBank under accession number CP019793. We have also screened the *C. gallinacea*-positive clinical samples (Additional file 1: Table S1) for plasmid presence by amplifying a 661 bp fragment of the plasmid's CDS1 (integrase) using primer pair plaF1 and plaR1 of the plasmid (Additional file 2: Table S2).

Macroscopic comparative genomic and phylogenetic analyses

The genome of *C. gallinacea* JX-1was compared in-depth to the reference genome of type strain 08-1274/3, as well as to publicly available genomes of other related chlamydial species, i.e. *C. avium* 10 DC88 (NZ_CP006571.1), *C. pecorum* E58 (CP002608), *C. psittaci* 6 BC (CP002586.1) and *C. abortus* S26/3 (CR848038.1). Pairwise genomic comparison was performed using the Artemis Comparison Tool (ACT) [15], and Geneious 9 [16] using alignments produced with progressive Mauve [17] and MAFFT [18]. The genomic regions of interest and/or loci were extracted from the analyzed genomes and aligned, in order to be used for further nucleotide and/or translated protein sequence analyses performed using DNASp 5.0 [19], as well as BLAST (https://blast.ncbi.nlm.nih.gov/Blast.cgi). In addition, we have also used open source TMHMM Server v. 2.0 (available from http://www.cbs.dtu.dk/services/TMHMM/) which predicts transmembrane helices in proteins, to predict chlamydial inclusion membrane proteins with a presence of bilobed hydrophobic domains using translated *C. gallinacea* hypothetical gene sequences. A mid-point rooted phylogenetic tree constructed from the alignment of all identified *C. gallinacea pmp* genes from both strains used in this study was generated with PhyML with 1000 bootstrap repetitions [20], as implemented in Geneious 9. Figures of the whole-genome comparison and specific genomic regions using blastn and tblastx algorithms were generated with Brig [21] and EasyFig [22], while the graphical representation of the *C. gallinacea* JX-1 genome and its elements was generated with the DnaPlotter [23].

The phylogenetic relationship of the two *C. gallinacea* strains was examined through comparison to each other and related chlamydial species using an 11.2 kbp alignment of concatenated sequences. The concatenated sequence consisted of 12 partial and full-length conserved chlamydial phylogenetic markers that were concatenated in the following order: six MLST house-keeping gene fragments (*gat*A, *hfl*X, *gid*A, *eno*A, *hem*N, *fum*C), and full-length major outer membrane protein gene *omp*A, DNA-directed RNA polymerase subunit beta gene *rpo*B, 50S ribosomal protein L3 gene *rpl*C, 50S ribosomal protein L4 gene *rpl*D, DNA recombination/repair protein gene *rec*A, and tyrosine—tRNA ligase gene *tyr*S. In addition to the two *C. gallinacea* genomes used in this study, each of these sequences were extracted from the genomes of the following related species: *C. avium* 10 DC88 (NZ_CP006571.1), *C. caviae* GPIC (NC_003361.3), *C. felis* F/C-56 (NC_007899.1), *Candidatus C. ibidis* 10-1398/6 (NZ_APJW00000000.1), *C. pneumoniae* LpColN (NC_017285.1), *C. pecorum* E58 (CP002608), *C. psittaci* 6 BC (CP002586.1), *C. trachomatis* AHAR-13 (CP000051.1), *C. muridarum* Nigg (NC_002620.2), *C. suis* MD56 (NZ_KI538658.1) and *C. abortus* S26/3 (CR848038.1). A mid-point rooted maximum-likelihood phylogenetic tree was constructed using PhyML with 1000 bootstrap repetitions, as integrated in Geneious 9.

MLST of *C. gallinacea*

In this study, we performed a complete MLST, based on a previously published scheme for chlamydiae [13]. The primers used to amplify seven *C. gallinacea*-specific house-keeping (HK) genes were designed in this study based on the sequence of *C. gallinacea* type strain 08-1274/3 and other related chlamydial species (Additional file 3: Table S3).

PCR amplification of the seven HK genes (as well as plasmid fragments) was performed in a LightCycler 480-II real-time PCR platform using a high-stringency 18-cycle step-down temperature protocol (Additional file 4: Table S4) as described [3, 4]. *C. gallinacea* JX-1 DNA was used as a positive control, while ultrapure H_2O was used as a negative control in each assay. The PCR products were electrophoresed through 2% agarose gel and purified using the QIAquick PCR Purification Kit (Qiagen) for automated DNA sequencing (GenScript, Jiangsu, Nanjing, China).

After optimization and development, the *C. gallinacea* MLST was applied to sequences of a total of 45 *C. gallinacea*-positive oral and cloacal swabs from chickens ($n = 20$), hens ($n = 4$), ducks ($n = 12$), pigeons ($n = 6$), and geese ($n = 3$), collected in different provinces of China (Additional file 1: Table S1).

C. gallinacea MLST and phylogenetic analysis were performed using DNASp 5.0 and Geneious 9. Briefly, forward and reverse chromatograms for each sequenced HK gene fragment were aligned and trimmed, and the fragment sequence for that allele was obtained. Allele and sequence type (ST) assignment for 25 *C. gallinacea* strains were determined and deposited at http://pubmlst.org/chlamydiales/ [24] (Additional file 5: Table S5).

Sequences of individual genes and concatenated gene sets were aligned using ClustalX. DnaSP 5.0 was used to analyze sequence polymorphisms by determining the number of synonymous (d_s) and non-synonymous (d_n) substitutions per site, Jukes-Cantor corrected, the number of polymorphic sites and haplotypes (Additional file 5: Table S5).

Best-fit models of nucleotide substitution for our data set were estimated by considering eleven substitution (nst = 11) models using jModelTest v.2.2. [25]. A Bayesian phylogenetic tree using concatenated MLST sequences of 25 *C. gallinacea* strains was constructed with MrBayes [26] with the HKY + I model, as implemented in Geneious 9. Run parameters included four Markov Chain Monte Carlo (MCMC) chains with a million generations, sampled every 1000 generations and with the first 10,000 trees were discarded as burn-in. The *C. avium* MLST sequence was used as an outgroup.

Results

Description of the *C. gallinacea* JX-1 genome

Using SMRT sequencing technology, we have completely sequenced and assembled the genome of *C. gallinacea* strain JX-1. The whole genome is sized 1,059,522 bp with an overall GC content of 37.93%, encompassing 957 predicted CDSs that account for 91.21% of the genome and a 7.49 kbp plasmid (Table 1, Fig. 1). Alignment to the reference genome of *C. gallinacea* type strain 08-1274/3 confirmed 100% chromosome coverage for our newly described JX-1 genome.

With the availability of the JX-1 genome, we were able to evaluate genome similarities and differences between *C. gallinacea* strains, but also to related chlamydial species (Fig. 2a). The two *C. gallinacea* genomes are virtually identical to each other (99.4% identity) and share the highest similarity with the *C. avium* genome (~78.9% sequence similarity) based on whole-genome MAFFT alignment. Comparison to the genomes of the other chlamydial species revealed more pronounced differences, mainly in the PZ and Pmp clusters, however with a remarkable overall chlamydial genome synteny (Fig. 2a). Phylogenetically, although closely related, the two *C. gallinacea* strains, grouped in a larger clade with their closest relative, *C. avium* (Fig. 2b).

The *C. gallinacea* genomes of field strain JX-1 and type strain 08-1274/3 exhibited high sequence similarity (99.4% identity) with ~6250 SNP differences between the two strains, while maintaining synteny and gene order. Both *C. gallinacea* genomes also contained the hallmark genomic features of chlamydiae, such as the highly conserved Type III Secretion System (T3SS), clusters of *pmp* genes, inclusion protein genes (*incs*), and a PZ (Fig. 1a) [27].

With SNPs evenly distributed along the chromosome, the major genetic differences between strains JX-1 and 08-1274/3 are presented in Table 2. In comparative genomic analysis, we have identified genes with high density of SNPs (with at least 5% total length sequence dissimilarity between the two strains), Interestingly, besides the previously recognized highly variable *ompA* gene, most of the remaining genes with high numbers of SNPs were annotated as metabolic genes (Table 2). Highly variable CDS GM000264, annotated as a hypothetical protein, appears to harbor a *C. gallinacea*-specific sequence based on BLAST searches, with only 20% similarity to a conserved hypothetical protein of *C. psittaci*. The putative product, however, does not seem to have a conserved domain. None of these identified genes appears to be under positive selection with the majority of accumulating SNPs being synonymous (Table 2). However, due to only two strains analyzed, at present we cannot accurately assess the selection on these genes.

The *C. gallinacea* strains possess a chlamydial T3SS comprised of a total of 36 genes encoding T3SS structural components, chaperones and secreted effectors (Additional file 6: Table S6). T3SS genes of the two *C. gallinacea* strains were highly conserved with 98–100% sequence similarity, in stark contrast to the previously described genetic diversity in these genes in related chlamydial species such as *C. psittaci* [28], *C. pecorum* [29] and human *C. trachomatis* [30].

We have also assessed the predicted Inclusion membrane proteins (Incs) for *C. gallinacea*, as during early infection the inclusion membrane modified by the insertion of a number of type III secreted effector proteins, and the inclusion proteins play a significant role [31]. Using open source prediction software TMHMM Server v. 2.0 using a cut-off of more than 40 amino acids in the bi-lobed hydrophobic domain, we have predicted a total of 29 putative Inc's with two transmembrane domains, besides the two annotated IncA, and IncB and IncC; five

Table 1 Description of *Chlamydia gallinacea* JX-1 genome

Strain	*Chlamydia gallinacea* JX-1
Clinical manifestation and anatomical site sample type	Asymptomatic/Cloacal swab
Host and country of origin	Chicken, China
Total No. of filtered reads	66,564
Average read length	11,688 bp
Average read depth	100 ×
Genome size (bp) and % GC	1,059,522 bp / 37.93% GC
No. of predicted CDSs	957
Chlamydial plasmid and size	Present: pJX-1 (7.49 Kbp)
nc RNAs	39 tRNAs, two 5S rRNA, and single copy of 16S rRNA and 23S rRNA
Pyrimidine genes	Present (*pyr*G, *pyr*E, *pyr*H, *ndk*)
Biotin operon	Present (*bio*ADFB, bioY)
Tryptophan operon	Absent
% DNA sequence identity *C. gallinacea* 08-1274/3	99.4%
Accession number	CP019792

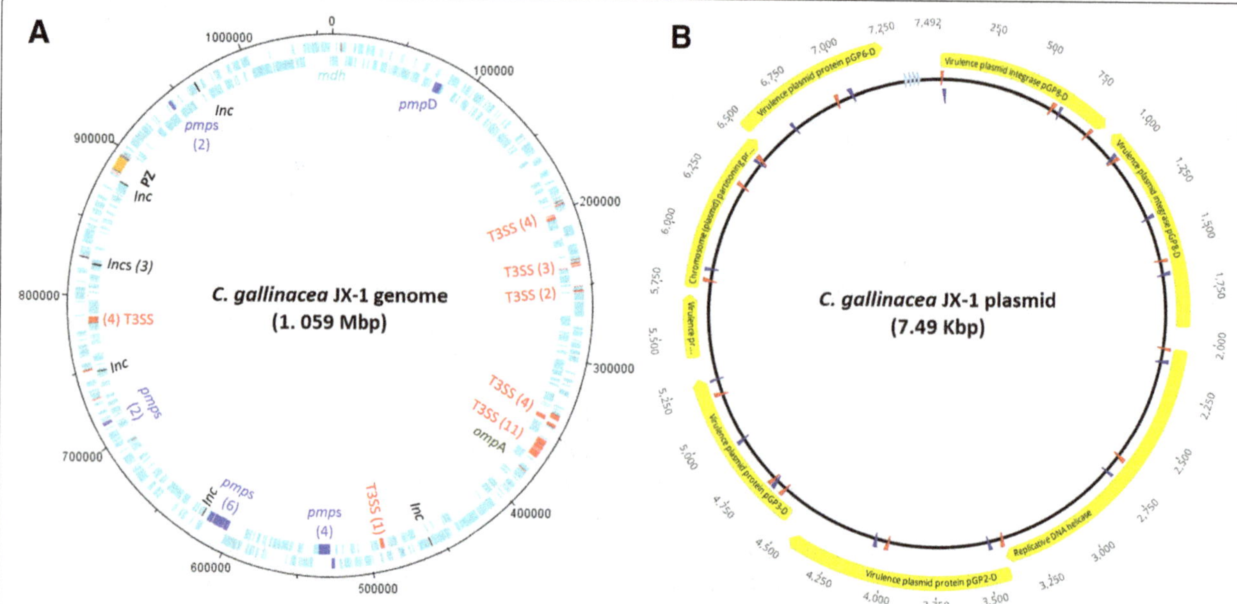

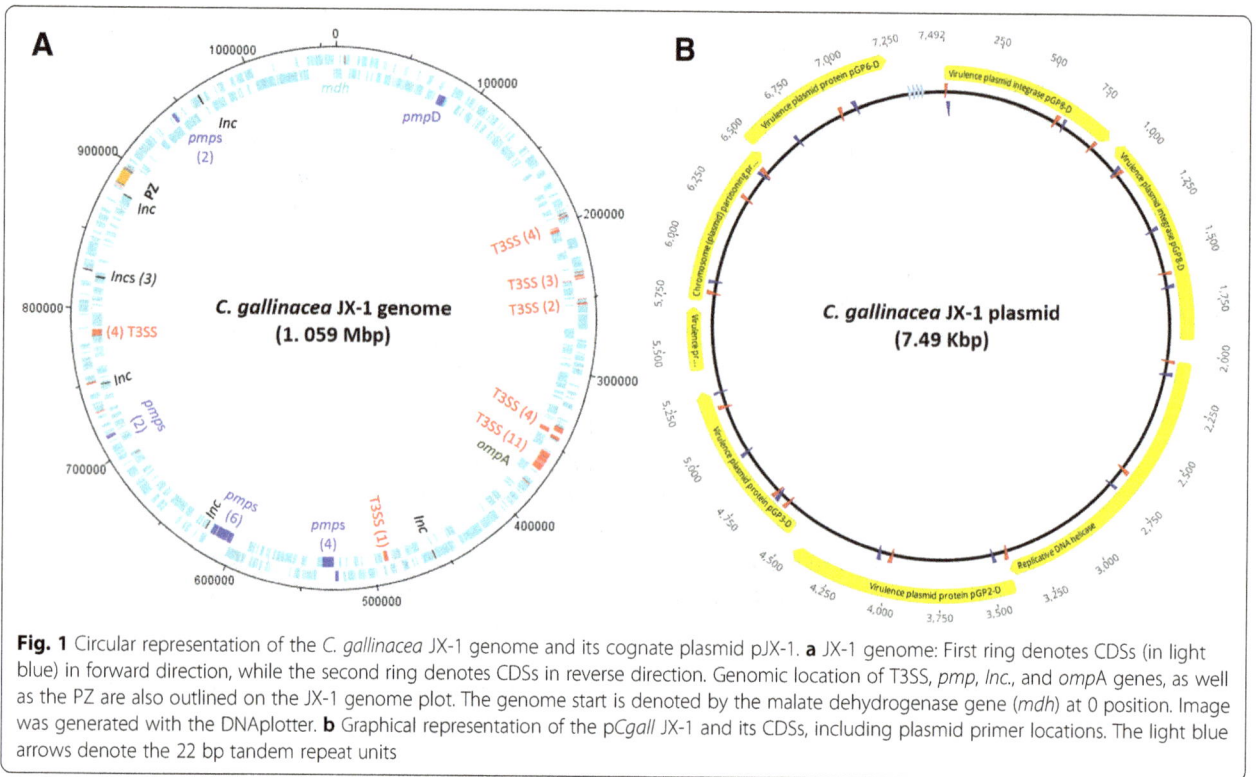

Fig. 1 Circular representation of the *C. gallinacea* JX-1 genome and its cognate plasmid pJX-1. **a** JX-1 genome: First ring denotes CDSs (in light blue) in forward direction, while the second ring denotes CDSs in reverse direction. Genomic location of T3SS, *pmp*, *Inc.*, and *ompA* genes, as well as the PZ are also outlined on the JX-1 genome plot. The genome start is denoted by the malate dehydrogenase gene (*mdh*) at 0 position. Image was generated with the DNAplotter. **b** Graphical representation of the p*Cgall* JX-1 and its CDSs, including plasmid primer locations. The light blue arrows denote the 22 bp tandem repeat units

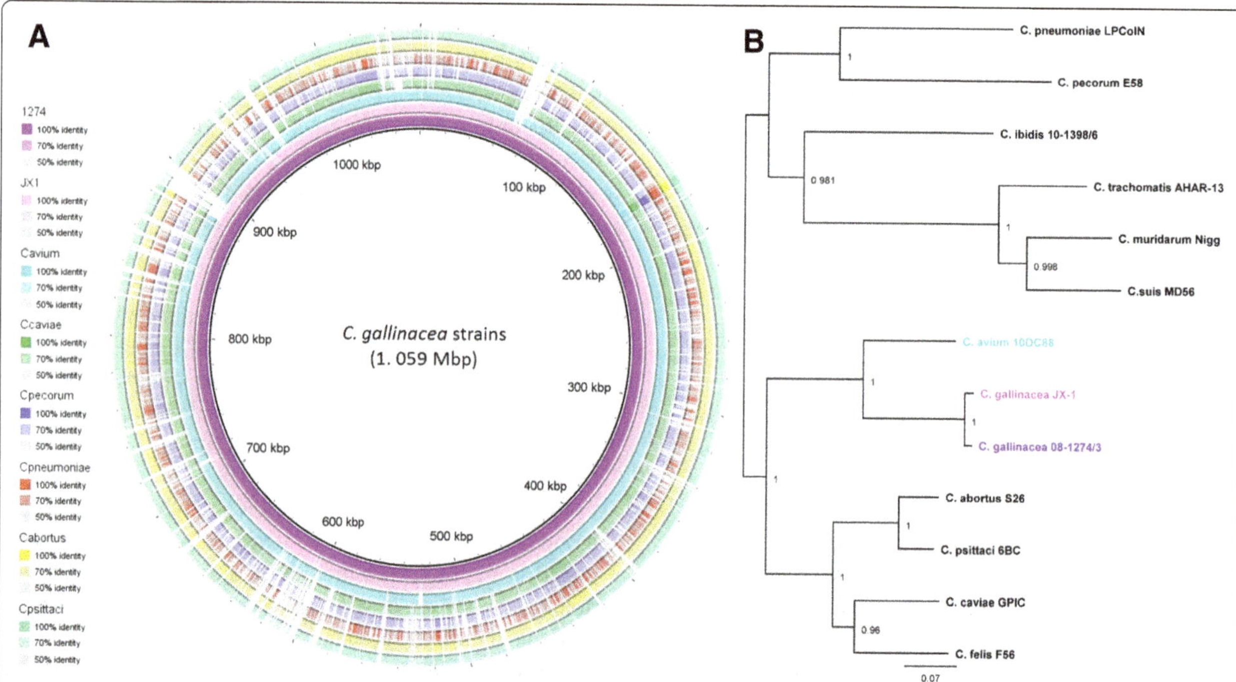

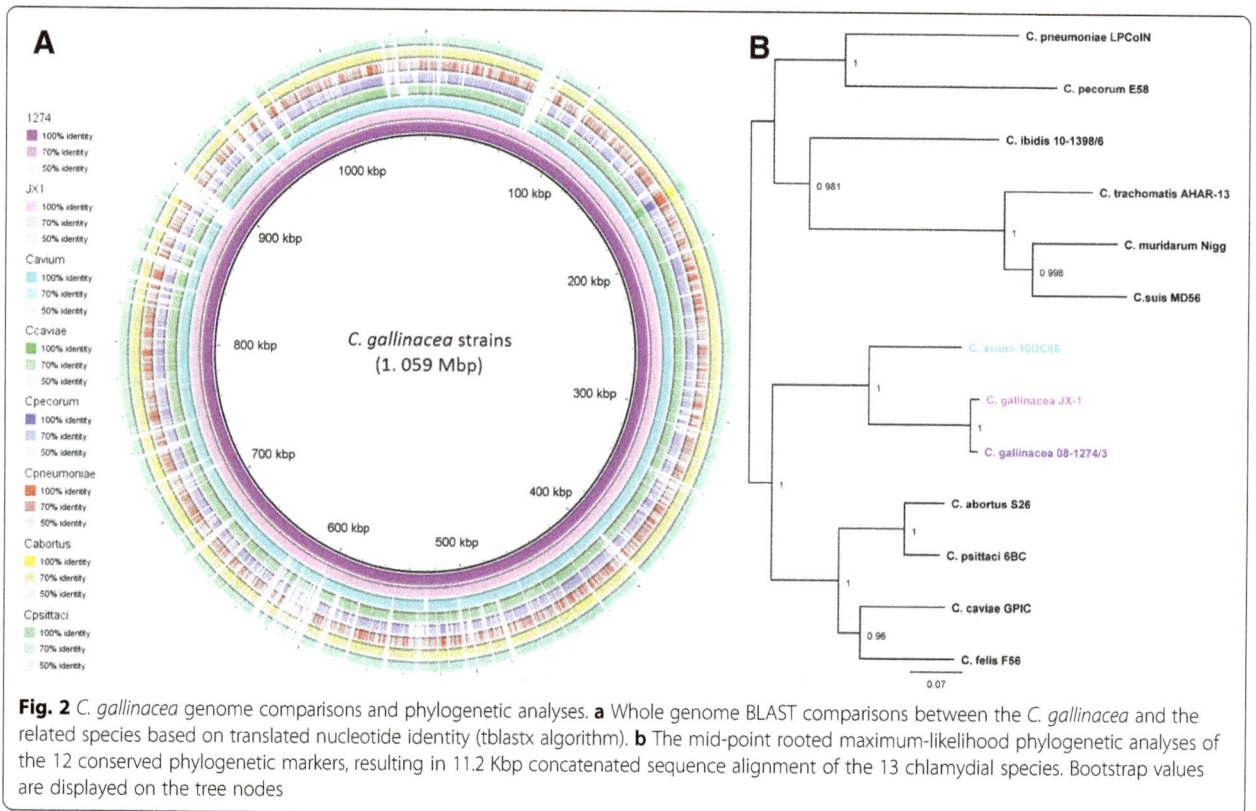

Fig. 2 *C. gallinacea* genome comparisons and phylogenetic analyses. **a** Whole genome BLAST comparisons between the *C. gallinacea* and the related species based on translated nucleotide identity (tblastx algorithm). **b** The mid-point rooted maximum-likelihood phylogenetic analyses of the 12 conserved phylogenetic markers, resulting in 11.2 Kbp concatenated sequence alignment of the 13 chlamydial species. Bootstrap values are displayed on the tree nodes

Table 2 Sequence analysis of *C. gallinacea* polymorphic genes

Locus tag and length (bp) in *C. gallinacea* JX-1	Predicted product	% DNA sequence similarity	Total No. of polymorphisms	No. of non-synonymous substitutions	No. of synonymous substitutions
GM000925 (1581 bp)	Lysine-tRNA ligase	95.1%	77 SNPs	11	65
GM000895 (873 bp)	Serine/threonine protein phosphatase	93.2%	59 SNPs	12	53
GM000890 (2631 bp)	Alanyl-tRNA synthetase	92.5%	198 SNPs	36	162
GM000889 (3252 bp)	Transcription-repair coupling factor	94.2%	190 SNPs	37	153
GM000888 (924 bp)	Uroporphyrinogen decarboxylase (hemE)	92.2%	72 SNPs	9	63
GM000887 (1374 bp)	Coproporphyrinogen oxidase (hemN)	93.0%	96 SNPs	31	65
GM000705 (1206 bp)	Major outer membrane protein, porin	86.7%	161 (104 SNPs and 57 indel)	38	60
GM000539 (1356 bp)	Sodium/alanine symporter family protein	95.2%	65 SNPs	12	53
GM000349 (1236 bp)	Cell wall hydrolase	91.7%	102 (87 SNPs and 15 indel)	41	45
GM000348 (1179 bp)	Phage T7 tail fiber family protein	80.3%	232 (133 SNPs and 99 indel)	59	74
GM000289 (2409 bp)	Glycogen phosphorylase	94.1%	142 SNPs	20	122
GM000288 (1290 bp)	Dihydrolipoamide acetyltransferase component	91.1%	115 SNPs	40	75
GM000264[a] (1765 bp)	Hypothetical protein	88.3%	202 (175 SNPs and 27 indel)	67	108

[a]C. gallinacea-specific sequence based on BLAST search

with four transmembrane domains; and three with six transmembrane domains (Additional file 7: Table S7; Additional file 8: Figure S1). This number of predicted Incs is comparable to that observed in the related chlamydial species [31].

Variation in the *pmp* genes: gene truncations rather than SNP accumulation

In the *C. gallinacea* genomes, the *pmp* genes were found to form two major clusters. In our analysis, we predicted a total of 15 *pmp* genes, however only B, A, E1, H, G1 and G2 appear to be intact in both *C. gallinacea* strains (Fig. 3). The *pmp*D gene of strain JX-1 was found to have a premature STOP codon (Fig. 3a). The remaining four *pmp* genes also had a premature STOP codon in both strains, which was predicted to truncate the encoded proteins before their respective C-terminal autotransporter domains as based on BLAST and CDD (Conserved domains) BLAST analysis. BLAST and phylogenetic analyses confirmed that the majority of the predicted *pmp* genes in the *C. gallinacea* genomes are paralogs of *pmp*G gene lineage (Fig. 3b).

In terms of the number of intact *pmp* genes, *C. gallinacea* was closest to *C. avium* ($n = 7$), while the related chlamydial species harbor between 14 (*C. abortus*) and 22 *pmp* genes (*C. psittaci*). In contrast to the related chlamydial species where these genes are major contributors of SNPs [29, 32], *pmp* genes of the two analysed *C. gallinacea* strains were conserved, with overall sequence similarity ranging from 95.5 to 99.6%. Comparable levels of similarity were observed when comparing the pmp amino acid sequences (Fig. 3c).

The *C. gallinacea* plasticity zone (PZ)

The PZ, notoriously known for harboring key virulence genes of chlamydiae, has been considered as the region of most extensive genetic differences between chlamydial genomes [27]. In comparison to the related chlamydial species where the PZ may be up to 50 kbp in size (e.g. in *C. muridarum* and *C. trachomatis*), *C. gallinacea* displayed gene content reduction in this region (14 kbp), similar to *C. abortus* (12 kbp). In our analysis, *C. avium* appeared to have the most reduced PZ (4.6 kbp) (Fig. 4). The PZ of two *C. gallinacea* strains included three hypothetical proteins, two acetyl-co-carboxylases and a single copy of the chlamydial *cytotoxin* (*tox*) gene, but remained highly conserved with 99.2% sequence similarity. Interestingly, *C. gallinacea* JX-1 had a premature STOP codon in the cytotoxin (*tox*) gene, while *tox* appeared to be intact in the type strain 08-1274/3 (Fig. 4).

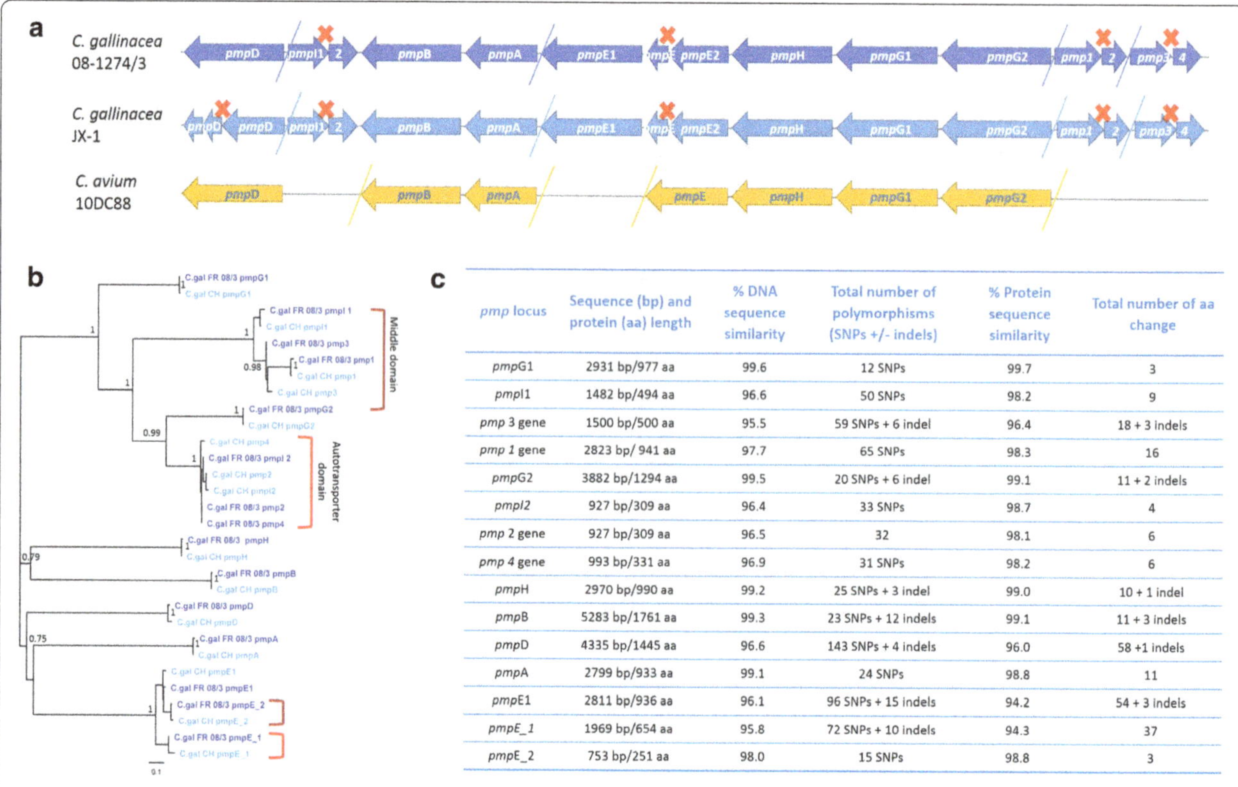

pmp locus	Sequence (bp) and protein (aa) length	% DNA sequence similarity	Total number of polymorphisms (SNPs +/- indels)	% Protein sequence similarity	Total number of aa change
pmpG1	2931 bp/977 aa	99.6	12 SNPs	99.7	3
pmpI1	1482 bp/494 aa	96.6	50 SNPs	98.2	9
pmp 3 gene	1500 bp/500 aa	95.5	59 SNPs + 6 indel	96.4	18 + 3 indels
pmp 1 gene	2823 bp/ 941 aa	97.7	65 SNPs	98.3	16
pmpG2	3882 bp/1294 aa	99.5	20 SNPs + 6 indel	99.1	11 + 2 indels
pmpI2	927 bp/309 aa	96.4	33 SNPs	98.7	4
pmp 2 gene	927 bp/309 aa	96.5	32	98.1	6
pmp 4 gene	993 bp/331 aa	96.9	31 SNPs	98.2	6
pmpH	2970 bp/990 aa	99.2	25 SNPs + 3 indel	99.0	10 + 1 indel
pmpB	5283 bp/1761 aa	99.3	23 SNPs + 12 indels	99.1	11 + 3 indels
pmpD	4335 bp/1445 aa	96.6	143 SNPs + 4 indels	96.0	58 +1 indels
pmpA	2799 bp/933 aa	99.1	24 SNPs	98.8	11
pmpE1	2811 bp/936 aa	96.1	96 SNPs + 15 indels	94.2	54 + 3 indels
pmpE_1	1969 bp/654 aa	95.8	72 SNPs + 10 indels	94.3	37
pmpE_2	753 bp/251 aa	98.0	15 SNPs	98.8	3

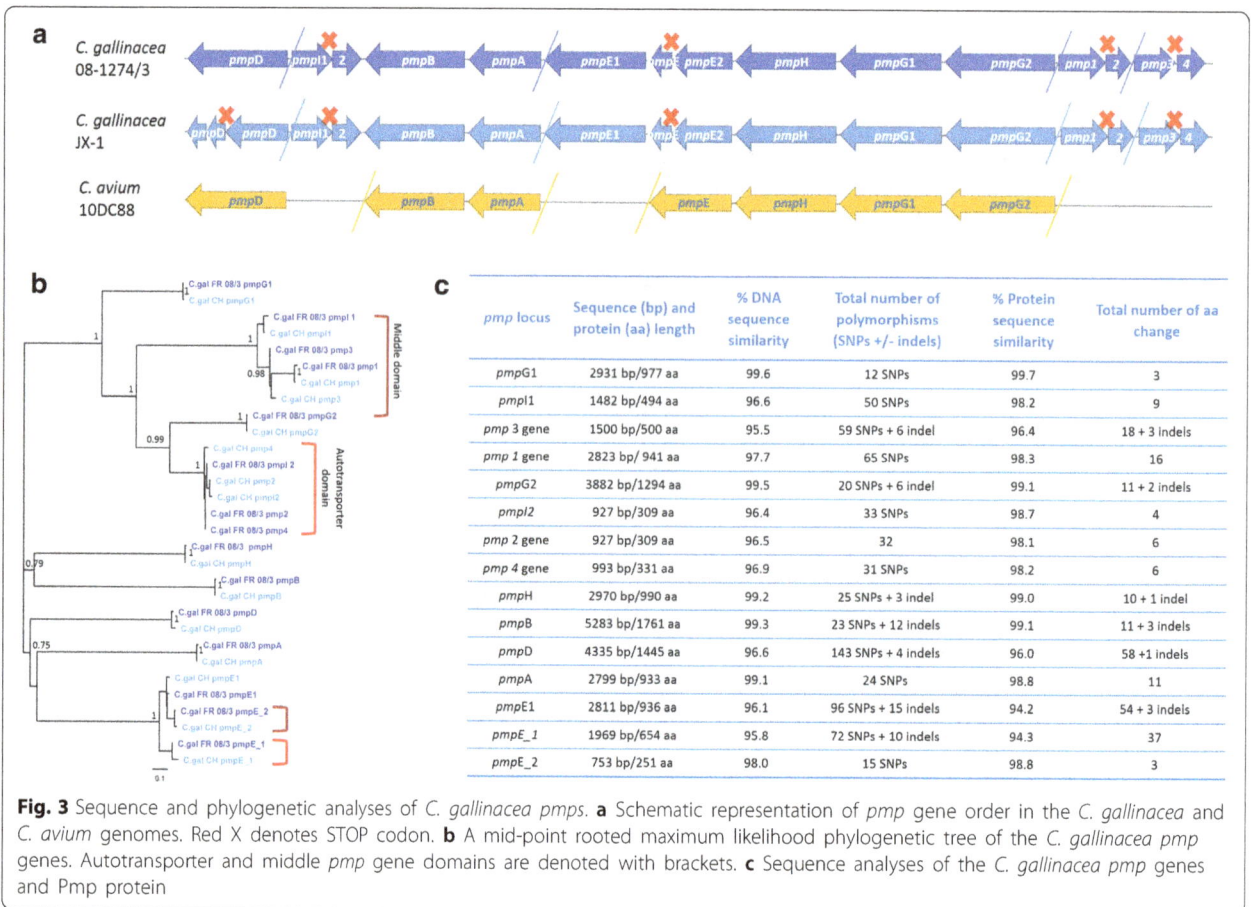

Fig. 3 Sequence and phylogenetic analyses of *C. gallinacea pmps*. **a** Schematic representation of *pmp* gene order in the *C. gallinacea* and *C. avium* genomes. Red X denotes STOP codon. **b** A mid-point rooted maximum likelihood phylogenetic tree of the *C. gallinacea pmp* genes. Autotransporter and middle *pmp* gene domains are denoted with brackets. **c** Sequence analyses of the *C. gallinacea pmp* genes and Pmp protein

Molecular characterization of plasmid pJX-1 and its distribution in *C. gallinacea* strains

Newly characterised plasmid pJX-1 of *C. gallinacea* was 7.49 kbp in length, sharing an identical annotation with eight CDSs and four 22 bp tandem repeats with plasmid p1274 of the type strain 08-1274/3 (Fig. 1b). Briefly, in pJX-1, CDSs 1 (pGP8), 2 (pGP8), 3 (pGP1), and 7 (*par*A) were denoted as putative integrase, helicase, and a partitioning plasmid proteins, respectively, while the CDSs 4 (pGP2), 5 (pGP3), 6 (pGP4), and 8 (pGP6) were denoted as putative chlamydia-specific plasmid virulence proteins, as previously described for related chlamydial plasmids [33, 34]. The sequence of pJX-1 was 99.9% identical to p1274, except for two point mutations at positions 6573 (C changed to T) and 7170 (C changed to A) in CDS 8 (pGP6). Further PCR examination of *C. gallinacea*-positive clinical samples revealed that pJX-1 was detected in all the 24 oral/cloacal swab samples from chickens, while clinical samples from other avian hosts remained negative (data not shown).

MLST of *C. gallinacea* reveals genetic diversity among strains in chickens

In order to obtain a snapshot of the genetic diversity of this emerging pathogen, we have applied our newly developed *C. gallinacea*-specific MLST to a range of *C. gallinacea*-positive clinical samples collected from birds (Additional file 1: Table S1). Unfortunately, due to insufficient amounts of *C. gallinacea* DNA, MLST was only successful in 23 cloacal and oral clinical samples from 3 hens and 20 chickens.

Using a set of 25 sequences, including the MLST sequences obtained from the two sequenced *C. gallinacea* genomes, sequence analysis of individual as well as concatenated gene fragments confirmed the evolutionary conservation of the HK genes, as all alleles were under purifying selection with dn/ds ratios < 1(Table 3). The highest number of mutations was noted in *hem*N (41 SNPs) and *eno*A (25 SNPs), resulting mainly in synonymous substitution, whereas *gat*A had none. *eno*A and *opp*A3 were the most diverse loci as they both occurred in six allelic variants (Table 3). The concatenated sequences (further used for phylogenetic analysis) and its cognate MLST allelic profiles resulted in a total of 15 haplotypes or *C. gallinacea* STs (Table 3).

To examine the genetic relationships between the *C. gallinacea* strains typed using our MLST method, a mid-point rooted Bayesian phylogenetic tree was constructed using the concatenated HK gene sequences

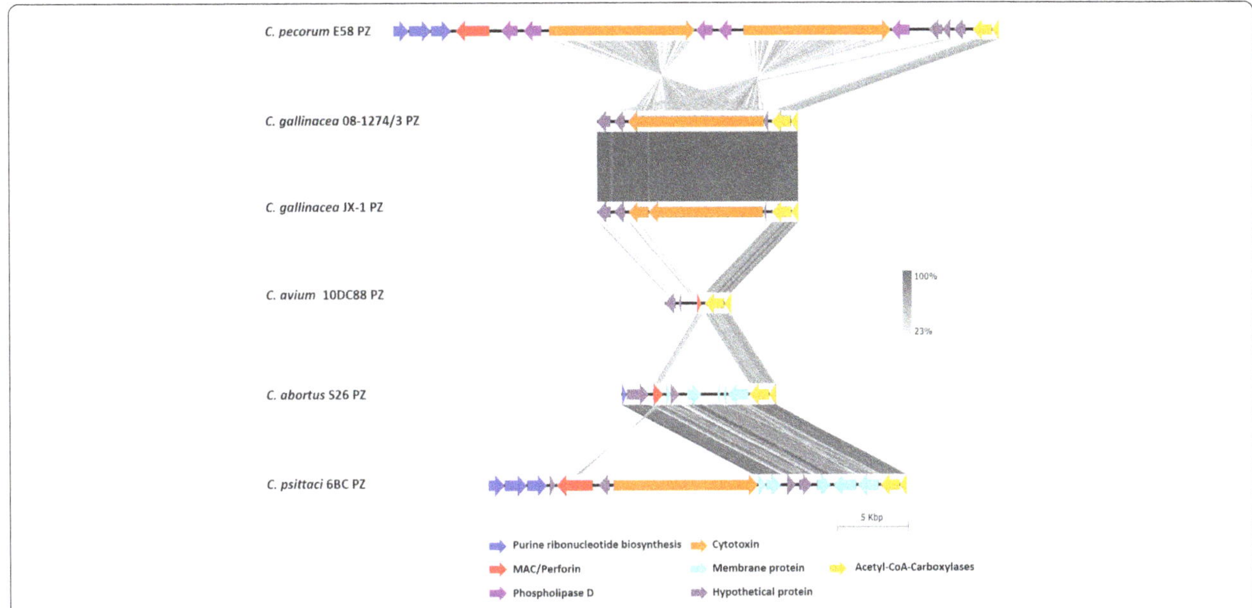

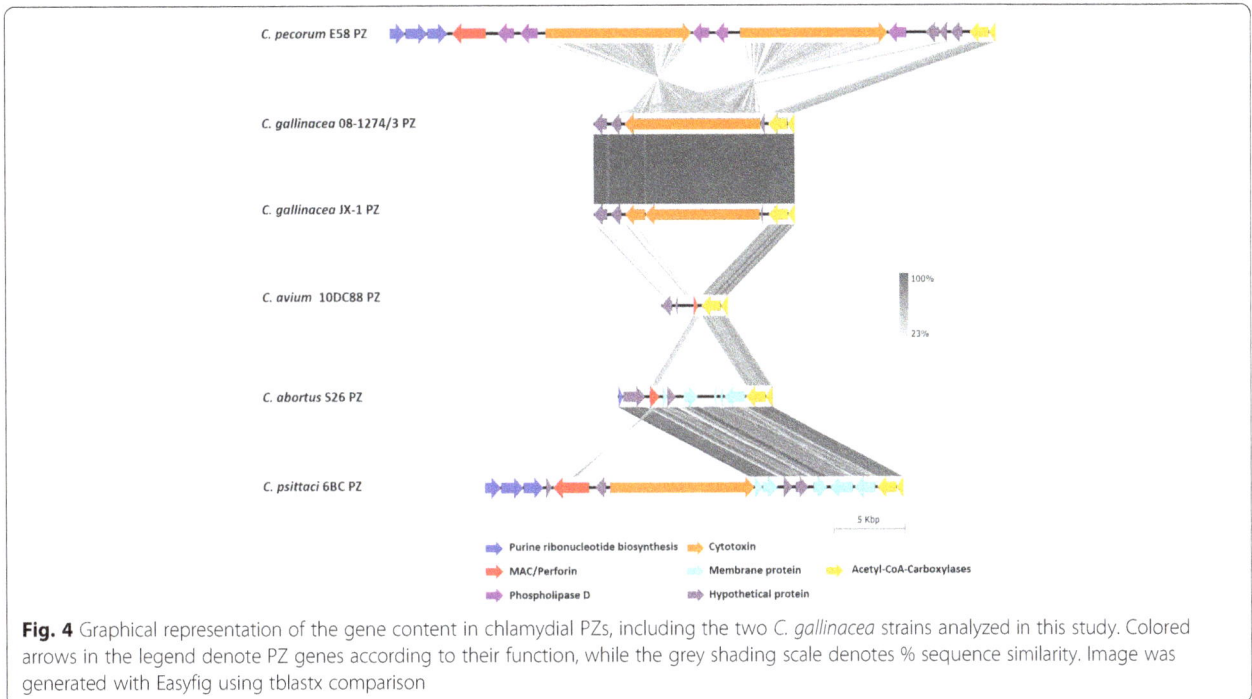

Fig. 4 Graphical representation of the gene content in chlamydial PZs, including the two *C. gallinacea* strains analyzed in this study. Colored arrows in the legend denote PZ genes according to their function, while the grey shading scale denotes % sequence similarity. Image was generated with Easyfig using tblastx comparison

amplified from the 25 strains included. Using the concatenated MLST sequence of the closest relative *C. avium* as an out-group, the phylogenetic tree separated the *C. gallinacea* type strain 08-1274/3 from all Chinese *C. gallinacea* strains into two distinct clades (Fig. 5a). Although in the same well-supported larger clade, *C. gallinacea* strains from China could be further sub-divided into 14 genetically distinct lineages. The phylogenetic tree constructed from the present *C. gallinacea* STs also revealed that: a) The same strain can infect two different anatomical sites in a single host (e.g. A/Ch_40: oral and cloaca); b) the same strain can be found in different animals from the same area (e.g. J/Hen 31, 12 and 4 strains) or geographically distinct areas (e.g. JX-1, J/ChA2432, J/ChA2360 and A/Ch29-1 strains); and c) closely related strains can also be found in geographically distinct areas (e.g. Ha/ChA3274 and Gx/ChA612) (Fig. 5).

Discussion

In this study, we present the first detailed analysis of *C. gallinacea* genomes. The two *C. gallinacea* genomes studied are compact, syntenic and highly conserved between them, while sharing some of the classical genomic features of *Chlamydia* spp., such as the highly conserved T3SS, *pmp* gene clusters, and a potentially virulence-associated plasmid. The two strains analyzed in this study, JX-1 from China and type strain 08-1274/3 from France form separate phylogenetic lineages within a clade with its closest relative *C. avium*. Using information derived from these genomes and previously described *Chlamydiales* MLST scheme [13], we also adapted a complete *C. gallinacea* MLST scheme and applied it to clinical strains from chickens of various Chinese provinces. The latter analysis revealed that this organism is genetically diverse, indicating the potential

Table 3 Sequence analysis of *C. gallinacea* MLST alleles for samples denoted in bold in Additional file 1: Table S1

Allele	Total number of mutations (Δnt)	No. of non-synonymous substitutions	No. of synonymous substitutions	N alleles
*gat*A	0	0	0	1
*opp*A_3	5	2	3	6
*hfl*X	3	0	3	4
*gid*A	2	1	1	2
*eno*A	25	1	24	6
*hem*N	41	13	28	3
*fum*C	7	1	6	5
Concatenated (3098 bp)	83	18	65	15

Δnt: No. of polymorphic sites; *d_s and d_n: the average number of synonymous substitutions per synonymous site and non- synonymous substitutions per non-synonymous site, respectively (Jukes – Cantor corrected); N Alleles: No. of unique sequences according to the gene

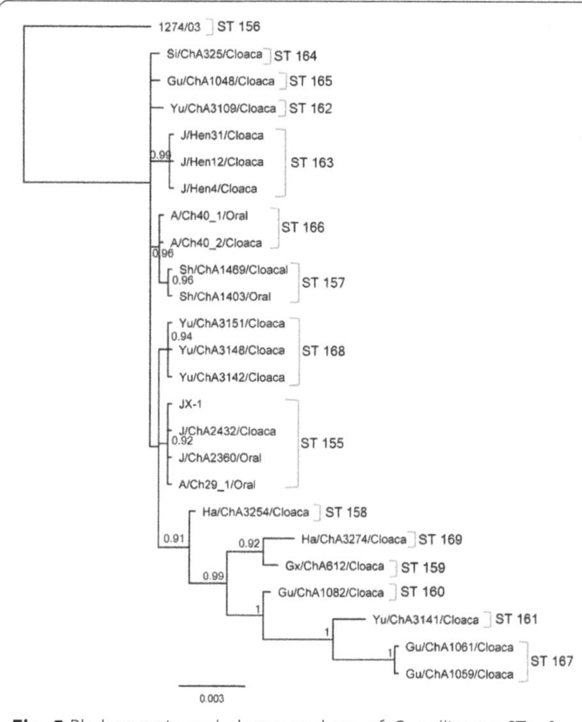

Fig. 5 Phylogenetic and cluster analyses of *C. gallinacea* STs. **A**: Bayesian phylogenetic analysis of the concatenated sequences of seven MLST fragments of 25 *C. gallinacea* genotypes. Posterior probabilities > 0.75 are displayed on tree nodes, and STs in a clade are indicated by brackets

for a complex epidemiology similar to other chlamydial species found in animals.

The previously described typical synteny and gene order for chlamydial organisms [9] was also observed in the *C. gallinacea* genomes. The two strains differed in approximately 6250 SNPs, mostly synonymous, which were distributed evenly along the chromosome. The extent of genetic variation in the *omp*A locus was not surprising, as a previous study on *C. gallinacea omp*A typing identified at least 13 genotypes [4]. The *omp*A variation at the variable domains in this species is also consistent with *omp*A diversity seen in other chlamydial pathogens [35–37].

As outlined in Table 2, the majority of SNPs accumulated in genes associated with metabolism, most notably in *hem*E and *hem*N genes that are involved in heme metabolism [38]. In prokaryotes, heme is an integral part of proteins involved in multiple electron transport chains for respiration, and a cofactor of many enzymes including catalases, peroxidases, and P(450) class cytochromes [39]. Further analysis including genome sequences of more strains from different hosts and regions will be necessary to elucidate possible functional consequences of the present observation. Similarly, the present dataset cannot provide an explanation why gene GM000264, which is coding for a hypothetical protein of unknown function, was found to accumulate SNPs and have the highest dn/ds ratio (0.526) among all analyzed genes.

In contrast to the high number of SNPs observed in the metabolic genes outlined in Table 2, the T3SS and *pmp* genes were unexpectedly conserved between the two *C. gallinacea* strains with less than 1% sequence dissimilarity. Such congruence and conservation in these genomic regions is in stark contrast to the high genetic variation (up to 15%) in these genes of other chlamydial species [28, 29]. The chlamydial T3SS, a system of structural, chaperone and secreted effector proteins, is considered as "virulence machinery" with a function to deliver effector proteins in order to subvert host cellular processes [40, 41]. As such, genetic variation and polymorphisms in these genes are probably associated with differences in virulence and host and/or tissue tropism, as previously observed for *C. pecorum*, *C. trachomatis*, *C. psittaci* and other chlamydial species [28, 42, 43]. Whether the sequence conservation in T3SS genes is present throughout the *C. gallinacea* taxon remains to be investigated, as well as their role in virulence of this organism. Equally, we further need to investigate the role in infection of the 37 putative *C. gallinacea* Incs identified in this study, and how genetically diverse they will be throughout the taxon.

Highly polymorphic chlamydial *pmp* gene families account for the majority of chromosomal SNPs in other related species. Generally speaking, the *pmp* loci constitute almost 4% of the chlamydial genome, thus suggesting an important biological function due to their adhesive as well as antigenic properties [32]. In this study, the *pmp* genes of *C. gallinacea* somewhat surprisingly displayed sequence conservation, although we observed non-synonymous changes and indels between the two strains. Proteins *Pmp*B, A, E1, H, G1 and G2 appear to be intact in both *C. gallinacea* strains, while strain JX-1 harbored a premature STOP codon in *pmp*D, which will require further investigations to assess whether the protein's function has been impaired. The remaining *pmp* genes, although truncated, appear to be PmpG paralogs based on our blast and phylogenetic analyses. Based on these (early) observations, our current hypothesis is that *pmp*G expansion may not be necessary for virulence or biology of *C. gallinacea*. Studies on *C. pneumoniae*, where Pmps (including at least 13 *pmp*G subtypes) represent major proteins in the outer membrane, showed that all *pmp* genes are transcribed and expressed during the infectious cycle [32, 44]. In *C. psittaci*, several *pmp*G genes that may be related to host tropism and virulence were identified. It is possible that the *pmp*G group plays a major role in host range, tissue tropism and virulence for different *C. psittaci* strains due to their high diversity and rapid evolution [45, 46]. Considering the importance of *pmps* in chlamydial genomes, the observed non-synonymous changes and indels between the two analyzed strains render future investigations on *pmp* diversity within the species of *C. gallinacea* highly pertinent.

The PZs of the two *C. gallinacea* strains proved compact and highly conserved, with 99.2% sequence similarity, with Mac/Perforin and Phospholipase D genes missing, but harboring the full-length chlamydial cytotoxin (*tox*) gene [47]. *C. gallinacea* JX-1 had a premature STOP codon in the N-terminal regions of the *tox* gene, a region that contain a catalytic glycosyltransferase domain, while *tox* was intact in *C. gallinacea* type strain 08-1274/3. However, whether this has an impact on the function of the *tox* gene remains to be elucidated. Chlamydial *tox* genes are considered important virulence factors and were associated with acute infection and disease [48]. They appear to be a species and niche-specific feature due to their full-length presence in only five related species, including *C. psittaci*, *C. felis*, *C. caviae* with a single gene copy, and *C. pecorum* and *C. muridarum* with two and three copies, respectively [9, 27]. In contrast, *tox* is partially truncated in *C. trachomatis*, while *C. avium*, *C. abortus* and *C. pneumoniae* do not have a *tox* gene. It will be interesting to see how the presence or absence of *tox* gene(s) in chlamydial pathogens can be correlated with virulence properties in the future.

Both *C. gallinacea* strains were found to carry the highly conserved plasmid. In the present study, the plasmid was only detected in 24 *C. gallinacea*-positive samples from chickens out of total 45 samples tested from different avian. Its absence in strains infecting other hosts could be due to a) the plasmid being host specific; and b) more likely, failure to detect it because of low DNA concentration and rapid degradation, considering that chlamydial plasmids are present in low copy numbers (1–10 per chromosome) [49]. Another possibility leading to failure of plasmid detection in clinical samples might be associated with high temperature during DNA extraction. In order to increase the efficiency of DNA extraction from clinical samples, we usually give high temperature (72 °C) and a long time for proteinase K incubation and for the multiple elution steps. This approach increases sensitivity for PCR-based clinical diagnostics, but may induce DNA breakage which probably result in failure of detecting the 7.49 kbp plasmid. The presence of a plasmid is a frequent feature of *Chlamydia* spp. genomes, but naturally occurring plasmid-less strains can also be found [29]. Species like *C. abortus* and (human) *C. pneumoniae* do not carry plasmids at all [50]. Both the role and distribution of the plasmid in *C. gallinacea* strains warrant further investigations.

Previous data from *ompA* genotyping indicated considerable genetic diversity of *C. gallinacea* field strains [4]. The present MLST confirmed this by identifying 15 novel STs among 25 strains.

An analysis of the *C. gallinacea* MLST data allows us to make some speculation on the epidemiology of this pathogen. Identification of the same sequence type in cloacal and oral sites (e.g. A/Ch_40: oral and cloaca, both ST 166) indicates that at least one possible transmission route could be fecal-oral (cloacal). The observations above also indicated transmission among different avian hosts in the same area (e.g. J/Hen 31, 12 and 4 strains, all ST 163) or geographically distinct areas (e.g. JX-1, J/ChA2432, J/ChA2360 and A/Ch29-1 strains, all ST 155). Moreover, there is probably transmission across different hosts, as *C. gallinacea* strains have been frequently isolated from birds and were also found in livestock that have been in contact [2].

While we could not apply MLST on *C. gallinacea* strains from ducks, geese and pigeons, nor to European and livestock strains, we do anticipate that future typing studies will shed more light on the complex epidemiology and genetic diversity of this chlamydial agent.

Conclusions

Awareness is growing that *C. gallinacea* infections in avian hosts are globally highly prevalent. In the present study, we have provided new evidence that this pathogen is genetically diverse, even though it is still open how these genetic differences among strains infecting chickens and other hosts translate into its biology and pathogenicity. Future studies should include analysis of strains from a variety of avian and other livestock hosts to enable us to understand the evolution and host adaption of this enigmatic pathogen.

Additional files

Additional file 1: Table S1. *C. gallinacea*-positive samples used for MLST and plasmid detection. (DOCX 13 kb)

Additional file 2: Table S2. Primers used for generating *C. gallinacea* plasmid fragments in this study. (DOCX 14 kb)

Additional file 3: Table S3. Primers used for MLST of *C. gallinacea* in this study. (DOCX 14 kb)

Additional file 4: Table S4. PCR conditions for amplifying seven HK genes in this study. (DOCX 12 kb)

Additional file 5: Table S5. Original data for MLST analysis of the *C. gallinacea*-positive samples. (XLSX 10 kb)

Additional file 6: Table S6. Original data for T3SS of *C. gallinacea*, including the locus-tags, and their annotation/putative function, length and direction. (XLS 36 kb)

Additional file 7: Table S7. List of the predicted transmembrane helices in the analyzed putative *C. gallinacea* inclusion (Inc) proteins from this study. (XLSX 10 kb)

Additional file 8: Figure S1. Graphical representation of the predicted transmembrane helices in the analyzed putative *C. gallinacea* inclusion (Inc) proteins from this study. (DOCX 423 kb)

Abbreviations
MLST: Multilocus sequence typing; *pmp*: Polymorphic membrane protein; PZ: Plasticity zone; *tox*: Cytotoxin

Acknowledgements
This publication made use of the PubMLST website (http://pubmlst.org/) developed by Keith Jolley (24) and hosted at the University of Oxford. The development of this website was funded by the Wellcome Trust.

Funding
This project is funded by the National Key R & D Program of China (2016YFD0500804), A Key Project of Anhui Province Education Department (No: KJ2016A822) and the Program of Anhui Science and Technology University (No: ZRC2016497).

Authors' contributions
CW, WG, MJ, KS, AP and BK designed the study and interpreted the data. WG, JL, JG, and JY performed the DNA extraction, MLST and DNA sequencing. WG, MJ and YP analyzed the MLST data and drew the figure. CW, WG, MJ, KS, AP, BK and YP were major contributors in writing the manuscript. All authors read and approved the final manuscript.

Competing interests
The authors declare that they have no competing interests.

Author details
[1]Jiangsu Co-Innovation Center for Prevention and Control of Important Animal Infectious Diseases and Zoonoses, Yangzhou University College of Veterinary Medicine, Yangzhou, Jiangsu 225009, People's Republic of China. [2]College of Animal Science, Anhui Science and Technology University, Maanshan, Anhui, China. [3]Centre for Animal Health Innovation, Faculty of Science, Health, Education and Engineering, University of the Sunshine Coast, Maroochydore, QLD, Australia. [4]Institute of Bioinformatics, Friedrich-Schiller-Universität Jena, Jena, Germany. [5]Department of Microbiology, University of Amsterdam, Amsterdam, The Netherlands. [6]College of Veterinary Medicine, Auburn University, Auburn, AL, USA. [7]Poultry Institute, Chinese Academy of Agricultural Sciences, Jiangsu Co-Innovation Center for Prevention and Control of Important Animal Infectious Diseases and Zoonoses, Yangzhou, Jiangsu, China.

References
1. Sachse K, Laroucau K. Avian chlamydiosis: two more bacterial players discovered. Vet J. 2014;200(3):347–8.
2. Li J, Guo W, Kaltenboeck B, Sachse K, Yang Y, Lu G, et al. *Chlamydia pecorum* is the endemic intestinal species in cattle while *C. gallinacea*, *C. psittaci* and *C. pneumoniae* associate with sporadic systemic infection. Vet Microbiol. 2016;193:93–9.
3. Li L, Luther M, Macklin K, Pugh D, Li J, Zhang J, et al. *Chlamydia gallinacea*: a widespread emerging *Chlamydia* agent with zoonotic potential in backyard poultry. Epidemiol Infect. 2017;145(13):2701–3.
4. Guo W, Li J, Kaltenboeck B, Gong J, Fan W, Wang C. *Chlamydia gallinacea*, not *C. psittaci*, is the endemic chlamydial species in chicken (*Gallus gallus*). Sci Rep. 2016;6:19638.
5. Zocevic A, Vorimore F, Marhold C, Horvatek D, Wang D, Slavec B, et al. Molecular characterization of atypical *Chlamydia* and evidence of their dissemination in different European and Asian chicken flocks by specific real-time PCR. Environ Microbiol. 2012;14(8):2212–22.
6. Krautwald-Junghanns ME, Stolze J, Schmidt V, Bohme J, Sachse K, Cramer K. Efficacy of doxycycline for treatment of chlamydiosis in flocks of racing and fancy pigeons. Tierarztl Prax Ausg K Kleintiere Heimtiere. 2013;41(6):392–8.
7. Laroucau K, Aaziz R, Meurice L, Servas V, Chossat I, Royer H, et al. Outbreak of psittacosis in a group of women exposed to *Chlamydia psittaci*-infected chickens. Euro Surveill. 2015;20(24)
8. Laroucau K, Vorimore F, Aaziz R, Berndt A, Schubert E, Sachse K. Isolation of a new chlamydial agent from infected domestic poultry coincided with cases of atypical pneumonia among slaughterhouse workers in France. Infect Genet Evol. 2009;9(6):1240–7.
9. Bachmann NL, Polkinghorne A, Timms P. *Chlamydia* genomics: providing novel insights into chlamydial biology. Trends Microbiol. 2014;22(8):464–72.
10. Taylor-Brown A, Vaughan L, Greub G, Timms P, Polkinghorne A. Twenty years of research into *Chlamydia*-like organisms: a revolution in our understanding of the biology and pathogenicity of members of the phylum *Chlamydiae*. Pathog Dis. 2015;73(1):1–15.
11. Holzer M, Laroucau K, Creasy HH, Ott S, Vorimore F, Bavoil PM, et al. Whole-Genome Sequence of *Chlamydia gallinacea* Type Strain 08-1274/3. Genome Announc. 2016;4(4)
12. Sachse K, Laroucau K, Riege K, Wehner S, Dilcher M, Creasy HH, et al. Evidence for the existence of two new members of the family *Chlamydiaceae* and proposal of *Chlamydia avium* sp. nov. and *Chlamydia gallinacea* sp. nov. Syst Appl Microbiol. 2014;37(2):79–88.
13. Pannekoek Y, Morelli G, Kusecek B, Morre SA, Ossewaarde JM, Langerak AA, et al. Multi locus sequence typing of *Chlamydiales*: clonal groupings within the obligate intracellular bacteria *Chlamydia trachomatis*. BMC Microbiol. 2008;8:42.
14. Aziz RK, Bartels D, Best AA, DeJongh M, Disz T, Edwards RA, et al. The RAST server: rapid annotations using subsystems technology. BMC Genomics. 2008;9:75.
15. Carver T, Berriman M, Tivey A, Patel C, Bohme U, Barrell BG, et al. Artemis and ACT: viewing, annotating and comparing sequences stored in a relational database. Bioinformatics. 2008;24(23):2672–6.
16. Kearse M, Moir R, Wilson A, Stones-Havas S, Cheung M, Sturrock S, et al. Geneious basic: an integrated and extendable desktop software platform for the organization and analysis of sequence data. Bioinformatics. 2012; 28(12):1647–9.
17. Darling AE, Mau B, Perna NT. Progressive Mauve: multiple genome alignment with gene gain, loss and rearrangement. PLoS One. 2010;5(6):e11147.
18. Katoh K, Standley DM. MAFFT: iterative refinement and additional methods. Methods Mol Biol. 2014;1079:131–46.
19. Librado P, Rozas J. DnaSP v5: a software for comprehensive analysis of DNA polymorphism data. Bioinformatics. 2009;25(11):1451–2.
20. Guindon S, Dufayard JF, Lefort V, Anisimova M, Hordijk W, Gascuel O. New algorithms and methods to estimate maximum-likelihood phylogenies: assessing the performance of PhyML 3.0. Syst Biol. 2010;59(3):307–21.
21. Alikhan NF, Petty NK, Ben Zakour NL, Beatson SA. BLAST ring image generator (BRIG): simple prokaryote genome comparisons. BMC Genomics. 2011;12:402.
22. Sullivan MJ, Petty NK, Beatson SA. Easyfig: a genome comparison visualizer. Bioinformatics. 2011;27(7):1009–10.
23. Carver T, Thomson N, Bleasby A, Berriman M, Parkhill J. DNAPlotter: circular and linear interactive genome visualization. Bioinformatics. 2009;25(1):119–20.
24. Jolley KA, Maiden MC. BIGSdb: scalable analysis of bacterial genome variation at the population level. BMC Bioinformatics. 2010;11:595.
25. Darriba D, Taboada GL, Doallo R, Posada D. jModelTest 2: more models, new heuristics and parallel computing. Nat Methods. 2012;9(8):772.
26. Huelsenbeck JP, Ronquist F. MRBAYES: Bayesian inference of phylogenetic trees. Bioinformatics. 2001;17(8):754–5.
27. Nunes A, Gomes JP. Evolution, phylogeny, and molecular epidemiology of *Chlamydia*. Infect Genet Evol. 2014;23:49–64.
28. Wolff BJ, Morrison SS, Pesti D, Ganakammal SR, Srinivasamoorthy G, Changayil S, et al. *Chlamydia psittaci* comparative genomics reveals intraspecies variations in the putative outer membrane and type III secretion system genes. Microbiology. 2015;161(7):1378–91.
29. Jelocnik M, Bachmann NL, Kaltenboeck B, Waugh C, Woolford L, Speight KN, et al. Genetic diversity in the plasticity zone and the presence of the chlamydial plasmid differentiates *Chlamydia pecorum* strains from pigs, sheep, cattle, and koalas. BMC Genomics. 2015;16:893.
30. Abdelsamed H, Peters J, Byrne GI. Genetic variation in *Chlamydia trachomatis* and their hosts: impact on disease severity and tissue tropism. Future Microbiol. 2013;8(9):1129–46.

31. Mital JMN, Dorward DW, Dooley CA, Hackstadt T. Role for chlamydial inclusion membrane proteins in inclusion membrane structure and biogenesis. PLoS One. 2013;8(5):e63426.
32. Vasilevsky S, Stojanov M, Greub G, Baud D. Chlamydial polymorphic membrane proteins: regulation, function and potential vaccine candidates. Virulence. 2016;7(1):11–22.
33. Gong S, Yang Z, Lei L, Shen L, Zhong G. Characterization of *Chlamydia trachomatis* plasmid-encoded open reading frames. J Bacteriol. 2013;195(17): 3819–26.
34. Jelocnik M, Bachmann NL, Seth-Smith H, Thomson NR, Timms P, Polkinghorne AM. Molecular characterisation of the *Chlamydia pecorum* plasmid from porcine, ovine, bovine, and koala strains indicates plasmid-strain co-evolution. PeerJ. 2016;4:e1661.
35. Legione AR, Patterson JL, Whiteley PL, Amery-Gale J, Lynch M, Haynes L, et al. Identification of unusual *Chlamydia pecorum* genotypes in Victorian koalas (Phascolarctos cinereus) and clinical variables associated with infection. J Med Microbiol. 2016;65(5):420–8.
36. Sachse K, Laroucau K, Hotzel H, Schubert E, Ehricht R, Slickers P. Genotyping of *Chlamydophila psittaci* using a new DNA microarray assay based on sequence analysis of *ompA* genes. BMC Microbiol. 2008;8:63.
37. Carlson JH, Porcella SF, McClarty G, Caldwell HD. Comparative genomic analysis of *Chlamydia trachomatis* oculotropic and genitotropic strains. Infect Immun. 2005;73(10):6407–18.
38. Heinemann IU, Jahn M, Jahn D. The biochemistry of heme biosynthesis. Arch Biochem Biophys. 2008;474(2):238–51.
39. Panek H, O'Brian MR. A whole genome view of prokaryotic haem biosynthesis. Microbiology. 2002;148(Pt 8):2273–82.
40. Nans A, Kudryashev M, Saibil HR, Hayward RD. Structure of a bacterial type III secretion system in contact with a host membrane in situ. Nat Commun. 2015;6:10114.
41. Peters J, Wilson DP, Myers G, Timms P, Bavoil PM. Type III secretion a la *Chlamydia*. Trends Microbiol. 2007;15(6):241–51.
42. Bachmann NL, Fraser TA, Bertelli C, Jelocnik M, Gillett A, Funnell O, et al. Comparative genomics of koala, cattle and sheep strains of *Chlamydia pecorum*. BMC Genomics. 2014;15:667.
43. Borges V, Gomes JP. Deep comparative genomics among *Chlamydia trachomatis* lymphogranuloma venereum isolates highlights genes potentially involved in pathoadaptation. Infect Genet Evol. 2015;32:74–88.
44. Grimwood J, Olinger L, Stephens RS. Expression of *Chlamydia pneumoniae* polymorphic membrane protein family genes. Infect Immun. 2001;69(4):2383–9.
45. Read TD, Joseph SJ, Didelot X, Liang B, Patel L, Dean D. Comparative analysis of *Chlamydia psittaci* genomes reveals the recent emergence of a pathogenic lineage with a broad host range. MBio. 2013;4(2)
46. Voigt A, Schofl G, Saluz HP. The *Chlamydia psittaci* genome: a comparative analysis of intracellular pathogens. PLoS One. 2012;7(4):e35097.
47. Read TD, Brunham RC, Shen C, Gill SR, Heidelberg JF, White O, et al. Genome sequences of *Chlamydia trachomatis* MoPn and *Chlamydia pneumoniae* AR39. Nucleic Acids Res. 2000;28(6):1397–406.
48. Rajaram K, Giebel AM, Toh E, Hu S, Newman JH, Morrison SG, et al. Mutational analysis of the *Chlamydia muridarum* plasticity zone. Infect Immun. 2015;83(7):2870–81.
49. Pickett MA, Everson JS, Pead PJ, Clarke IN. The plasmids of *Chlamydia trachomatis* and *Chlamydophila pneumoniae* (N16): accurate determination of copy number and the paradoxical effect of plasmid-curing agents. Microbiology. 2005;151(Pt 3):893–903.
50. Rockey DD. Unraveling the basic biology and clinical significance of the chlamydial plasmid. J Exp Med. 2011;208(11):2159–62.

10

Chronic wounds alter the proteome profile in skin mucus of farmed gilthead seabream

Héctor Cordero[1,2,3], Monica F. Brinchmann[3*], Alberto Cuesta[1] and María A. Esteban[1]

Abstract

Background: Skin and its mucus are known to be the first barrier of defence against any external stressors. In fish, skin wounds frequently appear as a result of intensive culture and also some diseases have skin ulcers as external clinical signs. However, there is no information about the changes produced by the wounds in the mucosae. In the present paper, we have studied the alterations in the proteome map of skin mucus of gilthead seabream during healing of experimentally produced chronic wounds by 2-DE followed by LC-MS/MS. The corresponding gene expression changes of some identified skin proteins were also investigated through qPCR.

Results: Our study has successfully identified 21 differentially expressed proteins involved in immunity and stress processes as well as other metabolic and structural proteins and revealed, for the first time, that all are downregulated in the skin mucus of wounded seabream specimens. At transcript level, we found that four of nine markers (*ighm*, *gst3*, *actb* and *krt1*) were downregulated after causing the wounds while the rest of them remained unaltered in the wounded fish. Finally, ELISA analysis revealed that IgM levels were significantly lower in wounded fish compared to the control fish.

Conclusions: Our study revealed a decreased-expression at protein and for some transcripts at mRNA levels in wounded fish, which could affect the functionality of these molecules, and therefore, delay the wound healing process and increase the susceptibility to any infection after wounds in the skin of gilthead seabream.

Keywords: Proteome, Wounds, Skin mucus, Gilthead seabream (*Sparus aurata*), Teleosts, IgM

Background

Teleost is the largest and most variable vertebrate taxon and most importantly, the earliest group of vertebrates possessing both an innate and adaptive immune system. Gilthead seabream (*Sparus aurata*; *Sparidae*; *Perciformes*; *Teleostei*) is a hermaphroditic protandrous marine species and one of the most farmed fish not only in Europe, but also worldwide with a global production of around 160,000 t in 2014 [1]. Intensive fish farming increases the occurrence of injuries and diseases, commonly associated with the appearance of wounds or ulcers in the skin, causing major economic losses [2, 3]. These injuries and diseases in the skin such as the white nodules from lymphocystis disease [4–6] or the physical wounds that increase the susceptibility of bacterial vibriosis [7] are devastating to farmed fish populations.

Skin mucus is mainly secreted by goblet cells in the skin of fish, protecting as a mechanical, physical, chemical, biological and immunological barrier against any external stressors [3, 8]. In recent years, skin mucus has become a hot topic as a faithful mirror of the immune status of fish [9]. Thus, many humoral immune activities such as proteases, antiproteases, peroxidases, esterases, alkaline phosphatase, lysozyme or immunoglobulins have been evaluated in skin mucus [10–12]. Apart from the individual characterization of antimicrobial peptides [13], immunoglobulins [14] or lectins [15], the recent advances in high throughput proteomics research methods have been used for identification and quantification of proteins [16]. Homology-driven proteomics is a major approach for identification of proteins in species where the sequences are not available [17]; however, identification of unknown proteins often relies on the similarity (rather than identity) when comparing with homologous protein sequences from phylogenetically related species [18], especially for the gilthead seabream,

* Correspondence: Monica.F.Brinchmann@nord.no
[3]Faculty of Biosciences and Aquaculture, Nord University, 8049 Bodø, Norway
Full list of author information is available at the end of the article

when the specific genome is not publically available and/or the transcriptome data are scarce.

Through this approach, the proteomic map of skin mucus has recently been studied in several fish species such as Atlantic cod [19], lumpsucker [20], European sea bass [21], and gilthead seabream [22, 23]. These studies have allowed the discovery of new molecules involved in protection and immunity of this mucosal surface. Besides changes in the skin mucus proteome, i.e. differentially expressed proteins have been studied after infection [24–27], from handling or crowding stress [28, 29], after parental care [30] and more recently after administration of different dietary supplements [29, 31]. However, despite the relevancy to fish health, there are no studies regarding the changes on the skin mucosae following injury so far.

The aim of this work was to study the alteration of the skin mucus proteome after inducing chronic wounds in gilthead seabream. This study was done using 2-DE followed by LC-MS/MS and provides a first idea about the changes of specific proteins involved in immunity, stress and metabolism, as well as structural proteins related to regeneration and healing processes present in skin mucus of gilthead seabream. Finally, we hypothesize that the proteomic levels in mucus and transcriptomic levels in skin are correlated as indicated by these markers as well as concentrations of IgM, which was the main systemic adaptive immune molecule found in skin mucus in our study.

Results

The differential proteome of skin mucus of gilthead seabream after causing chronic wounds was studied through 2-DE (Fig. 1) followed by LC-MS/MS approach (Tables 1 and 2). The total differentially expressed proteins were clustered in four groups: immune-related (I), stress-related (II), structural (III) and metabolic (IV) proteins as described below.

Immune-related molecules

The differential skin mucus proteome of gilthead seabream showed a general decrease of some proteins involved in several immune routes (Tables 1 and 2). One of the most important components of both innate and adaptive immunity, the complement molecule C3 (spot H26), was identified and down-regulated after chronic wounding. Similarly, APOA1 was identified in different parts of the gels (spots H2, H7 and H9) and also showed down-regulated expression in all the analysed protein spots.

It is well-known that some histones may act as antimicrobial peptides [32]. We have identified H2A (spot H5), H2B (spot H3) and H4 (spot H1) to be differentially down-regulated in the skin mucus from wounds of gilthead seabream. Finally, the main component of the adaptive immunity, IgM (spot 28), identified for the first time in skin mucus after 2-DE methodology, showed an interesting down-regulation after chronic wounding in skin mucus compared to control fish.

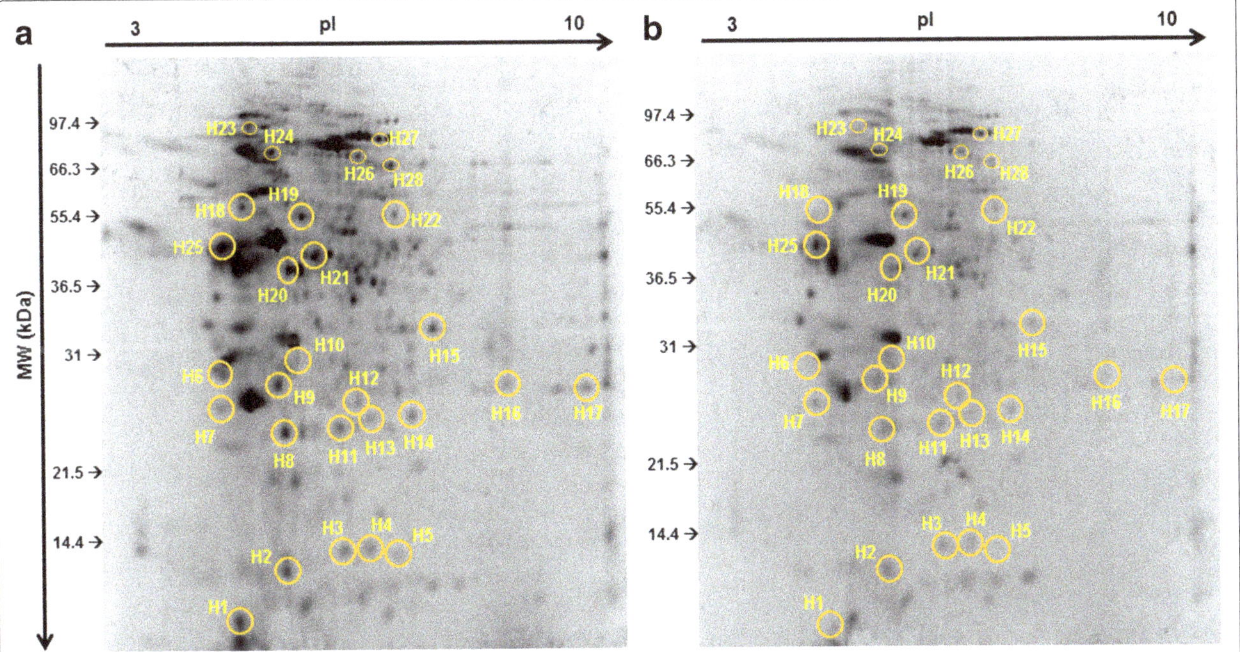

Fig. 1 Representative 2-DE gels of skin mucus of control (**a**) and wounded (**b**) *S. aurata* specimens. Skin mucus proteins were isoelectrically focused on 17 cm IPG strips (*pI* 3–10) and subjected to 12.5% SDS-PAGE. The 2-DE gels were stained with SYPRO® Ruby protein gel stain and the spots identified in (**a**–**b**) were annotated using the data from LC-MS/MS. The spot numbers represented in gels correspond to the protein identities mentioned in Table 2

Table 1 Details of the differentially expressed protein spots in skin mucus of *S. aurata* after chronic wounds

SN[a]	Protein name	Organism AN[b]	pI/MW[c]	S/C[d]	M_p/U_p^e	Peptide sequence and e-value[f]
H1	Histone H4	*Oncorhynchus mykiss* P62797	11.4/11.4	76/19	2/2	**VFLENVIR** (2.9*10^{-5}) **TVTAMDVVYALK** (0.002)
H2	Apolipoprotein A1	*Sparus aurata* AAT45246	5.3/15.9	151/39	3/3	**LLNLLSQAQTASGPMVEQASQDGR** (0.0068) **EYAETLQAKPEFQAFVK** (0.025) **VATALGEEASPLVDK** (0.016)
H3	Histone H2B	*Danio rerio* Q5BJA5	10.4/13.6	28/7	1/1	**LLLPGELAK** (0.0016)
H4	Cu/Zn Superoxide dismutase	*S. aurata* CAI79044	5.4/7.0	66/44	2/2	**HVGDLGNVTAGADNVAK** (4) **MLTLSGPLSIIGR** (0.14)
H5	Histone H2A	*D. rerio*	10.6/13.5	49/7	1/1	**AGLQFPVGR** (0.00014)
H6	14-3-3 protein beta/alpha-1	*O. mykiss* Q6UFZ9	4.6/27.7	143/15	2/5	**YLSEVASGDSK** (2.6*10^{-8}) **YLSEVASGDSKK** (0.35) NLLSVAYK (8.3/10^{-5}) VISSIEQK (1.3) DSTLIMQLLR (1.5*10^{-5})
H7	Apolipoprotein A1	*S. aurata* O42175	5.2/29.6	232/29	5/5	**AVLDVYLTQVK** (0.02) **AVNQLDDPQYAEFK** (0.0032) **IEEMYTQIK** (0.00025) **SSLAPQNEQLK** (0.00099) **TLLTPIYNDYK** (0.0014) **EVVQPYVQEYK** (0.092) **ITPLVEEIK** (0.0024)
H8	Phosphatidylethanolamine-binding protein 1	*S. aurata* FM145015	9.1/29.7	174/13	3/2	**LYDQLAGK** (28) **LYTLALTDPDAPSR** (0.0019) YGSVEIDELGK (0.00074)
H9	Apolipoprotein A1	*S. aurata* O42175	5.2/29.6	183/19	5/5	**IEEMYTQIK** (1.2) **SSLAPQNEQLK** (3.5) **TLLTPIYNDYK** (0.14) **EVVQPYVQEYK** (0.42) **ITPLVEEIK** (0.87)
H10	Actin cytoplasmic 1	*Ctenopharyngodon idella* P83751	5.3/42.1	53/7	3/3	**AGFAGDDAPR** (0.085) **DLTDYLMK** (0.089) **GYSFTTTAER** (6*10^{-5})
H11	Natural killer enhancing factor 2	*Larimichthys crocea* XP_010732927	5.9/21.8	278/26	6/2	**DYGVLKEDDGIAYR** (0.22) **EDDGIAYR** (21) IPLVADLTK (1.3*10^{-5}) GLFVIDDK (0.41) QITINDLPVGR (0.00085) LVQAFQHTDK (0.34)
H12	ADP-ribosylation factor 3	*Takifugu rubripes* P61207	6.8/20.7	106/24	4/4	**ILMVGLDAAGK** (4*10^{-7}) **MLAEDELR** (3*10^{-5}) **DAVLLVFANK** (0.056) **QDLPNAMNAAEITDK** (0.17)
H13	Natural killer enhancing factor 1	*Osmerus mordax* ACO 09982	5.8/22.3	102/14	3/3	**LAPDFTAK** (26) **AVMPDGQFK** (18) **QITINDLPVGR** (0.0028)
H14	Glutathione S-transferase 3	*S. aurata* AFV39802	6.9/25.5	206/19	5/3	**FTGILGDFR** (0.00069) **MTEIPAVNR** (0.1) **TVMEVFDIK** (2.2) YLPVFEK (11) AILNYIAEK (0.79)
H15	Triosephosphate isomerase A	*S. aurata* FG266106	8.7/28.8	203/18	5/4	**IIYGGSVTGATCK** (0.3) **NVSEAVANSVR** (0.0059) **KNVSEAVANSVR** (1200) **GAFTGEISPAMIK** (4.9) FGVAAQNCYK (11)
H16	Triosephosphate isomerase B	*D. rerio* Q90XG0	6.5/27.1	76/12	3/3	**FFVGGNWK** (0.065) **GAFTGEISPAMIK** (5.7*10^{-7}) **WVILGHSER** (0.037)

Table 1 Details of the differentially expressed protein spots in skin mucus of *S. aurata* after chronic wounds *(Continued)*

SN[a]	Protein name	Organism AN[b]	pI/MW[c]	S/C[d]	M_p/U_p[e]	Peptide sequence and e-value[f]
H17	Triosephosphate isomerase B	*D. rerio* Q90XG0	6.5/27.1	131/23	5/5	**FFVGGNWK** (6.1) **GAFTGEISPAMIK** ($1.7*10^{-6}$) **WVILGHSER** (0.001) **HVFGESDELIGQK** ($2.9*10^{-6}$) **VVLAYEPVWAIGTGK** (0.022)
H18	ATP synthase subunit beta	*Cyprinus carpio* Q9PTY0	5.1/55.3	317/28	10/10	**TIAMDGTEGLVR** (0.0043) **VLDTGAPIR** ($1.8*10^{-6}$) **IPVGPETLGR** ($7.4*10^{-8}$) **IMNVIGEPIDER** ($1.1*10^{-6}$) **VVDLLAPYAK** ($3*10^{-5}$) **IGLFGGAGVGK** ($6.8*10^{-6}$) **TVLIMELINNVAK** (0.022) **VALVYGQMNEPPGAR** ($5.4*10^{-5}$) **IPSAVGYQPTLATDMGTMQER** (0.0006) **AIAELGIYPAVDPLDSTSR** (0.0045)
H19	Actin-related protein	*T. rubripes* O73723	5.6/47.9	33/8	3/3	**FSYVCPDLVK** (0.062) **DYEEIGPSICR** (0.0066) **EVGIPPEQSLETAK** (0.14)
H20	Actin cytoplasmic 1	*Oreochromis mossambicus* P68143	5.3/42.1	144/15	5/5	**AGFAGDDAPR** ($3.8*10^{-8}$) **VAPEEHPVLLTEAPLNPK** (0.0038) **DLTDYLMK** (0.024) **GYSFTTTAER** ($1.5*10^{-5}$) **EITALAPSTMK** (0.066)
H21	Macrophage-capping protein	*L. crocea* XP_010735467	5.8/38.7	185/12	5/4	**TQVEILPQGK** (0.022) **MKTQVEILPQGK** (0.45) **MPELAESTPEEDSK** (0.16) **EIASLIR** (10) EGGVESGFR (1.8)
H22	Citrate synthase	*Katsuwonus pelamis* Q6S9V7	8.5/52.4	95/8	4/4	**DVLSDLIPK** (0.25) **ALGFPLERPK** (0.061) **VVPGYGHAVLR** ($3.7*10^{-5}$) **IVPNVLLEQGK** ($1.1*10^{-6}$)
H23	Heat shock cognate 71 kDa	*Oryzias latipes* Q9W6Y1	5.8/76.6	476/20	13/3	**NQVAMNPTNTVFDAK ($1.8*10^{-7}$)** **SFYPEEVSSMVLTK ($1.2*10^{-5}$)** **GQIHDIVLVGGSTR (0.0077)** VEIIANDQGNR ($5.8*10^{-6}$) MKEIAEAYLGK ($7.2*10^{-5}$) EIAEAYLGK (0.02) DAGTISGLNVLR ($3.6*10^{-5}$) IINEPTAAAIAYGLDKK ($1*10^{-6}$) STAGDTHLGGEDFDNR (0.0014) ARFEELNADLFR ($5.5*10^{-5}$) FEELNADLFR ($7.6*10^{-7}$) LLQDFFNGK ($9.2*10^{-6}$) NGLESYAFNMK (0.00053)
H24	Heat shock cognate 71 kDa	*Ictalurus punctatus* P47773	5.2/71.6	306/14	8/2	**TTPSYVAFTDSER ($1.8*10^{-6}$)** **FELTGIPPAPR (0.00019)** VEIIANDQGNR ($2.3*10^{-7}$) MKEIAEAYLGK (0.096) DAGTISGLNVLR ($7.8*10^{-7}$) STAGDTHLGGEDFDNR (0.00012) FEELNADLFR ($7.2*10^{-8}$) LLQDFFNGK (0.0026)
H25	Keratin type I	*O. mykiss* NP_001117848	5.2/51.9	521/14	9/0	KLEAANAELELK ($1.7*10^{-9}$) LEAANAELELK (0.00012) LAADDFR (0.0068) TKYENELAMR (0.041) QSVEADIAGLKR (43) SDLEMQIEGLK ($9.2*10^{-5}$) NHEEELLAMR (1.6) TRLEMEIAEYR (0.18) LEMEIAEYR (0.029)
H26	Complement component 3	*S. aurata*	8.1/186.9	152/4	7/6	**TLYTPESTVLYR** (18)

Table 1 Details of the differentially expressed protein spots in skin mucus of *S. aurata* after chronic wounds *(Continued)*

SN[a]	Protein name	Organism AN[b]	pI/MW[c]	S/C[d]	M_p/U_p^e	Peptide sequence and e-value[f]
		ADM13620				**DITYLILSR** (0.87) **VTGDPEATVGLVAVDK** (62) **SVPFIIIPMK** (13) **DSSLNDGIMR** (21) **VVPQGVLIK** (11) EYVLPSFEVK (100)
H27	Gelsolin	*S. aurata* HS984154	6.0/31.6	548/45	9/7	**QPGLQVWR** (0.035) **GGVASGFQHVVTNDMSAK** (13) **GDSFILDLGK** (0.059) **LHMVEEGEEPK** (25) **AFTEALGPK** (2.1) **TAIAPSTPDDEKADISNK** (0.00049) **GALYMISDASGTMK** (0.0044) **VSSVAPSSPFK** (0.0033) QAMLSPEECYILDNGVDK (1600) IENLDLKPVPK (54)
H28	Immunoglobulin M heavy chain	*S. aurata* AFN20639	6.1/51.2	50/2	1/1	**GFSPNSFQFK** (0.039)

[a]Spot number
[b]Accession number in NCBI or SwissProt databases
[c]Theoretical isoelectric point and molecular weight (kDa)
[d]Total score and coverage (%)
[e]Total matched peptides (Mp)/total unique peptides (Up)
[f]Unique peptides are in bold. Expect value (e-value) is noted for each peptide sequence

Stress-related molecules

Chronic wounds in the skin also altered some stress-related proteins in the mucus of gilthead seabream (Tables 1 and 2). Peroxiredoxins are a family of antioxidant enzymes that protect cells from oxidative damage [33]. Some of the most studied peroxiredoxins, identified here such as NKEF1 (spot 13) and NKEF2 (spot 11), were down-regulated after chronic wounding in skin mucus of gilthead seabream. Furthermore, we have identified SOD (spot H4), GST3 (spot 14) and HSC70 (it was identified in two parts of the proteome map, spots H23 and H24), which were also down-regulated after chronic wounding (Table 3).

Structural molecules

Our study indicated that structural proteins also play a major role in chronic injury of skin. We have identified ACTB (spots H10 and H20), ARP (spot 19), CAPG (spot H21), KRT1 (spot H25) and GSN (spot H27), and shown down-regulation in all cases with the lowest levels in KRT1 (Tables 1 and 2).

Metabolism molecules

Important proteins involved in several metabolic routes were identified in the present study. We found differential expression of YWHAZ (spot H6), PEBP1 (spot H8), ARF3 (spot H12), TPIA (spot H15), TPIB (spot H16 and H17), ATPB5B (H18) and CS (H22). All of these were down-regulated after chronic wounds in skin mucus of gilthead seabream (Tables 1 and 2).

Functional level of IgM

Our ELISA study with specific antibodies for total IgM of gilthead seabream showed a significant decrease of total IgM levels detected in skin mucus after chronic wounds compared to the levels detected in the skin mucus of control group (Fig. 2).

Transcript levels

Due to the importance of the skin mucus markers in the processes of immunity, inflammation, stress, skin regeneration and wound healing, we have selected and studied the gene expression profile of several immune-related (*ighm*, *c3* and *h2b*), stress-related (*hsc70*, *sod* and *gst3*) and finally structural-related molecules (*gsn*, *actb* and *krt1*) (Fig. 3). Regarding immune-related genes, *ighm* was significantly down-regulated in the wounded group, while the increase and decrease observed in *c3* and *h2b*, respectively, were not significant compared to the control group. Little variations were observed at transcript level in the case of stress-related genes, where only *gst3* showed a significant down-regulation in the wounded group, while *hsc70* and *sod* remained unaltered compared to the control group. Finally, the structural genes were the most affected by chronic wounds, as all of them the trend were down-regulation, with significant changes in the case of *actb* and *krt1*, the latter being the most affected molecule at transcript level in the wounded group compared to the control groups.

Discussion

Many factors such as stress by temperature, hypoxia, transportation, crowding, seasonal or dietary changes,

Table 2 List of proteins that are differentially expressed in skin mucus of *S. aurata* after chronic wounds

Spot	Protein name	Fold change	Previously detected in skin mucus?	References
H1	Histone H4 (H4)	↓ 0.01	Yes	[21, 23]
H2	Apolipoprotein A1 (APOA1)	↓ 0.04	Yes	[19–22, 29, 40]
H3	Histone H2B (H2B)	↓ 0.12	Yes	[20]
H4	Cu/Zn Superoxide dismutase (SOD)	↓ 0.09	Yes	[21–23]
H5	Histone H2A (H2A)	↓ 0.06	Yes	[23]
H6	14-3-3 protein beta/alpha 1	↓ 0.02	Yes	[19–23, 29]
H7	Apolipoprotein A1 (APOA1)	↓ 0.44	Yes	[19–22, 29]
H8	Phosphatidylethanolamine-binding protein 1 (PEBP1)	↓ 0.09	Yes	[22, 23]
H9	Apolipoprotein A1 (APOA1)	↓ 0.06	Yes	[19–22, 29, 40]
H10	Actin cytoplasmic 1 (ACTB)	↓ 0.32	Yes	[20–23, 29, 40]
H11	Natural killer enhancing factor 2 (NKEF2)	↓ 0.13	Yes	[21–23]
H12	ADP-ribosylation factor 3 (ARF3)	↓ 0.06	Yes	[29, 40]
H13	Natural killer enhancing factor 1 (NKEF1)	↓ 0.17	Yes	[20–23]
H14	Glutathione S-transferase 3 (GST3)	↓ 0.11	No	None
H15	Triosephosphate isomerase A (TPIA)	↓ 0.01	Yes	[19, 23, 29]
H16	Triosephosphate isomerase B (TPIB)	↓ 0.02	Yes	[19, 21, 23]
H17	Triosephosphate isomerase B (TPIB)	↓ 0.01	Yes	[19, 21, 23]
H18	ATP synthase subunit beta (ATB5B)	↓ 0.07	Yes	[20, 22, 23, 40]
H19	Actin-related protein (ARP)	↓ 0.46	Yes	[23, 24, 40]
H20	Actin cytoplasmic 1 (ACTB)	↓ 0.25	Yes	[19, 21–23, 40]
H21	Macrophage-capping protein (CAPG)	↓ 0.18	Yes	[22]
H22	Citrate synthase (CS)	↓ 0.09	Yes	[19]
H23	Heat shock cognate 71 kDa (HSC70)	↓ 0.12	Yes	[20, 22, 23, 40]
H24	Heat shock cognate 71 kDa (HSC70)	↓ 0.27	Yes	[20, 22, 23, 40]
H25	Keratin type I (KRT1)	↓ 0.11	Yes	[19–23, 29]
H26	Complement component 3 (C3)	↓ 0.36	Yes	[21, 29, 40]
H27	Gelsolin (GSN)	↓ 0.41	Yes	[21, 22, 40]
H28	Immunoglobulin M heavy chain (IgM)	↓ 0.07	No	None

↓ indicates under-expression of the proteins at $p < 0.01$. In addition, a literature-based comparison about presence of these proteins in skin mucus of other fish species after 2-DE spot detection is included

can affect directly the skin integrity in farmed fish. Most of the available studies have tried to improve the skin healing by dietary supplementation of diets with vitamin C [34], β-Glucans [35, 36] and minerals with different combinations of vitamins and glucans [37]. But curiously, the global molecular changes produced by wounds have scarcely been studied in fish. Only the transcriptomic changes using micro-array technology in the skin after skin and scale re-generation was reported [38]. The present study represents the first proteomic approach in the study of fish skin wounds.

From our own studies on fish skin mucus [11, 12, 39] and with proteomic tools [19, 26, 29], we provide evidence that 2-DE followed by LC-MS/MS provides good resolution and high performance for protein detection. One of the limitations of this approach could be the limited range of molecular weights available, thus mucins and other high molecular weight proteins have been undetected in these works. A recently published proteome map of gilthead seabream with more than 2000 proteins used 1-DE gels and mass spectrometry and any mucin was identified [40].

In the present study both protein levels and transcript levels were studied. In general one must have transcripts to make proteins, however due to, among others, RNA turnover rate, RNA localisation and protein turnover rate the changes in protein amount and RNA amount do not need to be the same. We found that *ighm*, *gst3*, *actb*, *krt1* transcripts were changed, whilst other transcripts were not significantly changed even if changes in proteins were observed.

Table 3 Information of primers used for qPCR study

Gene names	Accession number	Amplicon size	Sequence (5'→3')
Immunoglobulin mu heavy chain	JQ811851	113	F: CAACATGCCCAATTGATGAG R: GGCACGACACTCTAGCTTCC
Complement component 3	HM543456	106	F: CGCTCTTCTTGCTCTGGTGA R: CTGAGTTGATCCGTAGCCCC
Histone 2b	AM953480	174	F: AGACGGTCAAAGCACCAAAG R: AGTTCATGATGCCCATAGCC
Heat shock cognate 71 kDa	HS987272	124	F: GCCATGAACCCAACCAACAC R: GGCGGGTGTTGTCATTGATG
Superoxide dismutase	AJ937872	103	F: TCACGGACAAGATGCTCACT R: TCCTCGTTGCCTCCTTTTCC
Glutathione s-transferase	JQ308828	111	F: AGCGCTACCTTCCAGTGTTC R: CCTCCAACATCAGGGTGCAT
Gelsolin	HS984154	105	F: GCCATCAGAGCAACAGAGGT R: CTCACTGCCACACCACTGAT
Actin beta	AF316854	352	F: GGCACCACACCTTCTACAATG R: GTGGTGGTGAAGCTGTAGCC
Keratin 1	FJ744592	105	F: AGAGATCAATGACCTGCGGC R: CCCTCTGTGTCTGCCAATGT
Elongation factor 1 alpha	AF184170	115	F: TGTCATCAAGGCTGTTGAGC R: GCACACTTCTTGTTGCTGGA
Ribosomal protein s18	AM490061	109	F: CGAAAGCATTTGCCAAGAAT R: AGTTGGCACCGTTTATGGTC

Skin mucus is the first barrier of defense in fish, which contains immune components involved in both innate and adaptive immunity. In the present study we have demonstrated the presence of C3, APOA1, H2A, H2B, H4 and IgM. C3 can, upon cleavage, act as a chemoattractant (recruit immune cells), as opsonin (coat pathogens) to increase phagocytosis or as an agglutinin (coagulate pathogens) [21]. C3 was previously found in skin mucus of European sea bass [21]. While in the present study C3 was under-expressed in chronic wound specimens, in another study C3 was over-expressed after crowding stress in skin mucus of gilthead seabream [29].

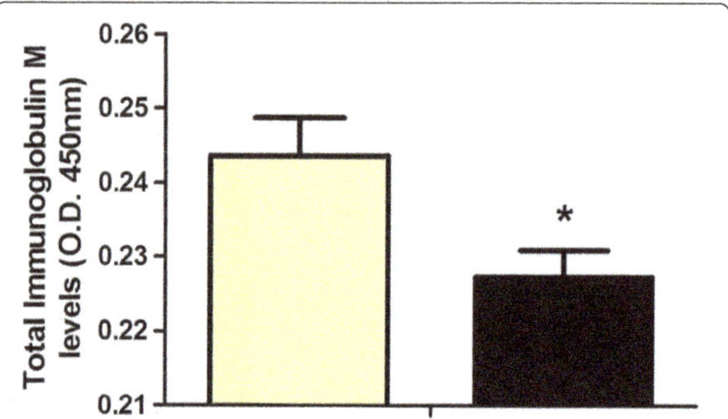

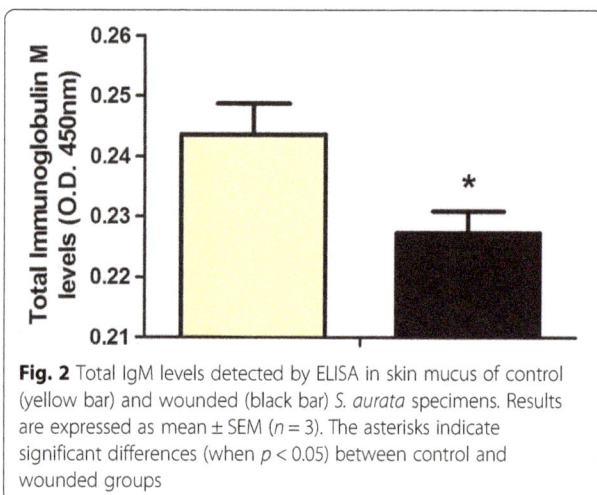

Fig. 2 Total IgM levels detected by ELISA in skin mucus of control (yellow bar) and wounded (black bar) *S. aurata* specimens. Results are expressed as mean ± SEM ($n = 3$). The asterisks indicate significant differences (when $p < 0.05$) between control and wounded groups

At transcriptional level, no changes in *c3* expression are reported in the skin of gilthead seabream after chronic wounds. Accordingly, in our previous study the transcript levels of *c3* were also unaltered in skin after crowding stress despite the protein differential expression in skin mucus of gilthead seabream [29].

APOA1 is the major component of high density lipoprotein in serum [41], which also act as a negative acute phase protein [42], and possesses bactericidal activity in vitro [43]; however, despite the previous finding of APOA1 as a conserved marker in skin mucus of European sea bass [21], Atlantic salmon [27], lumpsucker [20], Atlantic cod [19, 26] and gilthead seabream [29], its role in mucus is still unknown. Our study suggests that it plays a role as a negative acute phase protein may also occur in skin mucus as we found that APOA1 was under-expressed after chronic injury.

In addition to their classical role as histones folding DNA into chromatin, H2A, H2B and H4 are also known as antimicrobial peptides [32, 44], a role especially notable for H2A and H2B in skin mucus of fish [45, 46]. The histone H4 deserves more attention since previous studies have found this histone in the skin mucus [21], but little is known about its role as antimicrobial peptide. The under-expression of these three histones in skin mucus after chronic wounds may facilitate the entry of potential pathogens resulting in loss of immune defense. However, in sharp contrast with other studies where *h2b* was mostly up-regulated after virus and/or bacterial infections [44], in our study, *h2b* showed no

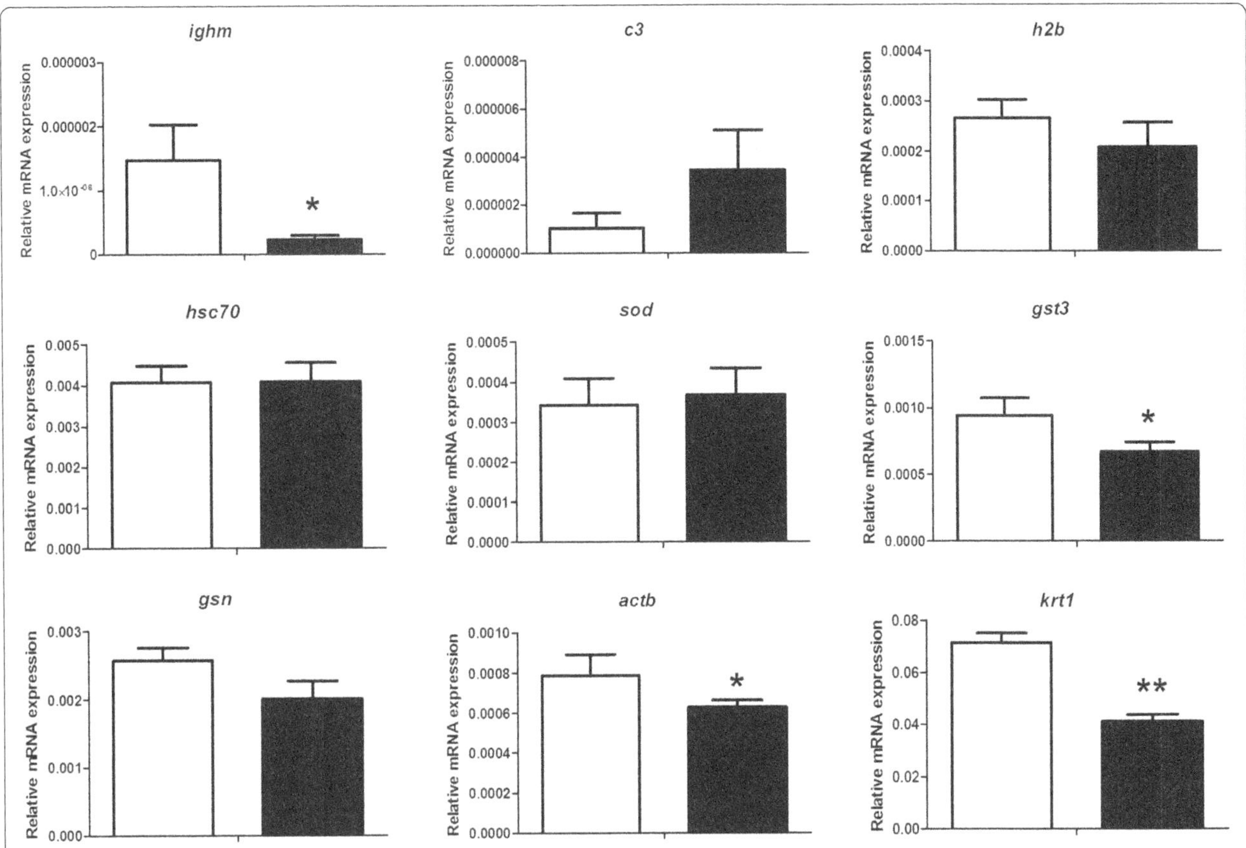

Fig. 3 Expression levels of some immune-related genes such as immunoglobulin *mu* heavy chain (*ighm*), complement component 3 (*c3*), histone 2b (*h2b*); some stress-related genes such as heat shock cognate 71 kDa protein (*hsc70*), superoxide dismutase (*sod*), glutathione s-transferase 3 (*gst3*); and some structural-related genes such as gelsolin (*gsn*), actin beta (*actb*), keratin 1 (*krt1*) in the skin of control (white bars) or wounded (black bars) *S. aurata* specimens. Transcripts were quantified by qPCR and normalised using the geometric average of the reference genes elongation factor 1 alpha (*ef1a*) and ribosomal protein S18 (*rps18*). The values are presented as mean ± SEM ($n = 6$). The asterisks indicate significant differences (*$p < 0.05$ or **$p < 0.01$) between control and wounded groups

differences at transcript level between control and wounded groups.

The main effector of the humoral systemic adaptive immunity, IgM, has been widely studied by ELISA in skin mucus of fish maintained under many different conditions and in several fish species [11, 12, 39]. However, in the present study we have identified IgM in a fish skin mucus proteome using 2-DE technology for the first time. IgM was under-expressed in skin mucus after chronic wounds. At transcript level, the down-regulation of *ighm* demonstrated the key role of this immunoglobulin in this type of stress. In many cases, the down-regulation of one gene or even the protein level are not correlated with the activity, but importantly in our study the IgM levels were also decreased when specific antibody was used. Further studies on this topic will help to characterize and elucidate the IgM functions in skin mucus as adaptive immunity players.

Here we hypothesize that the lower levels of these immunological proteins could promote the entry of pathogens into the fish body since the epidermis was removed and the skin was, therefore, interrupted. However, these lower levels could be related to the fact that abrasion promotes overproduction of immature mucus high in mucins unmeasurable in 2D gel analyses, which could lead to underestimation of the detected proteins. In this context, the knowledge on the production of mucins during the wound healing process would be essential.

On the other hand, it has been previously reported an increased inflammatory response ie. changed cytokine expression profile in wound healing on day 14 after wounding [36]. By contrast, in the present paper, we have not detected any cytokine, which does not necessarily mean absence of inflammation, but cytokines could be undetectable in our study because of their low molecular sizes and/or their limited presence in skin mucus. The differences in results could also be because the inflammatory response was detected mainly after 14 days [36], whilst our results were from 5 days of wound healing. There is a close relation between stress and immunity, especially in lower vertebrates such as fish, in which, for instance, cytokines and neuropeptides are performing roles in both neuroendocrine and immune system [47]. Another example of this relationship between stress and immunity are peroxiredoxins, which may act as modulators of inflammation in pathogen infection and in protection against cell death, tissue repair after damage, and tumour progression [48]. According to our results, in which NKEF1 and NKEF2 are under-expressed in skin mucus after chronic wounds, fish NKEFs expression, at either gene or protein level, is regulated by LPS treatment and pathogens including bacteria, viruses and parasites [33]. Concretely, NKEFs have been previously found in skin mucus of gilthead seabream [23], and over-expressed after crowding stress [29]. Our results indicate the opposite expression regulation when fish were stressed by crowding or damage and chronic wounds.

Also in close relation with the immunity, SOD is an enzyme that protects the tissue against oxidative stress by regulating various reactive oxygen (ROS) and reactive nitrogen species molecules [49]. In addition, T cell activation induces the secretion of SOD [50]. SOD was also identified previously in skin mucus of gilthead seabream [22, 23], however, this is the first time that this protein was demonstrated to be differentially expressed in skin mucus, but curiously no changes were found at transcript levels of *sod* in the skin after causing the wounds. In sharp contrast with our data, *sod* was up-regulated after in vitro exposure with different metals in gilthead seabream erythrocytes [51] as well as in gilthead seabream SAF-1 cell line [52].

GSTs are the superfamily of phase II detoxification enzymes that play crucial roles in cellular defense [21]. Some members of this superfamily have been previously identified in skin mucus of fish [9], reducing the amount of proteins in Atlantic cod after *V. anguillarum* infection [26] or increasing the amount of protein in gilthead seabream after probiotic intake [29]. In the present study GST3 was identified for first time in skin mucus, and was under-expressed after chronic wounds. At the transcriptional level, *gst3* was the only stress marker which was significantly down-regulated in skin of gilthead seabream after chronic wounds. By contrast, a previous study also in gilthead seabream reported an up-regulation of *gst3* in the liver after nanoparticle exposure [53]. However, there is no further information is available on the effects of *gst3* in the skin of teleost fish.

HSPs are part of a superfamily of stress proteins, highly conserved across species, often classified based on their molecular weight [21]. Both HSP70 and HSC70 may have similar cellular roles and have been previously found in skin mucus [20–23]. HSC70 can be mildly modulated by stressors such as heat [54], pathogens [55], and heavy metals [56]. According to these previous studies, at protein level, the present study demonstrated the under-expression of HSC70 in skin mucus after chronic wounds. By contrast, at transcript level, *hsc70* remains unaltered after wounding.

Some metabolic proteins have also been found to be under-expressed in skin mucus after chronic injury. PEBP1 was found in the mapping of gilthead seabream skin mucus [22], similar to YWHAZ [22, 23]. Moreover, YWHAZ was found in skin mucus of other fish species such as Atlantic cod [19], lumpsucker [20] and Atlantic salmon [24]. In agreement with the present study, it was reported that YWHAZ, ARF and TPIA were under-expressed after crowding stress in skin mucus of gilthead seabream [29]. CS and ATP5B were previously found in the skin mucus of Atlantic cod [19] and gilthead seabream [23], but this is the first time that these proteins were found differentially expressed in skin mucus of fish.

Beta actin (ACTB) is a multifunctional protein involved in cell motility and phagocytosis. It has been reported that ACTB can be fragmented after stress [27]. This fact could explain the under-expression of ACTB found in our study. In agreement with this result, ATCB was also under-expressed after crowding stress [29]. At transcript level, *actb* was also down-regulated after chronic wounds in gilthead seabream. The variations of *actb* in both skin and skin mucus in the present and other studies demonstrate that this molecule is highly influenced by the different stimuli, and therefore, its use as reference gene should be avoided, or at least reconsidered, in this tissue and fish species. In close relation with ACTB, ARP, CAPG and GSN were previously found in skin mucus of gilthead seabream [22, 23], however little is known about the interaction of all these proteins in stress processes since this is the first time that ARP and CAPG were found differentially expressed in skin mucus of fish. On the other hand, GSN was also expressed in skin mucus of gilthead seabream after stress stimuli [40]. The transcript levels of *gsn* were studied in gilthead seabream for first time in the present article, reporting no changes in the expression of *gsn* in the skin of gilthead seabream after chronic wounds. The importance of *gsn* in the skin remains unknown since most of the studies were focused in the corneal development and embryogenesis of zebrafish [57, 58].

KRTs are heteropolymeric intermediate filaments containing type I (KRT1) and type II (KRT2) keratins. These molecules have been reported in skin mucus of many fish species [9]. In the present study KRT1 was under-expressed after chronic wound in a similar fashion than KRT2 was under-expressed in skin mucus after infection [26]. In contrast, KRT1 was over-expressed in skin mucus after crowding stress [29]. It has been reported that KRTs play a role in the regulation stress-resistance in epithelial cells [59]. In addition, KRTs have been associated with pore-formation activities in skin mucus of fish [60]. A recent article reported an overexpression of KTR2 in skin mucus after different chronic stressors such as shaking, sounds and light flashes [40]. At transcript level, the present study revealed a great down-regulation of *krt1* after chronic wounding. Despite of the diversity of keratins reported in fish [61], there is very little information about the changes produced by these molecules at transcript levels in fish. Overall, it seems that KRTs are essential to maintain the proper function of skin mucus. The present findings of KRT1/*krt1* at both protein and transcript levels suggest an important role of this molecule after chronic wounds in the skin mucosae that it deserves to be studied in depth.

Conclusion

This study shows for first time the fish skin mucus proteome map of wounds. Thus, chronic wounding leads to a down-regulation in skin mucus proteins which are immune-related (C3, APOA1, H2A, H2B, H4 and IGM) and stress-related (NKEF1, NKEF2, SOD, GST3 and HSC-70), but also molecules involved in metabolism (PEBP1, YWHAZ, TPIA, TPIB, ARF, CS and ATP5B) and structural proteins (ATCB, ARP, CAPG, GSN and KRT1). The chronic wounding also leads a down-regulation of the transcripts corresponding to four of these proteins found in the skin of wounded specimens. These early alterations after chronic wounds could increase the susceptibility to pathogen infection due to the decrease in immune-related proteins as immune barrier and because of the decrease in structural proteins of the physical barrier, allowing for penetration of pathogens and, therefore, increasing the vulnerability of the fish.

Methods

Animal care

Forty specimens of gilthead seabream (*S. aurata*) (4.7 ± 1.3 g and 7.4 ± 0.6 cm), obtained from a local farm (Murcia, Spain), were kept in running seawater aquaria of 250 L (water flow 900 l h^{-1}) at 28 ‰ salinity, 22 °C and a photoperiod of 12 h light: 12 h dark. Fish were fed daily at 2% rate of fish biomass per day with commercial diet (Skretting). All the fish handling procedures were approved by the Ethical Committee of the University of Murcia (Permit Number: A13150104).

Chronic wounds

Fish were anesthetized with 100 mg L^{-1} of MS-222 (tricaine methanesulfonate; Sigma-Aldrich). Chronic wounds with a diameter of 8 mm and around 50 μm of depth were induced in the skin with an electric toothbrush (PRIMO) used for 30 s in each body side of the fish specimens (Fig. 4). The procedure was repeated twice each two days and sampled two days after the last abrasion (Fig. 4). The control group was handled in a similar manner as control fish without triggering wounds.

Mucus and tissues samples

Twenty fish per group were anesthetized as described above prior to sampling. Mucus was gently scraped off from the skin surface, avoiding blood, urine and faeces during collection [62]. In order to obtain a large enough amount of mucus, mucus samples from 10 fish were pooled as described elsewhere [29] resulting in two pools/groups. Mucus was transferred into tubes of 15 ml and stored at -80 °C until use. Skin tissue was collected in QIAzol lysis reagent (Qiagen) and stored at -80 °C for subsequent RNA extraction.

Histological analysis

Skin samples (n = 6) were collected and processed as described elsewhere [63]. Skin samples were sectioned at 5 μm and stained with periodic acid–Schiff (PAS; Merck) according to the manufacturer's instructions. Images were obtained under a light microscope (Leica DM6000B) with a digital camera (Leica DFC280) and processed by Leica Application Suite V 2.5.0. Software.

Mucus protein purification

Each sample was solubilised with 1 mM DTT and 1.5 mM EDTA, which serves to act as a mild mucolytic agent [64]. Next, after two rounds of sonication for 6 s followed by cooling for 1 min, samples were centrifuged at 20,000 g for 30 min at 4 °C. The supernatant containing the soluble mucus proteins was desalted with proteomic grade water (G Biosciences) using centrifugal filters of 3 KDa (VWR) by spinning 3 times at 14,000 g at 4 °C with 0.2 ml of ice cold water each time. The dialysed protein solution was further purified by 2D clean-up kit (Bio-Rad) following the manufacturer's instructions.

2-DE

The samples obtained after the 2D clean-up were resuspended in 2D lysis buffer (Bio-Rad) containing 7 M urea, 2 M thiourea, 1% (*w*/*v*) ASB-14, 40 mM Tris base, 0.001% bromophenol blue and 50 mM DTT (Sigma-Aldrich) and 0.5% (*v*/v) Biolytes 3–10 ampholyte (Bio-Rad). The protein content of the solubilised samples was estimated using Qubit protein assay (Life Technologies). Two hundred μg of proteins for each sample were rehydrated in 17 cm 3–10 IPG strips (Bio-Rad) and isoelectric focusing (IEF) was carried out using protean IEF cell (Bio-Rad). After IEF, the electro-focused IPG strips were reduced and alkylated for 15 min each in equilibration buffer containing 6 M urea (Sigma-Aldrich), 0.375 M Tris-HCl pH 8.8 (Bio-Rad), 2% (w/v) SDS (Sigma-Aldrich), 20% (v/v) glycerol (Merck) with 0.2% (w/v) DTT

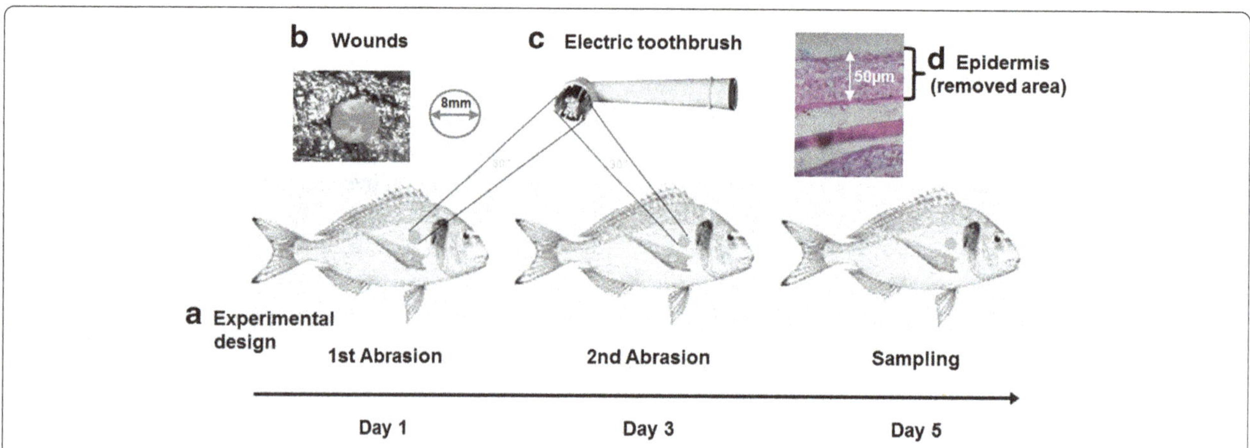

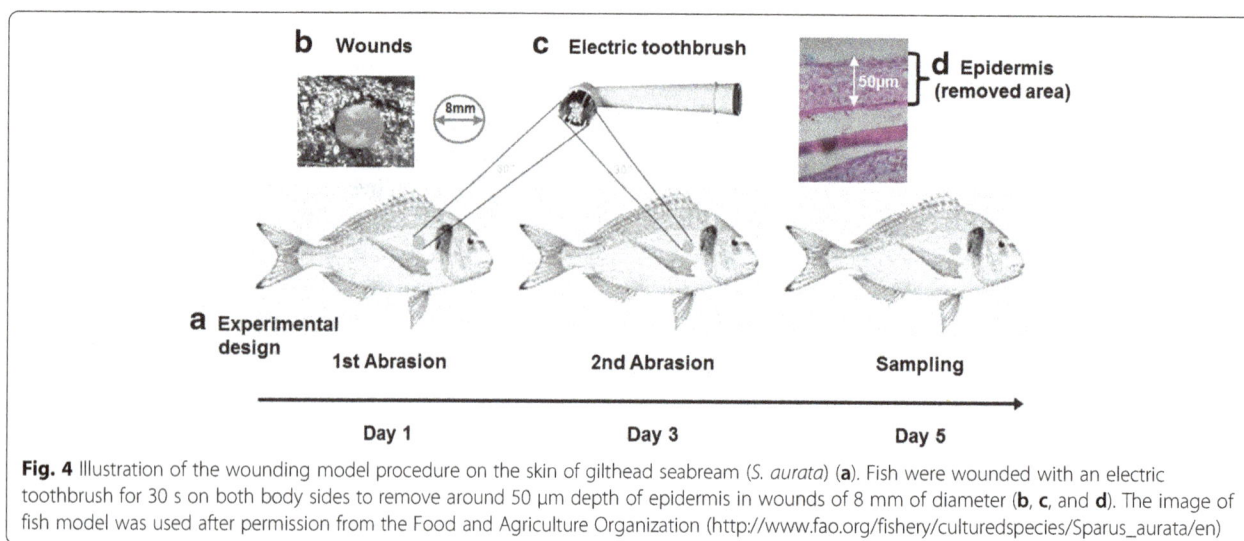

Fig. 4 Illustration of the wounding model procedure on the skin of gilthead seabream (*S. aurata*) (**a**). Fish were wounded with an electric toothbrush for 30 s on both body sides to remove around 50 μm depth of epidermis in wounds of 8 mm of diameter (**b**, **c**, and **d**). The image of fish model was used after permission from the Food and Agriculture Organization (http://www.fao.org/fishery/culturedspecies/Sparus_aurata/en)

(Sigma-Aldrich) or 0.3% (w/v) iodoacetamide (Bio-Rad), respectively. The equilibrated strips were loaded on 12.5% polyacrylamide gels to perform SDS-PAGE [65], run on PROTEAN II system (Bio-Rad). The gels were stained overnight with SYPRO® Ruby Protein Gel Stain (Life Technologies) according to the supplier's protocol. Gel image documentation was carried out using ChemiDocTM XRS imaging system (Bio-Rad). Raw pictures were analysed using PDQuest Advanced software version 8.0.1 (Bio-Rad) including detection of spots, normalization using local regression, spot matching and differential expression analysis. Protein spots were considered as differentially expressed when expression level was at least 1.5-fold different compared to the control group and when the differences were detected as significant at $p < 0.01$ by two tailed Student's t-test.

LC-MS/MS analysis

Spots from SYPRO-stained gilthead seabream skin mucus 2-DE gels ($n = 6$) were picked, excised and subjected to in-gel reduction, alkylation, and tryptic digestion using 2–10 ng/μl trypsin (V511A; Promega) as described elsewhere [66]. Peptide mixtures containing 0.1% formic acid were loaded onto a nanoACQUITY UltraPerformance LC (Waters), containing a 5 μm Symmetry C18 Trap column (180 μm × 20 mm; Waters) in front of a 1.7 μm BEH130 C18 analytical column (100 μm × 100 mm; Waters). Peptides were separated with a gradient of 5–95% acetonitrile, 0.1% formic acid, with a flow of 0.4 μl min^{-1} eluted to a Q-TOF Ultima mass spectrometer (Micromass/Waters). The samples were run in data dependent tandem mass spectrophotometry (MC/MC) mode. Peak lists were generated from MS/MS by Mascot Distiller Workstation and submitted to MASCOT search engine (version 2.5.1) and searched against NCBInr with the following parameters: maximum one missed cleavage by trypsin, peptide mass tolerance 100 ppm, MS/MS ion tolerance set to 0.1 Da, carbamidomethylation of cysteine selected as fixed modification and methionine oxidation as variable modification. Protein hits not satisfying a significance threshold ($p < 0.05$) or with low sequence coverage were further searched against Swissprot and vertebrate EST (expressed sequence tags) databases, taxonomy *Actinopterygii*.

Primer design

Primers were designed by OligoPerfect™ Designer (Life Technologies) from *S. aurata* sequences that are available in NCBInr database. Details regarding oligonucleotide primers and their attributes are given in Table 3.

Gene expression analysis

The mRNA levels corresponding to nine differentially expressed immune-related, stress-related and structural proteins in the skin of the experimental fish were analysed by real-time PCR (qPCR). RNA was extracted individually from 50 mg of skin from six specimens of gilthead seabream from both ulcered and control groups using QIAzol lysis reagent method (Qiagen) as described elsewhere [67]. The quality of total RNA was checked on a 1.2% agarose gel, followed by quantification using the Qubit® RNA assay kit and Qubit® 2.0 fluorometer (Life Technologies). The complementary DNA (cDNA) was synthetised from 1 μg of RNA using QuantiTec Reverse Transcription Kit (Qiagen). Ten times diluted cDNA was used to conduct qPCR on a ABI PRISM 7500 instrument (Applied Biosystems) as described elsewhere [21], using SYBR Green PCR Core Reagents (Applied Biosystems) and the $2^{-\Delta\Delta Ct}$ method [68]. Each plate subjected to qPCR contained a negative control for cDNA template (water) as well as a control for reverse transcription. No amplification product was observed in negative controls and neither primer-dimer formation nor secondary structures were observed in any case. All qPCR reactions were carried out in duplicate and quantification cycle (Ct) values of each gene (target) were converted into relative quantities. Normalization factors were calculated as the geometric mean of relative quantities of reference genes elongation factor 1 alpha (*ef1a*) and ribosomal protein S18 (*rps18*) using the BestKeeper© algorithm [69], which have been previously reported to be suitable reference genes in the skin of gilthead seabream as well [29].

Data are expressed as relative gene expression of each target gene (mean ± SEM). Statistical analysis (t-test) was performed using Statistical Package for the Social Sciences (SPSS) software v19.0. One or two asterisks denote significant differences when $p < 0.05$ or $p < 0.01$, respectively.

ELISA assay

Total mucus IgM levels were analysed by ELISA as described elsewhere [70]. First, 100 μl per well of 1/5 diluted mucus were placed in flat-bottomed 96-well plates in triplicate and the protein coating was performed by overnight incubation at 4 °C with 200 μl carbonate–bicarbonate buffer (35 mM $NaHCO_3$ and 15 mM Na_2CO_3, pH 9.6). After three rinses with phosphate buffered saline (PBS; Sigma-Aldrich) containing 0.05% Tween 20 (PBT, pH 7.3) the plates were blocked for 2 h at room temperature with blocking buffer containing 3% bovine serum albumin (BSA; Sigma-Aldrich) in PBT, followed by three rinses with PBT. The plates were then incubated for 1 h with 100 μl per well of mouse anti-gilthead seabream IgM monoclonal antibody (Aquatic Diagnostics Ltd.) (diluted 1/100 in blocking buffer), washed and incubated with secondary antibody anti–mouse IgG-HRP (diluted 1/1000 in blocking buffer; Sigma-Aldrich). After exhaustive rinsing with PBT, the plates were developed using 100 μl 0.42 mM 3,3,5,5-tetramethylbenzidine hydrochloride (Sigma-Aldrich) solution, freshly prepared in distilled water containing 0.01% H_2O_2 (Merck). The reaction was allowed to proceed for 10 min and stopped by the addition of 50 μl 2 M H_2SO_4 and the plates were read at 450 nm in a plate reader (BMG, Fluostar Omega). Negative controls were wells without mucus and wells without primary antibody, both in triplicates, whose OD values were subtracted for each sample value.

ELISA data were analysed by using t-test. Data are expressed as mean ± SEM. Statistical test was performed using SPSS software v19.0. Asterisks denote significant differences between groups when $p < 0.05$.

Abbreviations

ACTB: Actin cytoplasmic; APOA1: Apolipoprotein a1; ARF: Actin-related protein; ARF3: ADP-ribosylation factor 3; ATP5B: Adenosine triphosphate synthase subunit beta; C3: Complement component 3; CAPG: Macrophage-capping protein; CS: Citrate synthase; ELISA: Enzyme-linked immunosorbent assay; GSN: Gelsolin; GST3: Glutathione s-transferase 3; H2A: Histone 2a; H2B: Histone 2b; H4: Histone 4; HSC70: Heat shock cognate 71 kDa protein; IEF: Isoelectric focusing; IgM: Immunoglobulin mu heavy chain; KRT: Keratin; NKEF: Natural killer enhancing factor; PEBP1: Phosphatidylethanolamine-binding protein 1; PRDX: Peroxiredoxin; SOD: Cu/Zn superoxide dismutase; TPI: Triosephosphate isomerase; YWHAZ: 14–3-3 protein beta/alpha-1

Acknowledgements

The technical assistance of Dr. J-A Bruun from University of Tromsø is acknowledged.

Funding

The work of H.C. was funded by Spanish Ministry of Economy and Competitiveness (MINECO) with both FPI grant (BES-2012-052742) and stay abroad grant (EEBB-1-2016-10,533). The research was co-funded by a national project of the MINECO and European Regional Development Funds (FEDER/ERDF) (grant number AGL2014–51839-C5–1-R) and *Fundación Séneca de la Región de Murcia* (*Grupo de Excelencia* grant number 19883/GERM/15).

Authors' contributions

HC performed all the experiments, the data analysis and wrote the manuscript. MFB, AC and MAE conceived the project design, performed experimental design and wrote the manuscript. All authors have read and approved the final version of this manuscript.

Competing interests

The authors declare that they have no competing interests.

Author details

[1]Fish Innate Immune System Group, Department of Cell Biology and Histology, Faculty of Biology, Regional Campus of International Excellence "Campus Mare Nostrum", University of Murcia, 30100 Murcia, Spain. [2]Department of Microbiology and Immunology, Rosalind Franklin University of Medicine and Science, North Chicago, IL 60064, USA. [3]Faculty of Biosciences and Aquaculture, Nord University, 8049 Bodø, Norway.

References

1. FAO. The state of world fisheries and aquaculture 2014; 2014. p. 223.
2. Balcázar JL, De Blas I, Ruiz-Zarzuela I, Cunningham D, Vendrell D, Múzquiz JL. The role of probiotics in aquaculture. Vet Microbiol. 2006;114:173–86.
3. Esteban MA. An overview of the immunological defenses in fish skin. ISRN Immunol. 2012;2012:1–29.
4. Sarasquete C, González de Canales ML, Arellano J, Pérez-Prieto S, García-Rosado E, Borrego JJ. Histochemical study of lymphocystis disease in skin of gilthead seabream, *Sparus aurata* L. Histol Histopathol. 1998;13:37–45.
5. Borrego JJ, Valverde EJ, Labella AM, Castro D. Lymphocystis disease virus: its importance in aquaculture. Rev Aquacult. 2015;1:1–15.
6. Cordero H, Cuesta A, Meseguer J, Esteban MA. Characterization of the gilthead seabream (*Sparus aurata* L.) immune response under a natural lymphocystis disease virus outbreak. J Fish Dis. 2016;39:1467–76.

7. Balebona MC, Andreu MJ, Bordas MA, Zorrilla I, Moriñigo MA, Borrego JJ. Pathogenicity of Vibrio alginolyticus for cultured gilt-head sea bream (*Sparus aurata* L.). Appl Environ Microbiol. 1998;64:4269–75.
8. Shephard K. Functions for fish mucus. Rev Fish Biol Fisheries. 1994;4:401–29.
9. Brinchmann MF. Immune relevant molecules identified in the skin mucus of fish using -omics technologies. Mol BioSyst. 2016;12:2056–63.
10. Ross NW, Firth KJ, Anping W, Burka J, Johnson SC. Changes in hydrolytic enzyme activities of naïve Atlantic salmon *Salmo salar* skin mucus due to infection with the salmon louse *Lepeophtheirus salmonis* and cortisol implantation. Dis Aquat Org. 2000;41:43–51.
11. Guardiola FA, Cuesta A, Abellán E, Meseguer J, Esteban MA. Comparative analysis of the humoral immunity of skin mucus from several marine teleost fish. Fish Shellfish Immunol. 2014;40:24–31.
12. Cordero H, Cuesta A, Meseguer J, Esteban MA. Changes in the levels of humoral immune activities after storage of gilthead seabream (*Sparus aurata*) skin mucus. Fish Shellfish Immunol. 2016;58:500–7.
13. Rakers S, Niklasson L, Steinhagen D, Kruse C, Sundell K, Paus R. Antimicrobial peptides (AMPs) from fish epidermis: perspectives for investigative dermatology. J Invest Dermatol. 2013;133:1140–9.
14. Xu Z, Parra D, Gómez D, Salinas I, Zhang Y-A, Von Gersdorff JL, et al. Teleost skin, an ancient mucosal surface that elicits gut-like immune responses. Proc Natl Acad Sci U S A. 2013;110:13097–102.
15. Ng TB, Chi R, Cheung F, Cheuk C, Ng W, Fang EF, et al. A review of fish lectins. Curr Protein Pept Sci. 2015;16:337–51.
16. Gstaiger M, Aebersold R. Applying mass spectrometry-based proteomics to genetics, genomics and network biology. Nat Rev Genet. 2009;10:617–27.
17. Junqueira M, Spirin V, Balbuena TS, Thomas H, Adzhubei I, Sunyaev S, et al. Protein identification pipeline for the homology-driven proteomics. J Proteome. 2008;71:346–56.
18. Liska AJ, Shevchenko A. Expanding the organismal scope of proteomics: cross-species protein identification by mass spectrometry and its implications. Proteomics. 2003;3:19–28.
19. Rajan B, Fernandes JMO, Caipang CMA, Kiron V, Rombout JHWM, Brinchmann MF. Proteome reference map of the skin mucus of Atlantic cod (*Gadus morhua*) revealing immune competent molecules. Fish Shellfish Immunol. 2011;31:224–31.
20. Patel DM, Brinchmann MF. Skin mucus proteins of lumpsucker (*Cyclopterus lumpus*). Biochem Biophys Rep. 2017;9:217–25.
21. Cordero H, Brinchmann MF, Cuesta A, Meseguer J, Esteban MA. Skin mucus proteome map of European sea bass (*Dicentrarchus labrax*). Proteomics. 2015;15:4007–20.
22. Sanahuja I, Ibarz A. Skin mucus proteome of gilthead sea bream: a non-invasive method to screen for welfare indicators. Fish Shellfish Immunol. 2015;46:426–35.
23. Jurado J, Fuentes-Almagro CA, Guardiola FA, Cuesta A, Esteban MA, Prieto-Álamo MJ. Proteomic profile of the skin mucus of farmed gilthead seabream (*Sparus aurata*). J Proteome. 2015;120:21–34.
24. Provan F, Jensen LB, Uleberg KE, Larssen E, Rajalahti T, Mullins J, et al. Proteomic analysis of epidermal mucus from sea lice-infected Atlantic salmon, *Salmo salar* L. J Fish Dis. 2013;36:311–21.
25. Valdenegro-Vega V, Crosbie P, Bridle A, Leef M, Wilson R, Nowak BF. Differentially expressed proteins in gill and skin mucus of Atlantic salmon (*Salmo salar*) affected by amoebic gill disease. Fish Shellfish Immunol. 2014; 40:69–77.
26. Rajan B, Lokesh J, Kiron V, Brinchmann MF. Differentially expressed proteins in the skin mucus of Atlantic cod (*Gadus morhua*) upon natural infection with *Vibrio anguillarum*. BMC Vet Res. 2013;9:103.
27. Easy RH, Ross NW. Changes in Atlantic salmon (*Salmo salar*) epidermal mucus protein composition profiles following infection with sea lice (*Lepeophtheirus salmonis*). Comp Biochem Physiol - Part D. 2009;4:159–67.
28. Easy RH, Ross NW. Changes in Atlantic salmon *Salmo salar* mucus components following short- and long-term handling stress. J Fish Biol. 2010;77:1616–31.
29. Cordero H, Morcillo P, Cuesta A, Brinchmann MF, Esteban MA. Differential proteome profile of skin mucus of gilthead seabream (*Sparus aurata*) after probiotic intake and/or overcrowding stress. J Proteome. 2016;132:41–50.
30. Chong K, Joshi S, Jin LT, Shu-Chien AC. Proteomics profiling of epidermal mucus secretion of a cichlid (*Symphysodon aequifasciata*) demonstrating parental care behavior. Proteomics. 2005;5:2251–8.
31. Micallef G, Cash P, Fernandes JMO, Rajan B, Tinsley JW, Bickerdike R, et al. Dietary yeast cell wall extract alters the proteome of the skin mucous barrier in Atlantic Salmon (*Salmo salar*): increased abundance and expression of a calreticulin-like protein. PLoS One. 2017;12:e0169075.
32. Valero Y, Chaves-Pozo E, Meseguer J, Esteban MA, Cuesta A. Biological role of fish antimicrobial peptides. Antimicrobial Peptides. Nova Science Publishers; 2013. p. 31–60.
33. Valero Y, Martínez-Morcillo FJ, Esteban MA, Chaves-Pozo E, Cuesta A. Fish peroxiredoxins and their role in immunity. Biology. 2015;4:860–80.
34. Wahli T, Verlhac V, Girling P, Gabaudan J, Aebischer C. Influence of dietary vitamin C on the wound healing process in rainbow trout (*Oncorhynchus mykiss*). Aquaculture. 2003;225:371–86.
35. Vetvicka V, Vetvickova J. β(1-3)-D-glucan affects adipogenesis, wound healing and inflammation. Orient Pharma Exp Med. 2011;11:169–75.
36. Przybylska-Diaz D, Schmidt J, Vera-Jiménez N, Steinhagen D, Nielsen M. β-glucan enriched bath directly stimulates the wound healing process in common carp (*Cyprinus carpio* L.). Fish Shellfish Immunol. 2013;35:998–1006.
37. Jensen LB, Wahli T, McGurk C, Eriksen TB, Obach A, Waagbø R, et al. Effect of temperature and diet on wound healing in Atlantic salmon (*Salmo salar* L.). Fish Physiol Biochem. 2015;41:1524–43.
38. Vieira FA, Gregório SF, Ferraresso S, Thorne MAS, Costa R, Milan M, et al. Skin healing and scale regeneration in fed and unfed sea bream, *Sparus auratus*. BMC Genomics. 2011;12:490.
39. Guardiola FA, Cuesta A, Arizcun M, Meseguer J, Esteban MA. Comparative skin mucus and serum humoral defence mechanisms in the teleost gilthead seabream (*Sparus aurata*). Fish Shellfish Immunol. 2014;36:545–51.
40. Pérez-Sánchez J, Terova G, Simó-Mirabet P, Rimoldi S, Folkedal O, Calduch-Giner JA, et al. Skin mucus of gilthead sea bream (*Sparus aurata* L.). Protein mapping and regulation in chronically stressed fish. Front Physiol. 2017;8:34.
41. Breslow J, Ross D, McPherson J, Williams H, Kurnit D, Karathanasis SK, et al. Isolation and characterization of cDNA clones for human apolipoprotein A-I. Proc Natl Acad Sci U S A. 1982;79:6861–5.
42. Villarroel F, Bastías A, Casado A, Amthauer R, Concha MI, Apolipoprotein A-I. An antimicrobial protein in *Oncorhynchus mykiss*: evaluation of its expression in primary defence barriers and plasma levels in sick and healthy fish. Fish Shellfish Immunol. 2007;23:197–209.
43. Concha MI, Smith VJ, Castro K, Bastías A, Romero A, Amthauer RJ. Apolipoproteins A-I and A-II are potentially important effectors of innate immunity in the teleost fish *Cyprinus carpio*. Eur J Biochem. 2004;271:2984–90.
44. Valero Y, Arizcun M, Esteban MA, Cuesta A, Chaves-Pozo E. Transcription of histones H1 and H2B is regulated by several immune stimuli in gilthead seabream and European sea bass. Fish Shellfish Immunol. 2016;57:107–15.
45. Bergsson G, Agerberth B, Jörnvall H, Gudmundsson GH. Isolation and identification of antimicrobial components from the epidermal mucus of Atlantic cod (*Gadus morhua*). FEBS J. 2005;272:4960–9.
46. Fernandes JMO, Molle G, Kemp G, Smith VJ. Isolation and characterisation of oncorhyncin II, a histone H1-derived antimicrobial peptide from skin secretions of rainbow trout, *Oncorhynchus mykiss*. Dev Comp Immunol. 2004;28:127–38.
47. Tort L. Stress and immune modulation in fish. Dev Comp Immunol. 2011;35: 1366–75.
48. Ishii T, Warabi E, Yanagawa T. Novel roles of peroxiredoxins in inflammation, cancer and innate immunity. J Clin Biochem Nutr. 2012;50:91–105.
49. Break TJ, Jun S, Indramohan M, Carr KD, Sieve AN, Dory L, et al. Extracellular superoxide dismutase inhibits innate immune responses and clearance of an intracellular bacterial infection. J Immunol. 2012;188:3342–50.
50. Terrazzano G, Rubino V, Damiano S, Sasso A, Petrozziello T, Ucci V, et al. T cell activation induces CuZn superoxide dismutase (SOD)-1 intracellular re-localization, production and secretion. Biochim Biophys Acta. 2014;1843:265–74.
51. Morcillo P, Romero D, Meseguer J, Esteban MÁ, Cuesta A. Cytotoxicity and alterations at transcriptional level caused by metals on fish erythrocytes *in vitro*. Environ Sci Pollut Res. 2016;23:12312–22.
52. Morcillo P, Esteban MÁ, Cuesta A. Heavy metals produce toxicity, oxidative stress and apoptosis in the marine teleost fish SAF-1 cell line. Chemosphere. 2016;144:225–33.
53. Teles M, Fierro-Castro C, Na-Phatthalung P, Tvarijonaviciute A, Trindade T, Soares AMVM, et al. Assessment of gold nanoparticle effects in a marine teleost (*Sparus aurata*) using molecular and biochemical biomarkers. Aquat Toxicol. 2016;177:125–35.
54. Jesus TF, Inácio Â, Coelho MM. Different levels of *hsp70* and *hsc70* mRNA expression in Iberian fish exposed to distinct river conditions. Genet Mol Biol. 2013;36:61–9.

55. Das S, Mohapatra A, Sahoo PK. Expression analysis of heat shock protein genes during *Aeromonas hydrophila* infection in rohu, *Labeo rohita*, with special reference to molecular characterization of Grp78. Cell Stress Chaperon. 2015;20:73–84.
56. Morcillo P, Cordero H, Meseguer J, Esteban MA, Cuesta A. *vitro* immunotoxicological effects of heavy metals on European sea bass (*Dicentrarchus labrax* L.) head-kidney leucocytes. Fish Shellfish Immunol. 2015;47:245–54.
57. Jia S, Nakaya N, Piatigorsky J. Differential expression patterns and developmental roles of duplicated scinderin-like genes in zebrafish. Dev Dynam. 2009;238:2633–40.
58. Kanungo J, Kozmik Z, Swamynathan SK, Piatigorsky J. Gelsolin is a dorsalizing factor in zebrafish. Proc Natl Acad Sci U S A. 2003;100:3287–92.
59. Marceau N, Loranger A, Gilbert S, Daigle N, Champetier S. Keratin-mediated resistance to stress and apoptosis in simple epithelial cells in relation to health and disease. Biochem Cell Biol. 2001;79:543–55.
60. Molle V, Campagna S, Bessin Y, Ebran N, Saint N, Molle G. First evidence of the pore-forming properties of a keratin from skin mucus of rainbow trout (*Oncorhynchus mykiss*, formerly *Salmo gairdneri*). Biochem J. 2008;411:33–40.
61. Schaffeld M, Höffling S, Haberkamp M, Conrad M, Markl J. Type I keratin cDNAs from the rainbow trout: independent radiation of keratins in fish. Differentiation. 2002;70:282–91.
62. Palaksha KJ, Shin GW, Kim YR, Jung TS. Evaluation of non-specific immune components from the skin mucus of olive flounder (*Paralichthys olivaceus*). Fish Shellfish Immunol. 2008;24:479–88.
63. Cordero H, Ceballos-Francisco D, Cuesta A, Esteban MA. Dorso-ventral skin characterization of the farmed fish gilthead seabream (*Sparus aurata*). PLoS One. 2017;12:e0180438.
64. Reddy VM, Suleman FG, Hayworth DA. *Mycobacterium avium* binds to mouse intestinal mucus aldolase. Tuberculosis. 2004;84:303–10.
65. Laemmli UK. Cleavage of structural proteins during the assembly of the head bacteriophage T4. Nature. 1970;227:680–5.
66. Shevchenko A, Wilm M, Vorm O, Mann M. Mass spectrometric sequencing of proteins from silver-stained polyacrylamide gels. Anal Chem. 1996;68: 850–8.
67. Lokesh J, Fernandes JMO, Korsnes K, Bergh O, Brinchmann MF, Kiron V. Transcriptional regulation of cytokines in the intestine of Atlantic cod fed yeast derived mannan oligosaccharide or β-glucan and challenged with *Vibrio anguillarum*. Fish Shellfish Immunol. 2012;33:626–31.
68. Livak KJ, Schmittgen TD. Analysis of relative gene expression data using real-time quantitative PCR and the 2(–Delta Delta C(T)) method. Methods. 2001;25:402–8.
69. Pfaffl MW, Tichopad A, Prgomet C, Neuvians TP. Determination of stable housekeeping genes, differentially regulated target genes and sample integrity: BestKeeper – excel-based tool using pair-wise correlations. Biotechnol Lett. 2004;26:509–15.
70. Cuesta A, Meseguer J, Esteban MA. Total serum immunoglobulin M levels are affected by immunomodulators in seabream (*Sparus aurata* L.). Vet Immunol Immunopathol. 2004;101:203–10.

11

DNA methylation regulates discrimination of enhancers from promoters through a H3K4me1-H3K4me3 seesaw mechanism

Ali Sharifi-Zarchi[1,2,3,4†], Daniela Gerovska[5†], Kenjiro Adachi[6], Mehdi Totonchi[3], Hamid Pezeshk[7,8], Ryan J. Taft[9], Hans R. Schöler[6,10], Hamidreza Chitsaz[2], Mehdi Sadeghi[8,11], Hossein Baharvand[3,12*] and Marcos J. Araúzo-Bravo[5,13,14*]

Abstract

Background: DNA methylation at promoters is largely correlated with inhibition of gene expression. However, the role of DNA methylation at enhancers is not fully understood, although a crosstalk with chromatin marks is expected. Actually, there exist contradictory reports about positive and negative correlations between DNA methylation and H3K4me1, a chromatin hallmark of enhancers.

Results: We investigated the relationship between DNA methylation and active chromatin marks through genome-wide correlations, and found anti-correlation between H3K4me1 and H3K4me3 enrichment at low and intermediate DNA methylation *loci*. We hypothesized "seesaw" dynamics between H3K4me1 and H3K4me3 in the low and intermediate DNA methylation range, in which DNA methylation discriminates between enhancers and promoters, marked by H3K4me1 and H3K4me3, respectively. Low methylated regions are H3K4me3 enriched, while those with intermediate DNA methylation levels are progressively H3K4me1 enriched. Additionally, the enrichment of H3K27ac, distinguishing active from primed enhancers, follows a plateau in the lower range of the intermediate DNA methylation level, corresponding to active enhancers, and decreases linearly in the higher range of the intermediate DNA methylation. Thus, the decrease of the DNA methylation switches smoothly the state of the enhancers from a primed to an active state. We summarize these observations into a rule of thumb of one-out-of-three methylation marks: "In each genomic region only one out of these three methylation marks {DNA methylation, H3K4me1, H3K4me3} is high. If it is the DNA methylation, the region is inactive. If it is H3K4me1, the region is an enhancer, and if it is H3K4me3, the region is a promoter". To test our model, we used available genome-wide datasets of H3K4 methyltransferases knockouts. Our analysis suggests that CXXC proteins, as readers of non-methylated CpGs would regulate the "seesaw" mechanism that focuses H3K4me3 to unmethylated sites, while being repulsed from H3K4me1 decorated enhancers and CpG island shores.

(Continued on next page)

* Correspondence: baharvand@royaninstitute.org; mararabra@yahoo.co.uk
†Equal contributors
[3]Department of Stem Cells and Developmental Biology, Cell Science Research Center, Royan Institute for Stem Cell Biology and Technology, ACECR, Tehran, Iran
[5]Computational Biology and Systems Biomedicine, Biodonostia Health Research Institute, 20014 San Sebastián, Spain
Full list of author information is available at the end of the article

(Continued from previous page)

Conclusions: Our results show that DNA methylation discriminates promoters from enhancers through H3K4me1-H3K4me3 seesaw mechanism, and suggest its possible function in the inheritance of chromatin marks after cell division. Our analyses suggest aberrant formation of promoter-like regions and ectopic transcription of hypomethylated regions of DNA. Such mechanism process can have important implications in biological process in where it has been reported abnormal DNA methylation status such as cancer and aging.

Keywords: DNA methylation, Histone modifications, Promoters, Enhancers, H3K4me1, H3K4me3, Computational epigenomics, Next generation sequencing

Background

Multicellular organisms need to establish tissue- and temporal-specific transcriptional programs from a single genome sequence. Such programs coordinate transcription factors (TFs), chromatin-remodeling, chromatin-modifying enzymes, DNA methylation and DNA functional elements such as promoters, insulators, and enhancers. In a previous study on the interaction between DNA methylation and TFs, we found that the methylation-resistant CpG methylation motifs (CpGMMs) are in crosstalk with TFs in gene expression regulation [1]. Such crosstalk could be explained by two mechanisms. One, proposed by Schübeler's group [2], according to which the TFs binding to DNA regions protect them from being methylated. Another mechanism [1] might be that the methylation-resistant CpGMMs signal the TFs to recruit DNA sequence-specific unmethylation machinery. The two mechanisms are not exclusive and might apply cooperatively. Enhancers, making up 10% of the human genome [3, 4] are the most abundant class of regulatory elements. They up-regulate transcription independently of their orientation or distance to the Transcription Start Sites (TSSs), which makes the comprehensive identification of enhancers more difficult than that of other regulatory elements such as promoters (characterized by 5′-sequencing of genes), or insulators (generally bound by the CCCTC-binding factor, CTCF).

Since the first reports on the presence of methyl groups on some genomic cytosines, huge effort has been made to decrypt the function of DNA methylation, focused mostly on promoters, CpG islands and gene bodies, whereas open questions remain about the role of DNA methylation in enhancers [5]. Additionally, DNA methylation has a determinant role in regulating cell fate at distal regulatory regions rather than promoters and gene bodies [6]. Thus, a better understanding of DNA methylation depletion over enhancers is a crucial, yet cumbersome task due to the genomic and epigenomic complexities of the eukaryotic genomic structure.

Some chromatin modifications are employed in addition to the DNA sequence for a more accurate discrimination between promoters and enhancers [7]. Enhancers and promoters can be distinguished by the methylation status at H3K4. Enhancers are enriched for monomethylation of the 4th lysine of histone 3 (H3K4me1) [8], whereas high levels of trimethylation (H3K4me3) predominantly mark active or poised promoters [9]. However, H3K4me1 alone is not a definitive predictor of enhancer [10–12]. Additional chromatin features at enhancers specify three subcategories of enhancers: (i) Active enhancers: They have activation marks (H3K4me1 and H3K27ac), are bound by the Mediator complex [13], and exert regulatory function to increase the transcription of target genes and produce RNA. (ii) Primed enhancers: Enhancers can exist in a primed state prior to activation, they are marked with activation histone modifications (H3K4me1), which do not yield RNA. (iii) Poised enhancers: They are similar to primed enhancers, but distinguished by the presence of the repression mark (H3K27me3), which must be removed for the transition to an active enhancer state [9, 14].

Most of the genome-wide DNA methylation and histone modification studies on mammalian cells show inverse correlation between DNA methylation and histone H3K4 methylation [15–22]. Specifically, DNA methylation is associated with the absence of H3K4 methylation (H3K4me0) [23]. The interaction between DNA and histone methylation is regulated by a cross-talk in the cell between DNA Methyl-Transferases (DMTs) that can contain domains recognizing methylated histones and Histone Methyl-Transferases (HMTs) containing domains recognizing non-methylated DNA. These interactions involve DNA Methyl-CpG-Binding Domains (MBDs) recognizing DNA methylated CpGs, and zinc finger CXXC domains recognizing non-methylated DNA. Thus, several mechanisms based on the interaction between protein-H3K4me recognizing domains (ADD) [24] and protein-DNA methylation recognizing domains (CXXC and MBD) have been discovered that explain the cross-talk between H3K4 (mono-, di- and tri-) methylation and DNA methylation:

(i) The DMTs activity is regulated by the chromatin-interacting ATRX-DNMT3-DNMT3L (ADD) domain of Dnmt3a that recognizes H3K4me specifically. The ADD domain binds to the histone H3 tail that is unmethylated at lysine 4 [25, 26] and the chromatin methylation activity of Dnmt3a and Dnmt3a/3l is guided by interaction of the ADD domain with the histone H3 tail [27].

(ii) In mammals, there are six lysine-specific HMTs, of the COMPASS (COMplex of Proteins Associated with Set1) MLL/SET1 family, namely four Mixed Lineage Leukemia (MLL1 through 4) and two SET domain containing proteins (SET1A and SET1B) [9]. The MLL1/2 contain a CXXC domain, and use it to recognize DNA unmethylated CpG-rich regions [28–30] whereas the MLL3/4 lack the CXXC domain [31–33]. SET1B and SET1B also lack the CXXC domain. They make a complex with the CXXC finger protein 1 (CFP1), and use the CXXC domain of CFP1 to recognize DNA unmethylated CpG-rich regions [34–38]. CFP1 organizes genome-wide H3K4me3 in embryonic stem cells (ESCs) [39]. Although MLL/SET1 family proteins contain similar HMT catalytic SET domains and are capable of mono-, di-, and tri-methylation of H3K4, the transition from mono- to higher methylation states requires additional subunits [40]. Specifically, the robust tri--methylation activity appears to be mediated by the accessory subunit of the tryptophan-aspartic acid (WD) repeat domain 82 (WDR82) protein that binds SET1A/B but not the MLL proteins [41, 42].

The CXXC domain of CFP1 allows preferential binding of CFP1 to H3K4me3 at promoters. In contrast, other HMTs, such as the Trr/MLL3/MLL4 complex, lacking the CXXC domain [31–33], are likely responsible for deposition of H3K4me1 at enhancers [9]. This enhancer-promoter discrimination can be explained by differences in DNA sequence, with high number of CpG islands (usually hypo-methylated) observed at most promoters, but not at enhancers [43].

(iii) Alongside with these zinc finger CXXC domain recognizing non-methylated DNA is the MBD family of proteins recognizing DNA methylated CpGs. The MBD domain of the MBD1 protein binds more efficiently to methylated DNA within a specific sequence context, and a functional MBD domain is necessary and sufficient for recruitment of MBD1 to these *loci*, while DNA binding by the CXXC domain is largely dispensable [44].

While the use of ChIP-seq improved our knowledge of enhancer chromatin states, many questions related to chromatin state and enhancer function remain unanswered, such as the prediction and functional validation of putative enhancers, the determination of the genes associated with enhancers on a large scale, the disclosing of the mechanism that maintains histone marks at enhancers, the determination of whether poised enhancers contact their target promoters, and the defining of the direction of the flow of influence between the enhancer chromatin state and the target DNA promoter state: whether histone marks define enhancers, or histone marks are rather a consequence of the establishment of the enhancer state [3].

To find the interplay of DNA methylation with other epigenetic marks, we integrated high throughput profiles of DNA methylation, histone modification, DNA binding proteins and gene transcription in several mouse cell types (Table 1). After estimating the correlation of DNA methylation with different histone marks within different DNA regulatory regions, we demonstrated that H3K4me1 has different deposition from other active chromatin marks in regard to DNA methylation. We compared the impact of the 5-methylcytosine (5mC) and 5-hydroxymethylcytosine (5hmC) forms of DNA methylation on the regulation of H3K4 methylation, and uncovered their biological consequences. Most importantly, we developed a hypothesis that explains the role of DNA methylation in regulating a seesaw mechanism between H3K4me1 and H3K4me3, and provided additional evidence for the existence of such mechanism by integrating high throughput datasets of functional analyses obtained from gene knockout (KO) experiments.

Results

H3K4me1, in contrast to all other active chromatin marks, is positively correlated with DNA methylation within hypomethylated regions at enhancers and promoters

The correlation between specific chromatin marks and DNA methylation has already been studied in promoters and gene coding regions [1, 20], but with insufficient focus on enhancers. Therefore, we compiled a set of 210,048 genomic sites, each of length 1 k base (kb), centered over Promoters-TSSs (+/– 500 bp of the TSS), as well as the cross-tissue putative enhancers (reported in 19 mouse cell types). We calculated the average DNA methylation of each genomic site in mouse ESCs, and split the list of genomic sites into two groups based on their DNA methylation level: hypermethylated sites (DNA methylation >50%, $N = 186{,}564$) and hypomethylated sites (DNA methylation ≤50%, $N = 23{,}484$). Hyper- and hypomethylation usually refer to increased or decreased DNA methylation without a specific boundary, and we also use these terms to simplify the presentation of our results. The 50% is not a sharp boundary and slight changes in its value do not affect our conclusions.

Within each DNA methylation group, we analyzed the correlation with DNA methylation of promoters and enhancers. While the promoters are easy to determine since they are around the TSS, the enhancers can occur in any genomic region including repeat-associated regions {Short Interspersed Nuclear Element (SINE),

Table 1 High throughput profiles of DNA methylation, histone modification, transcription and protein binding data sets analyzed in this study. ID represents accession identifiers of the datasets in the National Center for Biotechnology Information (NCBI) Gene Expression Omnibus (GEO) [45, 92] or the European Molecular Biology Laboratory - European Bioinformatics Institute (EMBL-EBI) ArrayExpress [73, 93] databases, along with the reference publications. Target marks are the types of molecular targets represented in the datasets. Cell type notation: mouse embryonic stem cell (ESC), mouse embryonic fibroblast (MEF), wild type (WT), knockout (KO). Sequencing methods notation: Whole-Genome Bisulfite Sequencing (WGBS), Tet-Associated Bisulfite sequencing (TAB-seq), Reduced Representation Bisulfite Sequencing (RRBS), Chromatin Immunoprecipitation sequencing (ChIP-seq), RNA sequencing (RNA-seq)

Row	GEO ID	Target marks	Cell type/tissue	Method	Reference
1	GSE30206	DNA methylation	ESC	WGBS	Stadler et al. [56]
2	GSE44760	DNA methylation	MEF (WT, *Dnmt1* KO)	RRBS	Reddington et al. [59]
3	GSE36173	DNA hydroxymethylation	ESC	TAB-seq	Yu et al. [79]
4	GSE29218	H3K4me1, H3K4me3, Pol2, CTCF, H3K27ac, P300	ESC, MEF, Cortex, Liver	ChIP-seq	Shen et al. [45]
5	GSE12241	H3, H4K20me3, H3K36me3, H3K9me3	ESC, MEF	ChIP-seq	Mikkelsen et al. [38]
6	GSE28254	H3K27me3	ESC	ChIP-seq	Brinkman et al. [94]
7	GSE29413	H3K9me3	ESC	ChIP-seq	Karimi et al. [95]
8	E-ERAD-79	H3K4me(1,3)	ESC (WT, *Cfp1* KO)	ChIP-seq	Clouaire et al. [39]
9	GSE41440	H3K4me1, H3K27me3	MEF (WT, *Mll1* KO)	ChIP-seq	Herz et al. [33]
10	GSE44393	H3K4me3, H3K27me3	MEF (WT, *Dnmt1* KO)	ChIP-seq	Reddington et al. [59]
11	GSE39610	MBD (1A,1B,2,3,4), MECP2	ESC	ChIP-seq	Baubec et al. [16]
12	GSE34094	CTCF	ESC	ChIP-seq	Sleutels et al. [96]
13	GSE37338	Transcription	ESC	RNA-seq	Livyatan et al. [97]
14	GSE44733	Transcription	MEF (WT, *Dnmt1* KO)	RNA-seq	Reddington et al. [59]
15	GSE42836	DNA methylation	Liver, Cortex	WGBS	Hon et al. [98]

Long Interspersed Nuclear Element (LINE), Simple repeat, Long Terminal Repeat (LTR), DNA Transposon, Low complexity, DNA Transposon}, Intergenic, Intron, coding regions {Exon, 3'UTR, Transcription Termination Site (TTS)}, Non-coding, CpG island, and Others (merging the cases with less than 100 members, see Methods).

For each of the resulting 14 classes (one promoter and 13 enhancer classes), we calculated the correlation of DNA methylation with 9 chromatin marks {H3K4me1, H3K4me2, H3K4me3, H3K9ac, H3K9me3, H3K27ac, H3K20me3, H3K27me3, H3K36me3}, the repressive histone 3 (H3), the gene transcription marker RNA polymerase 2 (Pol2), the enhancer marker histone acetyltransferase P300, and the binding of the insulator marker CTCF in mouse ESCs (Fig. 1, and Table 1, rows 1, 4–7, 12).

The active chromatin marks (H3K4me1, H3K9ac, H3K4me3, H3K27ac) show negative correlations with DNA hypermethylated classes, while some repressive marks, including H3K9me3 and H3K20me3, are positively correlated (Fig. 1a). Among all the genomic regions in our study, the negative correlation with DNA hypermethylated is especially strong in Promoter-TSSs. The DNA hypomethylated sites represent a similar pattern, particularly for the active chromatin marks (H3K9ac, H3K4me3, and H3K27ac) (Fig. 1b). H3K4me1, however, exhibits an opposite pattern between DNA hyper- and hypomethylated regulatory regions: its correlation with DNA methylation is negative within all hypermethylated classes (Fig. 1a), but positive for DNA hypomethylated classes, especially within Promoter-TSSs, and within putative enhancers in CpG islands, Exons and 3'UTRs (Fig. 1b). The latter result is unexpected, since DNA methylation is generally regarded as a repressive epigenetic mark, and H3K4me1 is a hallmark of active or poised enhancers [4], hence a negative correlation is more likely between them.

According to their distance to the TSSs, the enhancers are usually classified into proximal and distal enhancers. The flanking regions of promoters are usually enriched with H3K4me1 defining proximal enhancers; however, the study of DNA methylation regulating histone methylation is more relevant in distal regions. Therefore, in order to focus our analysis on distal enhancers we excluded from the list of putative enhancers those located inside genes (promoters, exons and introns) or within a distance of less than 3 kb from the closest TSS. We found a clear anti-correlation between H3K4me1 and DNA methylation over hypermethylated distal enhancers (Additional file 1: Figure S1a), whereas the distal enhancers over genomic regions with DNA methylation lower than 50% are positively correlated between H3K4me1 and DNA methylation (Additional file 1: Figure S1b). Additionally, it is noteworthy in Additional file 1: Figure S1b the very high correlation of H3K4me1 with distal enhancers lying over CpG islands. Hence, H3K4me1 exhibits positive correlation with DNA hypermethylated enhancers in general and with DNA hypermethylated distal enhancers in particular.

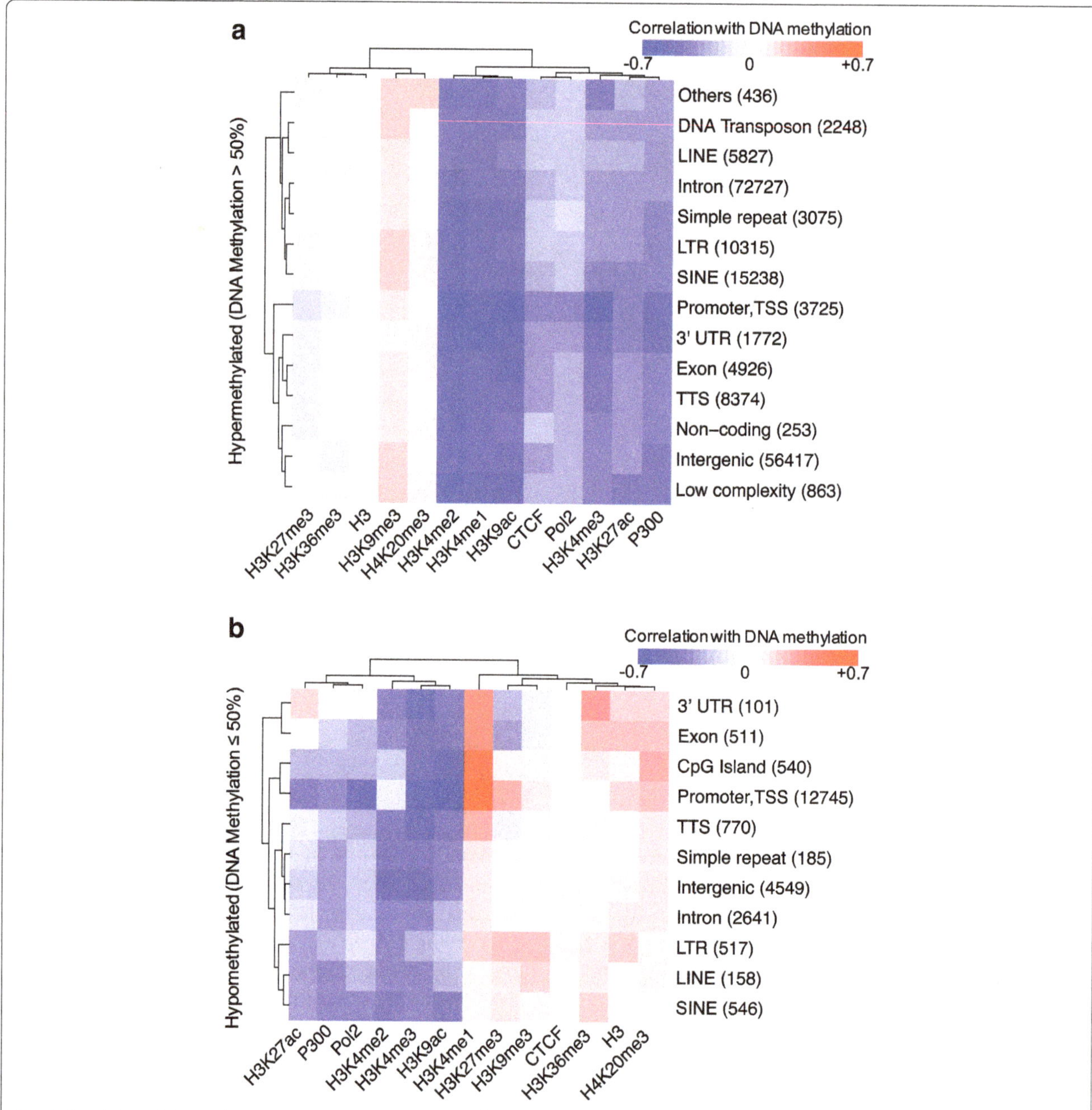

Fig. 1 Correlation of chromatin marks and gene transcription regulators with DNA methylation in promoters and putative enhancers. The promoters are labeled as Promoter, TSS. The putative enhancers are distributed across different classes including repeat-associated regions {Short Interspersed Nuclear Element (SINE), Long Interspersed Nuclear Element (LINE), Simple repeat, Long Terminal Repeat (LTR), DNA Transposon, Low complexity, DNA Transposon}, Intergenic, Intron, coding regions {Exon, 3'UTR, Transcription Termination Site (TTS)}, Non-coding regions, CpG island, and Others. The promoters and different classes of enhancers are split into (**a**) DNA hypermethylated (DNA methylation >0.5) and (**b**) DNA hypomethylated (DNA methylation ≤0.5) groups. In each DNA methylation group, regulatory sites are classified based on their genomic location (rows). For each class, Spearman's rank correlations, ρ, between DNA methylation of ESCs and 9 different chromatin marks, the repressive histone 3 (H3), the gene transcription marker RNA polymerase 2 (Pol2), the enhancer marker histone acetyltransferase P300, and insulator marker CCCTC-binding factor are presented in columns. Red, white and blue colors show positive, null and negative correlations, respectively

Since enhancers are often shorter than 1 kb and both H3K4me1 and DNA methylation could localize to the same 1 kb element, but not necessary with local overlap, this could drive correlations between DNA methylation and H3K4me1. Therefore, we performed the above correlation analysis with window sizes of +/− 100 bp (total size 200 bp, Additional file 1: Figure S2) and +/− 200 bp (total size 400 bp, Additional file 1: Figure S3). The smaller window sizes decrease the number of promoters/enhancers, since many of them lack required number of CpGs in the smaller window to measure the DNA methylation level. Nonetheless, these results confirm the correlation between H3K4me1/H3K4me3 and DNA methylation, independently of the window size.

H3K4me1, in contrast to all other active chromatin marks, is enriched at intermediate DNA methylation level

To analyze the distinct deposition of H3K4me1 over the DNA methylation landscape, we sorted the list of regulatory regions based on their DNA methylation level, and averaged the enrichments of each chromatin mark (Fig. 2a). We found that repressive chromatin marks such as H3K9me3, H4K20me3 and histone 3 (H3) are statistically significantly overrepresented in hypermethylated regions, while active chromatin marks are enriched at DNA hypomethylated promoters and enhancers (*p*-value <1e-15), i.e., the regulatory regions with DNA methylation >95% are 5-fold more enriched of H3K9me3 and simultaneously 10-fold less enriched of H3K4me3, compared to the <5% DNA methylated regions.

H3K4me1 enrichment is clearly distinct from all the other active chromatin marks (Fig. 2b). It is most enriched (0.9) at intermediate DNA methylation levels (25 - 75%), and is enrichment diminished at DNA methylation levels below 25% or above 75%, whereas H3K27ac, whose enrichment distinguishes the active from primed enhancers, is enriched in the lower range (25 - 35%) of the same intermediate DNA methylation level and decreases linearly in the higher range (35 - 75%) of the intermediate DNA methylation (Fig. 2b). Thus, when the DNA methylation of the enhancers decreases, the enhancers switch from a primed to an active state.

We studied the correlation of the signal of the three methylation states of H3K4 {me1, me2, me3} with the DNA methylation level, and found that while H3K4me2 and H3K4me3 signals anticorrelate with DNA methylation level across the whole DNA methylation range, H3K4me1 correlates positively with DNA methylation in the 0 - 50% range and negatively in the 50 - 100% range (Fig. 2f-h). We observed that DNA methylation affects RNA expression differentially promoters and enhancers. Whereas in the case of promoters, RNA expression was depleted for the middle range of DNA methylation (Fig. 2c), for the case of enhancers RNA expression was less affected for DNA methylation levels of more than 75%. We searched for non-canonically expressed enhancers, i.e., those that being highly methylated (DNA methylation >75%) are nevertheless expressed. Among them we found multiple enzymes, such as the three *loci* of the muscle pyruvate kinase (*Pkm*), lactate dehydrogenase C (*Ldhc*), glycogen synthase 2 (*Gys2*), prolyl 4-hydroxylase subunit β (*P4hb*), two *loci* of the protein phosphatase 4, catalytic subunit (*Ppp4c*), the epigenetic regulators tet methylcytosine dioxygenase 1 (*Tet1*), and jade family PHD finger 1 (*Jade1*); and transcriptional regulators such as the transcriptional repressor pro-apoptotic WT1 regulator (*Pawr*); the transcriptional and translational initiators: basic transcription factor 3 (*Btf3*) and eukaryotic translation initiation factor 4, gamma 2 (*Eif4g2*) among others (Fig. 2e).

Next, we validated our finding that in contrast to the other active chromatin marks (H3K9ac, H3K4me3, H3K4me2, H3K27ac), H3K4me1 is less enriched in both unmethylated and highly methylated regulatory regions, but overrepresented in regions with intermediate levels of DNA methylation (Fig. 2a and b), by co-localizing the DNA methylation level and histone mark signals with the known enhancer coordinates of the *Myc/c-Myc* and *Sox2* pluripotent genes in ESCs [45, 46] (Fig. 2i). In the case of *Myc*, the three known enhancers co-localize with peaks of high H3K4me1 signal and intermediate DNA methylation level. In the case of *Sox2*, two enhancers (5 and 6) co-localize with peaks of high H3K4me1 signal and intermediate DNA methylation level, and four enhancers (1, 2, 3 and 4) co-localize with peaks of the P300 and very low DNA methylation level.

Neither methyl-binding proteins, nor cytosine hydroxymethylation can explain the distinct H3K4me1/3 deposition

To search for possible molecular mechanisms that explain the positive correlation between DNA methylation and H3K4me1 at hypo- to intermediate DNA methylation level at regulatory sites, we examined two conjectures: (i) Proteins with MBDs could be potential mediators of the distinct H3K4me1/3 deposition. (ii) The transition of cytosine methylation towards unmethylation through the cytosine hydroxymethylation transitory state could be associated with the H3K4me1 enrichment at intermediate DNA methylation level.

MBD proteins link to DNA through binding DNA methylated sites to some histone modifications, i.e. MBD1 forms a complex with the H3K9 methylase SETDB1, which is suggested to form stable heterochromatin histone marks over methylated DNA [47, 48]. Additionally, MBD3 is enriched at active promoters (with a positive correlation with H3K4me3) and at the enhancers of active genes that are usually H3K4me1 marked [49, 50]. Indirect interactions between MBDs and H3K4 methylation can also be hypothesized, i.e. ZIC2, an enhancer-binding factor which co-localizes H3K4me1 and the other enhancer marks (P300, H3K27ac) is shown to interact with MBD3/NURD in mouse ESCs [51]. Thus, the MBDs could be effectors of the crosstalk between DNA methylation and the H3K4me1 and H3K4me3 interaction observed here.

Therefore, to check the MBD effectors hypothesis we compared the chromatin immunoprecipitation sequencing (ChIP-seq) profiles of the MBD proteins for which data is available: MBD1A/B, MBD2, MBD3, MBD4 and

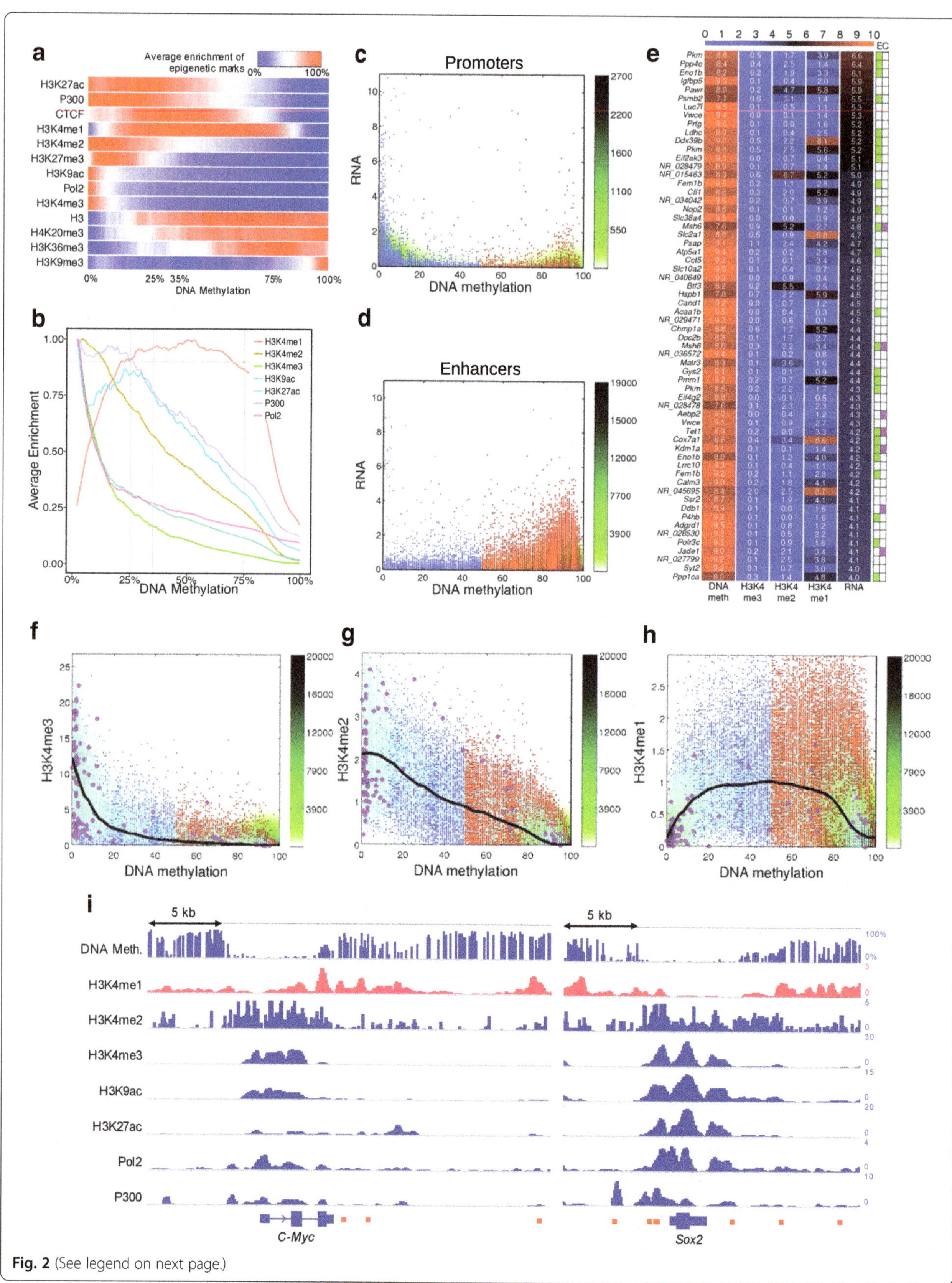

Fig. 2 (See legend on next page.)

(See figure on previous page.)
Fig. 2 Distinct deposition of H3K4me1 from the other active chromatin marks. The regulatory sites are sorted according to their DNA methylation level in ESCs from 0 to 100% methylated. Average enrichment of different chromatin marks (rows) over sites of the same DNA methylation level are shown with (**a**) color bars and (**b**) lines (for the seven active chromatin marks). Average enrichments are scaled to have equal maximum for different marks. Pairwise scatter plots of DNA methylation versus RNA transcription for promoters (**c**) and enhancers (**d**). The scattering density is shown in green. Red and blue dots show sites with DNA methylation lower or higher that 50%, respectively. Cyan spreads show promoter sites and magenta circles show the promoters whose transcription is more than 4 in log_2 scale. (**e**) Heat map of hypermethylated enhancers (DNA methylation >75%) and expressed transcripts (transcription >4 in log_2 scale). To adjust the color codification, the DNA methylation, percentages are multiplied by 0.1, and H3K4me2 and H3K4me2 peaks by 5, the RNA-seq values are in log_2 scale. Higher values correspond to redder color. The table to the right annotates Gene Ontology (GO) terms: E (Enzymatic activity) in green and C (Chromatin organization regulation) in magenta. H3K4 methylation, me3 (**f**), me2 (**g**) and me1 (**h**), enrichments within regulatory sites versus DNA methylation. Each point represents a single regulatory site. Each point represents a single regulatory site. The scattering density is shown in green. Red and blue dots show sites with DNA methylation lower or higher that 50%, respectively. Cyan spreads show promoter sites and magenta circles show the promoters whose transcription is more than 4 in log_2 scale. The over imposed black lines mark the median of the H3K4 methylations smoothed using a robust loess regression. (**i**) DNA methylation and enrichment of the seven active chromatin marks around the *Myc* and *Sox2* gene *loci*. The location of all known putative *Myc* and *Sox2* enhancers taken from the supplemental material of Shen et al. [45] and from PHANTOM5 [46], are marked by red bars at the bottom. The y-axis represents the DNA methylation measured as the percentage of reads that support the methylated state of each CpG (estimated methylation level). For each histone mark track and for the Pol2 and P300 tracks, the y-axis represents the normalized level of ChIP-seq signal over the genomic regions

MECP2 (Table 1, row 11) with enrichment sites of H3K4me1 and H3K4me3 in mouse ESCs (Table 1, row 4). In this analysis we included all genomic sites that showed a statistically significant peak of the chromatin marks or of the protein binding, regardless of whether such genomic sites are located within promoter/enhancer regions or not. H3K4me1 peaks occur at intermediate to high DNA methylation level, median DNA methylation (Med) = 76%, whereas the MBD proteins binding *loci* are very highly DNA methylated (Med > 90%), with the exception of MBD3 (Med = 52%) and MBD2 (Med = 81%). H3K4me3 enrichment occurs at low DNA methylation level (Med = 24%) (Fig. 3a). Such results point out lack of correlation between H3K4me3 deposition and MBD protein binding DNA methylation over all the DNA methylation ranges (low, intermediate and high), and not so obvious lack of correlation between H3K4me1 deposition and MBD protein binding DNA methylation. To resolve this case, we zoomed into the intermediate to high range of DNA methylation (50 - 100%) to check some possible correlation of MBD binding and H3K4me1 enrichment. For this purpose, we calculated the fraction of the highly methylated peaks (DNA methylation >95%) among all peaks of H3K4me1 and H3K4me3, and MBD binding regions (Fig. 3b). 10 - 20% of the MBD binding peaks populate the over 95% DNA methylation range, in contrast to only 2% H3K4me1 marks populating the same range, which rejects the possibility of overlap direct interaction between methyl-binding proteins and H3K4 methylation. We analyzed further this possibility through computing the number of all possible pairwise overlaps between peaks of two signals (chromatin marks or methyl-binding proteins) (Fig. 3c). We found that for all methyl-binding proteins there were more peak overlaps with H3K3me3 than with H3K3me1. The methyl-binding protein with highest number of overlaps with H3K4me1 is MBD3, i.e. it has a 21% of peaks overlapping with the H3K4me1 (accounting for 7592 peaks), and a 23% peaks overlapping with H3K3me3 (amounting to 8524 peaks). The other methyl-binding proteins have even less overlaps with H3K4me1 peaks (5 to 13%). These results abrogate the hypothesis of a possible connection between methyl-binding proteins and H3K4me1 deposition.

We observed that H3K4me1 is enriched at intermediate DNA methylation level, leading to the conjecture that such intermediary level might correspond to bidirectional DNA high ↔ low methylation transitions. Since it has been considered that DNA cytosine hydroxymethylation (5hmC) is an intermediate state in the process of active DNA cytosine demethylation [52], is conceivable to hypothesize that the observed intermediate DNA cytosine methylation associated with H3K4me1 enrichment might also correlate with DNA cytosine hydroxymethylation. Therefore, it is worth to study whether there is a correlation between DNA cytosine hydroxymethylation and the dynamics of the distinct H3K4me1/3 states when DNA methylation is in the transitory way to be reduced during the intermediary DNA cytosine hydroxymethylation.

To check the DNA cytosine hydroxymethylation hypothesis, we designed a method to find out which one of the DNA cytosine methylations (5mC or 5hmC), has stronger impact on the level of H3K4 methylation (H3K4me1 and H3K4me3). For this purpose, we compared alternations between a present (+) and an absent (−) state of one form of cytosine methylation (5mC or 5hmC) while the other form remains constant at a background level. Since WGBS (Whole-Genome Bisulfite Sequencing) data cannot discriminate directly between 5mC and 5hmC levels of a CpG but the sum of both DNA methylation types, we designed a method to infer 5mC from WGBS and TAB-seq (Tet-assisted bisulfite sequencing), see Eq. 1 in Methods section. We identified

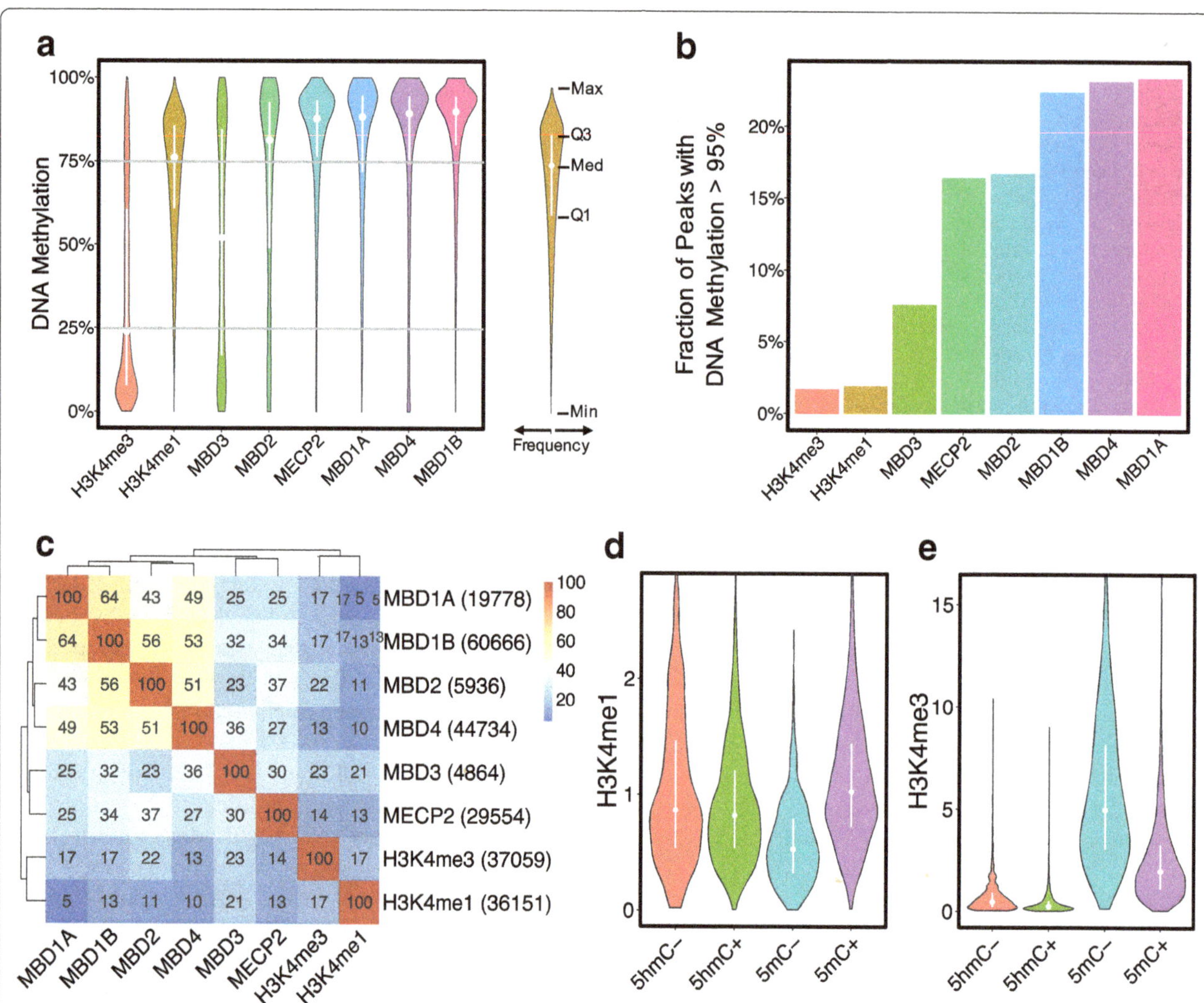

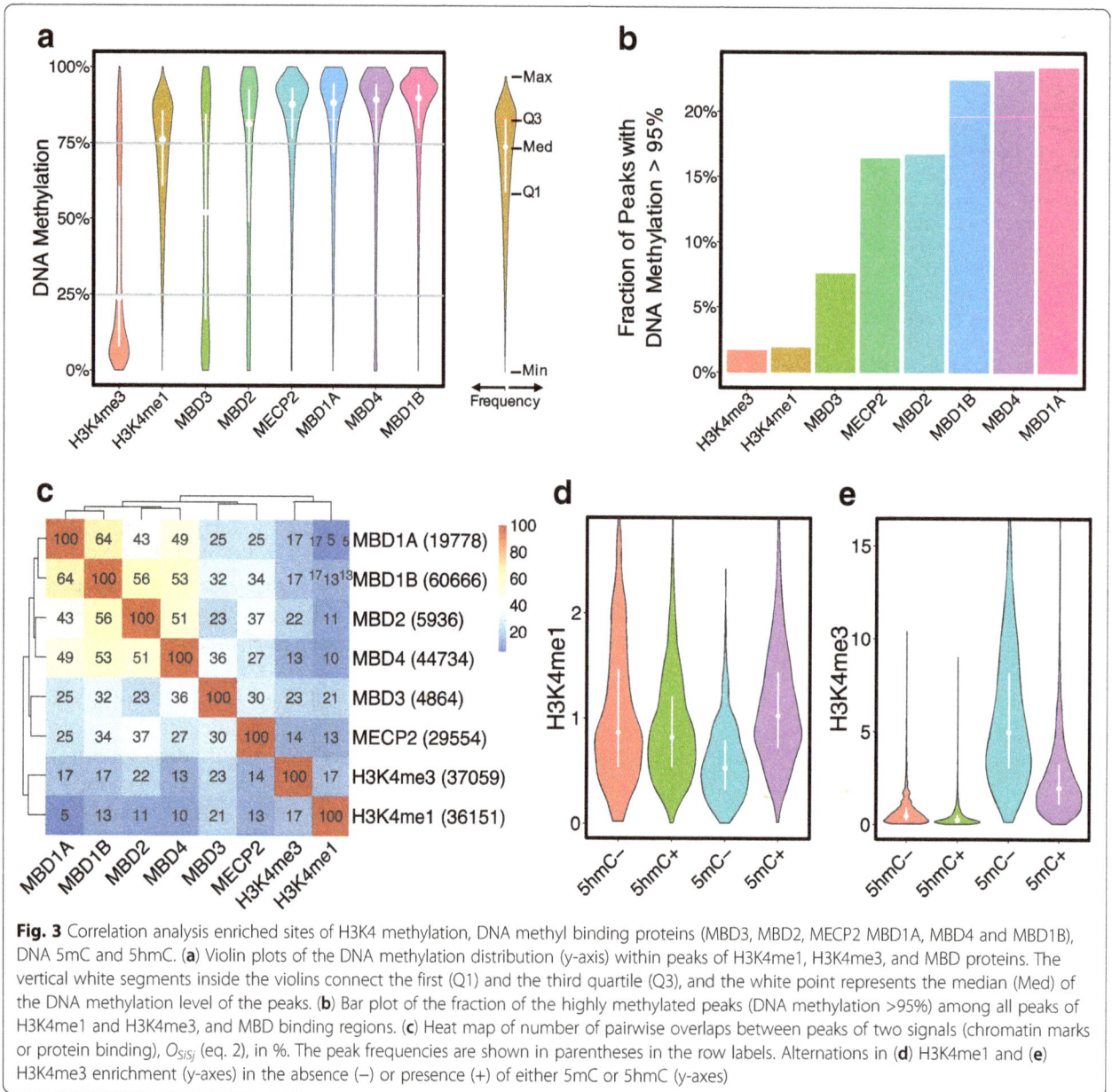

Fig. 3 Correlation analysis enriched sites of H3K4 methylation, DNA methyl binding proteins (MBD3, MBD2, MECP2 MBD1A, MBD4 and MBD1B), DNA 5mC and 5hmC. (**a**) Violin plots of the DNA methylation distribution (y-axis) within peaks of H3K4me1, H3K4me3, and MBD proteins. The vertical white segments inside the violins connect the first (Q1) and the third quartile (Q3), and the white point represents the median (Med) of the DNA methylation level of the peaks. (**b**) Bar plot of the fraction of the highly methylated peaks (DNA methylation >95%) among all peaks of H3K4me1 and H3K4me3, and MBD binding regions. (**c**) Heat map of number of pairwise overlaps between peaks of two signals (chromatin marks or protein binding), $O_{s_i s_j}$ (eq. 2), in %. The peak frequencies are shown in parentheses in the row labels. Alternations in (**d**) H3K4me1 and (**e**) H3K4me3 enrichment (y-axes) in the absence (–) or presence (+) of either 5mC or 5hmC (y-axes)

two groups of putative enhancers for each form of cytosine methylation (5mC or 5hmC). Each of these two groups has two subgroups, each subgroup with a similar distribution of one form of cytosine methylation working as a background but with altered level into two states (present +, or absent -) of the other form of cytosine methylation. Thus, the 5hmC alteration group consists of two subgroups {5hmC+, 5hmC-} of enhancers with significantly different 5hmC (present +, or absent -) but equal 5mC distributions, while the 5mC alteration group consists of two subgroups {5mC+, 5mC-} of enhancers with significantly different 5mC (present +, or absent -) but equal 5hmC distributions. Hence we could study the effect of the "altered" (present +, or absent -) form of DNA cytosine methylation, independently from the "background" (equal) form of methylation. We calculated the enrichment of H3K4me1 and H3K4me3 for each of the identified groups to study whether the hydroxymethylation of cytosines (5hmC) is the cause of the positive correlation between DNA methylation and H3K4me1 on DNA hypomethylated regulatory sites (Figs. 3d and e, and Table 1, rows 1, 3 and 4). We found out that alternation in 5mC levels coincides with a significant change in both H3K4me1 and H3K4me3 enrichment of regulatory sites, the H3K4me1 level increases from the group of 5mC- to the group of 5mC+ enhancers whereas the H3K4me3 level decreases from the group of 5mC- to the group of 5mC+. However, both H3K4me1 and H3K4me3 enrichments of enhancers having similar 5mC but different 5hmC are almost the same. Hence, a possible role of cytosine hydroxymethylation on H3K4me1/3 regulation is rejected and the role of cytosine methylation on H3K4me1/3 regulation is reinforced.

DNA methylation regulates H3K4me1 - H3K4me3 seesaw

Since our previous conjectures for explaining the molecular mechanisms ruling the enrichment of H3K4me1 within DNA methylated regulatory sites were rejected, we asked the reverse question: Why H3K4me1 is not increased at DNA unmethylated regulatory sites (promoters and putative enhancers) as it could be expected for an active mark? We have already observed elevated H3K4me3 over diminished H3K4me1on DNA unmethylated regulatory sites, Particularly, the enrichment of H3K4me3 has the highest fold-change between DNA hypo- and hypermethylated regulatory sites among all active chromatin marks in this study. Hence, we hypothesized the existence of a seesaw between H3K4me1 and H3K4me3 occupancy within regulatory sites, which is controlled by DNA methylation. While both chromatin marks are depleted at DNA hypermethylated regions, the activation of this seesaw mechanism is restricted to the regulatory sites with zero to intermediate levels of DNA methylation.

We checked this hypothesis in mouse pluripotent ESCs (Fig. 4 and Table 1, rows 1 and 4). Regulatory regions with the highest H3K4me3 enrichments were DNA unmethylated and H3K4me1 decreased (Fig. 4a). In contrast, the regions with elevated H3K4me1 enrichment had higher DNA methylation but less H3K4me3. A similar analysis of cortex and liver cells (Table 1, row 15) confirmed that our finding is also true for differentiated cells (Figs. 4b and c).

The regulation of the H3K4me1 - H3K4me3 seesaw by DNA methylation is mediated through protein CXXC DNA binding domains

The MLL1/2 and SET1A/B protein complexes responsible for deposition of H3K4me3 to the nucleosomes [42, 53, 54] share homologous CXXC subunits (CXXC7 in MLL1/2 and CXXC1 in the CFP1 component of the SET1A/B complex). These CXXC subunits are missing in the H3K4me1 depositing histone methyltransferase MLL3/4 complex [31–33]. CXXC binding domains are known to bind unmethylated CpG rich genomic regions, particularly CpG islands [39, 55]. To obtain mechanistic insights into the seesaw mechanism here proposed, we studied the influence of the presence or absence of CXXC domains on the performance of the seesaw mechanism through the computational analysis of knock out (KO) of such domains in pluripotent and differentiated mouse cells.

An expected consequence of the seesaw mechanism would be the elevation of H3K4me1 after the block of H3K4me3 in DNA hypomethylated regions. We counted the number of H3K27me3 and H3Kme1 peaks in wild type (WT) and CXXC7 (MLL1) KO from mouse embryonic fibroblast (MEF) (Table 1, row 9) and the results confirmed such prediction: the number of H3K4me1 peaks in the MLL1 KO is 31% higher (p-value <1e-15, binomial test) than in the WT cells (Fig. 4d). The frequency of H3K27me3 peaks had a minor (< 4%) difference that showed the analysis was not biased (Fig. 4d).

Next, we studied how the influence on the H3K4 methylation exerted by the CXXC1 (CFP1) component of SET1A/B in ESCs is related to DNA methylation. We used the ESC WT and *Cfp1* KO H3K4me1/3 ChIP-seq peaks from Clouaire et al. [39] (Table 1, row 8) and we co-localized them with ESC DNA methylation from Stadler et al. [56] (Table 1, row 1). We identified 8409 H3K4me3 peaks specific for WT cells and 13,184 peaks specific for *Cfp1* KO, in addition to the 78,847 common peaks between WT and *Cfp1* KO cells. The number of distal (i.e. > 5 kb distance from a TSS) H3K4me3 peaks is significantly increased in *Cfp1* KO cells (11,352 peaks in *Cfp1* KO, and 4663 in WT).

To study how DNA methylation influences the lack of unmethylated CpG CXXC binding domains (*Cfp1* KO) on H3K4me3, we selected the peaks with at least 2% CpG content. We found a significant change between the DNA methylation of the WT and *Cfp1* KO specific H3K4me3 peaks, median DNA methylation of 13 and 79%, respectively (Fig. 4e). Since WT peaks are restricted to DNA hypomethylated regions, this finding suggests that the ablation of *Cfp1* allows the appearance of H3K4me3 peaks in DNA hypermethylated regions. This suggestion is in agreement with previous studies [39, 55]. We do not exclude, however, the possibility of reduced activity of DNA methyltransferases and global hypomethylation in Cfp1-KO cells as reported previously [57].

Additionally, we identified 7638 H3K4me1 peaks specific to WT, 8234 specific to *Cfp1* KO cells, and 116,373 H3K4me1 peaks in both cell types. Since we hypothesized that there is a seesaw mechanism between H3K4me1 and H3K4me3 within low to intermediate DNA methylation, we focused our analysis on peaks with DNA methylation below 50%. The WT-specific H3K4me1 peaks have significantly higher H3K4me1, but lower H3K4me3 enrichment than the *Cfp1* KO specific peaks, and vice versa (Fig. 4f, g, p-value <1e-15). Particularly, H3K4me1 enrichment shows a significantly negative correlation (Pearson's correlation coefficient $r = -0.71$) with H3K4me3 enrichment, i.e. within low to intermediate DNA methylation increased H3K4me1 levels encompassed with reduced H3K4me3

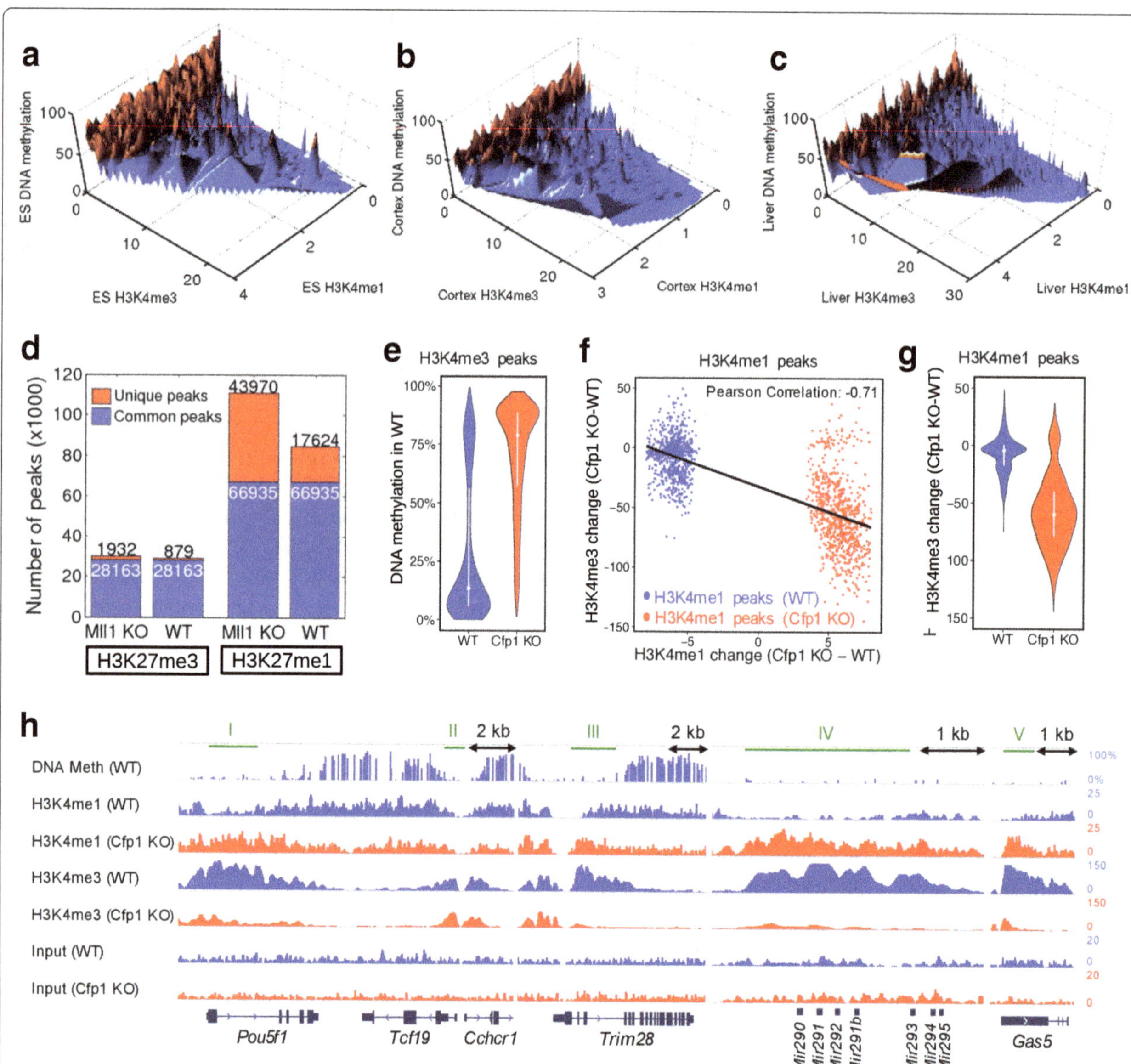

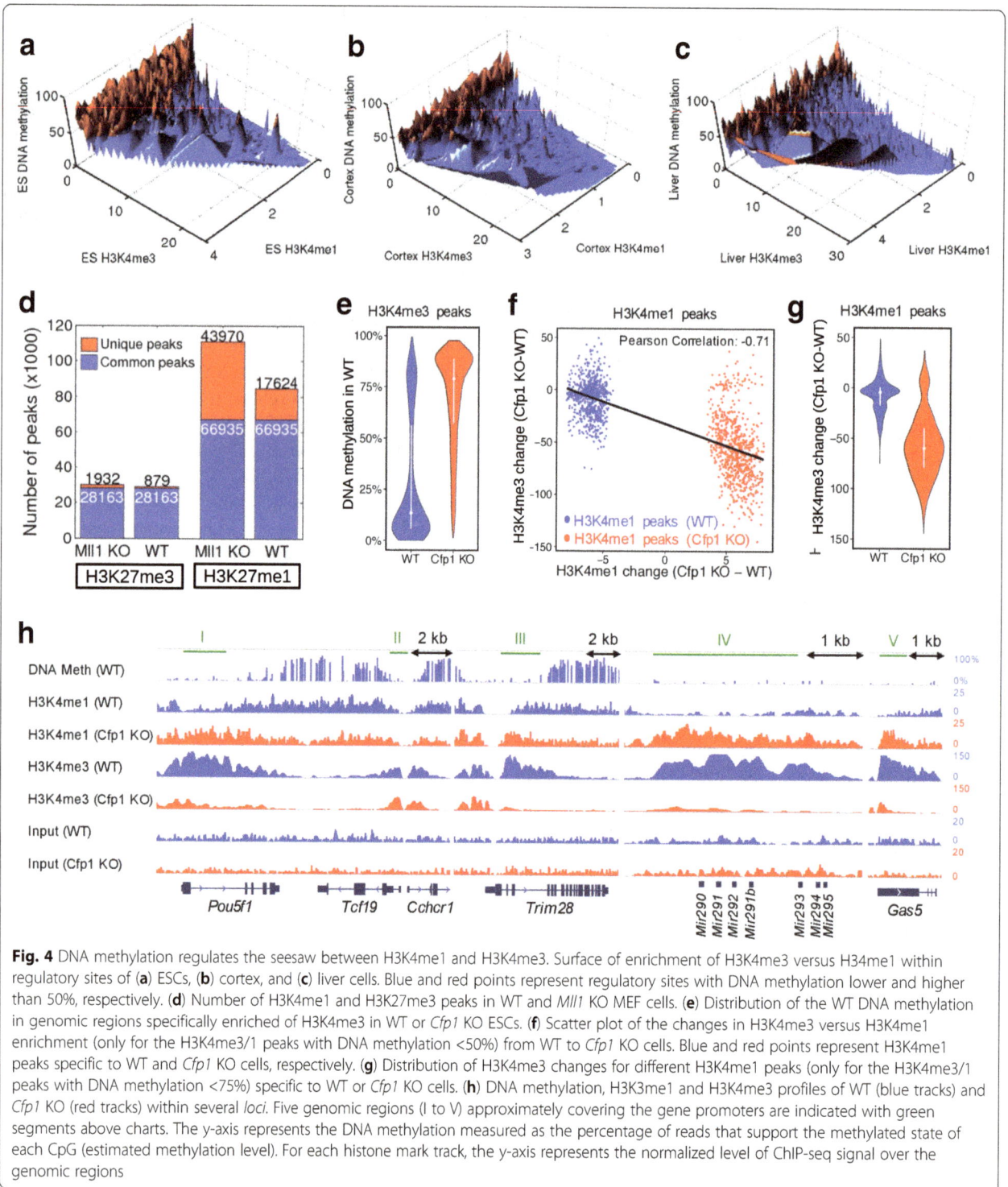

Fig. 4 DNA methylation regulates the seesaw between H3K4me1 and H3K4me3. Surface of enrichment of H3K4me3 versus H34me1 within regulatory sites of (**a**) ESCs, (**b**) cortex, and (**c**) liver cells. Blue and red points represent regulatory sites with DNA methylation lower and higher than 50%, respectively. (**d**) Number of H3K4me1 and H3K27me3 peaks in WT and *Mll1* KO MEF cells. (**e**) Distribution of the WT DNA methylation in genomic regions specifically enriched of H3K4me3 in WT or *Cfp1* KO ESCs. (**f**) Scatter plot of the changes in H3K4me3 versus H3K4me1 enrichment (only for the H3K4me3/1 peaks with DNA methylation <50%) from WT to *Cfp1* KO cells. Blue and red points represent H3K4me1 peaks specific to WT and *Cfp1* KO cells, respectively. (**g**) Distribution of H3K4me3 changes for different H3K4me1 peaks (only for the H3K4me3/1 peaks with DNA methylation <75%) specific to WT or *Cfp1* KO cells. (**h**) DNA methylation, H3K3me1 and H3K4me3 profiles of WT (blue tracks) and *Cfp1* KO (red tracks) within several *loci*. Five genomic regions (I to V) approximately covering the gene promoters are indicated with green segments above charts. The y-axis represents the DNA methylation measured as the percentage of reads that support the methylated state of each CpG (estimated methylation level). For each histone mark track, the y-axis represents the normalized level of ChIP-seq signal over the genomic regions

levels when *Cfp1* is knocked out which further confirms the seesaw model (Fig. 4f), thus reduced H3K4me3 (due to Cfp1KO) elevates the seesaw towards H3K4me1.

To illustrate how the co-localization of the H3K4me1 and H3K4me3 signals is influenced by the disruption of CFP1, we studied the genomic region around the master of pluripotency transcription factor *Pou5f1/Oct4* (Fig. 4h). The unmethylated promoter of *Pou5f1* (region I) is depleted of H3K4me1 and enriched of H3K4me3 in WT cells, while the *Cfp1* KO cells are enriched of H3K4me1 and depleted of H3K4me3 in the same *loci*. Similarly, the transcriptional intermediary factor *Trim28*, the pluripotency-associated *Mir290* cluster of microRNAs, and the non-coding RNA gene *Gas5* (regions III-IV) show elevated H3K4me1 coinciding with depleted H3K4me3 after *Cfp1* KO. These results show how the disruption of CFP1, alters the balance between H3K4me1 and H3K4me3. The promoter shared between *Tcf19* (Transcription factor 19) and *Cchcr1* (Coiled-coil α-helical rod protein 1) transcribed in opposite directions (region II), however, it shows almost similar chromatin patterns in WT and *Cfp1* KO cells.

DNA hypomethylation causes H3K4me3 enrichment and aberrant gene expression

We have provided several lines of evidence showing that the seesaw mechanism between H3K4me1 and H3K4me3 is regulated by DNA methylation. However, the biological impact of such regulation still needs to be identified. It is also important to determine whether this regulatory function of DNA methylation is a specific property of pluripotent cells or whether it exists also in differentiated cells. Therefore, we studied MEF cells in absence (KO) or presence (WT) of *Dnmt1*, the key maintainer of DNA methylation after cell division (Fig. 5 and Table 1, rows 2, 10 and 14). In addition to 23,859 common H3K4me3 peaks in *Dnmt1* WT and KO cell types, we found a gain of 8648 (30%) of genomic *loci* which were H3K4me3-enriched specifically in the *Dnmt1*-KO cells. This is almost twice the number of specific peaks of WT cells (4515 WT-specific) (Fig. 5a). Furthermore, the number of H3K27me3 peaks had a modest change (3%) between cell types, which confirms that the results were not cell type dependent. Similar to *Cfp1* KO cells, there were significantly more frequent distal H3K4me3 peaks specific for *Dnmt1* KO cells ($N = 5652$) compared to the WT specific ($N = 1516$) (Additional file 1: Fig. S4).

The DNA methylation pattern is significantly different between the specific peaks for each cell type, WT and *Dnmt1* KO (Fig. 5b, Additional file 1: Fig. S1). The WT-specific H3K4me3 peak locations are hypomethylated in both WT and *Dnmt1* KO cells (21 and 18% median DNA methylation, respectively). In contrast, *Dnmt1* KO-specific H3K4me3 peaks show significant loss of DNA methylation after *Dnmt1* KO (28%), while they are hypermethylated in WT (77% median DNA methylation).

Presence of H3K4me3 peaks coincides with a major shift in transcription (Fig. 5c). Among *Dnmt1* KO-specific H3K4me3 peaks, up-regulated transcribed regions in the same cell type (compared to the WT cells) are 21 times more frequent than down-regulated ones ($N = 2931$ and 139 respectively, minimum 2-fold change in transcription). Similarly, WT- specific H3K4me3 peaks with more than 2-fold up-regulation in the same cells (compared to *Dnmt1* KO) are 7.5 times more frequent than down-regulated regions ($N = 1064$ and 141, respectively).

We analyzed how the H3K4me3 peaks for each cell type, WT and *Dnmt1* KO, split between enhancers and promoters. At DNA methylomics level, the WT-specific H3K4me3 peaks are under-methylated in *Dnmt1* KO samples in relation to WT samples both in promoters (Fig. 5d) and enhancers (Fig. 5f). Interestingly, in the case of enhancers, there is a depression of DNA methylation in the *Dnmt1* KO samples for the DNA methylation level around 75% of the WT samples (Fig. 5f). The *Dnmt1* KO-specific peaks are slightly more under-methylated over promoters (Fig. 5e) than over enhancers (Fig. 5g), with a high dispersion of DNA methylation in *Dnmt1* KO samples in the same *loci* of enhancers in which the WT samples are highly DNA methylated (Fig. 5g). At transcriptomics level, the expression behavior in WT-specific H3K4me3 peaks over promoters (Fig. 5h) and over enhancers (Fig. 5j) is very similar. In both cases the transcription in WT samples is up-regulated in relation to the transcription in *Dnmt1* KO samples. Interestingly, in the case of *Dnmt1* KO-specific H3K4me3 peaks there is a strong dichotomy in the transcription behavior of enhancers (Fig. 5i) and promoters (Fig. 5k). In both cases, the expression is very similar in *Dnmt1* KO and WT samples for expression level higher than 4 (in $\log_2$ scale). However, for low transcription levels, the transcription is up-regulated in *Dnmt1* KO samples in relation to WT samples in enhancers (Fig. 5k).

We studied how this observation at genomics level translates into the co-localization of signals at *loci* of four cell-type specific genes (*Tex19.1*, *Hspb2*, *Capn11*, *En1*) and one house housekeeping gene, *Gapdh* (Fig. 5i). Both epigenetic (DNA methylation and H3K4me3) and transcriptional patterns of *Gapdh* (region I) are similar in WT and *Dnmt1* KO cells. In contrast, the pluripotency-associated gene *Tex19.1* that is specifically active in ES, placenta and germ cells [58] loses DNA methylation (from 100% to 25–75%) in its promoter in *Dnmt1* KO cells. This is supported by the fact that the number of CpGs with 75–100% methylation is reduced to almost zero at genomics scale in *Dnmt1* KO cells [59]. The DNA methylation loss is coincident with H3K4me3 enrichment and downstream ectopic expression in MEF cells (region II). Same scenario develops at the Heat Shock Protein Family B (Small) Member 2 coding *Hspb2* gene, normally expressed in muscle and heart (region III). Region IV is an intronic long terminal repeat (LTR) located within the spermatogenic-specific Calpain 11 coding gene *Capn11*. It is silent in WT MEFs, H3K4me3 enriched and transcribed after being hypomethylated in *Dnmt1* KO cells, although the *Capn11* gene itself is silent in both cell types. An intergenic region upstream of the neural specific Engrailed Homeobox *En1* coding gene is also shown to undergo DNA hypomethylation, H3K4me3 enrichment and active transcription in *Dnmt1* KO cells (region V).

We compared the genomic location of H3K4me3 peaks specific to WT and *Dnmt1* KO cells (Fig. 5m). The *Dnmt1* KO-specific H3K4me3 peaks were overrepresented within

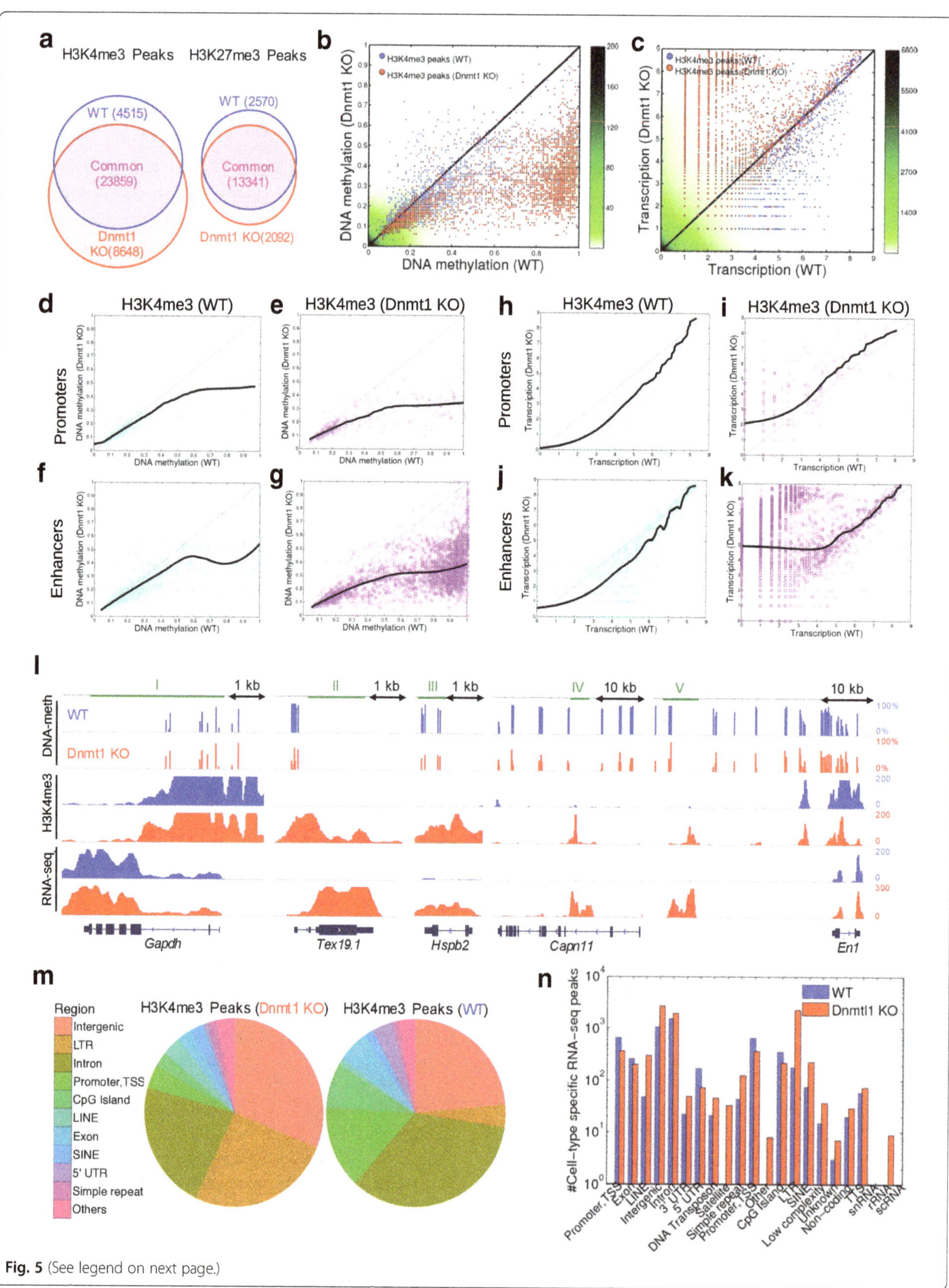

Fig. 5 (See legend on next page.)

(See figure on previous page.)
Fig. 5 DNA hypomethylation is followed by H3K4me3 enrichment and activates transcription. (**a**) Venn diagrams of number of H3K4me3 (left) and H3K27me3 (right) peaks in WT and *Dnmt1* KO MEF cells, in blue and red, respectively. (**b**) Scatter plot of DNA methylation profiles in *Dnmt1* KO versus WT cells. (**c**) Scatter plot of RNA-Seq transcription profiles of *Dnmt1* KO versus WT. The scattering density is shown in green. (**d**) Scatter plot of DNA methylation profiles in *Dnmt1* KO versus WT cells for H3K4me peaks over promoters of WT cells. (**e**) Scatter plot of DNA methylation profiles in *Dnmt1* KO versus WT cells for H3K4me peaks over promoters of Dnmt1 KO cells. (**f**) Scatter plot of DNA methylation profiles in *Dnmt1* KO versus WT cells for H3K4me peaks over enhancers of WT cells. (**g**) Scatter plot of DNA methylation profiles in *Dnmt1* KO versus WT cells for H3K4me peaks over enhancers of Dnmt1 KO cells. (**h**) Scatter plot of transcriptomics profiles in *Dnmt1* KO versus WT cells for H3K4me peaks over promoters of WT cells. (**i**) Scatter plot of transcriptomics profiles in *Dnmt1* KO versus WT cells for H3K4me peaks over promoters of Dnmt1 KO cells. (**j**) Scatter plot of transcriptomics profiles in *Dnmt1* KO versus WT cells for H3K4me peaks over enhancers of WT cells. (**k**) Scatter plot of transcriptomics profiles in *Dnmt1* KO versus WT cells for H3K4me peaks over enhancers of Dnmt1 KO cells. (**l**) DNA methylation, H3K4me3 and transcription in several *loci* of WT (blue tracks) and *Dnmt1* KO (red tracks) MEF cells. Green bars above the gene map locate the CpG islands. The y-axis represents the DNA methylation measured as the percentage of reads that support the methylated state of each CpG (estimated methylation level). For each histone mark track, the y-axis represents the normalized level of ChIP-seq signal over the genomic regions. (**m**) Pie charts of the genomic structural composition of the H3K4me3 peaks *loci* specific to WT and *Dnmt1* KO cells. (**n**) Number of specific RNA-seq peaks in WT and *Dnmt1* KO cells

retroelements including LTR, LINE (long intergenic non-coding elements) and SINE (short intergenic non-coding elements). Exons, promoters and distal CpG-rich regions were elevated for WT peaks. This finding was confirmed by profiling RNA-seq peaks specific to WT and *Dnmt1* KO cells. LTR, LINE and SINE elements were significantly overrepresented in KO cells, while WT cells showed transcription enrichment within introns, intergenic regions and LTRs (Fig. 5n).

Discussion

We analyzed the crosstalk between DNA methylation and different chromatin marks over a broad range of regulatory regions including putative enhancers and promoters. Intriguingly, in contrast to the expected significantly negative correlation between DNA methylation and active chromatin marks, we found that H3K4me1 enrichment has significantly positive correlation with intermediate (in the range between 25 and 75%) DNA methylation at regulatory regions. Existing reports about H3K4me1 and DNA methylation claim both positive [21, 60] and negative [15–17] correlations. Our results re-conciliate the two seemingly contradictory observations zooming into the less studied fuzzy intermediate range between the wide-used extreme hyper and hypo DNA methylation states.

We observed anti-correlation between H3K4me1 and H3K4me3 enrichment at low (0 - 25%) and intermediate (25 - 75%) DNA methylation. While a negative correlation between an active epigenetic mark (H3K4me1) and a repressive one (DNA methylation) at high DNA methylation (>75%) seems acceptable, we tried to uncover the mechanism responsible for the anti-correlation between H3K4me1 and H3K4me3 within low and intermediate (0 - 75%) DNA methylated regulatory regions. We hypothesized "seesaw" dynamics between H3K4me1 and H3K4me3 in the 0 - 75% DNA methylation range: while the enrichment of one mark rises up, the enrichment of the other drops down. DNA methylation discriminates between enhancers and promoters, marked by H3K4me1 and H3K4me3, respectively: low methylated regions are H3K4me3 enriched, while those with intermediate DNA methylation levels are progressively H3K4me1 enriched. Additionally, the enrichment of H3K27ac, distinguishing active from primed enhancers, follows a plateau in the lower range of the intermediate DNA methylation level (25 - 35% DNA methylation), corresponding to active enhancers, and decreases linearly in the higher range of the intermediate DNA methylation (35 - 75%). Thus, the decrease of the DNA methylation switches smoothly the state of the enhancers from a primed to an active state.

Although simultaneous mono- and trimethylation of a single H3K4 are mutually excluded, different cells of a population can have different chromatin marks at the same genomic region, and such marks are dynamically changed through the enzymatic activity of methylases and demethylases. Therefore, we use the term "seesaw" rather than "mutual exclusion" to define such mechanism, which includes a balanced state with both marks enriched at lower levels.

The H3K4me1-H3K4me3 seesaw mechanism controlled by DNA methylation is valid for both pluripotent and differentiated cells, i.e. it is not cell type-specific. We scrutinized whether DNA methylation has a mechanistic function in the discrimination of H3K4me1 from H3K4me3 marked regulatory sites. While low and intermediate DNA methylated regions of WT ESCs are depleted of H3K4me3 peaks, knocking out of the H3K4me3 methyltransferase *CxxC1* domain of *Cfp1* increases the H3K4me3 enrichment in these regions. This can be linked to reduced DNA methylation of these regions after decreased DNA methyltransferase level in *Cfp1* KO cells [61]. Additionally, unmethylated CpG-rich regions are shown to be sufficient for CFP1 binding and H3K4me3 enrichment [55], hence SET1A/B complex can be deficient of unmethylated CpG recognition sites in absence of *Cfp1* that can result in H3K4me3 enrichment of even hypermethylated regions. Both possibilities suggest an active function of DNA methylation in regulating H3K4me3 deposition.

Reports implicitly confirm that blocking H3K4me3 would result in enriched H3K4me1. The WD (glycine-histidine) repeat domain 5, *Wdr5*, a core member of mammalian H3K4me3 methyltransferases, interacts with H3K4me2 and mediates transition to H3K4me3 [62]. Immunoblot of *Wdr5* knockdown ESCs shows enriched H3K4me1 in response to depleted H3K4me3, which is due to increased DNA demethylation of H3K4me2 [63]. Furthermore, H3K4me3 enrichment coincides with H3K4me1 depletion after knock down of the histone demethylase *Kdm5c* [64]. The same report demonstrates that H3K4me1 is depleted at H3K4me3 peak summits. H3K4me1 peaks have higher frequency of in absence of the H3K4me3 methyltransferase domain CXXC7 of Mll1 [33]. Analysis of MEF cells in absence of *Dnmt1* provides further evidence of the functional role of DNA methylation in differentiated cells. Significant loss of DNA methylation within H3K4me3 peaks of *Dnmt1* KO cells compared to the WT MEF shows that DNA hypomethylation is a precondition for H3K4me3 deposition.

Switching between H3K4me1 and H3K4me3 discloses the role of DNA methylation in discriminating promoters and enhancers. H3K4me3 is shown to facilitate access and assembly of the RNA polymerase 2, Pol2, as well as to promote transcriptional initiation through binding of TFIID [64]. On one hand, transcriptional activity is also shown to influence H3K4me3 enrichment [65]. Our results indicate a dramatic transcriptional activity coincident with H3K4me3 enrichment in consequence of DNA hypomethylation in *Dnmt1* KO cells, which is not limited to gene coding regions but also overrepresented within non-coding and intergenic regions, particularly the retroelements. On the other hand, H3K4me1 is specifically recognized by a number of chromatin-interacting proteins [66] and is also shown to guide a pioneer transcription factor *Foxa1* for initiating enhancer complex formation. Depleted H3K4me1 by overexpression of H3K4 demethylase is followed by abrogated recruitment of the transcription factor [67], suggesting a causal role for H3K4me1 in enhancer priming.

We propose that the seesaw mechanism operates as follows: DNA unmethylated CpG-rich regions provide the basis for H3K4me3 methyltransferases (i.e. SET1A/B, MLL1/2) to bind and functionally mark the area as a promoter. This can be done by increasing the conversion of H3K4me1/2 to H3K4me3 that results in increased H3K4me3 in parallel with decreased H3K4me1, which leads to the seesaw mechanism (Fig. 6). The intermediate DNA methylation levels can reduce the binding of the CpG sensitive CXXC domain of H3K4me3 methylases, while giving access to CXXC-free H3K4me1 methylases, which results in limited conversion of H3K4me1 to H3K4me3, and marks the *locus* as enhancer by H3K4me1 enrichment. This mechanism driven by H3K4 methyltransferases is complementary to the regulatory role of H3K4 demethylases Kdm5b/c in the discrimination between promoters and enhancers [64, 68]. Thus the H3K4me3 methyltransferases and the H3K4 demethylases make possible the reversible seesaw between enhancers and promoters.

The activation of the seesaw mechanism occurs for low to medium DNA methylation levels. When DNA methylation is high, both H3K4me1 and H3K4me3 are low, marking an inactive genomic region. When DNA methylation decreases to intermediate level, H3K4me1 is high and H3K4me3 is low, marking an enhancer region. Finally, when DNA methylation decreases to a low level, H3K4me1 is low and H3K4me3 is high, marking a promoter region. We can summarize these observations into a rule of thumb of one-out-of-three methylation marks: "In each genomic region only one out of the following three methylation marks {DNA methylation, H3K4me1, H3K4me3} is high: if it is DNA methylation, the region is inactive, if it is H3K4me1, the region is an enhancer, and if it is H3K4me3, the region is a promoter".

Conclusions

To explain H3K4me1 depletion at high levels of DNA methylation, we suggest two possible mechanisms: (I) A passive mechanism, in which the heterochromatin structure of the genome that is incorporated with stable hypermethylation [69] can make chromatin inaccessible for many DNA or chromatin-bound proteins, TFs and potentially the H3K4me1 histone methyltransferases. (II) An active mechanism in which the recruitment of TFs by H3K4me1 leads to DNA hypomethylation and enhancer priming [67]. Interestingly, binding of some TFs causes DNA hypomethylation at low to intermediate levels in the population [1, 56], which is in agreement with our observation of enriched H3K4me1 at intermediate DNA methylation.

Additionally, our findings suggest a potential mechanism for inheritance of histone codes, particularly H3K4 methylation, during cell division: While the machinery maintaining DNA methylation during cell division is well-studied [70], little is known about how histone codes are inherited by the daughter cells. H3K4me3 methylase is shown to remain associated with the newly replicated DNA through unknown mechanisms during cell division, although the histones carrying the H3K4me3 mark are replaced by unmethylated histone 3 (H3) after DNA replication [71]. We suggest that re-established DNA methylation of the nascent DNA could provide complementary information on how H3K4 methylases transmit the H3K4 methylation patterns from the parent

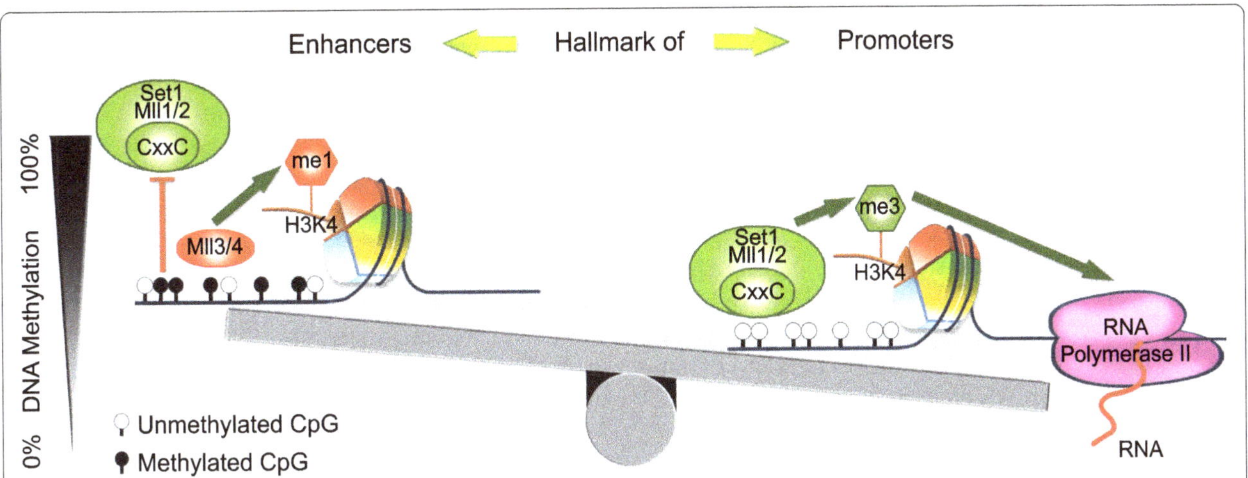

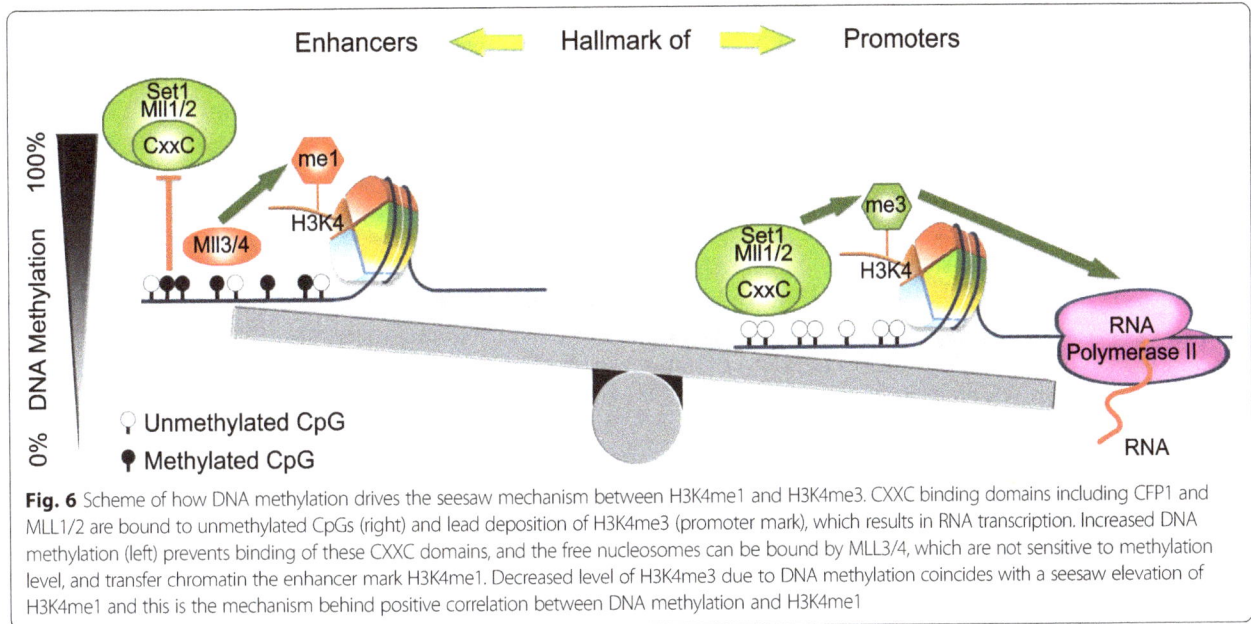

Fig. 6 Scheme of how DNA methylation drives the seesaw mechanism between H3K4me1 and H3K4me3. CXXC binding domains including CFP1 and MLL1/2 are bound to unmethylated CpGs (right) and lead deposition of H3K4me3 (promoter mark), which results in RNA transcription. Increased DNA methylation (left) prevents binding of these CXXC domains, and the free nucleosomes can be bound by MLL3/4, which are not sensitive to methylation level, and transfer chromatin the enhancer mark H3K4me1. Decreased level of H3K4me3 due to DNA methylation coincides with a seesaw elevation of H3K4me1 and this is the mechanism behind positive correlation between DNA methylation and H3K4me1

to the daughter cells in a more reliable manner. This would help to establish a framework for the inheritance of chromatin marks and a genomic map of promoters and enhancers that are inherited by DNA methylation.

The DNA methylation regulating a H3K4me1 - H3K4me3 seesaw mechanism has implications in developmental biology, cellular reprogramming, cancer and aging. It changes the balance in differentiation- and pluripotency-related genes. The promiscuous DNA hypomethylation of cancer cells can disrupt the normal deposition pattern of promoter and enhancer chromatin marks, followed by aberrant transcription of silent genes and non-coding regions. The disturbance in DNA methylation can change also the balance between enhancers and promoters in aging related genes.

Methods

Data sources

ChIP-seq data of genome-wide maps of chromatin marks, Pol2 and gene expression regulators, RNA-seq and different forms of bisulfite sequencing (Bis-seq) including Whole-Genome Bisulfite Sequencing (WGBS), Reduced Representation Bisulfite Sequencing (RRBS) and Tet-assisted bisulfite sequencing (TAB-seq), used for measurement of DNA hydroxymethylation 5hmC) were obtained from several GEO or ArrayExpress datasets (Table 1). The coordinates of putative enhancers of 19 mouse tissues and cell types were taken from Tan et al. [72].

NGS data preprocessing

The mouse reference genome assembly mm9 was used for the whole analysis, and the University of California, Santa Cruz (UCSC) liftOver tool was applied for address conversion of some datasets already aligned to mm8. We developed a pipelined script to download next-generation sequencing data in SRA or other available formats, converted them to fastq format, aligned them using Bowtie2 [73] and identified statistically significant peaks compared to the whole cell extract (WCE) inputs when available. The processing of the data has been performed as follows: The raw fastq files were aligned to the reference genome using Bowtie2, and then converted to the genome coverage wiggle format using a pipeline of several commands including bamToBed, genomeCoverageBed, bedGraphToBigWig and finally bigWigToWig commands of BEDTools [74] and UCSC Genome Browser toolkits [75]. We then used MACS2 [76] for peak finding and MAnorm [77] for normalization of genome coverage data. Hence the genome coverage values that are depicted in the figures are the normalized total number of NGS reads that are aligned to each genomic region.

DNA methylation data processing

To assess the degree of DNA methylation of each CpG, we used our parallel processing pipeline software for automatic analysis of bisulfite sequencing data (P3BSseq) [78]. We define a CpG as 100% methylated when all the reads that are aligned to this CpG in a genomic region are 100% methylated (the reads in the CpG *loci* are of the form CpG rather than TpG that is a result of C → T conversion for unmethylated CpGs following Sodium-bisulfite treatment). We set a minimum read CpG coverage criterion, keeping only the reads having at least 10 CpG dinucleotides with coverage of minimum 5× in the bisulfite sequencing (Bis-seq) data. The CpG dinucleotides with minimum 5-reads coverage were kept in each Bis-seq data, and the average methylation ratios over CpGs inside the 1 kb window were assigned as DNA methylation level of each site.

WGBS data cannot discriminate directly between 5mC and 5hmC levels of a CpG but the sum $5{\times}mC_{WGBS}$ of 5mC and 5hmC, i.e. $5{\times}mC_{WGBS} = 5mC + 5hmC$. However, since TAB-seq measures specifically the 5hmC level of each CpG, we estimate the 5mC level of a CpG by subtracting the TAB-seq measured $5hmC_{TAB}$ ratios [45, 72, 79–81] from the total DNA methylation level (5mC + 5hmC measured in WGBS experiments) as

$$5mC = 5xmC_{WGBS} - 5hmC_{TAB} \quad (1)$$

which allows us to evaluate specifically 5mC ratios at single-base resolution. To discriminate CpGs with 5hmC from those that do not have 5hmC (those that have only 5mC), we performed (Eq. 1) calculation only on CpGs with significantly reliable 5hmC levels (False Discovery Rate (FDR) < 0.05 and Phred quality score ≥ 20).

Assessment of the enrichment of epigenetic marks in promoters and enhancers

When studying the difference of enrichment of the several epigenetic marks between promoters and enhancers, promoter regions are relatively easy to define based on the position of the transcription start sites (TSSs). However, enhancers do not have well defined positions, and they can occur in almost any genomic region. Therefore, in order to obtain a comprehensive picture of the discrimination of epigenetic marks between promoters and enhancers, in our results we re-annotated the enhancer positions into 20 different genomic categories. Thus, we created a draft list of 428,297 non-overlapped 1 kb genomic sites centered over TSSs (–/+ 500 bp of the TSS) of genes of the National Center for Biotechnology Information (NCBI) Reference Sequence Database (RefSeq), and cross-tissue putative enhancers of 19 mouse cell types [45, 72, 79–81].

Since the enhancers can occur in any genomic region, we furthermore mapped them into 20 different categories: including repeat-associated regions {Short Interspersed Nuclear Element (SINE), Long Interspersed Nuclear Element (LINE), Simple repeat, Long Terminal Repeat (LTR), DNA Transposon, Low complexity, DNA Transposon, Satellite}, Intergenic, Intron, non-coding, CpG island, and coding regions {Exon, 5'UTR, 3'UTR, and transcription termination site (TTS)}, regions associated with different types RNA species {rRNA, scRNA, snRNA, tRNA}, and regions with "Unknown" annotation. The TSSs are 1 kb genomic sites centered (–/+ 500 bp) over the TTS. Additionally, we have created the category "Others", that appears across the different results. In this category, we merged the cases with less than 100 members. From the initial non-overlapped 1 kb genomic sites, we filtered 210,048 sites having at least 10 CpG dinucleotides with minimum 5x coverage in the Bis-seq data. We established a feature $s \times r$matrix M with the s = 210,048 sites in the rows, and different r = 13 gene regulation features including epigenetic marks {H3K27me3, H3K36me3, H3K9me3, H3K20me3, H3K4me2, H3K4me1, H3K9ac, H3K4me3, H3K27ac}, protein bindings {P300} and other genomic features {H3, CTCF, Pol2} in the columns. Enrichments of chromatin marks within each site were calculated as the average depth of reads within the 1 kb window.

Correlation analysis of epigenetic marks with DNA methylation

All regulatory sites used in this study were classified based on the genomic structure using the annotatePeaks.pl script of the HOMER suite [73, 82]. For each of the 21 classes (one promoter and 20 enhancers classes) we computed the Spearman's rank correlation coefficients ρ between the DNA methylation and the enrichment of 13 gene regulation features {H3K27me3, H3K36me3, H3K9me3, H3K20me3, H3K4me2, H3K4me1, H3K9ac, H3K4me3, H3K27ac, P300, H3, CTCF, Pol2}, creating a correlation $c \times r$ matrix R with the c = 21 classes (one corresponding to promoters and 20 to enhancers) in the rows, and different r = 13 gene regulation features in the columns. Next, all sites were sorted into 100 bins according to DNA methylation levels (i.e., bin_1-included those sites with DNA methylation level between 0 and 1%) and split into two matrices of correlations, one, R^{Hyper} with DNA hypermethylated sites (DNA methylation >50%), and another, R^{Hypo} with DNA hypomethylated sites (DNA methylation ≤50%). The results were represented in heat maps after hierarchical clustering of rows and columns of the matrices of correlation R^{Hyper} and R^{Hypo}. For each of the 13 gene regulation features, the enrichments were averaged on all sites assorted in the same bin, and the results were linearly scaled between 0 and 1. The Integrative Genomics Viewer (IGV) was used for *locus*-specific representation of ChIP-seq and DNA methylation data [83].

Peak analysis of methyl-binding proteins and chromatin marks

We used MACS [84] to calculate the fraction of peaks with DNA methylation level above 95% over the total number of peaks. Additionally, peaks of each pair of signals were compared to find overlaps. Two peaks p_{Si} and p_{Sj} of two different signals Si and Sj, were considered overlapped if some genomic region (even as small as a single nucleotide) was included in both. Thus we define an overlap binary variable o_{SiSj}, equal to 1, if $p_{Si} \cap p_{Sj} \geq 1$, and 0, otherwise. For each pair of signals S_i, S_j, with $\#p_{Si}$

and $\#p_{Sj}$ number of peaks, respectively, we calculated their percentage of overlap O_{SiSj} as the number of overlapped peaks $\#o_{SiSj}$ divided by the number of peaks of the signal with smaller number of peaks, in %, i.e.

$$O_{SiSj} = 100\#o_{SiSj}/\min\left(\#p_{Si}, \#p_{Sj}\right) \quad (2)$$

and represented it in a hierarchical clustered heatmap.

Discrimination between the impact of DNA 5mC and 5hmC on H3K4 methylation

To study which of the DNA cytosine methylations (5mC or 5hmC) have stronger impact on the level of H3K4 methylation, we modeled such impact with probability theory. We observed initially that the 5hmC level (measured by TAB-seq) is gained on putative enhancers that have also higher 5mC levels (estimated by Eq. 1), hindering to consider 5mC or 5hmC as independent variables. Assuming *H3K4me1* and *H3K4me3* to be the probabilistic events of significant alternations in H3K4me1 and H3K4me3, respectively, and *5mC* and *5hmC* as the events of change in 5mC and 5hmC levels, respectively, we compared the conditional probabilities *P(H3K4me1|5mC)*, *P(H3K4me3|5mC)*, *P(H3K4me1|5hmC)*, and *P(H3K4me3|5hmC)*. Therefore, we computed the conditional probability of either H3K4me1 or H3K4me3 as a response of the 5hmC as the variable, under fixed 5mC distribution, and vice versa, 5mC as the variable, under fixed 5hmC distribution. Namely, to discriminate the possible relationship between the H3K4me1 and H3K4me3 chromatin marks and 5mC versus 5hmC, we compared alternations of one form of cytosine methylation (5mC or 5hmC) when the other form was constant (5hmC or 5mC). This is a challenging task since alternations in 5hmC is usually coincident with changes in 5mC level. To address this issue, we used a probabilistic approach to identify two groups of putative enhancers to compare for each form of cytosine methylation, (4 groups in total). Each pair of enhancer groups had a similar distribution of one form of methylation (called "background"), but altered level of the other form of methylation ("altered"). Hence, we could study the possible effect of the "altered" form of methylation independently from the "background" form of methylation.

To study the interplay between H3K4me1 or H3K4me3, and 5hmC as the altered methylation (with 5mC as the background), we built the 5hmC altered group considering two groups of enhancers {5hmC+, 5hmC-} with significantly different 5hmC but equal 5mC distributions. The first group, representing the presence of the 5hmC signal called 5hmC+, consists of 2501 putative enhancers with a minimum of 20 CpG dinucleotides and an average of 5hmC between 15 and 30% within a 1 kb window. The second group, representing the absence of 5hmC signal is called 5hmC-, has the same number of putative enhancers as the 5hmC+ group, and the same minimum of 20 CpGs but with an average of 5hmC 0% within a 1 kb window. To eliminate the possible effect in the 5mC background of 5mC alternations between the two groups, for each enhancer in the 5hmC+ group we selected an enhancer with the constraints of the other group (20 CpGs and 0% 5hmC) in such a way that the 5mC levels of the two enhancers were the closest possible, thus ensuring similar 5mC background distribution in the two groups. Thus, both groups have mean 5mC equal to 68% (p-value = 1). Still, 5hmC levels were significantly different. The mean 5hmC level in 5hmC+ and 5hmC-group was 17 and 0%, respectively (p-value <1e-15). We did not select the 5hmC + group from higher levels of 5hmC due to the lack of sufficient number of enhancers fulfilling the criteria for comparison of the two group.

A similar approach was used to study the interplay of either H3K4me1 or H3K4me3 with 5mC, under fixed 5hmC. The 5mC-presence group, 5mC+, consisted of putative enhancers with 5mC between 15 and 30% (the same range used for the 5hmC+ group), while for the 5mC-absence group, 5mC-, we used a slightly relaxed criterion of 5% as the maximum 5mC level, since there were too few enhancers with absolutely 0% 5mC within a 1 kb window. Both 5mC+ and 5mC- groups had zero level for the 5hmC background. The mean 5mC levels in the 5mC+ and 5mC- group were 22 and 3%, respectively (p-value <1e-15). There were 1791 and 1365 putative enhancers in the 5mC + and 5mC- groups, respectively, which were all putative enhancers that met the above criteria. The distribution of H3K4me1 and H3K4me3 enrichments were estimated for each of the four groups (5hmC+, 5hmC-, 5mC+ and 5mC-) of enhancers.

Graphical representation of 3-dimensional information

To better represent 3-dimensional genomics data, we developed R functions to produce scatter plots with automatic conversion of the third dimension to the color spectrum of data points. This substantially improved the insight into the data. These functions first eliminated outliers or incomplete data. We used the following criterion to remove data points as outliers to ensure at least 98% of the data are kept for the analysis. If the lower 1% and upper 99% quartiles were Q_1 and Q_{99}-respectively, we defined IQ as the length of the interval between them: $IQ = Q_{99} - Q_1$. A data point x is considered an outlier if either $x\text{-}Q_{99} > 1.1IQ$, or $Q_1\text{-}x > 1.1IQ$, i.e. x is outside the interval between Q_1 and Q_{99} at a distance higher than 10% of such interval. The incomplete data are those lacking required CpGs in the window around a genomic site to infer the DNA methylation level. We then sorted all data points according to the third dimension into a particular number of bins (identified as function argument) to produce equal-width bins in the whole range of the 3rd dimension. Data points of each bin were subsequently assigned the same color of the whole spectrum. The data points were interpolated with a triangle-based linear method and projected them onto a 3-dimensional surface to ensure that the visual representation was not biased to the points overlaying other points. The scatter plots representing H3K4me1, H3K4me3 and DNA methylation in different cell types were produced by this method.

Peak intersection analysis between WT and KO cells

Cross-normalization of processed ChIP-seq data was performed with the MAnorm software [77]. The same software was used to identify the common and specific peaks for pairs of cell types (WT versus KO). A peak was called specific to one cell type if the normalized enrichment value had more than a 2-fold change between the two cell types with a p-value <1e-5. Transcriptional activity on each peak was estimated by $\log_2(1 + m)$, where m is the maximum number of RNA-seq reads aligned to the same genomic position inside the peak. Peaks specifically enriched of H3K4me3 and transcription were classified into 15 categories (Intron, LTR, Exon, Intergenic, 3'UTR, SINE, CpG island, Promoter, TSS, LINE, Simple repeat, 5'UTR, Non-coding, Low complexity, DNA Transposition) according to the genomic region using the HOMER suite. The small classes with less than 100 peaks were merged and labeled as an additional 16th category "Others".

R packages used in the analysis

We used the following R packages in our analysis: scales, intervals, modeest, bioDist, Hmisc, e1071, rpart, data.table, abind, plyr [85], raster, gplots, ggplot2 [86], pheatmap, reshape [87], multicore, zoo [88], directlabels, Biobase [89], GEOquery [90], limma [91].

Abbreviations

5hmC: 5-hydroxymethylcytosine; 5mC: 5-methylcytosine; ADD: ATRX-DNMT3-DNMT3L; Bis-seq: Bisulfite sequencing; *Btf3*: Basic transcription factor 3; *Cchcr1*: Coiled-coil α-helical rod protein 1; CFP1: CXXC finger protein 1; ChIP-seq: Chromatin Immunoprecipitation sequencing; ChIP-seq: Chromatin immunoprecipitation sequencing; COMPASS: COMplex of Proteins Associated with Set1; CpGMM: CpG methylation motif; DMT: DNA Methyl-Transferase; *Eif4g2*: Eukaryotic translation initiation factor 4, gamma 2; EMBL-EBI: European Molecular Biology Laboratory - European Bioinformatics Institute; ESC: Embryonic stem cell; FDR: False Discovery Rate; GEO: Gene Expression Omnibus; GO: Gene Ontology; *Gys2*: Glycogen synthase 2; H3: Histone 3; HMT: Histone Methyl-Transferase; IGV: Integrative Genomics Viewer; *Jade1*: Jade family PHD finger 1; Kb: Kilo base; KO: Gene knockout; *Ldhc*: Lactate dehydrogenase C; LINE: Long Interspersed Nuclear Element; LTR: Long Terminal Repeat; MBD: Methyl-CpG-Binding Domain; Med: Median DNA methylation; MEF: Mouse embryonic fibroblast; NCBI: National Center for Biotechnology Information; *P4hb*: Prolyl 4-hydroxylase subunit β; *Pawr*: Pro-apoptotic WT1 regulator; *Pkm*: Muscle pyruvate kinase; Pol2: RNA polymerase 2; *Ppp4c*: Protein phosphatase 4, catalytic subunit; RNA-seq: RNA sequencing; RRBS: Reduced Representation Bisulfite Sequencing; SINE: Short Interspersed Nuclear Element; TAB-seq: Tet-assisted bisulfite sequencing; *Tcf19*: Transcription factor 19; *Tet1*: The epigenetic regulators tet methylcytosine dioxygenase 1; TF: Transcription factor; TSS: Transcription Start Site; TTS: Transcription Termination Site; WCE: Whole cell extract; WGBS: Whole-Genome Bisulfite Sequencing; WT: Wild type

Acknowledgments

We thank Farzane Emami for creating the artwork of the scientific concept. Data analysis was performed using computing cluster facilities of the Institute for Research in Fundamental Sciences, Tehran, and of the Biodonostia Health Research Institute, San Sebastian.

Funding

DG and MJ A-B. have been supported by grants DFG10/15, DFG15/15 and DFG141/16 from Diputación Foral de Gipuzkoa, Spain, Ministry of Economy and Competitiveness, Spain, MINECO grants PI16/01430 and BFU 2016–7798-P and funds from IKERBASQUE, Basque Foundation for Science, Spain. Also, this study was funded by grants from Royan Institute, the Iran National Science Foundation (INSF), and Iran Science Elites Federation to HB.

Authors' contributions

AS-Z, MS, and MJA-B designed the project. KA, MT, RJT, DG, and HRS provided biological insights and checked the results. HP and DG contributed to the statistical analysis. AS-Z performed the computational analysis. HC checked the plots. HB and MS supervised the project. AS-Z, DG and MJA-B wrote the manuscript. All authors have read and approved the manuscript.

Competing interests

The authors declare that they have no competing interests.

Author details

[1]Department of Bioinformatics, Institute of Biochemistry and Biophysics, University of Tehran, Tehran, Iran. [2]Computer Science Department, Colorado State University, Fort Collins, CO, USA. [3]Department of Stem Cells and Developmental Biology, Cell Science Research Center, Royan Institute for Stem Cell Biology and Technology, ACECR, Tehran, Iran. [4]Department of Computer Engineering, Sharif University of Technology, Tehran, Iran. [5]Computational Biology and Systems Biomedicine, Biodonostia Health Research Institute, 20014 San Sebastián, Spain. [6]Department of Cell and Developmental Biology, Max Planck Institute for Molecular Biomedicine, Münster, Germany. [7]School of Mathematics, Statistics and Computer Science, College of Science, University of Tehran, Tehran, Iran. [8]School of Biological Sciences, Institute for Research in Fundamental Sciences (IPM), Tehran, Iran. [9]Illumina Inc., San Diego, USA. [10]Medical Faculty, University of Münster, Münster, Germany. [11]National Institute of Genetic Engineering and Biotechnology (NIGEB), Tehran, Iran. [12]Department of Developmental Biology, University of Science and Culture, Tehran, Iran. [13]Computational Biology and Bioinformatics Group, Max Planck Institute for Molecular Biomedicine, Münster, Germany. [14]IKERBASQUE, Basque Foundation for Science, 48011 Bilbao, Spain.

References

1. Luu P-L, Schöler HR, Araúzo-Bravo MJ. Disclosing the crosstalk among DNA methylation, transcription factors, and histone marks in human pluripotent cells through discovery of DNA methylation motifs. Genome Res. 2013; 23(12):2013–29.
2. Lienert F, Wirbelauer C, Som I, Dean A, Mohn F, Schübeler D. Identification of genetic elements that autonomously determine DNA methylation states. Nat Genet. 2011;43(11):1091–7.
3. Zentner GE, Scacheri PC. The chromatin fingerprint of gene enhancer elements. J Biol Chem. 2012;287(37):30888–96.
4. Bulger M, Groudine M. Functional and mechanistic diversity of distal transcription enhancers. Cell. 2011;144(3):327–39.
5. Jones PA. Functions of DNA methylation: islands, start sites, gene bodies and beyond. Nat Rev Genet. 2012;13(7):484–92.
6. Koh KP, Rao A. DNA methylation and methylcytosine oxidation in cell fate decisions. Curr Opin Cell Biol. 2013;25(2):152–61.
7. Heintzman ND, Stuart RK, Hon G, Fu Y, Ching CW, Hawkins RD, et al. Distinct and predictive chromatin signatures of transcriptional promoters and enhancers in the human genome. Nat Genet. 2007;39(3):311–8.
8. Banerji J, Olson L, Schaffner W. A lymphocyte-specific cellular enhancer is located downstream of the joining region in immunoglobulin heavy chain genes. Cell. 1983;33(3):729–40.
9. Calo E, Wysocka J. Modification of enhancer chromatin: what, how, and why? Mol Cell. 2013;49(5):825–37.
10. Heintzman ND, Hon GC, Hawkins RD, Kheradpour P, Stark A, Harp LF, et al. Histone modifications at human enhancers reflect global cell-type-specific gene expression. Nature. 2009;459(7243):108–12.
11. Wang Z, Zang C, Rosenfeld JA, Schones DE, Barski A, Cuddapah S, et al. Combinatorial patterns of histone acetylations and methylations in the human genome. Nat Genet. 2008;40(7):897–903.
12. Barski A, Cuddapah S, Cui K, Roh T-Y, Schones DE, Wang Z, et al. High-resolution profiling of histone methylations in the human genome. Cell. 2007;129(4):823–37.
13. Kagey MH, Newman JJ, Bilodeau S, Zhan Y, Orlando DA, van Berkum NL, et al. Mediator and cohesin connect gene expression and chromatin architecture. Nature. 2010;467(7314):430–5.
14. Heinz S, Romanoski CE, Benner C, Glass CK. The selection and function of cell type-specific enhancers. Nat Rev Mol Cell Biol. 2015;16(3):144–54.
15. Gifford CA, Ziller MJ, Gu H, Trapnell C, Donaghey J, Tsankov A, et al. Transcriptional and epigenetic dynamics during specification of human embryonic stem cells. Cell. 2013;153(5):1149–63.
16. Baubec T, Ivánek R, Lienert F, Schübeler D. Methylation-dependent and -independent genomic targeting principles of the MBD protein family. Cell. 2013;153(2):480–92.
17. Xie R, Everett LJ, Lim H-W, Patel NA, Schug J, Kroon E, et al. Dynamic chromatin remodeling mediated by polycomb proteins orchestrates pancreatic differentiation of human embryonic stem cells. Cell Stem Cell. 2013;12(2):224–37.
18. Ong C-T, Corces VG. Enhancers: emerging roles in cell fate specification. EMBO Rep. 2012;13(5):423–30.
19. Weber M, Hellmann I, Stadler MB, Ramos L, Pääbo S, Rebhan M, et al. Distribution, silencing potential and evolutionary impact of promoter DNA methylation in the human genome. Nat Genet. 2007;39(4):457–66.
20. Meissner A, Mikkelsen TS, Gu H, Wernig M, Hanna J, Sivachenko A, et al. Genome-scale DNA methylation maps of pluripotent and differentiated cells. Nature. 2008;454(7205):766–70.
21. Zhang X, Bernatavichute YV, Cokus S, Pellegrini M, Jacobsen SE. Genome-wide analysis of mono-, di- and trimethylation of histone H3 lysine 4 in Arabidopsis Thaliana. Genome Biol. 2009;10(6):R62.
22. Hodges E, Smith AD, Kendall J, Xuan Z, Ravi K, Rooks M, et al. High definition profiling of mammalian DNA methylation by array capture and single molecule bisulfite sequencing. Genome Res. 2009;19(9):1593–605.
23. Laurent L, Wong E, Li G, Huynh T, Tsirigos A, Ong CT, et al. Dynamic changes in the human methylome during differentiation. Genome Res. 2010;20(3):320–31.
24. Cheng X, Blumenthal RM. Introduction–epiphanies in epigenetics. Prog Mol Biol Transl Sci. 2011;101:1–21.
25. Ooi SKT, Qiu C, Bernstein E, Li K, Jia D, Yang Z, et al. DNMT3L connects unmethylated lysine 4 of histone H3 to de novo methylation of DNA. Nature. 2007;448(7154):714–7.
26. Otani J, Nankumo T, Arita K, Inamoto S, Ariyoshi M, Shirakawa M. Structural basis for recognition of H3K4 methylation status by the DNA methyltransferase 3A ATRX–DNMT3–DNMT3L domain. EMBO Rep. 2009; 10(11):1235–41.
27. Zhang Y, Jurkowska R, Soeroes S, Rajavelu A, Dhayalan A, Bock I, et al. Chromatin methylation activity of Dnmt3a and Dnmt3a/3L is guided by interaction of the ADD domain with the histone H3 tail. Nucleic Acids Res. 2010;38(13):4246–53.
28. Birke M, Schreiner S, García-Cuéllar M-P, Mahr K, Titgemeyer F, Slany RK. The MT domain of the proto-oncoprotein MLL binds to CpG-containing DNA and discriminates against methylation. Nucleic Acids Res. 2002;30(4):958–65.
29. Ayton PM, Chen EH, Cleary ML. Binding to nonmethylated CpG DNA is essential for target recognition, transactivation, and myeloid transformation by an MLL oncoprotein. Mol Cell Biol. 2004;24(23):10470–8.
30. Allen MD, Grummitt CG, Hilcenko C, Min SY, Tonkin LM, Johnson CM, et al. Solution structure of the nonmethyl-CpG-binding CXXC domain of the leukaemia-associated MLL histone methyltransferase. EMBO J. 2006;25(19): 4503–12.
31. van Nuland R, Smits AH, Pallaki P, Jansen PWTC, Vermeulen M, Timmers HTM. Quantitative dissection and stoichiometry determination of the human SET1/MLL histone methyltransferase complexes. Mol Cell Biol. 2013; 33(10):2067–77.
32. Schuettengruber B, Martinez A-M, Iovino N, Cavalli G. Trithorax group proteins: switching genes on and keeping them active. Nat Rev Mol Cell Biol. 2011;12(12):799–814.
33. Herz H-M, Mohan M, Garruss AS, Liang K, Takahashi Y-H, Mickey K, et al. Enhancer-associated H3K4 monomethylation by Trithorax-related, the drosophila homolog of mammalian Mll3/Mll4. Genes Dev. 2012;26(23):2604–20.
34. Lee JH, Voo KS, Skalnik DG. Identification and characterization of the DNA binding domain of CpG-binding protein. J Biol Chem. 2001;276(48):44669–76.
35. Lee J-H, Skalnik DG. CpG-binding protein (CXXC finger protein 1) is a component of the mammalian Set1 histone H3-Lys4 methyltransferase complex, the analogue of the yeast Set1/COMPASS complex. J Biol Chem. 2005;280(50):41725–31.
36. Lee J-H, Tate CM, You J-S, Skalnik DG. Identification and characterization of the human Set1B histone H3-Lys4 methyltransferase complex. J Biol Chem. 2007;282(18):13419–28.
37. Guenther MG, Levine SS, Boyer LA, Jaenisch R, Young RAA. Chromatin landmark and transcription initiation at most promoters in human cells. Cell. 2007;130(1):77–88.
38. Mikkelsen TS, Ku M, Jaffe DB, Issac B, Lieberman E, Giannoukos G, et al. Genome-wide maps of chromatin state in pluripotent and lineage-committed cells. Nature. 2007;448(7153):553–60.
39. Clouaire T, Webb S, Skene P, Illingworth R, Kerr A, Andrews R, et al. Cfp1 integrates both CpG content and gene activity for accurate H3K4me3 deposition in embryonic stem cells. Genes Dev. 2012;26(15):1714–28.
40. Shilatifard A. The COMPASS family of histone H3K4 methylases: mechanisms of regulation in development and disease pathogenesis. Annu Rev Biochem. 2012;81:65–95.
41. Lee J-H, Skalnik DG. Wdr82 is a C-terminal domain-binding protein that recruits the Setd1A Histone H3-Lys4 methyltransferase complex to transcription start sites of transcribed human genes. Mol Cell Biol. 2008;28(2):609–18.
42. Wu M, Wang PF, Lee JS, Martin-Brown S, Florens L, Washburn M, et al. Molecular regulation of H3K4 trimethylation by Wdr82, a component of human Set1/COMPASS. Mol Cell Biol. 2008;28(24):7337–44.
43. Deaton AM, Bird A. CpG islands and the regulation of transcription. Genes Dev. 2011;25(10):1010–22.
44. Clouaire T, de Las Heras JI, Merusi C, Stancheva I. Recruitment of MBD1 to target genes requires sequence-specific interaction of the MBD domain with methylated DNA. Nucleic Acids Res. 2010;38(14):4620–34.
45. Shen Y, Yue F, McCleary DF, Ye Z, Edsall L, Kuan S, et al. A map of the cis-regulatory sequences in the mouse genome. Nature. 2012;488(7409):116–20.
46. Lizio M, Harshbarger J, Shimoji H, Severin J, Kasukawa T, Sahin S, et al. Gateways to the FANTOM5 promoter level mammalian expression atlas. Genome Biol. 2015;16:22.
47. Sarraf SA, Stancheva I. Methyl-CpG binding protein MBD1 couples histone H3 methylation at lysine 9 by SETDB1 to DNA replication and chromatin assembly. Mol Cell. 2004;15(4):595–605.
48. Li L, Chen B-F, Chan W-Y. An epigenetic regulator: methyl-CpG-binding domain protein 1 (MBD1). Int J Mol Sci. 2015;16(3):5125–40.

49. Günther K, Rust M, Leers J, Boettger T, Scharfe M, Jarek M, et al. Differential roles for MBD2 and MBD3 at methylated CpG islands, active promoters and binding to exon sequences. Nucleic Acids Res. 2013;41(5):3010–21.
50. Shimbo T, Du Y, Grimm SA, Dhasarathy A, Mav D, Shah RR, et al. MBD3 localizes at promoters, gene bodies and enhancers of active genes. PLoS Genet. 2013;9(12):e1004028.
51. Luo Z, Gao X, Lin C, Smith ER, Marshall SA, Swanson SK, et al. Zic2 is an enhancer-binding factor required for embryonic stem cell specification. Mol Cell. 2015;57(4):685–94.
52. Guibert S, Weber M. Functions of DNA methylation and hydroxymethylation in mammalian development. Curr Top Dev Biol. 2013;104:47–83.
53. Wang X, Xuan Z, Zhao X, Li Y, Zhang MQ. High-resolution human core-promoter prediction with CoreBoost_HM. Genome Res. 2009;19(2):266–75.
54. Hu D, Garruss AS, Gao X, Morgan MA, Cook M, Smith ER, et al. The Mll2 branch of the COMPASS family regulates bivalent promoters in mouse embryonic stem cells. Nat Struct Mol Biol. 2013;20(9):1093–7.
55. Thomson JP, Skene PJ, Selfridge J, Clouaire T, Guy J, Webb S, et al. CpG islands influence chromatin structure via the CpG-binding protein Cfp1. Nature. 2010;464(7291):1082–6.
56. Stadler MB, Murr R, Burger L, Ivanek R, Lienert F, Schöler A, et al. DNA-binding factors shape the mouse methylome at distal regulatory regions. Nature. 2011;480(7378):490–5.
57. Carlone DL, Lee J-H, Young SRL, Dobrota E, Butler JS, Ruiz J, et al. Reduced genomic cytosine methylation and defective cellular differentiation in embryonic stem cells lacking CpG binding protein. Mol Cell Biol. 2005;25(12):4881–91.
58. Kuntz S, Kieffer E, Bianchetti L, Lamoureux N, Fuhrmann G, Viville S. Tex19, a mammalian-specific protein with a restricted expression in pluripotent stem cells and germ line. Stem Cells Dayt Ohio. 2008;26(3):734–44.
59. Reddington JP, Perricone SM, Nestor CE, Reichmann J, Youngson NA, Suzuki M, et al. Redistribution of H3K27me3 upon DNA hypomethylation results in de-repression of Polycomb target genes. Genome Biol. 2013;14(3):R25.
60. Teng L, Tan K. Finding combinatorial histone code by semi-supervised biclustering. BMC Genomics. 2012;13:301.
61. Butler JS, Palam LR, Tate CM, Sanford JR, Wek RC, Skalnik DG. DNA Methyltransferase protein synthesis is reduced in CXXC finger protein 1-deficient embryonic stem cells. DNA Cell Biol. 2009;28(5):223–31.
62. Wysocka J, Swigut T, Milne TA, Dou Y, Zhang X, Burlingame AL, et al. WDR5 associates with histone H3 methylated at K4 and is essential for H3 K4 methylation and vertebrate development. Cell. 2005;121(6):859–72.
63. Ang Y-S, Tsai S-Y, Lee D-F, Monk J, Su J, Ratnakumar K, et al. Wdr5 mediates self-renewal and reprogramming via the embryonic stem cell core transcriptional network. Cell. 2011;145(2):183–97.
64. Outchkourov NS, Muiño JM, Kaufmann K, van Ijcken WFJ, Koerkamp MJG, Van Leenen D, et al. Balancing of histone H3K4 methylation states by the Kdm5c/SMCX histone demethylase modulates promoter and enhancer function. Cell Rep. 2013;3(4):1071–9.
65. Okitsu CY, Hsieh JCF, Hsieh C-L. Transcriptional activity affects the H3K4me3 level and distribution in the coding region. Mol Cell Biol. 2010;30(12):2933–46.
66. Eberl HC, Spruijt CG, Kelstrup CD, Vermeulen M, Mann M. A map of general and specialized chromatin readers in mouse tissues generated by label-free interaction proteomics. Mol Cell. 2013;49(2):368–78.
67. Sérandour AA, Avner S, Percevault F, Demay F, Bizot M, Lucchetti-Miganeh C, et al. Epigenetic switch involved in activation of pioneer factor FOXA1-dependent enhancers. Genome Res. 2011;21(4):555–65.
68. Kidder BL, Hu G, Zhao K. KDM5B focuses H3K4 methylation near promoters and enhancers during embryonic stem cell self-renewal and differentiation. Genome Biol. 2014;15(2):R32.
69. Cedar H, Bergman Y. Linking DNA methylation and histone modification: patterns and paradigms. Nat Rev Genet. 2009;10(5):295–304.
70. Nishiyama A, Yamaguchi L, Sharif J, Johmura Y, Kawamura T, Nakanishi K, et al. Uhrf1-dependent H3K23 ubiquitylation couples maintenance DNA methylation and replication. Nature. 2013;502(7470):249–53.
71. Petruk S, Sedkov Y, Johnston DM, Hodgson JW, Black KL, Kovermann SK, et al. TrxG and PcG proteins but not methylated histones remain associated with DNA through replication. Cell. 2012;150(5):922–33.
72. Tan L, Xiong L, Xu W, Wu F, Huang N, Xu Y, et al. Genome-wide comparison of DNA hydroxymethylation in mouse embryonic stem cells and neural progenitor cells by a new comparative hMeDIP-seq method. Nucleic Acids Res. 2013;41(7):e84.
73. Langmead B, Salzberg SL. Fast gapped-read alignment with bowtie 2. Nat Methods. 2012;9(4):357–9.
74. Quinlan AR, Hall IM. BEDTools: a flexible suite of utilities for comparing genomic features. Bioinforma Oxf Engl. 2010;26(6):841–2.
75. Kent WJ, Zweig AS, Barber G, Hinrichs AS, Karolchik D. BigWig and BigBed: enabling browsing of large distributed datasets. Bioinforma Oxf Engl. 2010;26(17):2204–7.
76. Zhang Y, Liu T, Meyer CA, Eeckhoute J, Johnson DS, Bernstein BE, et al. Model-based analysis of ChIP-Seq (MACS). Genome Biol. 2008;9(9):R137.
77. Shao Z, Zhang Y, Yuan G-C, Orkin SH, Waxman DJ. MAnorm: a robust model for quantitative comparison of ChIP-Seq data sets. Genome Biol. 2012;13(3):R16.
78. Luu P-L, Gerovska D, Arrospide-Elgarresta M, Retegi-Carrión S, Schöler HR, Araúzo-Bravo MJ. P3BSseq: parallel processing pipeline software for automatic analysis of bisulfite sequencing data. Bioinformatics. 2017; 33(3):428–31.
79. Yu M, Hon GC, Szulwach KE, Song C-X, Zhang L, Kim A, et al. Base-resolution analysis of 5-hydroxymethylcytosine in the mammalian genome. Cell. 2012;149(6):1368–80.
80. Booth MJ, Branco MR, Ficz G, Oxley D, Krueger F, Reik W, et al. Quantitative sequencing of 5-methylcytosine and 5-hydroxymethylcytosine at single-base resolution. Science. 2012;336(6083):934–7.
81. Sun Z, Terragni J, Jolyon T, Borgaro JG, Liu Y, Yu L, et al. High-resolution enzymatic mapping of genomic 5-hydroxymethylcytosine in mouse embryonic stem cells. Cell Rep. 2013;3(2):567–76.
82. Heinz S, Benner C, Spann N, Bertolino E, Lin YC, Laslo P, et al. Simple combinations of lineage-determining transcription factors prime cis-regulatory elements required for macrophage and B cell identities. Mol Cell. 2010;38(4):576–89.
83. Thorvaldsdóttir H, Robinson JT, Mesirov JP. Integrative genomics viewer (IGV): high-performance genomics data visualization and exploration. Brief Bioinform. 2013;14(2):178–92.
84. Feng J, Liu T, Qin B, Zhang Y, Liu XS. Identifying ChIP-seq enrichment using MACS. Nat Protoc. 2012;7(9):1728–40.
85. The Split-Apply-Combine Strategy for Data Analysis. Journal of Statistical Software. https://www.jstatsoft.org/article/view/v040i01
86. Wickham H. ggplot2: elegant graphics for data analysis. Springer New York: New York, NY; 2009.
87. Wickham HA. Practical tools for exploring data and models. ProQuest; 2008.
88. Zeileis A, Grothendieck G. zoo: S3 infrastructure for regular and irregular time series. ArXiv Prepr Math0505527. 2005. http://arxiv.org/abs/math/0505527
89. Gentleman RC, Carey VJ, Bates DM, Bolstad B, Dettling M, Dudoit S, et al. Bioconductor: open software development for computational biology and bioinformatics. Genome Biol. 2004;5(10):R80.
90. Davis S, Meltzer PS. GEOquery: a bridge between the gene expression omnibus (GEO) and BioConductor. Bioinforma Oxf Engl. 2007;23(14):1846–7.
91. Smyth GK. Limma: linear models for microarray data. In: Bioinformatics and computational biology solutions using R and Bioconductor. Springer; 2005. p. 397–420.
92. Barrett T, Wilhite SE, Ledoux P, Evangelista C, Kim IF, Tomashevsky M, et al. NCBI GEO: archive for functional genomics data sets–update. Nucleic Acids Res. 2013;41(Database issue):D991–5.
93. Rustici G, Kolesnikov N, Brandizi M, Burdett T, Dylag M, Emam I, et al. ArrayExpress update–trends in database growth and links to data analysis tools. Nucleic Acids Res. 2013;41(Database issue):D987–90.
94. Brinkman AB, Gu H, Bartels SJJ, Zhang Y, Matarese F, Simmer F, et al. Sequential ChIP-bisulfite sequencing enables direct genome-scale investigation of chromatin and DNA methylation cross-talk. Genome Res. 2012;22(6):1128–38.
95. Karimi MM, Goyal P, Maksakova IA, Bilenky M, Leung D, Tang JX, et al. DNA methylation and SETDB1/H3K9me3 regulate predominantly distinct sets of genes, retroelements, and chimeric transcripts in mESCs. Cell Stem Cell. 2011;8(6):676–87.
96. Sleutels F, Soochit W, Bartkuhn M, Heath H, Dienstbach S, Bergmaier P, et al. The male germ cell gene regulator CTCFL is functionally different from CTCF and binds CTCF-like consensus sites in a nucleosome composition-dependent manner. Epigenetics Chromatin. 2012;5(1):8.
97. Livyatan I, Harikumar A, Nissim-Rafinia M, Duttagupta R, Gingeras TR, Meshorer E. Non-polyadenylated transcription in embryonic stem cells reveals novel non-coding RNA related to pluripotency and differentiation. Nucleic Acids Res. 2013;41(12):6300–15.
98. Hon GC, Rajagopal N, Shen Y, McCleary DF, Yue F, Dang MD, et al. Epigenetic memory at embryonic enhancers identified in DNA methylation maps from adult mouse tissues. Nat Genet. 2013;45(10):1198–206.

12

Genetic relatedness of *Vibrio cholerae* isolates within and between households during outbreaks in Dhaka, Bangladesh

Christine Marie George[1*], Mahamud Rashid[2], Mathieu Almeida[3], K. M. Saif-Ur-Rahman[2], Shirajum Monira[2], Md. Sazzadul Islam Bhuyian[2], Khaled Hasan[4], Toslim T. Mahmud[2], Shan Li[3], Jessica Brubaker[4], Jamie Perin[4], Zillur Rahman[2], Munshi Mustafiz[2], David A. Sack[4], R. Bradley Sack[4], Munirul Alam[2] and O. Colin Stine[3]

Abstract

Background: Household contacts of cholera patients have a 100 times higher risk of developing a cholera infection than the general population. To compare the genetic relatedness of clinical and water source *Vibrio cholerae* isolates from cholera patients' households across three outbreaks, we analyzed these isolates using whole-genome-sequencing (WGS) and multilocus variable-number tandem-repeat analysis (MLVA).

Results: The WGS analyses revealed that 80% of households had source water isolates that were more closely related to clinical isolates from the same household than to any other isolates. While in another 20% of households an isolate from a person was more closely related to clinical isolates from another household than to source water isolates from their own household. The mean pairwise differences in single nucleotide-variant (SNV) counts for isolates from the same household were significantly lower than those for different households (2.4 vs. 7.7 $p < 0.0001$), and isolates from the same outbreak had significantly fewer mean pairwise differences compared to isolates from different outbreaks (mean: 6.2 vs. 8.0, $p < 0.0001$). Based on MLVA in outbreak 1, we observed that the majority of households had clinical isolates with MLVA genotypes related to other clinical isolates and unrelated to water source isolates from the same household. While in outbreak 3, there were different MLVA genotypes between households, however within the majority of households, the clinical and water source isolates had the same MLVA genotypes. The beginning of outbreak 2 resembled outbreak 1 and the latter part resembled outbreak 3. We validated our use of MLVA by comparing it to WGS. Isolates with the identical MLVA genotype had significantly fewer mean pairwise SNV differences than those isolates with different MLVA genotypes (mean: 4.8 vs. 7.7, $p < 0.0001$). Furthermore, consistent with WGS results, the number of pairwise differences in the five MLVA loci for isolates within the same household was significantly lower than isolates from different households (mean: 1.6 vs. 3.0, $p < 0.0001$).

Conclusion: These results suggest that transmission patterns for cholera are a combination of person-to-person and water-to-person cholera transmission with the proportions of the two modes varying within and between outbreaks.

Keywords: *Vibrio cholerae*, Bangladesh, Whole genome sequencing, Multilocus variable-number tandem-repeat analysis, Outbreak surveillance, Household transmission, Genetics

* Correspondence: cgeorg19@jhu.edu
[1]Department of International Health, Program in Global Disease Epidemiology and Control, Johns Hopkins Bloomberg School of Public Health, 615 N. Wolfe Street, Room E5535, Baltimore, MD 21205-2103, USA
Full list of author information is available at the end of the article

Background

The World Health Organization estimates that there are 3–5 million cholera cases worldwide per year resulting in more than 100,000 deaths [1]. Studies have identified water [2–5] and food borne contamination [6, 7] to be the main transmission routes for cholera. Household contacts are at a 100 times higher risk of developing a cholera infection than the general population [3, 8–10]. However, most previous studies among this high risk population were conducted before genetic identification of *Vibrio cholerae* strains was available.

Genetic identification of *V. cholerae* strains allows for sources of infection in a household to be identified. One genetic method, whole genome sequencing (WGS) distinguishes between isolates based on single nucleotide variants (SNVs). WGS data has revealed three phylogenetically distinct waves of cholera spreading around the world [11] and has been shown to be useful in outbreak investigation to identify separate outbreaks within a single time period [12]. A second genetic method, multilocus variable-number tandem-repeat analysis (MLVA) distinguishes between different strains of *V. cholerae* based on the number of short (6 to 9), repeating nucleotide sequences at five loci. An observational study of household contacts of cholera cases using MLVA in Dhaka, Bangladesh found that *V. cholerae* strains were genetically identical at five loci between index cases and household contacts for only 46% of pairs analyzed [13]. This result is very different than the nearly 90% matching by serogroup and serotype.

A recent study compared MLVA and WGS and found that they reflected the same genetic history [14], in contrast to two earlier reports [15, 16]. Rashid et al. found that isolates closely related by MLVA had significantly fewer nucleotides differences when compared to each other than when compared to isolates distantly related by MLVA [14]. In this study Rashid et al. reported about a shorter time scale (less than 1 year vs 38 years) than the first report [16] and did more extensive sampling than the second report [15].

We recently conducted a randomized controlled trial of a health facility based handwashing with soap and water treatment intervention for the household contacts of cholera patients (Cholera-Hospital-Based Intervention-for-7-Days (CHoBI7) Trial) to reduce cholera among this high risk population in Dhaka, Bangladesh [17]. In an attempt to investigate transmission patterns within cholera-patient households, we performed pulsed-field gel electrophoresis (PFGE) on clinical and water isolates collected from patient households in this trial. Of the 33 *V. cholerae* isolates analyzed by PFGE, 88% were found to have identical banding patterns [18]. The close similarity between clinical and water isolates within patient households is suggestive of the household's drinking water being the source of infecting inoculum in these homes.

Building on this previous work in our current study, we will compare the genetic relatedness of *V. cholerae* O1 isolates from cholera patients, their household members, and their household water sources by WGS and MLVA. These methods will allow us to differentiate between *V. cholerae* O1 isolates that are typically indistinguishable by PFGE [13, 19, 20]. Our objective is to investigate cholera transmission patterns in patient households over a 1 year period to determine if all outbreaks are the same or if there is variability in transmission patterns across outbreaks.

Results

Epidemiology

Our 1 year surveillance period identified a total of 136 culture confirmed cholera patients with three distinct case based outbreaks which were separated by a month with less than 4 cholera patients (Fig. 1). There was a summer peak (June –August 2013) with 33 cholera patients, a fall peak (September 2013 –January 2014) with 33 cholera patients, and a spring peak (March to June 2014) with 70 cholera patients. During these outbreaks, 19% of household contacts and 30% of drinking water sources used in cholera-patient households had detectable *V. cholerae* O1 by culture. A total of 621 isolates, 288 clinical and 333 water *V. cholerae* O1 isolates were analyzed from 31 households in which both patient and water samples were positive, 27 households had a patient and *source* water samples available. For each positive sample within the household, up to 10 *V. cholerae* O1 Ogawa isolates were collected. There was no significant difference in the location of cholera-patient households across outbreaks ($p = 0.251$) (Fig. 2). Average distances between households were 6.5, 5.4, and 6.8 km in outbreaks 1, 2, and 3, respectively. In addition, there was no apparent clustering of the households based on the outbreak.

Whole genome sequencing

Thirty-eight genomes of O1 isolates collected from 17 households were sequenced, 13 households had multiple isolates, and 10 had both water and clinical isolates (Additional file 1: Table S1). Thirteen genomes were from Outbreak 1, 12 from Outbreak 2, and 13 from Outbreak 3. On average, each genome was assembled into 80.9 scaffolds (Additional file 2: Table S2), with an average depth of 415 reads. After aligning the genomes as described in the methods section, we identified 66 SNVs distributed in 29 different scaffolds, with 81% of them corresponding to the chromosome I and mostly in regulatory genes.

We estimated the genetic relatedness of the 38 genomes using a maximum likelihood tree calculated from the variants detected in the alignments of the scaffolds bigger than 10 kb (Fig. 3). Only high quality SNVs were used in the bee swarm plot (Fig. 4). The

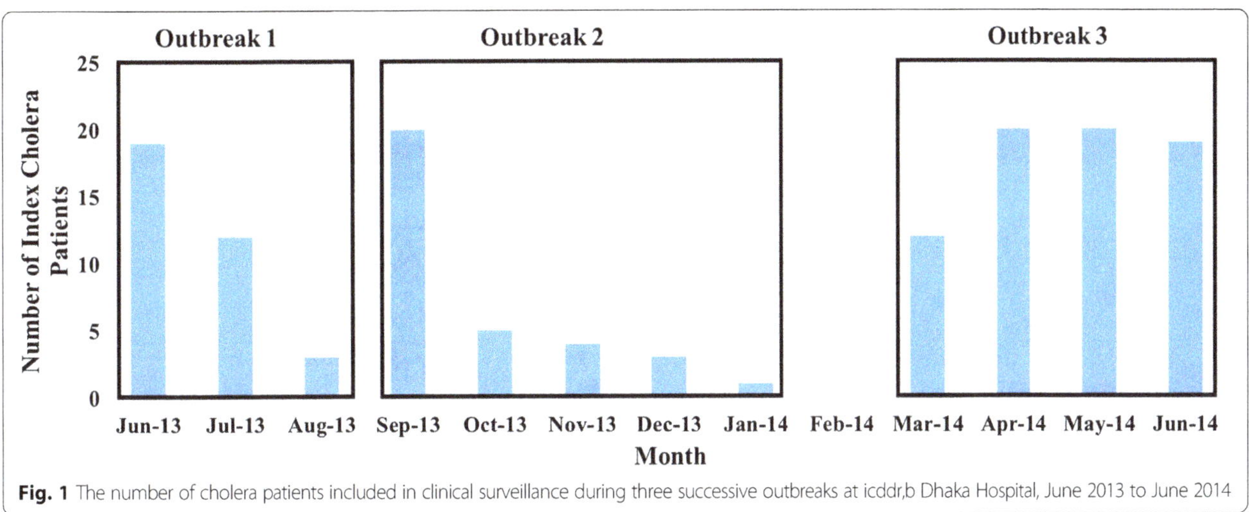

Fig. 1 The number of cholera patients included in clinical surveillance during three successive outbreaks at icddr,b Dhaka Hospital, June 2013 to June 2014

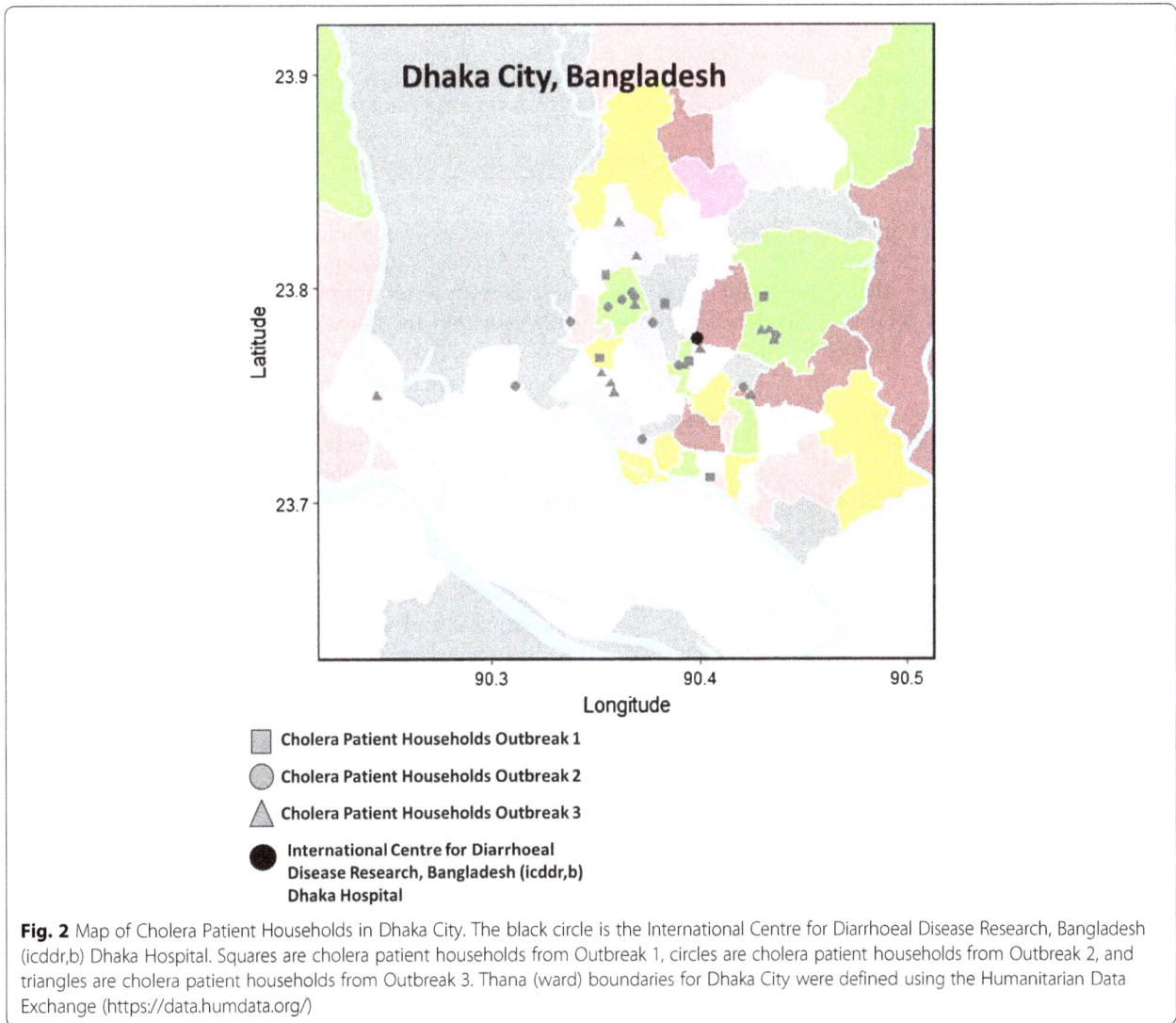

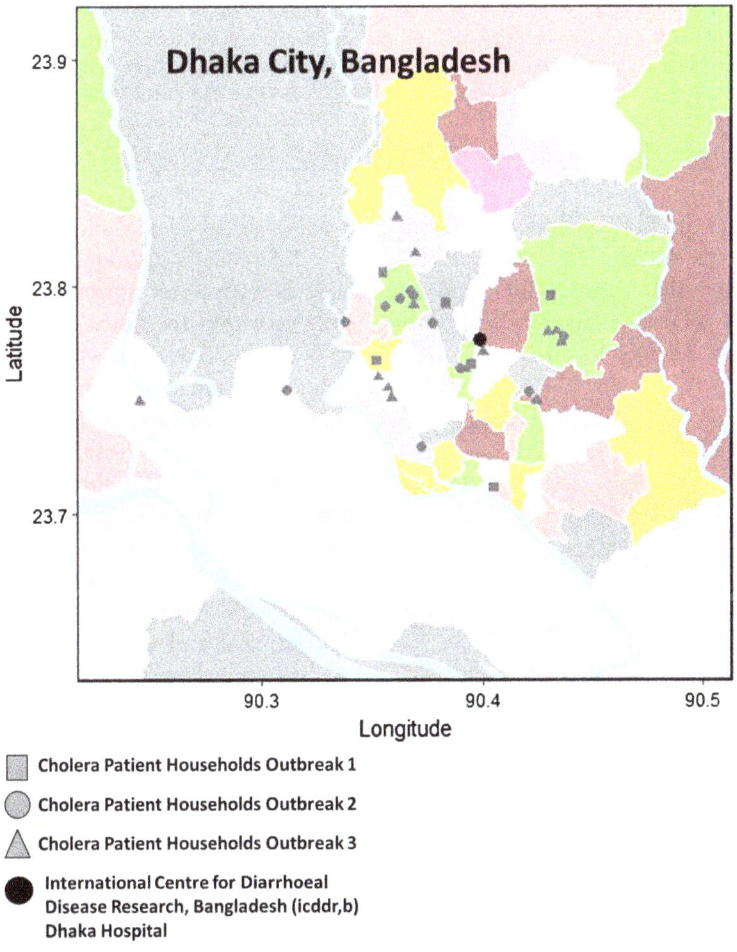

Fig. 2 Map of Cholera Patient Households in Dhaka City. The black circle is the International Centre for Diarrhoeal Disease Research, Bangladesh (icddr,b) Dhaka Hospital. Squares are cholera patient households from Outbreak 1, circles are cholera patient households from Outbreak 2, and triangles are cholera patient households from Outbreak 3. Thana (ward) boundaries for Dhaka City were defined using the Humanitarian Data Exchange (https://data.humdata.org/)

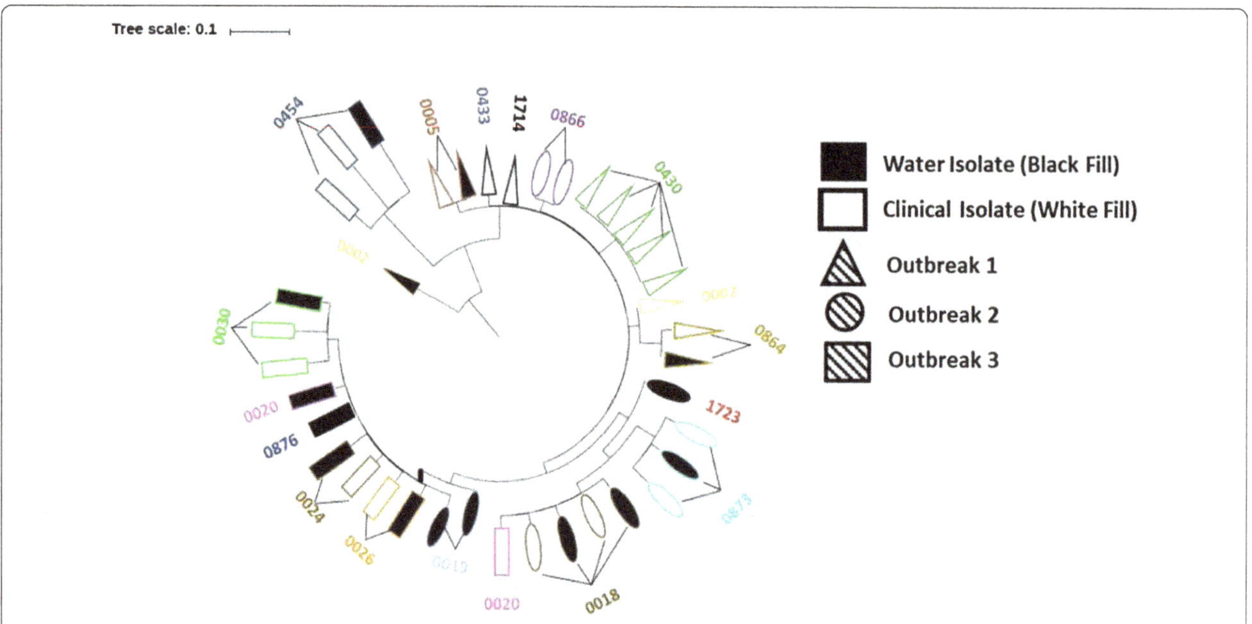

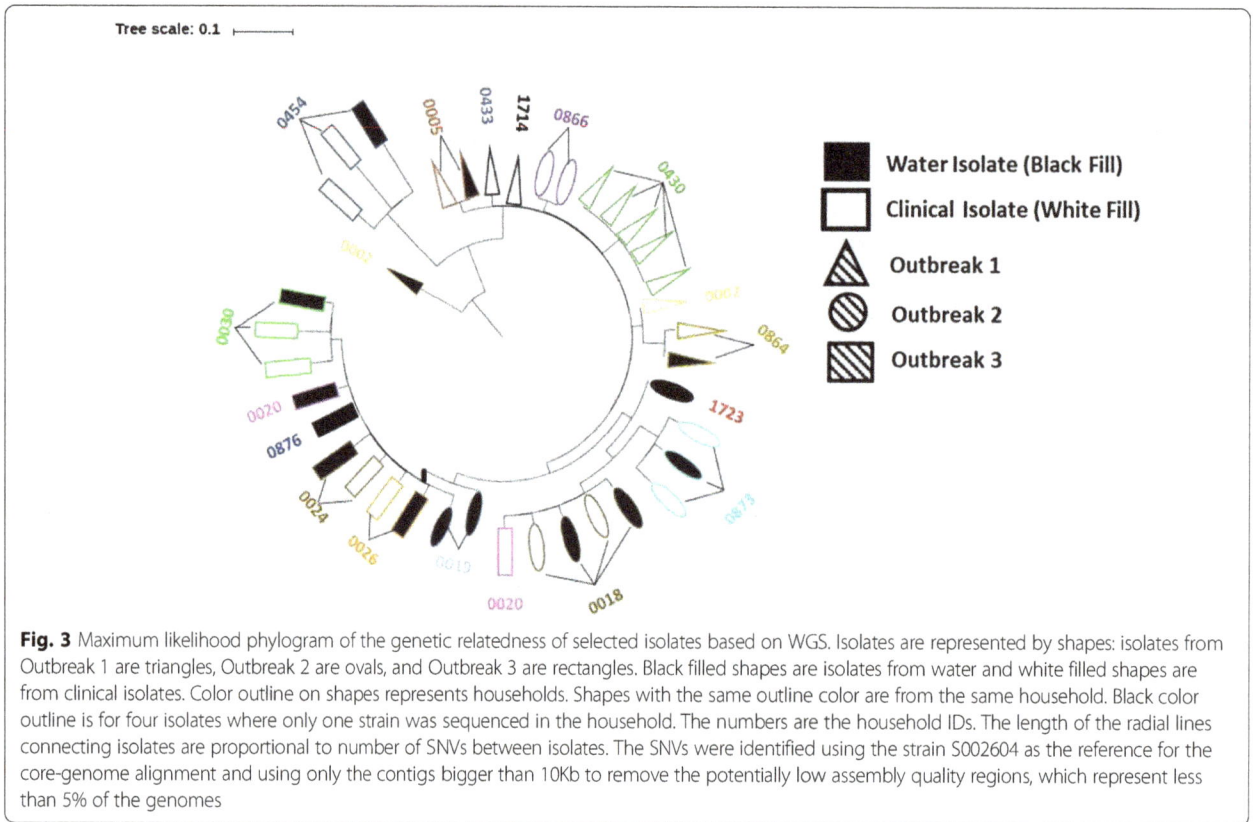

Fig. 3 Maximum likelihood phylogram of the genetic relatedness of selected isolates based on WGS. Isolates are represented by shapes: isolates from Outbreak 1 are triangles, Outbreak 2 are ovals, and Outbreak 3 are rectangles. Black filled shapes are isolates from water and white filled shapes are from clinical isolates. Color outline on shapes represents households. Shapes with the same outline color are from the same household. Black color outline is for four isolates where only one strain was sequenced in the household. The numbers are the household IDs. The length of the radial lines connecting isolates are proportional to number of SNVs between isolates. The SNVs were identified using the strain S002604 as the reference for the core-genome alignment and using only the contigs bigger than 10Kb to remove the potentially low assembly quality regions, which represent less than 5% of the genomes

mean pairwise differences in SNV counts for isolates from the same households (mean: 2.4 (range: 0–12)) were significantly lower than those for different households (7.7 (range: 1–21)) ($p < 0.0001$). Within the household, clinical isolates averaged less than or equal to 4 SNVs differences, while between households the average difference is greater than 4. The same is true when isolates from water sources are added to the comparison, isolates within the household average less than or equal to 4 SNVs differences, while for isolates between households the average difference is greater than 4. If an average of four differences is accepted as a threshold for determining within versus between household transmission of *V. cholerae*, then two households out of 10 households with clinical and source water isolates (20%) (0002 & 0020, see Fig. 3) had a person acquire their infecting *V. cholerae* outside the home or a source we did not measure. While for the other 8 out of 10 households with clinical and water isolates (80%), the genotype of the source water isolate in the household was more closely related to that of infected individuals in the same household than any other isolate.

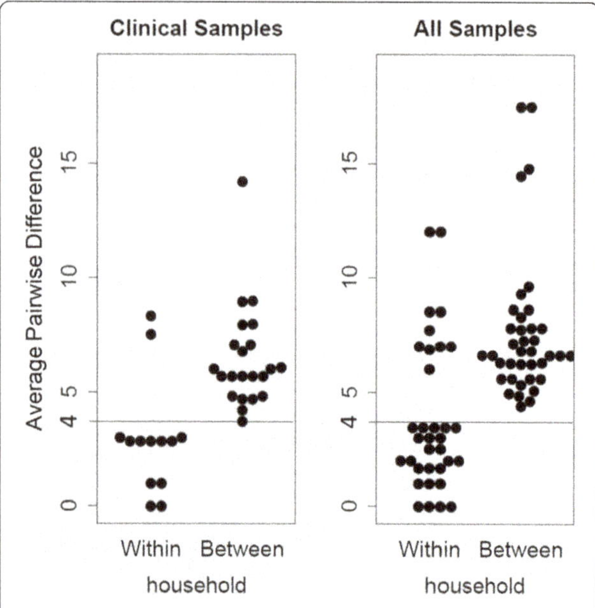

Fig. 4 Average Pairwise SNV differences within and between households for **a**) clinical samples from people and **b**) all samples

The differences between groups of isolates are shown in Table 1. Source water isolates had significantly greater mean pairwise differences compared to clinical isolates (mean: 9.0 vs. 6.2, $p < 0.0001$). Isolates from the same outbreak had significantly fewer mean pairwise differences compared to isolates from different outbreaks (mean: 6.2 vs. 8.0, $p < 0.0001$). When stratified by isolate type (clinical vs. water), isolates from the same outbreak had significantly fewer pairwise differences for clinical isolates ($p < 0.0001$), but not for water isolates ($p = 0.26$).

Table 1 Pairwise comparisons of single nucleotide variant (SNV) counts for *Vibrio cholerae* clinical and water isolates by WGS

	Pairwise Comparisons*	All Outbreaks Mean SNV ± SD (Min-Max)	*p*-value†
Inter vs. Intra Household Variability in SNV Counts			
Isolates from Same Households	33	2.4 ± 3.1 (0–12)	<0.0001
Isolates from Different Households	670	7.7 ± 4.8 (1–21)	
Same MLVA Genotype vs. Different MLVA Genotype Variability in SNV Counts			
Same MLVA Genotype	77	4.8 ± 4.9 (0–18)	<0.0001
Different MLVA Genotype	628	7.7 ± 4.8 (0–21)	
Single Locus Variants in MLVA Genotype vs. Different MLVA Genotype Variability in SNV Counts			
Single Locus Variants in MLVA Genotypes	89	4.7 ± 3.1 (0–15)	<0.0001
Different MLVA Genotype	539	8.2 ± 4.8 (1–21)	
Clinical vs. Water Isolate Variability in SNV Counts			
Clinical Isolates	231	6.2 ± 3.7 (1–18)	<0.0001
Water Isolates	120	9.0 ± 6.1 (1–20)	
Same vs. Different Outbreaks Variability in SNV Counts			
Isolates from Same Outbreak	232	6.1 ± 4.9 (0–21)	0.01
Isolates from Different Outbreak	473	8.0 ± 4.8 (1–21)	

†Permutation Tests
*Comparisons of Pairs of Isolates
SD Standard Deviation

Multilocus variable-number tandem-repeat analysis

Substantial genetic variation in the collected isolates was observed by MLVA. The genotype of each isolate was specified by the alleles at the five loci. There were 8 alleles at VC0147, 10 at VC0437, 9 at VC1650, 11 at VCA0171, and 9 at VCA0283. One hundred twenty-four distinct genotypes were identified: 25 genotypes from clinical isolates, 81 from water isolates, and 18 genotypes that were found in both clinical and water isolates.

Multiple MLVA genotypes were observed in isolates collected from a single clinical or water sample. The variation within a single sample was greater for water isolates (mean: 6 genotypes, range: 2–10 genotypes) compared to clinical isolate (mean: 2 genotypes, range: 1–5) ($p < 0.0001$). Three quarters (76%, 42/55) of clinical samples and all (100%, 46/46) of water samples had at least two isolates with different MLVA genotypes (Additional file 1: Table S1). For clinical specimens, the proportion of specimens with more than one MLVA genotype increased significantly over the course of the three outbreaks from 47% to 100% ($p < 0.0001$), no significant difference was observed for water samples (Table 2). Of note, one allele at VCA0171 did not amplify from 99 DNA sequences despite repeated attempts; subsequent WGS revealed the locus is present, but a mutation altered the last nucleotide of the primer binding site and interfered with amplification. Thirty percent (99/333) of water isolates had an unamplified allele at VCA0171 compared to none of clinical isolates (0/288) ($p < 0.0001$). The unamplified allele was present in 66% of the water isolates in the first outbreak and in only 1% of the water isolates in the third outbreak ($p < 0.0001$).

The majority of households had clinical and water isolates with identical MLVA genotypes. Eighty-two percent (9/11) of households had infected household members with identical MLVA genotypes. In 56% (15/27) of households, an isolate from the index case had the identical genotype as a *source* water isolate. In 58% (18/31) of households, there were infected household members *and* water isolates (stored and source water isolates) with identical MLVA genotypes (Table 2). Conversely, in 42% of households, there were no identical genotypes found in both household member and water samples.

The genetic relatedness of *V. cholerae* isolates in patients and water differed substantially between the three outbreaks (Fig. 5). In the first outbreak, there was a single predominate MLVA genotype found in 93% (13 of 14) of clinical specimens and the other seven clinical MLVA genotypes in this outbreak were single-locus variants (SLVs) of this genotype. In contrast, of the 44 MLVA genotypes found in the water isolates during this outbreak, only 3 were the same or SLVs of the genotype in infected individuals within these households. In the second outbreak, there were two MLVA genotype lineages found in the infected household members. One was the MLVA lineage from the first outbreak observed in

Table 2 Relatedness of clinical and water MLVA genotypes of *Vibrio cholerae* by outbreak

	All Outbreaks	Cholera Outbreaks			*p*-value†
		Outbreak 1	Outbreak 2	Outbreak 3	
		June to August 2013	September 2013 to January 2014	March to June 2014	
Samples with at least two isolates with different MLVA genotypes					
Clinical Samples	76% (42/55)	47% (8/17)	79% (15/19)	100% (19/19)	<0.0001
Water Samples	100% (46/46)	100% (8/8)	100% (15/15)	100% (23/23)	‡
Isolates with unamplified VCA0171 Allele					
Clinical Isolates	0% (0/288)	0% (0/139)	0% (0/109)	0% (0/40)	‡
Water Isolates	30% (99/333)	66% (52/79)	39% (46/118)	1% (1/136)	<0.0001
Household Characteristics					
Household member and water isolates with identical MLVA genotypes	58% (18/31)	50% (3/6)	55% (6/11)	64% (9/14)	0.09
Index cholera patient and *source water* isolate with identical MLVA genotypes	56% (15/27)	40%(2/5)	56% (5/9)	62% (8/13)	0.09
Household member and *source water isolates* with identical MLVA genotypes	52% (16/31)	50% (3/6)	46% (5/11)	57% (8/14)	0.09
Household member and *stored water* isolates with identical MLVA genotypes	40% (2/5)	0% (0/1)	50% (1/2)	50% (1/2)	0.4
Household member isolates with identical MLVA genotypes	82% (9/11)	66% (2/3)	75% (3/4)	100% (4/4)	0.2

†Fisher Exact Test
‡ Fisher exact couldn't be calculated because all proportions were the same across outbreaks

nine clinical specimens and six households. The second MLVA lineage was observed in twelve clinical samples and nine households. Of note, while only one of these two lineages was in any given individual prior to November 2013, after that date four of six infected household members had both lineages represented. There was a similar finding for water samples, up until November 2013 for Outbreak 2 there were 31 MLVA genotypes that were unrelated to the isolates in the clinical specimens. After November, all of the isolates observed in the water were related to the two predominate MLVA lineages. In the first household of Outbreak 3, a new MLVA lineage was observed in both patients and in the water, it also was found in the water of the second household. However the second and subsequent households look like the second portion of the second outbreak. The water from Outbreak 3 households contained the two previous lineages from outbreak 2 in ten households, a third lineage in two other households, and a fourth in three more households.

Validation of MLVA by WGS

Isolates with eighteen different MLVA genotypes were analyzed by WGS. Isolates with the identical MLVA genotype had significantly fewer mean pairwise SNV differences than those isolates with different MLVA genotypes (mean: 4.8 vs. 7.7, $p < 0.0001$). Furthermore, isolates that differed at a single MLVA locus were more closely related than those with greater MLVA allelic variation (mean: 4.70 vs. 8.23, $p < 0.001$).

Consistent with results from WGS, isolates collected within households were more closely related by MLVA than those from different households. The number of pairwise differences in the five MLVA loci for isolates within a household was significantly lower than isolates from different households (mean: 1.6 vs. 3.0, $p < 0.0001$). The lower number of pairwise differences was significant even when stratified by clinical (mean: 0.29 vs. 2.1, $p < 0.0001$) and water isolates (mean 1.7 vs. 3.3, $p < 0.0001$). Water isolates also had a significantly greater number of MLVA loci differences compared to clinical isolates (mean: 3.3 vs. 2.0, $p < 0.0001$). Isolates from the same outbreak had significantly fewer MLVA loci differences compared to isolates from different outbreaks (mean: 2.5 vs. 3.2, $p < 0.0001$).

Discussion

Our results are consistent with the presence of two modes of cholera transmission: person-to-person and water-to-person with the proportions of the two modes varying within and between outbreaks. Our WGS analyses of 38 isolates revealed that 80% of households had source water isolates that were more closely related to clinical isolates from the same household than to any other isolates. While in another 20% of households, an isolate from a person was more closely related to clinical isolates from another household than to source water isolates from their own household. We interpret the former to be water-to-person transmission and the latter to be person-to-person transmission. In order to expand

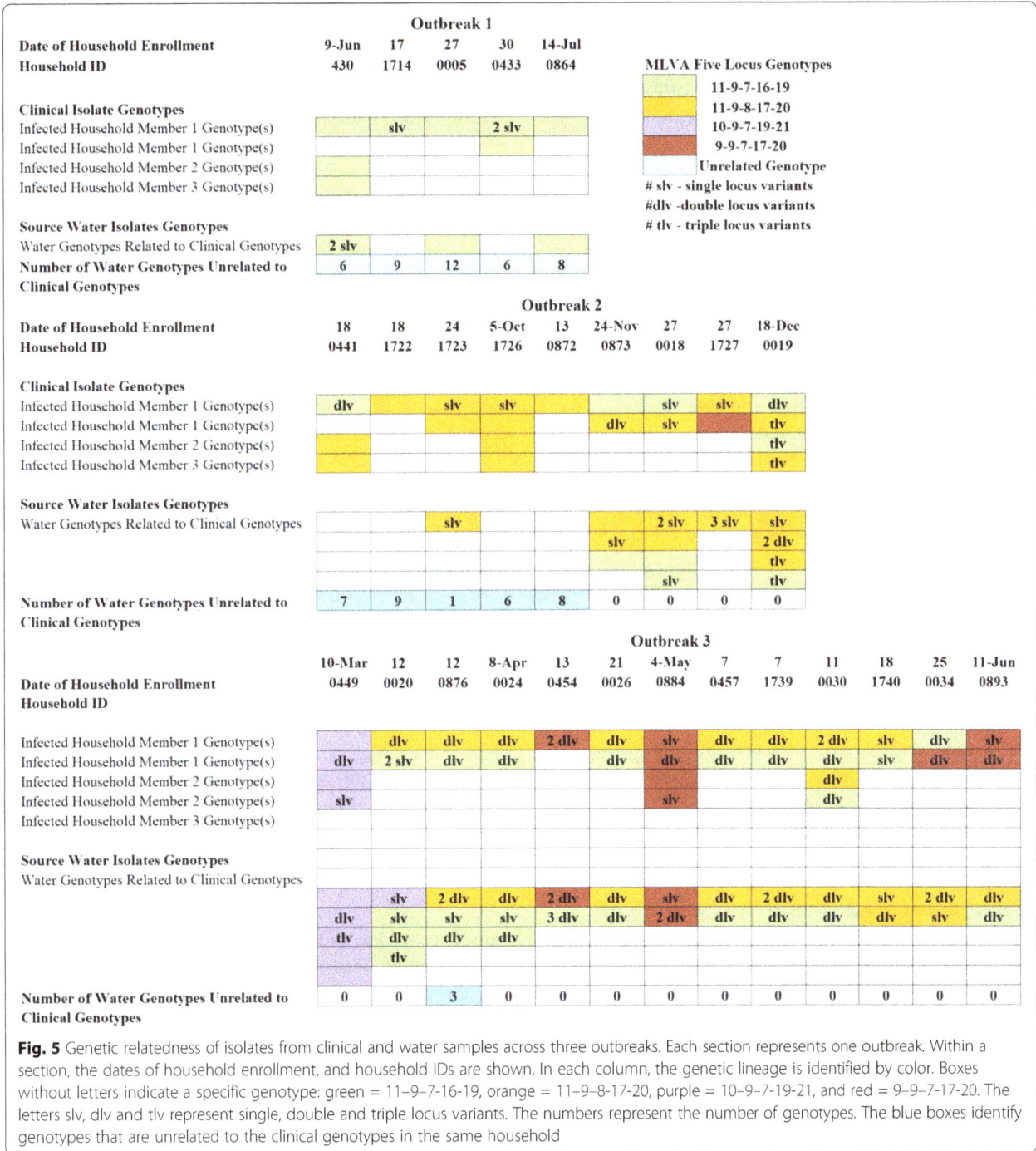

Fig. 5 Genetic relatedness of isolates from clinical and water samples across three outbreaks. Each section represents one outbreak. Within a section, the dates of household enrollment, and household IDs are shown. In each column, the genetic lineage is identified by color. Boxes without letters indicate a specific genotype: green = 11–9–7-16-19, orange = 11–9–8-17-20, purple = 10–9–7-19-21, and red = 9–9–7-17-20. The letters slv, dlv and tlv represent single, double and triple locus variants. The numbers represent the number of genotypes. The blue boxes identify genotypes that are unrelated to the clinical genotypes in the same household

the sample size, we used MLVA to estimate the genetic relatedness of 621 systematically collected isolates. Our estimates of relatedness by MLVA reflected the relatedness that was observed by WGS. Based on our MLVA, we observed differences in transmission patterns between and within outbreaks. In Outbreak 1, we observed that the majority of households had clinical isolates with MLVA genotypes unrelated to water source isolates while clinical isolates within households were all genetically related, consistent with person-to-person transmission. In contrast, in Outbreak 3, there were different MLVA genotypes between households, although within the household, clinical and water source isolates had the same MLVA genotypes consistent with water-to-person transmission. Outbreak 2 showed a combination of both modes of transmission with the beginning of the outbreak resembling Outbreak 1 and the latter part resembling Outbreak 3. Therefore, these findings indicate that in this urban setting cholera control programs should focus on both hygiene promotion, to reduce person-to-person transmission, and on municipal and point of use water treatment programs, to reduce water-to-person transmission.

The genotypes from isolates within a household were more closely related than those between households. We therefore suspect the spread of cholera within many households comes from a contaminated water source used for drinking water by household members, as evidenced by the WGS and MLVA data showing that water and clinical isolates within a household are more closely related than isolates outside the household. Alternatively, there were instances when the spread of cholera within the household came from other infected household members through poor hygiene practices, as evidenced in Outbreak 1 where household members had the same MLVA genotypes, while the water source isolates for the most part were unrelated to the clinical isolates.

We observed more water than clinical MLVA genotypes, while in previous studies, more clinical genotypes were observed than water genotypes [21, 20]. This is likely the result of study design and sampling issues with previous studies. In each of the previous studies, Rashed et al. [21] and Stine et al. [20] a single colony was picked from each water sample likely resulting in an underestimation of the variation in the water. When comparing only clinical samples, Kendall et al. [13] in another household contact study of cholera patients found 83 distinct MLVA clinical genotypes. However this study was conducted over a time span of more than 3 years compared to our 1 year period, perhaps allowing more time to accumulate additional clinical MLVA genotypes. In addition, our results revealed the majority of those infected with cholera had multiple MLVA genotypes present in their stool samples consistent with previous work [13, 21].

In our study, WGS and MLVA data reflected similar genetic relationships for comparisons made between *V. cholerae* isolates. Isolates with the same MLVA genotype were significantly more similar by WGS than isolates with different MLVA genotypes. In addition, isolates that differed at a single MLVA locus were also significantly more similar than those that differed at more than one MLVA locus. This finding is consistent with previous work by Rashid et al. who showed that distinct MLVA clonal complexes represent separate WGS lineages [14].

Our study has several strengths. The first was the environmental surveillance of the source and stored drinking water in the households of cholera patients. Second, we conducted WGS and MLVA which allowed us to complement our WGS analysis with a larger number of strains analyzed by MLVA. Third, we collected multiple isolates from all clinical and water samples which allowed us to investigate diversity of MLVA genotypes within samples.

Our study has a few limitations. First, the 621 water and clinical isolates collected were limited to 31 households spread over 3 outbreaks. Second, we only included households that had a water sample with detectable *V. cholerae*. Third, we only sequenced a fraction of the isolates we collected, although given the high correlation between MLVA and WGS changes in the conclusions seem unlikely.

Conclusion

These results provide evidence consistent with two modes of cholera transmission, water-to-person and person-to-person transmission, within and between the households of cholera patients during outbreaks in Dhaka, Bangladesh. Remarkably, person-to-person transmission was the dominate mode of transmission in one outbreak while water-to-person transmission was the dominate mode in subsequent outbreaks.

Methods

The 31 households from the CHoBI7 randomized controlled trial (RCT) [17] in Dhaka, Bangladesh included in this analysis were enrolled from June 2013 to June 2014. Cholera patients confirmed by bacterial culture presenting at the icddr,b Dhaka hospital Sunday to Thursday were enrolled. A description of the intervention is described elsewhere [17]. Case households were visited at Days 1, 3, 5, 7, and 9 after the index patient was identified. At each visit, rectal swab samples were collected from household contacts and a water sample was collected from the household's water source, a piped connection to the municipal water supply, and stored drinking water in the home to test for *V. Cholerae* by bacterial culture. Household contacts were defined as individuals sharing the same cooking pot with the cholera patient for the past 3 days.

Sample collection and processing

Stool samples from the index patient were collected at the hospital. Rectal swab samples from household contacts and water samples were collected from the households of enrolled cholera patients. All samples were processed and analyzed by bacterial culture and serotyped using published methods [17, 22]. For clinical samples, up to 5 colonies were selected and for each water sample up to 10 colonies.

Whole genome sequencing (WGS)

Genomic DNA was extracted from 38 isolates using published phenol-chloroform extraction methods [23]. Genome sequencing was performed as previously published on an Illumina HiSeq2500 (Illumina, San Diego, CA, USA) [24, 25].

High quality reads of the 101-base paired-end reads were selected (https://www.bioinformatics.babraham.ac.uk/projects/fastqc/), assembled with "Spades" software (v.3.6.2) [26], and annotated using the RAST server [27]. The assembled genomes were submitted to Genbank and are associated with the BioprojectID: PRJNA371610. We choose the wave 3 isolate genome, S002604, as the reference, in order to minimize the number of variable sites that were different between the reference genome and all the genomes in our sample since those variable sites do not contribute to the analysis. PARSNP (v1.2) [28] was used to extract and align the variable nucleotides using previously published options [24, 25]. The '.ggr' file was loaded in Gingr (v1.2) [28] to visualize the alignments. A '.vcf' file was used to remove all variants less than 1 kb from the end of the contigs in the '.mfa' file using an in-house script. FastTree2 (v2.1.9) [29] generated the maximum-likelihood newick tree file using the revised alignment file. iTOL (http://itol.embl.de/) [30] was employed to visualize the maximum-likelihood tree.

Multilocus variable-number tandem-repeat analysis (MLVA)

To complement the WGS analysis with a larger number of strains, MLVA was performed on DNA from 621 *V. cholerae* water and clinical isolates from patient households. DNA was isolated from 5 μl of culture using Prepman (ABI) according to the manufacturer's instructions. To perform MLVA, the DNA from the *V. cholerae* O1 isolates was genotyped at each of five previously identified MLVA loci (VC0147, VC0437, VC1650, VCA0171 & VCA0283) using previously published methods [13].

Statistical analysis

Fisher's exact, paired t-tests, permutation tests, and Poisson regression models were computed using SAS (version 9.3) to analyze WGS and MLVA data. For WGS, pairwise comparisons were made of the SNV counts for each isolate compared to the reference strain, S002604. For MLVA, the relatedness of *V. cholerae* isolates was assessed using all five MLVA loci. Pairwise comparisons were made based on the number of loci with different alleles (e.g. there would be one difference if a single locus varied, two if two loci varied). The statistical analysis of the cholera patient household locations was performed using R version 3.3.2 with package ggplot.

Abbreviations

icddr,b: International Centre for Diarrhoeal Disease Research, Bangladesh; MLVA: Multilocus variable-number tandem-repeat analysis; PFGE: Pulsed-field gel electrophoresis; SLVs: Single-locus variants; SNVs: Single nucleotide variants; WGS: Whole genome sequencing

Acknowledgments

We thank the study participants and the following research staff who conducted the field work for this study: Ismat Minhaz Uddin, Rafiqul Islam, Al-Mamun, Maynul Hasan, Kalpona Akhter, Khandokar Fazilatunnessa, Sadia Afrin Ananya, Akhi Sultana, Sohag Sarker Jahed Masud, Abul Sikder, Shirin Akter, and Laki Das.
This research was supported by the Center for Global Health at Johns Hopkins University and the National Institute of Allergy and Infectious Diseases, National Institutes of Health. icddr,b thanks the governments of Australia, Bangladesh, Canada, Sweden, and United Kingdom for providing core/unrestricted support.

Funding

This work was funded by the National Institute of Allergy and Infectious Disease (NIAID 1K01AI110526-01A1). The funding body had no role in the study design, data collection, analysis and interpretation of data, or in writing the manuscript.

Authors' contributions

CMG, OCS, DS, RBS, MR conceived and designed the research. CMG, MA, MR, MSIB, KH, TTM, SUR, SL, JB, JP, ZR, MM, OCS analyzed the data. CMG, KMS, OCS, SM, MA wrote the manuscript. All authors read and approved the final manuscript.

Competing interests

The authors declare that they have no competing interests.

Author details

[1]Department of International Health, Program in Global Disease Epidemiology and Control, Johns Hopkins Bloomberg School of Public Health, 615 N. Wolfe Street, Room E5535, Baltimore, MD 21205-2103, USA. [2]International Centre for Diarrhoeal Disease Research, Shaheed Tajuddin Ahmed Ave, Dhaka 1213, Bangladesh. [3]University of Maryland School of Medicine, 655 W Baltimore S, Baltimore, MD 21201, USA. [4]Johns Hopkins School of Public Health, 615 N. Wolfe Street, Baltimore, MD 21205-2103, USA.

References

1. Organization WH. WHO | Cholera - World Health Organization. http://www.who.int/cholera/en/.
2. Deb B, Sircar B, Sengupta P, De S, Mondal S, Gupta D, Saha N, Ghosh S, Mitra U, Pal S. Studies on interventions to prevent eltor cholera transmission in urban slums. Bull World Health Organ. 1986;64(1):127.
3. Hughes JM, Boyce JM, Levine RJ, Khan M, Aziz K, Huq M, Curlin GT. Epidemiology of eltor cholera in rural Bangladesh: importance of surface water in transmission. Bull World Health Organ. 1982;60(3):395.
4. Kumar P, Mishra DK, Deshmukh DG, Jain M, Zade AM, Ingole KV, Goel AK, Yadava PK. Vibrio cholerae O1 Ogawa el tor strains with the ctxB7 allele driving cholera outbreaks in south-western India in 2012. Infect Genet Evol. 2014;25:93–6.
5. Bhuyan SK, Vairale MG, Arya N, Yadav P, Veer V, Singh L, Yadava PK, Kumar P. Molecular epidemiology of Vibrio cholerae associated with flood in Brahamputra River valley, Assam, India. Infect Genet Evol. 2016;40:352–6.

6. Sinclair G, Mphahlele M, Duvenhage H, Nichol R, Whitehorn A, Küstner H. Determination of the mode of transmission of cholera in Lebowa. An epidemiological investigation. S Afr Med J. 1982;62(21):753.
7. Acosta CJ, Galindo CM, Kimario J, Senkoro K, Urassa H, Casals C, Corachán M, Eseko N, Tanner M, Mshinda H. Cholera outbreak in southern Tanzania: risk factors and patterns of transmission. Emerg Infect Dis. 2001;7(3 Suppl):583.
8. Weil AA, Khan AI, Chowdhury F, LaRocque RC, Faruque A, Ryan ET, Calderwood SB, Qadri F, Harris JB. Clinical outcomes in household contacts of patients with cholera in Bangladesh. Clin Infect Dis. 2009;49(10):1473–9.
9. Spira W, Khan MU, Saeed Y, Sattar M. Microbiological surveillance of intra-neighbourhood el tor cholera transmission in rural Bangaldesh. Bull World Health Organ. 1980;58(5):731.
10. Glass RI, Svennerholm AM, Khan MR, Huda S, Huq MI, Holmgren J. Seroepidemiological studies of el tor cholera in Bangladesh: association of serum antibody levels with protection. J Infect Dis. 1985;151(2):236–42.
11. Mutreja A, Kim DW, Thomson NR, Connor TR, Lee JH, Kariuki S, Croucher NJ, Choi SY, Harris SR, Lebens M, et al. Evidence for several waves of global transmission in the seventh cholera pandemic. Nature. 2011;477(7365):462–5.
12. Shah MA, Mutreja A, Thomson N, Baker S, Parkhill J, Dougan G, Bokhari H, Wren BW. Genomic epidemiology of Vibrio cholerae O1 associated with floods, Pakistan, 2010. Emerg Infect Dis. 2014;20(1):13–20.
13. Kendall EA, Chowdhury F, Begum Y, Khan AI, Li S, Thierer JH, Bailey J, Kreisel K, Tacket CO, LaRocque RC. Relatedness of Vibrio cholerae O1/O139 isolates from patients and their household contacts, determined by multilocus variable-number tandem-repeat analysis. J Bacteriol. 2010;192(17):4367–76.
14. Rashid MU, Almeida M, Azman AS, Lindsay BR, Sack DA, Colwell RR, Huq A, Morris JG Jr, Alam M, Stine OC. Comparison of inferred relatedness based on multilocus variable-number tandem-repeat analysis and whole genome sequencing of Vibrio cholerae O1. FEMS Microbiol Lett. 2016;363(12)
15. Abd El Ghany M, Chander J, Mutreja A, Rashid M, Hill-Cawthorne GA, Ali S, Naeem R, Thomson NR, Dougan G, Pain A. The population structure of Vibrio cholerae from the Chandigarh Region of Northern India. PLoS Negl Trop Dis. 2014;8(7):e2981.
16. Lam C, Octavia S, Reeves P, Wang L, Lan R. Evolution of seventh cholera pandemic and origin of 1991 epidemic, Latin America. Emerg Infect Dis. 2010;16(7):1130–2.
17. George CM, Monira S, Sack DA, Rashid MU, Saif-Ur-Rahman KM, Mahmud T, Rahman Z, Mustafiz M, Bhuyian SI, Winch PJ, et al. Randomized controlled trial of hospital-based hygiene and water treatment intervention (CHoBI7) to reduce cholera. Emerg Infect Dis. 2016;22(2):233–41.
18. Rafique R, Rashid MU, Monira S, Rahman Z, Mahmud MT, Mustafiz M, Saif-Ur-Rahman KM, Johura FT, Islam S, Parvin T, et al. Transmission of infectious Vibrio cholerae through drinking water among the household contacts of cholera patients (CHoBI7 trial). Front Microbiol. 2016;7:1635.
19. Danin-Poleg Y, Cohen LA, Gancz H, Broza YY, Goldshmidt H, Malul E, Valinsky L, Lerner L, Broza M, Kashi Y. Vibrio cholerae strain typing and phylogeny study based on simple sequence repeats. J Clin Microbiol. 2007; 45(3):736–46.
20. Stine OC, Alam M, Tang L, Nair GB, Siddique AK, Faruque SM, Huq A, Colwell R, Sack RB, Morris JG Jr. Seasonal cholera from multiple small outbreaks, rural Bangladesh. Emerg Infect Dis. 2008;14(5):831.
21. Rashed SM, Azman AS, Alam M, Li S, Sack DA, Morris JG Jr, Longini I, Siddique AK, Iqbal A, Huq A, et al. Genetic variation of Vibrio cholerae during outbreaks, Bangladesh, 2010-2011. Emerg Infect Dis. 2014;20(1):54–60.
22. Alam M, Sultana M, Nair GB, Siddique AK, Hasan NA, Sack RB, Sack DA, Ahmed KU, Sadique A, Watanabe H, et al. Viable but nonculturable Vibrio cholerae O1 in biofilms in the aquatic environment and their role in cholera transmission. Proc Natl Acad Sci U S A. 2007;104(45):17801–6.
23. Chowdhury NR, Chakraborty S, Ramamurthy T, Nishibuchi M, Yamasaki S, Takeda Y, Nair GB. Molecular evidence of clonal Vibrio Parahaemolyticus pandemic strains. Emerg Infect Dis. 2000;6(6):631–6.
24. Kachwamba Y, Mohammed AA, Lukupulo H, Urio L, Majigo M, Mosha F, Matonya M, Kishimba R, Mghamba J, Lusekelo J, et al. Genetic characterization of Vibrio cholerae O1 isolates from outbreaks between 2011 and 2015 in Tanzania. BMC Infect Dis. 2017;17(1):157.
25. Garrine M, Mandomando I, Vubil D, Nhampossa T, Acacio S, Li S, Paulson JN, Almeida M, Domman D, Thomson NR, et al. Minimal genetic change in Vibrio cholerae in Mozambique over time: multilocus variable number tandem repeat analysis and whole genome sequencing. PLoS Negl Trop Dis. 2017;11(6):e0005671.
26. Bankevich A, Nurk S, Antipov D, Gurevich AA, Dvorkin M, Kulikov AS, Lesin VM, Nikolenko SI, Pham S, Prjibelski AD, et al. SPAdes: a new genome assembly algorithm and its applications to single-cell sequencing. J Comput Biol. 2012;19(5):455–77.
27. Overbeek R, Olson R, Pusch GD, Olsen GJ, Davis JJ, Disz T, Edwards RA, Gerdes S, Parrello B, Shukla M. The SEED and the rapid annotation of microbial genomes using subsystems technology (RAST). Nucleic Acids Res. 2014;42(D1):D206–14.
28. Treangen TJ, Ondov BD, Koren S, Phillippy AM. The harvest suite for rapid core-genome alignment and visualization of thousands of intraspecific microbial genomes. Genome Biol. 2014;15(11):524.
29. Price MN, Dehal PS, Arkin AP. FastTree 2–approximately maximum-likelihood trees for large alignments. PLoS One. 2010;5(3):e9490.
30. Letunic I, Bork P. Interactive tree of life (iTOL) v3: an online tool for the display and annotation of phylogenetic and other trees. Nucleic Acids Res. 2016;44(W1):W242–5.

13

Transmissibility of intra-host hepatitis C virus variants

David S. Campo[1*], June Zhang[1,2], Sumathi Ramachandran[1] and Yury Khudyakov[1]

Abstract

Background: Intra-host hepatitis C virus (HCV) populations are genetically heterogeneous and organized in subpopulations. With the exception of blood transfusions, transmission of HCV occurs via a small number of genetic variants, the effect of which is frequently described as a bottleneck. Stochasticity of transmission associated with the bottleneck is usually used to explain genetic differences among HCV populations identified in the source and recipient cases, which may be further exacerbated by intra-host HCV evolution and differential biological capacity of HCV variants to successfully establish a population in a new host.

Results: Transmissibility was formulated as a property that can be measured from experimental Ultra-Deep Sequencing (UDS) data. The UDS data were obtained from one large hepatitis C outbreak involving an epidemiologically defined source and 18 recipient cases. *k*-Step networks of HCV variants were constructed and used to identify a potential association between transmissibility and network centrality of individual HCV variants from the source. An additional dataset obtained from nine other HCV outbreaks with known directionality of transmission was used for validation.

Transmissibility was not found to be dependent on high frequency of variants in the source, supporting the earlier observations of transmission of minority variants. Among all tested measures of centrality, the highest correlation of transmissibility was found with Hamming centrality ($r = 0.720$; $p = 1.57$ E-71). Correlation between genetic distances and differences in transmissibility among HCV variants from the source was found to be 0.3276 (Mantel Test, $p = 9.99$ E-5), indicating association between genetic proximity and transmissibility. A strong correlation ranging from 0.565–0.947 was observed between Hamming centrality and transmissibility in 7 of the 9 additional transmission clusters ($p < 0.05$).

Conclusions: Transmission is not an exclusively stochastic process. Transmissibility, as formally measured in this study, is associated with certain biological properties that also define location of variants in the genetic space occupied by the HCV strain from the source. The measure may also be applicable to other highly heterogeneous viruses. Besides improving accuracy of outbreak investigations, this finding helps with the understanding of molecular mechanisms contributing to establishment of chronic HCV infection.

* Correspondence: fyv6@cdc.gov
[1]Division of Viral Hepatitis, Molecular Epidemiology and Bioinformatics, Centers for Disease Control and Prevention, Atlanta, GA, USA
Full list of author information is available at the end of the article

Background

Hepatitis C virus (HCV) infects nearly 3% of the world's population and is a major cause of liver disease worldwide [1]. HCV infection is an important US public health problem, being the most common chronic blood-borne infection and the leading cause for liver transplantation [2]. Since 2007, HCV has surpassed Human Immunodeficiency Virus (HIV) as a cause of death in the United States [3]. It is estimated that 2.7–3.9 million people in the United States have chronic HCV infection and that > 15,000 die each year from HCV-related disease, with mortality expected to rise in the coming years [4]. Approximately 80% of patients who become infected with HCV develop chronic infections and are at risk for advanced liver disease; 15%–30% of these patients progress to liver fibrosis and cirrhosis and up to 5% die from liver failure due to cirrhosis or hepatocellular carcinoma [2]. Outbreaks of HCV infections are associated with unsafe injection practices, drug diversion, and other exposures to blood and blood products [5].

RNA viruses such as HCV exist as a heterogeneous population of closely related but genetically distinct variants in each infected person [6, 7]. During a transmission event, this variability creates an opportunity for differences in the genetic composition of variants found in the source and recipients. For HIV, it is known that, although source may have many viral variants, recipients are productively infected by a small number of them, often only one [8]. For HCV, several studies have shown that only a small number of HCV variants establish infection in a new host [9–12], commonly resulting in a profound founder effect [13].

Establishment of new infections originating from a small, random sample of intra-host variants from the source (often referred to as a bottleneck) may explain the genetic differences in viral populations found between the source and recipients. A fact that does not support this simple explanation is that a transmission/founder (T/F) virus is often different from major variants present in the source [8]. It is also important to note that due to a high rate of mutations, the viral population (often sampled weeks or months after the transmission event) may be greatly different from the population existing in the source at the time of transmission. Thus, there are a variety of transmission events that can contribute to detected genetic differences between viral populations identified in the source and recipients at the time of sampling.

Besides stochasticity of transmission, changes in genetic composition of the source and recipient populations may be explained by differences in biological properties among intra-host viral variants. It is conceivable that viral variants may differ in their capacity to be transmitted and to establish the first detectable viral population in a new host, with some being more likely to be T/F variants. Viral envelope glycoproteins, which are involved principally in virus attachment and entry into target cells, are likely to correlate with such transmission-related properties. Several recent HIV studies have given support to this active selection model by showing that newly acquired variants often have shorter glycosyl residues and/or less glycosylated envelope glycoproteins than those present in chronically infected persons [8]. For HCV, there are two main lines of evidence for the active selection model. First, Kell et al. [14] showed that T/F variants are recognized by retinoic acid-inducible gene I (RIG-I) in a manner dependent on length of the U-core motif of the poly-U/UC pathogen-associated molecular pattern (PAMP) and are recognized by RIG-I to induce innate immune responses that restrict acute infection. Second, as we have previously shown, physicochemical properties within the hypervariable region 1 (HVR1) of E2 envelope glycoprotein are significantly different between HCV variants found in acutely and chronically infected individuals [15]. This finding was recently confirmed in a study of three transmission pairs that found a common HVR1 amino acids pattern in transmitted HCV variants, with each transmission pair sharing specific patterns [11].

Finding specific phenotypic properties contributing to transmissibility of variants can help to understand the establishment of HCV chronic infection, improve accuracy of outbreak detection, and rationally design strategies for preventing HCV infection by limiting the search space for vaccine candidates. Here, we define transmissibility as a property that can be measured from experimental data obtained using ultra-deep sequencing (UDS) and identify properties of the source HCV variants that are associated with transmissibility.

Methods

Data

We conducted an in-depth analysis of one HCV infection outbreak (hereinafter referred to as the AW outbreak) where the source case of HCV infection was known. The AW outbreak investigation started with identification of two patients diagnosed with acute HCV infection from the same hospital. Further investigation implicated a drug-diverting; HCV-infected surgical technician as the source of the outbreak. Sera from the source and patients found to be serologically HCV positive were used to conduct HCV sequence analysis. In total, 5970 patients were notified of their possible exposure to HCV, 88% of whom were tested and had results reported to the state public health department. Ultimately, 18 patients had HCV sequences highly related to the surgical technician's virus [16]. The associations identified from the AW outbreak analysis were further tested using data from nine other transmission clusters with known sources identified from eight additional outbreak investigations [16–22]. All nine outbreaks we examined were serologically confirmed, epidemiologically defined, and reported to the Centers for Disease Control and Prevention between 2008-2013 [5].

Sequencing

The HCV strain identified in the AW outbreak was sequenced from all 19 cases using UDS. PCR products were pooled and subjected to pyrosequencing using GS FLX Titanium Sequencing Kit (454 Life Sciences, Roche, Branford, CT). The UDS files were processed using the error correction algorithms KEC and ET [23]. All sequence variants represented with a single UDS read were excluded from consideration. No sequences were obtained from one recipient despite several attempts. It must be noted that UDS was used only in the AW outbreak. Sequences for the other outbreaks were obtained using the End-Point Limiting-Dilution Real-Time PCR method for sequencing of multiple HVR1 clones [24–26]. For each sample of HCV sequences, a Multiple Sequence Alignment (MSA) was created using MAFFT 7.221 [27]. The primer sequences were removed and the final sequences were 264 nucleotides in length.

Transmissibility

We developed a new measure of transmissibility for each variant in a source. Let R denote a set of recipients, with N being a total number of recipients. Let R_j be a set of distinct variants in recipient j. A total number of unique variants in recipient j is $|R_j|$. For each variant i in the source, the average distance to recipient R_j is:

$$d(i, R_j) = \frac{1}{|R_j|}\sum_{k=1}^{|R_j|} d(i,k),$$

where $d(i,k)$ is the Hamming distance between a source's variant i and recipient's variant k. Rather than using a raw value $d(i,R_j)$, we used the rank of each source variant with respect to recipient:

$$r(i, R_j) = \mathrm{rank}(-d(i, R_j))$$

Finally, transmissivity T_i of each variant is the harmonic mean of its rank with respect to all recipients (Fig. 1):

$$T_i = \frac{N}{\sum_{j=1}^{|R|} \frac{1}{r(i,R_j)}}$$

We considered several transformation schemes such as the top quartile: for each recipient, the sequence was marked as 1 if it belongs to the top quartile of genetic distances, or 0 if otherwise, which highly correlated with the rank-based measure ($r = 0.8625$; p value = 5.82 E-132). However, we found ranking more suitable because, in the case of one or few recipients, the top quartile did not generate a continuous value for all sequences;

Networks

For the set of 6231 distinct variants found in the AW outbreak, we visualized the sequence similarity by means of a k-step network as previously described [7, 24, 28]; nodes in the k-step network correspond to variants. The k-step network is equivalent to the union of all possible Minimum Spanning Trees and allows one to efficiently visualize the genetic relatedness among all variants present in a sample. The networks were drawn using GEPHI software [29].

Variant properties

The read frequency of a variant is the number of cleaned experimental reads associated with that particular sequence from experiments. We previously showed that experimentally obtained frequencies are highly correlated with true proportions found in the sample [23]. Several topological properties of the k-step network were

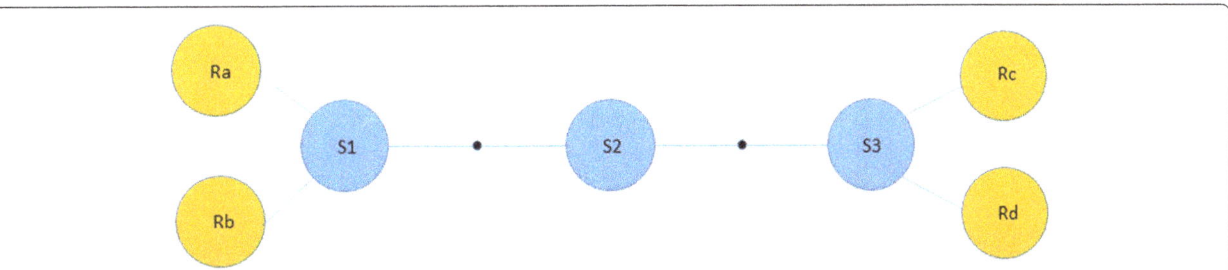

Fig. 1 Diagram illustrating multiple clusters of transmissibility. There are three variants in the source (S1, S2 and S3). The lines represent one-step mutation, showing that S1 is very close to recipients Ra and Rb, whereas S3 is very close to recipients Rc and Rd. If we use the arithmetic mean of the genetic distances, all three sequences would get the same value (3), which would assign an artificially low value to S2. If we use the harmonic mean instead, S1 and S3 have a smaller value than S2 (1.5 vs 2)

measured for each node: 1) degree centrality (number of links), 2) closeness centrality (reciprocal of the average shortest path) [30], 3) betweenness centrality (fraction of shortest path that goes through each node) [31], and 4) k-shell (a degree-based decomposition) [32]. Finally, the Hamming centrality of each sequence was calculated. Hamming centrality is analogous to closeness centrality [30] but it uses the matrix of Hamming distances among all pairs of sequences instead of the matrix of shortest paths in the network. Let N be the number of variants in a sample, the Hamming centrality (*HC*) of variant x is given by:

$$HC_x = \frac{1}{\sum_y d(y,x) N-1}$$

where $d(y, x)$ is the hamming distance between variants x and y.

Hypothesis testing

We performed a Mann-Whitney U test to compare the frequency distributions of shared and non-shared variants. In order to evaluate if location in sequence space was associated with Transmissibility, we performed a Mantel test [33]. The Mantel test is a statistical permutation test ($n = 10{,}000$) of the correlation between two dissimilarity matrices: the first matrix is the Hamming distance between any two variants, whereas the second matrix is the absolute difference in their transmissibility.

Results

Frequency of shared HCV variants

The data available from the AW outbreak were especially amenable to analysis of transmissibility of intra-host HCV variants: (1) an epidemiologically established transmission from a single source case to 18 recipients over a 5-week period [22]; and (2) a large number of intra-host HCV HVR1 variants sampled from each case using UDS [23]. Sequences were obtained from the source and all but one recipient. A total of 6231 HCV variants obtained from the outbreak cases were represented by 137,691 sequence reads. An average of 388.16 variants, ranging from 2 to 816, were sampled from each case (Fig. 2). The confirmed outbreak source is designated as AW02. Although the number of variants sampled from the source was not highest among all cases, the intra-host HCV population had the highest level of nucleotide diversity [24]. Of the 441 unique source variants, 119 were shared with the recipients. Out of the 17 recipients, 12 shared at least one variant with the source (Fig. 2). While most recipients had only a few variants in common with the source, patient AW05 shared 113.

HCV variant frequencies, defined by the number of sequence reads representing each variant, varied in a broad range. Shared variants had a 6.4 times higher average frequency than non-shared variants (43.23 vs 6.70, p value <0.0001). Despite this observation, correlation between the number of recipients, where a certain source variant was found, and its frequency in the source was very low ($r = 0.0949$; $p = 0.0464$). This correlation, however, improved when Log10 of frequency was used ($r = 0.4020$; $p < 0.0001$). In support of this finding, the most frequent source variant (35.30% of the whole population) was found in only one of the 17 recipients, whereas the most shared variant, found in 8 recipients, was represented in the source by only a few sequence reads (0.11% of the whole population, less than 24.71% of all source variants).

Genetic relatedness of HCV variants

The genetic relationships among HCV variants from the source and recipients was visualized using a k-step

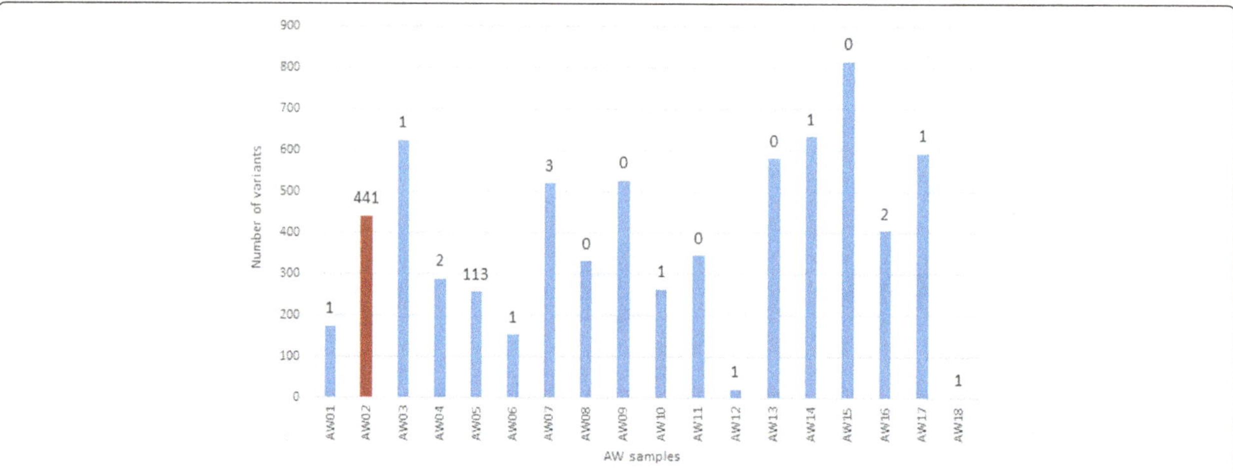

Fig. 2 Number of variants in each sample of the AW outbreak. The number above each bar indicates the number of HCV variants that are shared with the source (AW2, shown in red)

network, where each node is a distinct HCV variant ($n = 6231$) and the size of the node is proportional to the number of samples that share that variant (Fig. 3). There are several modules in the network, each representing a cluster of closely related HCV variants. The source variants are distributed among several clusters. The cluster containing a majority of HCV variants that were common among recipients was built around a single source variant. This particular source variant was shared with eight recipients, indicating a variable capacity of different source variants to serve as founders in these recipients.

Transmissibility

Transmissibility measured using intra-host variants shared between source and recipients is highly sensitive to stochastic disparities associated with variant sampling and variation in evolutionary history in infected persons. The transmissibility measure developed here reduces this sensitivity. A bimodal frequency distribution of transmissibility values for HCV variants (Fig. 4) reflects a complex modular organization of intra-host HCV population in the source (Fig. 3). This suggests the existence of more than one cluster, from which founders of the acute populations in recipients were recruited, with only very few variants having the highest values of transmissibility.

Pearson correlation was calculated between several sequence properties and transmissibility, though variant frequency did not correlate with transmissibility (Fig. 5).

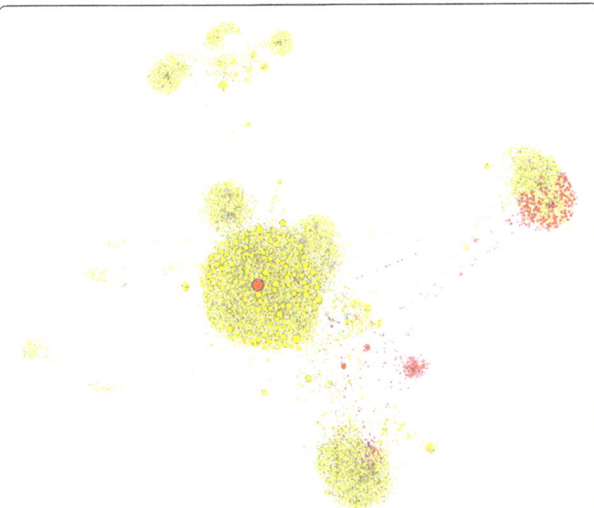

Fig. 3 k-step network of the AW outbreak. Each node is a distinct HCV variant (n = 6231) found among the 18 samples. The size of the node is proportional to the number of samples that share that variant. Red nodes are variants found in the source; yellow nodes are variants found exclusively among recipients. Out of 11,418 links, 94.35% have a Hamming distance of 1, 5.24% a distance of 2, 0.32% a distance of 3, 0.06% a distance of 4, and 0.03% a distance of 15

The source variants with high transmissibility had a moderate frequency (Fig. 6). Transmissibility was found to correlate with several measures of network centrality, including degree, closeness, and betweenness centrality. However, the highest correlation was found with Hamming centrality ($r = 0.720$; $p = 1.57$ E-71), which is calculated from the matrix of Hamming distances rather than from the k-step network.

We checked whether genetically close sequences have similar values of transmissibility. Correlation between matrices of genetic distances and differences in transmissibility among HCV variants was found to be low but significant ($r = 0.3276$; Mantel Test p value = 9.99 E-5). Figure 7 shows how the average difference in transmissibility increases with genetic distance. However, for large genetic distances, the differences are reduced, suggesting the existence of more than one cluster of high transmissibility.

Other HCV outbreaks

The strong association we identified in the AW outbreak between Hamming centrality and transmissibility was further tested using data from nine HCV transmission clusters from the eight other outbreak investigations. We obtained much lower sample sizes than in the AW outbreak here because of the alternate sequencing method used (average number of reads = 50.62). Nonetheless, a strong correlation was observed between Hamming centrality and transmissibility in 7 of the 9 transmission clusters (average $r = 0.8011$, ranging from 0.5650 to 0.947; p-value < 0.05).

Discussion

Stochastic transmission of a limited number of HCV variants from highly heterogeneous intra-host populations generates an opportunity for significant differences in genetic composition of HCV populations found in the source and recipients. Each intra-host HCV population is organized in subpopulations [34] which can be visualized using k-step networks [7]. Simulation experiments of transmission events have previously shown that stochastic sampling of a very small fraction of variants from the source results in picking representatives from each large subpopulation [7], indicating that each HCV subpopulation in the recipient has a very high probability to be founded by variants transmitted from the source. However, not all founded subpopulations increase their size at the beginning of infection in a new host to the level detectable by the available sequencing strategies, including UDS used here. Only one or few subpopulations grow in density early in infection, establishing a dominant subpopulation during the acute stage of HCV infection [34]. The other subpopulations become detectable later in infection after decline of the initial HCV

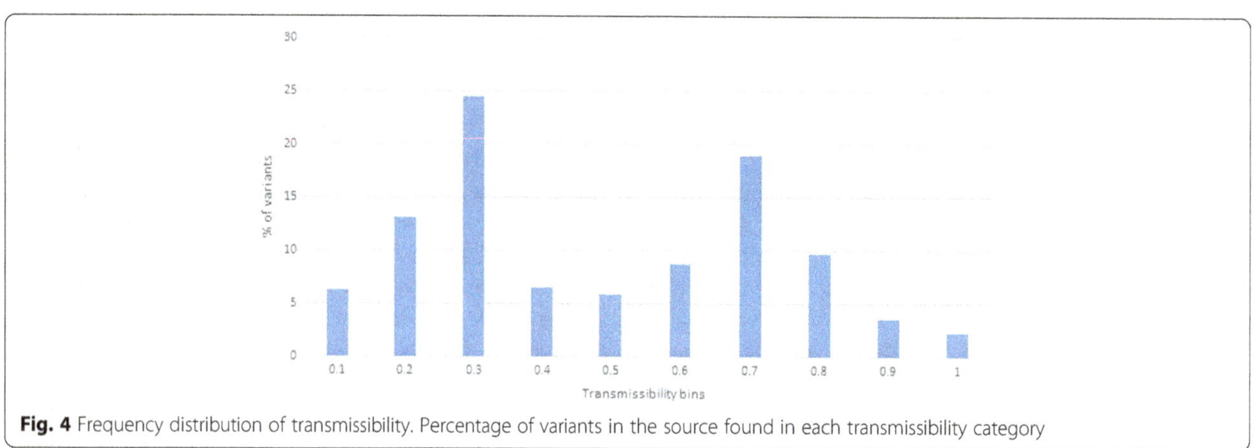

Fig. 4 Frequency distribution of transmissibility. Percentage of variants in the source found in each transmissibility category

variants [34]. Thus, HCV T/F variants from different subpopulations may differ in their biological capacity to rapidly establish a dominant population at the early stages of infection. The frequent finding that the dominant intra-host population in recipients is genetically closest to a minority variants from the source [8] argues against a simple stochasticity of transmission. In addition, identification of common patterns of HCV variants shared among linked individuals strongly suggests specific biological properties for T/F variants [11].

Here, we explored genetic data from one large outbreak involving many cases of HCV infection with epidemiologically confirmed transmission from a single case [16] to identify a differential capacity of the source HCV variants to facilitate establishment of the first dominant population in new hosts. We define transmissibility as a property that can be measured from experimental UDS data and can be applied to other genetically heterogeneous pathogens. This measure of transmissibility is based on: (i) use of average distance (from each sequence in the donor to all sequences in the recipient) instead of minimal distance because the latter created too many ties among sequences which was problematic given the need for ranking; (ii) use of ranks rather than raw distances to control for variation in the range of genetic distances for different recipients; (iii) a compound measure for all recipients was generated using the harmonic mean to reduce probability of producing a high transmissibility value for a single HCV variant located between source subpopulations, variants from which transmitted to different recipients (Fig. 1); (iv) the use of a harmonic mean gives more weight to variants that are very close to one or more recipients; although not used in this study, other schemes such as an expert-based voting system can be applied to generate this compound measure; and (v) use of hamming distance instead of other genetic or patristic distance is justified by the fact, based on our previous finding [24], that hamming distance performs slightly better in identifying closely related individuals linked by transmission.

Transmissibility measured here mitigates several problems associated with the use of shared variants (distance equal 0) rather than all sampled variants at the entire range of genetic distances from each other. For instance,

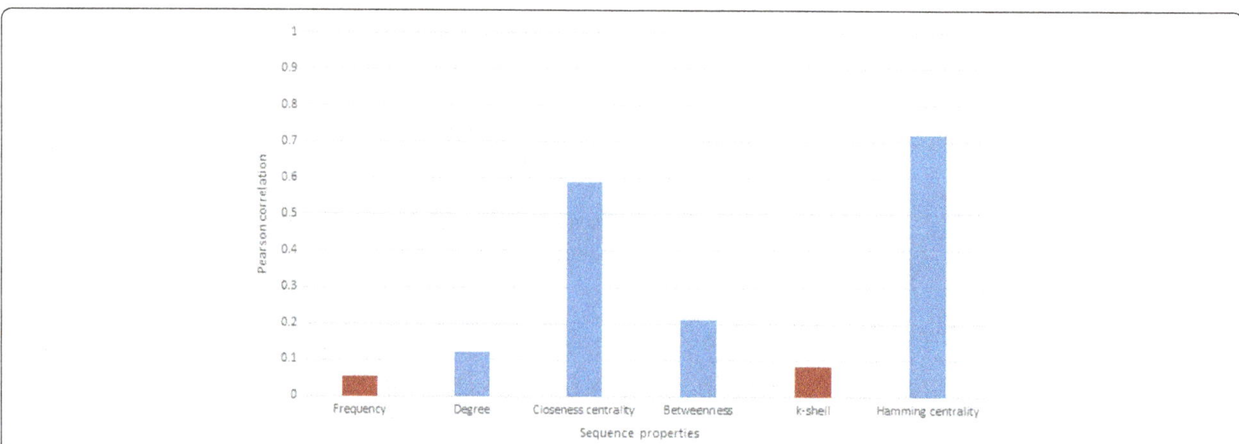

Fig. 5 Pearson correlation between transmissibility and different sequence properties. Blue bars correspond to *p*-values lower than 0.05, whereas red bars correspond to p-values higher than > 0.05

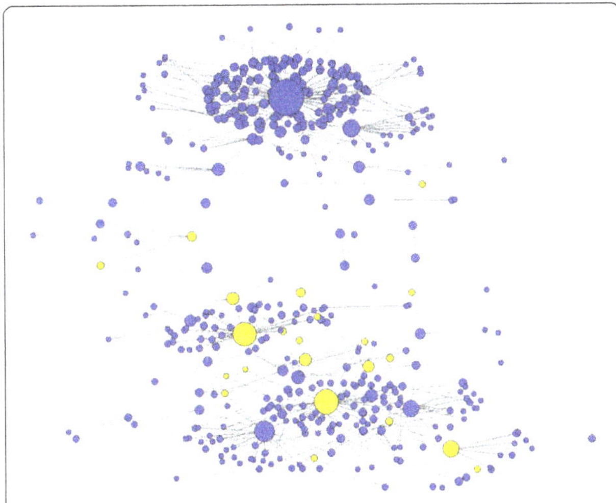

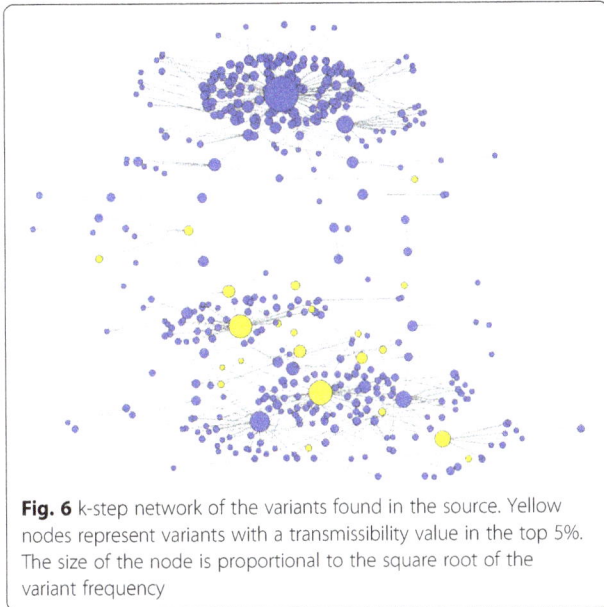

Fig. 6 k-step network of the variants found in the source. Yellow nodes represent variants with a transmissibility value in the top 5%. The size of the node is proportional to the square root of the variant frequency

in the AW outbreak, out of the 17 recipients, 5 did not share any variant and 1 shared many variants with the source. This variation in the number of shared variants is most probably associated with:

(i) Stochasticity of sampling: Due to the high level of intra-host HCV genetic heterogeneity, stochasticity of sampling is very high. Even when comparing two samples from the same individual obtained in the same UDS experiment, we found that the average level of variant sharing was approximately 53.80% (average of 576 pairwise comparisons among 24 samples obtained from the same chronically HCV-infected individual, data not shown).

(ii) Sampling times: Sampling from source and recipient cases does not occur at the time of transmission. Owing to the high rate of diversification of RNA viruses, genetic composition of the HCV population sampled weeks or months after the transmission event may not reflect what was present at the time of transmission in the source. In the AW outbreak studied here, all recipients were infected just within 5 weeks and the time between infection and sampling was ~30 weeks.

Both factors have a significant effect on the transmissibility value measured using shared variants. The measure formulated here is robust to variations in sampling, owing to the use of genetic distances from every source variant to all variants from every recipient. Even though the actual T/F variants may not be sampled from recipient because of their low frequency, the genetically proximal HCV variants of sufficient frequency may still be detected and used as a proxy for the T/F variant. The correlation observed here between Hamming distance and the difference in transmissibility among genetically close HCV variants indicates that these variants are similar in their transmissibility, which explains the robustness of our transmissibility measure to sampling stochasticity. In addition, the use of the entire range of distances creates a measure that is continuous rather than binary and enables more accurate modelling.

Although shared variants tend to have a higher average frequency than non-shared variants, there was a very low correlation between the number of recipients who carried a certain source variant, and the variant frequency ($r = 0.0949$; $p = 0.0464$). Transmissibility formulated here did not correlate with the variant frequency. Taken together, these observations suggest that a solely stochastic model accounts for a very small portion of the variance in transmissibility among the source HCV variants. It is conceivable that, to be physically transferred to a recipient, HCV variants must be of sufficient frequency in the source, but other phenotypic traits become important for selecting the variant(s) that establish the first dominant population upon transmission to a new host.

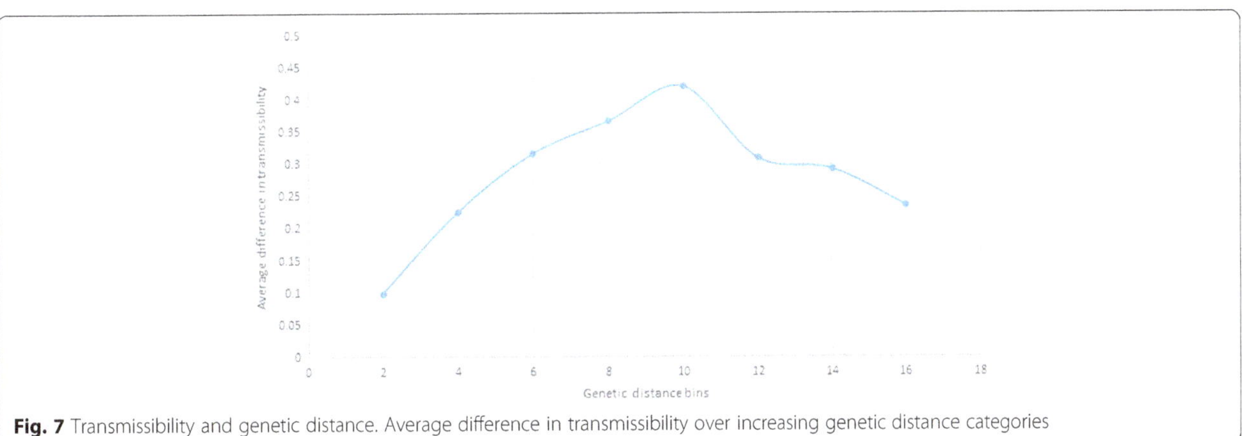

Fig. 7 Transmissibility and genetic distance. Average difference in transmissibility over increasing genetic distance categories

Given the high level of HCV genetic heterogeneity, a large variety of phenotypic traits may be responsible for transmissibility of variants. Depending on the strain and the genetic environments of the host and recipient, different combinations of traits may play a major role in defining each specific transmission event. Identification of common amino acid patterns among shared HCV HVR1 variants in transmission pairs suggests the existence of phenotypic properties associated with transmissibility [11]. However, these patterns were experimentally shown to confer no greater capacity for cell entry than the ones derived from non-transmitted variants [11].

In this study, we found that the Hamming centrality of HCV variants from the source of our outbreak was strongly associated with transmissibility of variants to recipients. This property has the potential to be applied to any highly heterogeneous virus since it is not defined by a motif, but by the location of variants in the sequence space occupied by the entire viral population from the source. A weak association of transmissibility with other measures of network centrality like closeness or betweenness centrality indicates that the Hamming centrality is less dependent than these other measures on comprehensive sampling of HCV variants. The association with the variant centrality in genetic space suggests existence of certain phenotypic traits that can affect transmissibility. One of the possible traits is mutational robustness [35], which has been experimentally observed for HCV [7] and results in generation of numerous viable mutant variants. Another beneficial trait for transmissibility is a weak cross-immunoreactivity with its mutant progeny so that antibodies against T/F variants would be inefficient in neutralizing the genetically proximal HCV variants. In order to play a role in transmission, this weak cross-immunoreactivity should be host independent, which is supported by recent findings regarding the important contribution of cross-immunoreactivity in directing intra-host HCV evolution [36] and limited cross-immunoreactivity among HCV variants in acute infection [37].

The main limitation of this study is that its analysis is based on genetic data from only ten transmission clusters, suggesting that the results should be cautiously applied to other datasets. However, it must be noted that HCV outbreaks are difficult to detect, or fully characterize, because HCV infections are asymptomatic in more than 70% of infected persons for years, which explains limited availability of genetic data from outbreaks. Further improvement of molecular surveillance using novel technologies, like Global Hepatitis Outbreak and Surveillance Technology (GHOST) (Longmire et al. in this issue), is required to develop advanced approaches for accurate tracking of viral transmissions.

Conclusions

Stochastic sampling does not comprehensively explain genetic differences between intra-host HCV populations found in a source and the recipients. Transmissibility, as formally measured in this study, is associated with certain biological properties that define location of variants in the HCV genetic space. Besides improving accuracy of outbreak investigations, this finding has important implications for understanding of molecular mechanisms contributing to establishment of HCV chronic infection.

Acknowledgements
Not applicable

Funding
All work and publication costs were funded by the Centers for Disease Control and Prevention.

About this supplement
This article has been published as part of *BMC Genomics* Volume 18 Supplement 10, 2017: Selected articles from the 6th IEEE International Conference on Computational Advances in Bio and Medical Sciences (ICCABS): genomics. The full contents of the supplement are available online at https://bmcgenomics.biomedcentral.com/articles/supplements/volume-18-supplement-10.

Authors' contributions
DSC and YK designed the study. SR conducted the laboratory experiments. DSC and JZ analysed the data. DSC and YK wrote the manuscript. All authors evaluated the final draft. All authors read and approved the final manuscript.

Competing interests
The authors declare that they have no competing interests.

Author details
[1]Division of Viral Hepatitis, Molecular Epidemiology and Bioinformatics, Centers for Disease Control and Prevention, Atlanta, GA, USA. [2]Department of Electrical Engineering, University of Hawaii, Manoa, HI, USA.

References

1. Mohd Hanafiah K, Groeger J, Flaxman AD, Wiersma ST. Global epidemiology of hepatitis C virus infection: new estimates of age-specific antibody to HCV seroprevalence. Hepatology. 2013;57(4):1333–42.
2. Alter M. Epidemiology of hepatitis C virus infection. World J Gastroenterol. 2007;13(17):2436–41.
3. Ly KN, Xing J, Klevens RM, Jiles RB, Ward JW, Holmberg SD. The increasing burden of mortality from viral hepatitis in the United States between 1999 and 2007. Ann Intern Med. 2012;156(4):271–8.
4. Ward JW. The hidden epidemic of hepatitis C virus infection in the United States: occult transmission and burden of disease. Topics Antiviral Med. 2013;21(1):15–9.
5. Healthcare-Associated Hepatitis B and C Outbreaks Reported to CDC in 2008-2013 [http://www.cdc.gov/hepatitis/Outbreaks/HealthcareHepOutbreakTable.htm].

6. Domingo E, Sheldon J, Perales C. Viral quasispecies evolution. Microbiol Mol Biol Rev. 2012;76(2):159–216.
7. Campo DS, Dimitrova Z, Yamasaki L, Skums P, Lau DT, Vaughan G, Forbi JC, Teo CG, Khudyakov Y. Next-generation sequencing reveals large connected networks of intra-host HCV variants. BMC Genomics. 2014;15(Suppl 5):S4.
8. Sagar M. HIV-1 transmission biology: selection and characteristics of infecting viruses. J Infect Dis. 2010;202(Suppl 2):S289–96.
9. Wang G, Sherrill-Mix S, Chang K, Quince C, Bushman F. Hepatitis C virus transmission bottlenecks analyzed by deep sequencing. J Virol. 2010;84(12): 6218–28.
10. Bull RA, Luciani F, McElroy K, Gaudieri S, Pham ST, Chopra A, Cameron B, Maher L, Dore GJ, White PA, et al. Sequential bottlenecks drive viral evolution in early acute hepatitis C virus infection. PLoS Pathog. 2011;7(9): e1002243.
11. D'Arienzo V, Moreau A, D'Alteroche L, Gissot V, Blanchard E, Gaudy-Graffin C, Roch E, Dubois F, Giraudeau B, Plantier JC, et al. Sequence and functional analysis of the envelope glycoproteins of hepatitis C virus variants selectively transmitted to a new host. J Virol. 2013;87(24):13609–18.
12. Li H, Stoddard MB, Wang S, Blair LM, Giorgi EE, Parrish EH, Learn GH, Hraber P, Goepfert PA, Saag MS, et al. Elucidation of hepatitis C virus transmission and early diversification by single genome sequencing. PLoS Pathog. 2012; 8(8):e1002880.
13. Preciado MV, Valva P, Escobar-Gutierrez A, Rahal P, Ruiz-Tovar K, Yamasaki L, Vazquez-Chacon C, Martinez-Guarneros A, Carpio-Pedroza JC, Fonseca-Coronado S, et al. Hepatitis C virus molecular evolution: transmission, disease progression and antiviral therapy. World J Gastroenterol. 2014; 20(43):15992–6013.
14. Kell A, Stoddard M, Li H, Marcotrigiano J, Shaw GM, Gale M Jr. Pathogen-associated molecular pattern recognition of hepatitis C virus transmitted/founder variants by RIG-I is dependent on U-Core length. J Virol. 2015;89(21):11056–68.
15. Astrakhantseva IV, Campo D, Araujo A, Teo C-G, Khudyakov Y, Kamili S: Variation in physicochemical properties of the hypervariable region 1during acute and chronic stages of hepatitis C virus infection. In: Bioinformatics and Biomedicine Workshops (BIBMW), 2011 IEEE International Conference. IEEE Xplore Digital Library; 2011: P. 72 - 78.
16. Warner AE, Schaefer MK, Patel PR, Drobeniuc J, Xia G, Lin Y, Khudyakov Y, Vonderwahl CW, Miller L, Thompson ND. Outbreak of hepatitis C virus infection associated with narcotics diversion by an hepatitis C virus-infected surgical technician. Am J Infect Control. 2015;43(1):53–8.
17. Noviello S, Smith P, Chai F, Nainan O, Genovese-Candela A, Armellino D, Farber B, Stricof R, Chang H, Birkhead G et al: Hepatitis C virus transmission by a cardiac surgeon in the United States. 2015.
18. Chai F, Xia G, Williams I, Rosenberg J, Janowski M, Ginsberg M, Alter M, Nainan O, Gunn R, Janowski M. Transmission of hepatitis C virus at a pain remediation clinic - San Diego, California 2003. In: 43rd annual meeting of the Infectious Diseases Society of America (IDSA): October 6 to 9. Arlington: Infectious Diseases Society of America; 2005. http://www.idsociety.org/Index.aspx.
19. Lee K, Scoville S, Taylor R, Baum S, Chai F, Bower W, Soltis M, Xia G, Hachey W, Betz T et al. Outbreak of acute hepatitis C virus (HCV) infections of two different genotypes associated with an HCV-infected anesthetist. In: 47th annual meeting of the Infectious Diseases Society of America (IDSA): October 29 to November 1 2009. Arlington: Infectious Diseases Society of America; 2009. http://www.idsociety.org/Index.aspx.
20. Thompson N, Novak R, White-Comstock M, Xia G, Ganova-Raeva L, Ramachandran S, Khudyakov Y, Bialek S, Williams I. Patient-to-patient hepatitis C virus transmissions associated with infection control breaches in a Hemodialysis unit. J Nephrol Therapeutics. 2012;S10:002.
21. Fischer GE, Schaefer MK, Labus BJ, Sands L, Rowley P, Azzam IA, Armour P, Khudyakov YE, Lin Y, Xia G, et al. Hepatitis C virus infections from unsafe injection practices at an endoscopy clinic in Las Vegas, Nevada, 2007-2008. Clin Infect Dis. 2010;51(3):267–73.
22. Moore ZS, Schaefer MK, Hoffmann KK, Thompson SC, Xia GL, Lin Y, Khudyakov Y, Maillard JM, Engel JP, Perz JF, et al. Transmission of hepatitis C virus during myocardial perfusion imaging in an outpatient clinic. Am J Cardiol. 2011;108(1):126–32.
23. Skums P, Dimitrova Z, Campo DS, Vaughan G, Rossi L, Forbi JC, Yokosawa J, Zelikovsky A, Khudyakov Y. Efficient error correction for next-generation sequencing of viral amplicons. BMC Bioinformatics. 2012;13(Suppl 10):S6.
24. Campo D, Xia G, Dimitrova Z, Lin Y, Ganova-Raeva L, Punkova L, Ramachandran S, Thai H, Sims S, Rytsareva I, et al. Accurate genetic detection of hepatitis C virus transmissions in outbreak settings. J Infect Dis. 2015;213(6):957–65.
25. Ramachandran S, Xia GL, Ganova-Raeva LM, Nainan OV, Khudyakov Y. End-point limiting-dilution real-time PCR assay for evaluation of hepatitis C virus quasispecies in serum: performance under optimal and suboptimal conditions. J Virol Methods. 2008;151(2):217–24.
26. Ramachandran S, Zhai X, Thai H, Campo DS, Xia G, Ganova-Raeva LM, Drobeniuc J, Khudyakov YE. Evaluation of intra-host variants of the entire hepatitis B virus genome. PLoS One. 2011;6(9):e25232.
27. Katoh K, Standley DM. MAFFT multiple sequence alignment software version 7: improvements in performance and usability. Mol Biol Evol. 2013; 30(4):772–80.
28. Quirin A, Cordón O, Guerrero-Bote V, Vargas-Quesada B, Moya-Anegón F: A quick MST-based algorithm to obtain pathfinder networks (∞, n − 1). J Am Soc Inf Sci Technol 2008, 59(12):1912-1924.
29. Bastian M, Heymann S, Jacomy M. Gephi: an open source software for exploring and manipulating networks. San Jose: International AAAI conference on weblogs and social media; 2009. https://www.aaai.org/.
30. Latora V, Marchiori M. Efficient behavior of small-world networks. Physics Review Letters. 2001;87(19):1–14.
31. Freeman L. A set of measures of centrality based on betweenness. Sociometry. 1977;40:35–41.
32. Alvarez-hamelin I, Dall'Asta L, Vespignani A. K-core decomposition: a tool for the visualization of large scale networks. Adv Neural Inf Proces Syst. 2006; 18(41):1–13.
33. Mantel N, Valand R. A technique of nonparametric multivariate analysis. Biometrics. 1970;26:547–58.
34. Ramachandran S, Campo DS, Dimitrova ZE, Xia GL, Purdy MA, Khudyakov YE. Temporal variations in the hepatitis C virus intrahost population during chronic infection. J Virol. 2011;85(13):6369–80.
35. van Nimwegen E, Crutchfield JP, Huynen M. Neutral evolution of mutational robustness. Proc Natl Acad Sci U S A. 1999;96(17):9716–20.
36. Skums P, Bunimovich L, Khudyakov Y. Antigenic cooperation among intrahost HCV variants organized into a complex network of cross-immunoreactivity. Proc Natl Acad Sci U S A. 2015;112(21):6653–8.
37. Campo DS, Dimitrova Z, Yokosawa J, Hoang D, Perez NO, Ramachandran S, Khudyakov Y. Hepatitis C virus antigenic convergence. Sci Rep. 2012;2:267–77.

14

Prediction of missing common genes for disease pairs using network based module separation on incomplete human interactome

Pakeeza Akram and Li Liao*

Abstract

Background: Identification of common genes associated with comorbid diseases can be critical in understanding their pathobiological mechanism. This work presents a novel method to predict missing common genes associated with a disease pair. Searching for missing common genes is formulated as an optimization problem to minimize network based module separation from two subgraphs produced by mapping genes associated with disease onto the interactome.

Results: Using cross validation on more than 600 disease pairs, our method achieves significantly higher average receiver operating characteristic ROC Score of 0.95 compared to a baseline ROC score 0.60 using randomized data.

Conclusion: Missing common genes prediction is aimed to complete gene set associated with comorbid disease for better understanding of biological intervention. It will also be useful for gene targeted therapeutics related to comorbid diseases. This method can be further considered for prediction of missing edges to complete the subgraph associated with disease pair.

Keywords: Disease module separation, Optimization, Interactome, Missing gene, Comorbidity

Background

Genetic cause for diseases is complex and complicated, and can rarely be attributed to a single gene. Instead, often, multiple factors are involved in manifestation of disease symptoms. Furthermore, as genes can take on more than one function and different pathways and processes are intertwined and can crosstalk to one another, it is therefore also quite common that one gene may be implicated in two or more diseases. As a result, it is sensible and informative to examine not only the associated genes of one disease to understand its pathology but also the overlap between the sets of associated genes of two diseases of high comorbidity risk in order to shed lights on the interplay of the two diseases [1–3]. Yet, the knowledge that can be gained from a list of genes, or their product proteins, would be quite limited if not putting them in the biological context, such as the signaling transduction pathways, regulatory and metabolic pathways in which they are involved.

Numerous efforts are being taken to identify relationship between two diseases [4, 5]. Comorbidity refers to the phenomenon that two (or more) diseases co-occur. Bar the pure coincidence, comorbidity would indicate that the two diseases are somehow pathologically similar. The similarity may reveal at various levels: from more phenotyptic ones, such as disease symptoms or coexpression of associated genes, to more genotypic ones, such as sharing common genes between the respective gene sets associated with the diseases. Indeed, there are reports on disease relationship which incorporates the fact of common

* Correspondence: LiLiao@udel.edu
Department of Computer & Information Sciences, University of Delaware, Newark, DE, USA

genetic origin of diseases [6, 7]. Recently emerged disease network theory has shifted focus to disease module [4]. Disease module contains those set of genes whose mutations have effect on phenotype, these set of genes are not scattered by chance in the interactome but they reside close to each other due to their interactions. These interactions form one or several connected subgraphs called as "disease module". Specifically, network based separation of a disease pair A and B (SAB) is introduced to compare shortest distances between proteins within each disease to the shortest distances between disease pair A and B. Relationships between a pair of diseases that have been revealed via other means, such as gene ontology (GO) term similarity and relative risk (RR) for comorbidity, are to correlate with the overlapping of two disease modules, supporting the hypothesis: cause of disruption leading to one disease may cause another disease sharing common characteristics. For example, [4] used the disease history of 30 million individuals aged 65 and older (U.S. Medicare) to determine for each disease pair the relative risk RR of disease comorbidity, finding that the relative risk drops from RR ≥10 for $S_{AB} < 0$ to the random expectation of RR ≈ 1 for $S_{AB} > 0$. While the esults show great promise, however, a significant challenge presents due to the limitation of having very few data. For example, there are only 7% of the disease pairs which overlap with each other and have negative S_{AB} value. At the system level, only 20% of the disease interaction network has been captured [4].

In this study, we attempt to address the issue by a novel method to predict the missing genes in the disease module using available information such as genes association to diseases, relative risk of cooccurrence of two diseases and human interactome. Our work starts out with the findings about disease module separation S_{AB} from [4] and explores its utility as a powerful indicator to determine comorbid diseases: smaller SAB indicates that two selected diseases are more closely located in the interactome, and hence may show comorbid behavior. To complete the set of gene associated with disease and contribute towards completing the interactome, it is critical to identify missing common genes. The method formulates the task of searching for missing common genes as an optimization problem to minimize a network based module separation between two subgraphs formed by mapping the disease associated genes onto the interactome. Tested on a dataset of more than 600 disease pairs using cross-validation, it is shown that the method achieves an average ROC score of 0.95.

Methods

In this section, we first briefly introduce the various concepts related to disease module on incomplete interactome, especially a quantity S_{AB}, called module separation, as given in [4], to measure relationship between two disease modules A and B. Then we explain in detail our method of finding missing common genes for a given pair of diseases formulated as an optimization problem to minimize S_{AB}.

Disease module on Interactome and module separation

Interactome contains all protein-protein interactions in the cell, and can be conveniently represented as a graph (or network), in which proteins are represented as nodes and interaction between two proteins is represented as an edge connecting the two corresponding nodes. Reconstructing the interactome is a central task in systems biology, which studies the cell as a system in a holistic way instead of simple ensemble of isolated items. Due to the limitation of the current technology, interactome for most organisms, even model organisms, is incomplete, with missing nodes and edges. Nonetheless, the incomplete interactome can already provide valuable insights into many biological processes which cannot be obtained otherwise. In [4], it is shown how to uncover disease-disease relationships through the incomplete interactome. Diseases with genetic causes have been studied widely, often with a focus to identify the culprit gene only, to find that in many cases the cause cannot be attributed to a single gene; instead it is very common that multiple genes involving in multiple cellular processes may be at play. Without putting these pieces in a bigger context, it is difficult to fully understand the pathological mechanisms. Work in [4] presents a systematic study to uncover disease-disease relationships by mapping the associated genes onto the interactome.

As mentioned by [4], given a pair of diseases A and B, the genes known to be associated with them are put into two separate sets G_A and G_B respectively. Let graph G be the interactome, with node set V, and edge set E. Let map the genes in G_A and G_B onto G with two different colors, say, nodes in G corresponding to genes in G_A are colored red and nodes in G corresponding to genes in G_B are colored blue. For any shared gene, i.e., a gene is known to be associated to both disease A and disease B, then the corresponding node will be colored half red and half blue. Although all the red nodes are genes associated with disease A, indicating relatedness among them, they may not form a single connected component (or subgraph) of graph G of the interactome; often they form several connected components. This may be due to either incompleteness of the interactome (i.e., missing edges) or unknown associated genes, or a combination of both. However, if the connected components are too fragmented, say not significantly different from what can be formed by randomly mapped genes, then it is difficult to reliably infer useful relationships. So, in [4], the size of the largest connected component, as a percentage of the total number of genes associated to a disease, must be maintained beyond a threshold, which is set based on percolation theory and the data used in the study. And the largest connected component, meeting the size requirement, is then called module as representative for the disease. For example, multiple sclerosis (MS) has 69 known associated genes and the largest connected component, which is qualified as a module with a size of 11, and rheumatoid arthritis (RA) has 51 associated gene and the largest connected component, which is qualified as module with a size of 9.

To uncover disease-disease relationships, a quantity called module separate S_{AB} is introduced as follows.

$$s_{AB} \equiv < d_{AB} > - \frac{< d_{AA} > + < d_{BB} >}{2} \tag{1}$$

where $<d_{AB}>$ is the average of the shortest distance for each gene of disease A to reach a gene of disease B and vice versa, $<d_{AA}>$ is the average of the shortest distance for every gene in disease A to reach another gene in disease A, and $<d_{BB}>$ the average of the shortest distance for genes of disease B to reach another gene in disease B. Figure 1 shows how S_{AB} is computed for a toy example. More comprehensive results in [4] demonstrate that this network-based measurement of disease module separation is more indicative of pathological manifestations of disease pairs than simply measuring the overlap between the associated gene sets, such as Jaccard Index:

$$J = |G_A \cap G_B| / |G_A \cup G_B| \tag{2}$$

It is reported in [4] that, when the disease history of 30 million individuals aged 65 and older is used to determine the relative risk RR of disease comorbidity for each disease pair, the relative risk drops from $RR \geq 10$ for $S_{AB} < 0$ to the random expectation of $RR \approx 1$ for $S_{AB} > 0$.

Detection of missing shared genes

To further explore the predictive power of the disease module separation, we use it to tackle the incompleteness of the data. Specifically, for disease pairs that are known to share high comorbidity and therefore are expected to have a small, preferably negative, module separation, but instead have large positive S_{AB}, we hypothesize that the discrepancy is due to some missing pieces of information, such as a missing shared gene, which if recovered should bring the two disease modules closer, i.e., to decrease S_{AB}. Therefore, we formulate the detection of missing common genes between two disease modules as an optimization problem as follows.

$$x^* = \text{argmin } SAB[+x] \tag{3}$$

$$x \in (G_A \cup G_B) - (G_A \cap G_B)$$

where x goes over genes distinctly associated to either disease A or disease B, and $S_{AB}[+x]$ is the module separation when x is added as a shared gene between disease A and B, and x* is the predicted missing shared gene which minimizes the module separation. The minimization can be achieved either by exhaustive search when the sets G_A and G_B are not very large or by some heuristics when the search space becomes huge. Note that, although Eq. (3) is formulated for finding a single (most probable) missing common gene, in practice, Eq. (3) can be applied sequentially multiple times

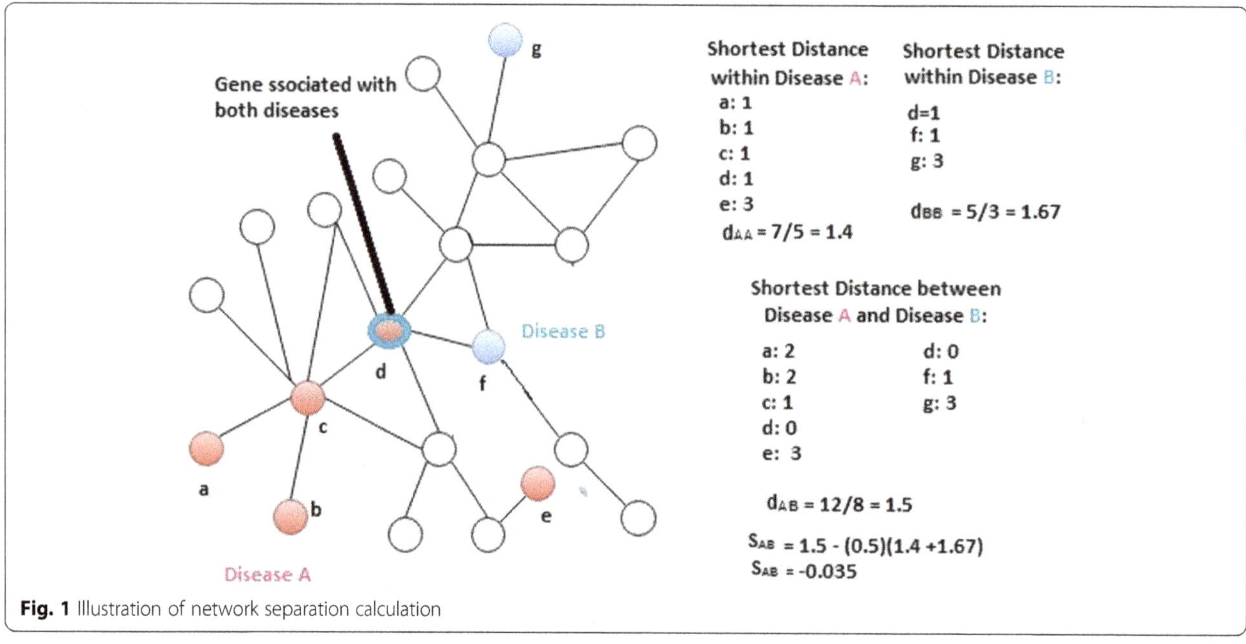

Fig. 1 Illustration of network separation calculation

for recovering multiple missing common genes. It is also worthwhile to note that the set of missing common genes recovered by using Eq. (3) iteratively one gene at a time may likely be different from a set of missing common genes should their candidacy as common gene be evaluated altogether, possibly due to the topology of the interactome and how these genes are located. So, if the number of missing common genes k is known, an alternative formulation of the optimization problem can be defined as follows.

$$X^* = \underset{X \in (G_A \cup G_B) - (G_A \cap G_B)}{\operatorname{argmin}} S_{AB}[+X] \quad (4)$$

where X^* is the optimal set of missing common genes, and X is any subset of size k from the genes that are distinctly associated with either disease A or disease B. This formulation, while theoretically sound and appealing, has two practical issues: a) the number of missing common genes k is not known a priori; and b) the increased computational complexity due to combinatorial in selecting k out n, where $n = |G_A \cup G_B| - |G_A \cap G_B|$. Because of these issues, we only tested Eq. (4) for $k = 2$ and $k = 3$, while the results reported in the next section are mainly based on Eq. (3).

Results

In this section, we tested our method for identifying missing genes with the data used in [4]. We first describe briefly the dataset, and then present the results which are evaluated using a cross validation scheme.

Dataset

The data, including Human interactome, disease gene association, network properties of disease pairs and comorbidity data, was used in the study from [4] and was downloaded from their website. Comorbidity (RR score) for several diseases using medicare data from USA has been calculated by [8]. The dataset contains 913 disease modules with negative S AB value and known RR score. The comorbidity value ranges from 0 to 6497. Comorbidity value 1.0 or above is considered high [4]. Among the 605 disease modules, 148 of them have comorbidity value ranging from 3.0 to 6497, and only 25 disease pairs with RR score above 100. Most of the disease modules have pairwise RR score between 0 and 3.0.

While the method is ultimately aimed at finding de novo missing common genes between a disease pair, for evaluation purpose, the method is tested, in a cross-validation scheme, at recovering known common genes. Therefore, a disease pair must have common genes to be used in the test. It was found that, out of 913 disease pairs, there are 605 disease pairs that satisfy the requirement, and the remaining 308 disease modules, either do not have any common gene or have all the genes common and hence are removed from the test dataset.

Cross-validation and performance

The cross-validation scheme is designed as follows. For a disease pairs A and B:

1. Randomly select multiple common genes and reserve them as positive test examples.
2. Randomly select multiple non-common genes from G_A and G_B respectively, and reserve them as negative test examples.
3. For each gene x in the test set, run the search algorithm as given in Eq. (3), and compute $S_{AB}[+x]$, the module separation when x is marked as shared, and x goes over all test examples associated with diseases A and B. Then compute prediction score $s(x) = S_{AB} - S_{AB}[+x]$.
4. Rank all the test examples x's by s(x) in a descending order: the higher the score s(x), the higher that x is ranked and hence more likely to be a common gene. Receiver operating characteristic (ROC) score is computed by comparing the ranked list and the ground truth of the test examples.

Note that in the experiments reported below, 10 common genes, if available, were selected from $G_A \cap G_B$ and 10 uncommon genes were selected from $(G_A \cup G_B) - (G_A \cap G_B)$ for cross validation.

The performance is evaluated by using receiver operating characteristic (ROC) score. From the list of the test examples ranked by their prediction score s(x), ROC curve plots the true positive rate as the function of false positive rate when a threshold moves from the top to the bottom of the ranked list – test examples with prediction score larger than or equal to the threshold are predicted as positive and otherwise as negative. ROC score is the area under the curve of ROC curve and thus has a range of [0, 1], with 0.5 corresponding to a random classifier and higher score corresponding to better predictive power. The average ROC score for the whole dataset is 0.947, as reported in Table 1. When Eq. (4) is used in place of Eq. (3), the average ROC score is 0.976 and 0.979 for k = 2 and 3 respectively. This confirms that considering candidate missing common genes as a subset can indeed achieve better prediction as compared to considering candidate missing common genes individually, though the gain in performance seems to be tapering as the value of k increases. Table 1 also lists the average ROC score for several cases: a) disease pairs with comorbidity in [0,1], b) disease pairs with comorbidity in [1, 2], c) disease pairs with comorbidity in [2, 3], and d) disease pairs with comorbidity >3.0, with case e) being all pairs included. It can be seen clearly that

Table 1 Average ROC Scores with standard deviation, precision and recall for various comorbidity ranges

	Comorbidity Range				
	0-8000	0-1	1-2	2-3	>3
Number of Disease Pairs	605	133	248	76	148
Average ROC Score (Shortest Distance)	0.947	0.966	0.950	0.952	0.920
Stddev (Shortest Distance)	0.094	0.063	0.089	0.072	0.124
Average ROC Score (Average Distance)	0.491	0.513	0.495	0.508	0.458
Stdev (Average Distance)	0.279	0.279	0.288	0.269	0.269
Average ROC Score (Randomization)	0.601	0.606	0.614	0.555	0.599
Stedev (Randomization)	0.278	0.282	0.287	0.258	0.2468
Average Precision (Shortest Distance)	0.88	0.88	0.85	0.89	0.96
Stddev (Shortest Distance)	0.27	0.28	0.31	0.25	0.15
Average Precision (Average Distance)	0.72	0.72	0.71	0.69	0.64
Stdev (Average Distance)	0.311	0.31	0.32	0.33	0.30
Average Precision (Randomization)	0.66	0.70	0.63	0.66	0.72
Stedev (Randomization)	0.29	0.28	0.29	0.30	0.29
Average Recall (Shortest Distance)	0.91	0.94	0.93	0.93	0.88
Stddev (Shortest Distance)	0.13	0.11	0.13	0.09	0.16
Average Recall (Average Distance)	0.69	0.72	0.70	0.70	0.64
Stdev (Average Distance)	0.30	0.28	0.30	0.31	0.30
Average Recall (Randomization)	0.78	0.80	0.79	0.73	0.76
Stedev (Randomization)	0.26	0.25	0.26	0.26	0.25

high average ROC scores are achieved for all cases, with case a) achieving marginally the highest. This finding is noteworthy as it suggests that S_{AB} is a useful indicator across all range of relative risk (RR) value whereas in [4] strong correlation was observed between RR drops and S_{AB} switching from negative to positive. Precision and recall reported in Table 1 are computed using a threshold on prediction score s(x) which is set as suggested in [9]. Essentially, the threshold is set by using ROC curve on the test data to determine the highest peak point of ROC curve from the diagonal line, i.e., the prediction score of the test example that corresponds the peak point is used as the threshold. Average precision and recall are reported as 0.88 and 0.91 respectively for comorbid disease pairs using shortest distance as method to measure module separation. Figure 2 represents a graphical representation of the evaluation metrics (roc score, precision and recall) used for two methods for calculating module separation and when used for randomized data.

In addition to the average ROC scores, the histogram plot of ROC scores is shown in Fig. 3. In the histogram, a point in a curve shows in the vertical axis the percentage of disease pairs that have a performance greater or equal than ROC score given in the horizontal axis. It also shows the random ROC score in yellow color.

We further examined how the prediction performance is affected by the number of common genes, i.e., the size of the training set. Specifically, we grouped disease pairs based on the range of overlap between associated genes: i) 5 ~ 10 common genes, ii) 10 ~ 15, and iii) 15 or more common genes.

The effect of the size of training set and the range of RR on prediction performance is reported in Table 2, which lists the number of disease pairs achieving a given ROC score range for different groups under different RR range. For example, 42 pairs with 0~ 5 common genes and RR between 0 and 1.0 have received ROC score in the range (0.9, 1.0) The results show that as the number of common genes increases, the prediction performance in terms of distribution over various ranges is quite stable, with slight improvement, suggesting the method is robust under various conditions. In each case we had all the results above ROC score 0.5. And, more than 80% of the disease modules provide missing gene prediction ROC score between 0.9-1.

Discussion

It should be noted that the missing common gene problem, despite of its apparent importance, has not yet been addressed elsewhere in the literature to our best knowledge. Still, in order to get a sense how well the proposed method does in comparison to a baseline, we randomize the common genes for each disease pair. Specifically, for each disease pair, the set of common gene is replaced with the same number of genes randomly selected from the whole set of genes in the interactome. The rationale

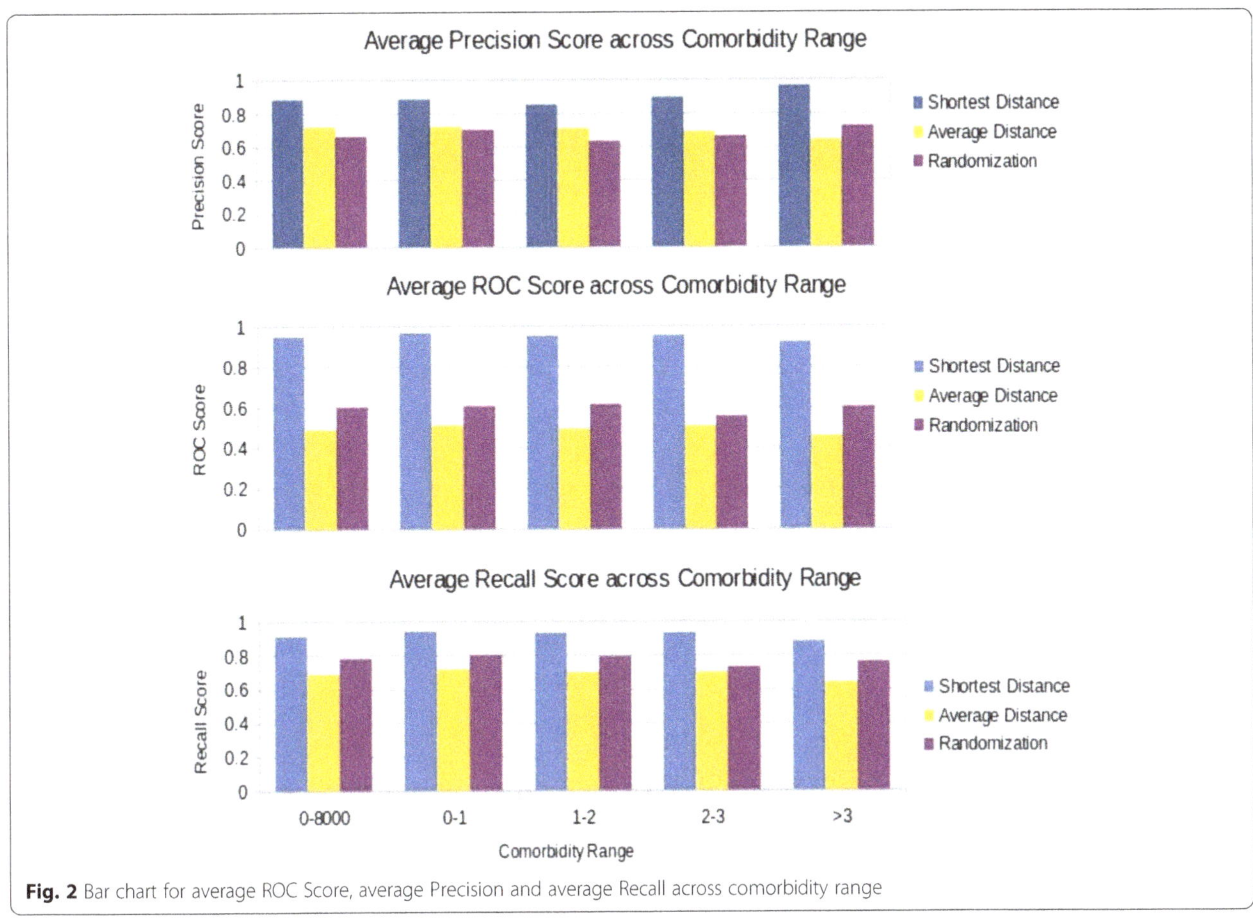

Fig. 2 Bar chart for average ROC Score, average Precision and average Recall across comorbidity range

for doing so is to keep the count of common genes for each disease pair unchanged and also maintain the topology of interactome and the overall relative locations of the two diseases in the pair. When everything else was kept the same, it was found that the average ROC score dropped to 0.601 for the 605 disease pairs with their common genes randomized. The detailed results for different comorbidity ranges 250 with respect to the randomized baseline are listed in Table 1, and the histogram of ROC scores for the baseline is shown as plot F in Fig. 2.

For comparison, we also modify how the module separation is calculated. Specifically, instead of the shortest distances used in Eq. (1), we replaced $<d_{AB}>$ with the average distance for all distinct A -B gene pairs, $<d_{AA}>$ is the average distance for all gene pairs within disease module A, and $<d_{BB}>$ the average distance for all gene pairs within disease module B. Use this modified module separation, let's call it all-pair-average based module separation $S_{<AB>}$, we get an average ROC score 0.49 for all 605 disease pairs. The histogram of the ROC scores is shown in Fig. 2 as plot G. One plausible explanation of why the all-pair-average based module separation performs poorly is that the module separation has become much less sensitive to swapping a single gene x's classification in Eq. (3) – from common gene to non-common gene and vice versa.

From comparison to the baseline of randomized data and an alternative definition of module separation, the results show that our proposed method performs very well, suggesting the optimization formulated in Eq. (3) as a viable solution to finding missing common genes for a given pair of diseases. Note that the predictive power is measured by ROC score, which does not require a pre-set threshold on the score s(x) when it is used for making prediction. Not requiring a pre-set threshold on the prediction score contributes to the popularity of ROC as a metric for assessing predictive power of a binary classifier: the ability to differentiating positive examples from negative examples when ranking on these examples by the prediction score. This is because in reality it is often difficult to set a priori threshold on the prediction score, although it can be set in certain ad hoc ways. In our situation, the score s(x), computed for each module separation sAB, depends on the interactome topology, the unknown number of missing genes and other factors, which makes it difficult to have a preset threshold for any give disease pair, least to say a common threshold for all disease pairs. Even if we

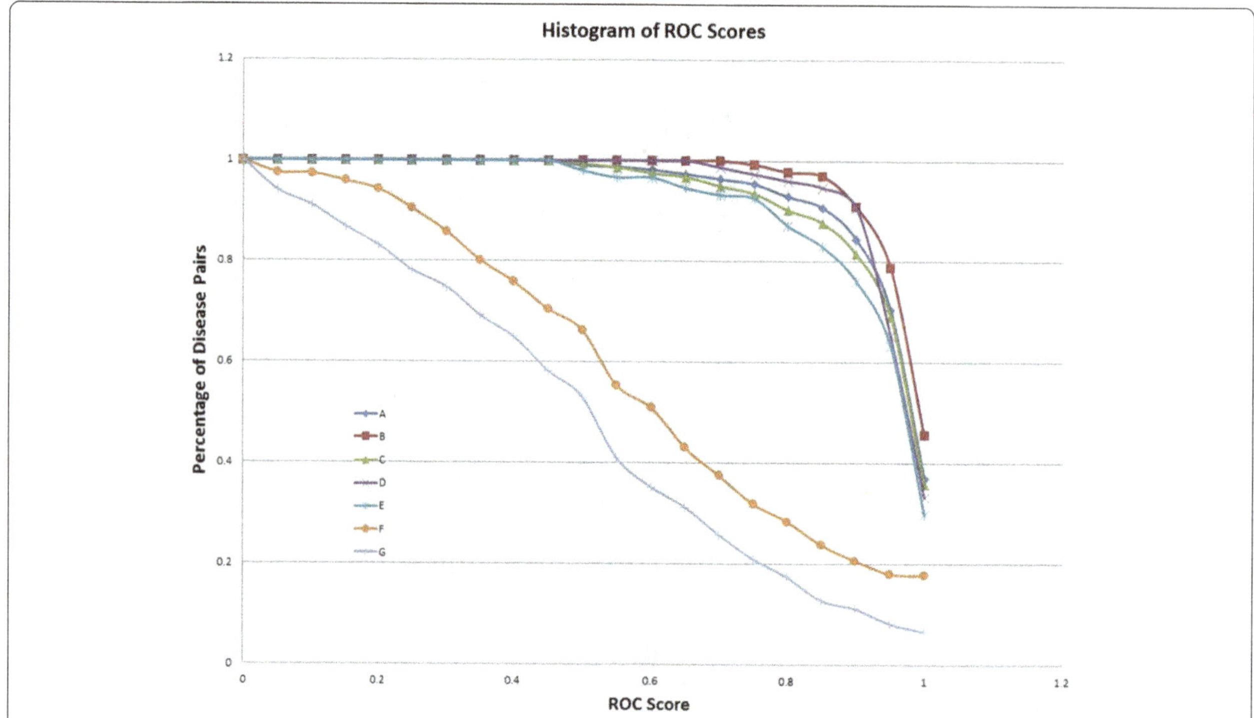

Fig. 3 Histogram of ROC Scores. A: comorbidity range 0 ~ 1; B: comorbidity range 1~2; C: comorbidity range 2 ~3; D: comorbidity range > 3; E: comorbidity range 0 ~ 10,000; F: randomized common genes; G: SAB based on average distance

Table 2 Effect of the size of training set and the range of RR on prediction performance

ROC Score Range	Comorbidity Range				
	0-8000	0-1	1-2	2-3	>3
i) 0 ~ 5 Common Genes					
0.5-0.6	2	0	2	0	0
0.7-0.8	5	2	3	0	0
0.9-1.0	174	46	81	18	29
Total	181	48	86	18	29
ii) 5 ~ 10 Common Genes					
0.5-0.6	0	0	0	0	0
0.7-0.8	2	0	2	0	0
0.9-1.0	121	36	48	15	22
Total	123	36	50	15	22
iii) 10 - 15 Common Genes					
0.5-0.6	0	0	0	0	0
0.7-0.8	1	0	0	0	1
0.9-1.0	46	12	21	4	9
Total	47	12	21	4	10
iv) 15 or more Common Genes					
0.5-0.6	10	1	3	0	6
0.7-0.8	24	1	6	4	13
0.9-1.0	220	35	82	35	68
Total	254	37	91	39	87

normalize the score s(x) as s(x) = (S_{AB} – S_{AB}[+x]) / S_{AB} it is unlikely that a threshold set for one disease pair (e.g., by using the method cited in Ref [9]) would be the same for another disease pair, because different disease pairs can have their genes residing on different locations of the network (and hence having different network topologies) and can have different number of missing common genes. So, for practical use of our method, we envision that, for any pair of diseases with high comorbidity yet a large module separation, biologists would suspect some common genes are missing and then use our method to suggest a short list of candidates (i.e., these with top ranking score s(x)) for further investigation.

In this work, we used brute force to search all genes associated to the disease pair, as our focus is on the viability of using module separation to detect missing common genes not on the speed. In the dataset, we used for this study, the average number of genes in a disease pair is 168 and it takes 2 min 43 s to search all genes in the disease pair for putative common genes on a desktop computer: 2.90Ghz intel core i7, 8.00Gb memory. While it is desirable as a future work to find a faster heuristic algorithm for search as the number of genes increase, the brute force approach seems to be acceptable for typical cases.

Conclusions

In this work, we developed a novel method to predict missing common genes for a given disease pairs. The method formulates the task as an optimization problem of minimizing network based module separation for subgraphs formed by associated genes on the interactome, with the hypothesis that correctly identified missing common genes would bring the two-module closer. The results of cross-validation from a benchmark dataset of more than 600 disease pairs show high prediction accuracy on average, measured as ROC score. The method provides a useful tool to infer better understanding of disease- disease interaction in terms of related genes. While the method is tested in cross-validation mode in this study, it can be easily deployed to predict de novo *m*issing genes, i.e., those genes that are not associated with any disease but have an impact on the phenotype of both diseases. It is worthwhile to note that the results reported in this study are based on incomplete Human interactome – protein interactions that exist but have not be detected by experiments and reported in literature, and thus are referred to as missing edges in the protein-protein interaction network. Therefore, the accuracy for missing gene prediction may change, likely for higher, as the interactome becomes more complete. In fact, as an effort to address the challenge presented by missing data, in future work this method could be extended for predicting missing edges in an incomplete interactome as well.

Acknowledgements
The authors are grateful to Joerg Menche for making the comorbidity data available. The authors would also like to thank the anonymous reviewers for their invaluable comments. The authors are thankful to Fulbright for funding the research.

Funding
PA is funded on a Fulbright scholarship. Publication charges for this article have been funded by a University of Delaware Professional and Development Award, awarded to PA for the dissemination of this work including travel and publication charges. The funding agency had no role in the design, collection, analysis, data interpretation and writing of this study.

About this supplement
This article has been published as part of *BMC Genomics* Volume 18 Supplement 10, 2017: Selected articles from the 6th IEEE International Conference on Computational Advances in Bio and Medical Sciences (ICCABS): genomics. The full contents of the supplement are available online at https://bmcgenomics.biomedcentral.com/articles/supplements/volume-18-supplement-10.

Authors' contributions
PA implemented data collection, method formulation and analysis and interpretation of results, and wrote the paper. LL designed the study, conceived method formulation, analyzed data and wrote the paper. All authors read and approved the final manuscript.

Competing interests
The authors declare that they have no competing interests.

References

1. Almaas E. Biological impacts and context of network biology. J Exp Biol. 2007;210:1548–58.
2. Alon U. Network motifs: theory and experimental approaches. Nat Rev Genet. 2007;8:450–61.
3. Yildrim MA, Goh K-I, Cusick ME, Barabási A-L, Vidal M. Drug-target network. Nat Biotechnol. 2007;25:1119–26.
4. Menche J, Sharma A, Kitsak M, Ghiassian S, Vidal M, Loscalzo J, Barabási AL. Uncovering disease-disease relationships through the incomplete human interactome. Science. 2015;347(6224):1257601.
5. Žitnik M, Janjić V, Larminie C, Zupan B, Pržulj N. Discovering disease-disease associations by fusing systems-level molecular data. Sci Rep. 2013;3:3202.
6. Goh K-I, Cusick ME, Valle D, Childs B, Vidal M, Barabási A-L. The human disease network. Proc Natl Acad Sci U S A. 104(21):8685–90.
7. Rzhetsky A, Wajngurt D, Park N, Zheng T. Probing genetic overlap among complex human phenotypes. Proc Natl Acad Sci USA. 2007;104(28):11694–9.
8. Park J, Lee D-S, Christakis NA, Barabási A-L. The impact of cellular networks on disease comorbidity. Mol Syst Biol. 2009;5:262.
9. Perkins NJ, Schisterman EF. The inconsistency of 'optimal' cut-points using two ROC based criteria. Am J Epidemio. 2006;163(7):670–75.

15

Significant differences in terms of codon usage bias between bacteriophage early and late genes: A comparative genomics analysis

Oriah Mioduser[1†], Eli Goz[1,2†] and Tamir Tuller[1,2,3*]

Abstract

Background: Viruses undergo extensive evolutionary selection for efficient replication which effects, among others, their codon distribution. In the current study, we aimed at understanding the way evolution shapes the codon distribution in early vs. late viral genes in terms of their expression during different stages in the viral replication cycle. To this end we analyzed 14 bacteriophages and 11 human viruses with available information about the expression phases of their genes.

Results: We demonstrated evidence of selection for distinct composition of synonymous codons in early and late viral genes in 50% of the analyzed bacteriophages. Among others, this phenomenon may be related to the time specific adaptation of the viral genes to the translation efficiency factors involved at different bacteriophage developmental stages. Specifically, we showed that the differences in codon composition in different temporal gene groups cannot be explained only by phylogenetic proximities between the analyzed bacteriophages, and can be partially explained by differences in the adaptation to the host tRNA pool, nucleotide bias, GC content and more. In contrast, no difference in temporal regulation of synonymous codon usage was observed in human viruses, possibly because of a stronger selection pressure due to a larger effective population size in bacteriophages and their bacterial hosts.

Conclusions: The codon distribution in large fractions of bacteriophage genomes tend to be different in early and late genes. This phenomenon seems to be related to various aspects of the viral life cycle, and to various intracellular processes. We believe that the reported results should contribute towards better understanding of viral evolution and may promote the development of relevant procedures in synthetic virology.

Keywords: Viral evolution, Codon usage bias (CUB), Bacteriophage genome evolution, Viral life cycle, Coding regions, Synthetic virology

Background

Deciphering the regulatory information encoded in the genomes of phages and other viruses, and the relation between the nucleotide composition of the coding regions and the viral fitness is of great interest in recent years.

Gene expression within different Deoxy ribonucleic Acid (DNA) viruses or viruses with DNA intermediate, such as herpeses, lenti-retro, polyoma, papilloma, adeno, parvo and various families of bacteriophages is regulated in a temporal fashion and can be divided into early and late stages with respect to the viral replication cycle [1–8].

The early genes are expressed following the entry into the host cell and code typically for non-structural proteins that are responsible for different regulatory functions in processes such as: viral DNA replication, activation of late genes expression, trans-nuclear

* Correspondence: tamirtul@post.tau.ac.il
†Equal contributors
[1]Department of Biomedical Engineering, Tel-Aviv University, Ramat Aviv, Israel
[2]SynVaccineLtd. Ramat Hachayal, Tel Aviv, Israel
Full list of author information is available at the end of the article

transport, interaction with the host cell, induction of the cell's DNA replication machinery necessary for viral replication, etc. [9, 10]. Late genes largely code for structural proteins required for virion assembly; they are generally highly expressed and their expression is usually induced or regulated by the early genes [9, 10].

Several studies have shown that viral codon frequencies tend to undergo evolutionary pressure for specific CUB; among others, it was suggested that viral CUB is under selection for improving the viral fitness, and in specifically the viral gene expression [11–33].

In particular, in [17] different trends of translation efficiency adaptation of the coding regions of the bacteriophage Lambda early and late genes were demonstrated. Specifically, it was shown that the preferences of codons in early genes, but not in the late genes, were similar to those of the bacterial host [17]. The analysis of ribosome profiling data revealed that the codon decoding rates of viral genes tend to correlate with their expression levels [17]. Interestingly, during the initial stages of phage development the decoding rates in early genes were found to be higher than the decoding rates in late genes; in more progressive viral cycles an opposite trend was demonstrated [17].

In this study we go further, and perform a comparative genomics analysis of the temporal differences in CUB in almost all known viruses with existing in the literature classification of their genes into early and late groups. Specifically basing on analysis of 14 bacteriophages and 11 human viruses we suggest that 50% of the analyzed bacteriophages tend to undergo an extensive evolutionary selection for distinct compositions of synonymous codons in early and late viral genes. We analyze the features of the genomes that undergo this type of selection and argue that the differential CUB can be related to various intracellular phenomena and processes, such as: translational selection and regulation [11, 12, 17, 21, 22, 28, 31], mutational bias and pressure [16, 20, 21, 23, 26, 27, 30, 32, 33], amino acids (AA) compositions [12, 16], and other genomic characteristics, some of which are still not fully understood [13, 14, 29, 34].

Finally, we discuss a possible application of our findings to synthetic virology. Specifically, we suggest using the temporally regulated CUB for controlling the viral gene expression at different time points during the life cycle for designing of optimized and/or deoptimized synthetic viruses which can be used in exploring novel strategies in vaccination (e.g. life attenuated vaccines) and cancer therapy (oncolytic viruses).

Results

The research outline of the study is described in Fig. 1. More details can be found in the following sections.

Bacteriophage early and late genes tend to have different compositions of synonymous codons

Genome level information about the different viruses analyzed in this study, like their hosts, number of genes, gene lengths and ENC, is displayed in Additional file 1: Table S1 and Figure S1.

In order to compare the synonymous codons usage in early and late genes, each coding sequence was represented by its relative synonymous codons frequencies (RSCF) - a 61 dimensional vector expressing each sense codon by its frequency in that sequence normalized relative to the frequencies of other synonymous codons coding for the same AA. We then performed a clustering analysis, assuming that RSCF vectors that are closer with respect to Euclidian metric correspond to genes with a more similar content of synonymous codons (see Materials and Methods).

Our results suggest that early and late genes in 50% of the analyzed bacteriophages tend to exploit different synonymous codons. Specifically, in 7 of the 14 analyzed bacteriophages, early and late genes were found to be significantly (p-value ≤0.05) separated according to the frequencies of their synonymous codons (Figs. 2, 3a, b, Additional file 1: Figure S2 and Figure S3 in Section 1.2). Our analysis provide evidence that different sets of synonymous codons in early vs. late genes are selected for in the course of viral evolution; these differences may be related to the optimization of bacteriophage fitness in different phases of the viral lifecycles.

In addition, 6 out of 14 bacteriophages were also found to be significantly (p-value <0.05) separated according to the AA composition of their early and late genes (Fig. 3b, Additional file 1: Figure S4 and Figure S5 in Section 1.2). 4 viruses were characterized both by a differential synonymous codon usage and by a differential AA usage in their early and late genes. These findings suggest that among others, the different codon distribution in early and late genes may be partially related to the functionality of the encoded proteins via their AA content and possibly protein folding [35].

To check if bacteriophages with significant differences in synonymous codons usage in temporal genes tend to have more similar genomic sequences (usually related to smaller evolutionary distances), we reconstructed a phylogenetic tree of the bacteriophage proteomes based on Average Repetitive Subsequences (ARS) distance matrix and neighbor joining method as described in Materials and Methods section and in references therein (Fig. 3a). We then performed a statistical analysis in order to investigate the relation between the differences in temporal regulation of synonymous codons in different viruses and their evolutionary distances. We did not find such a relation

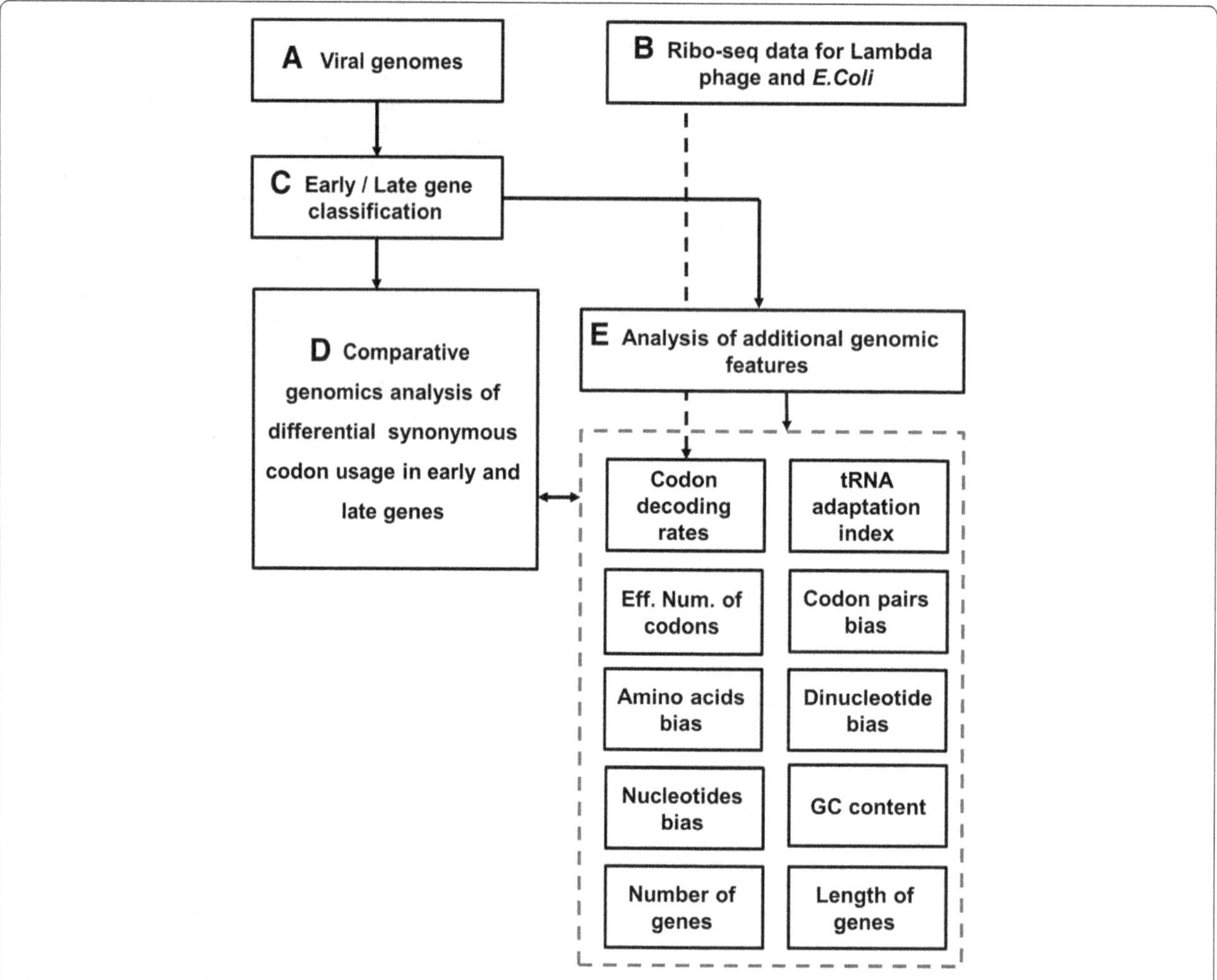

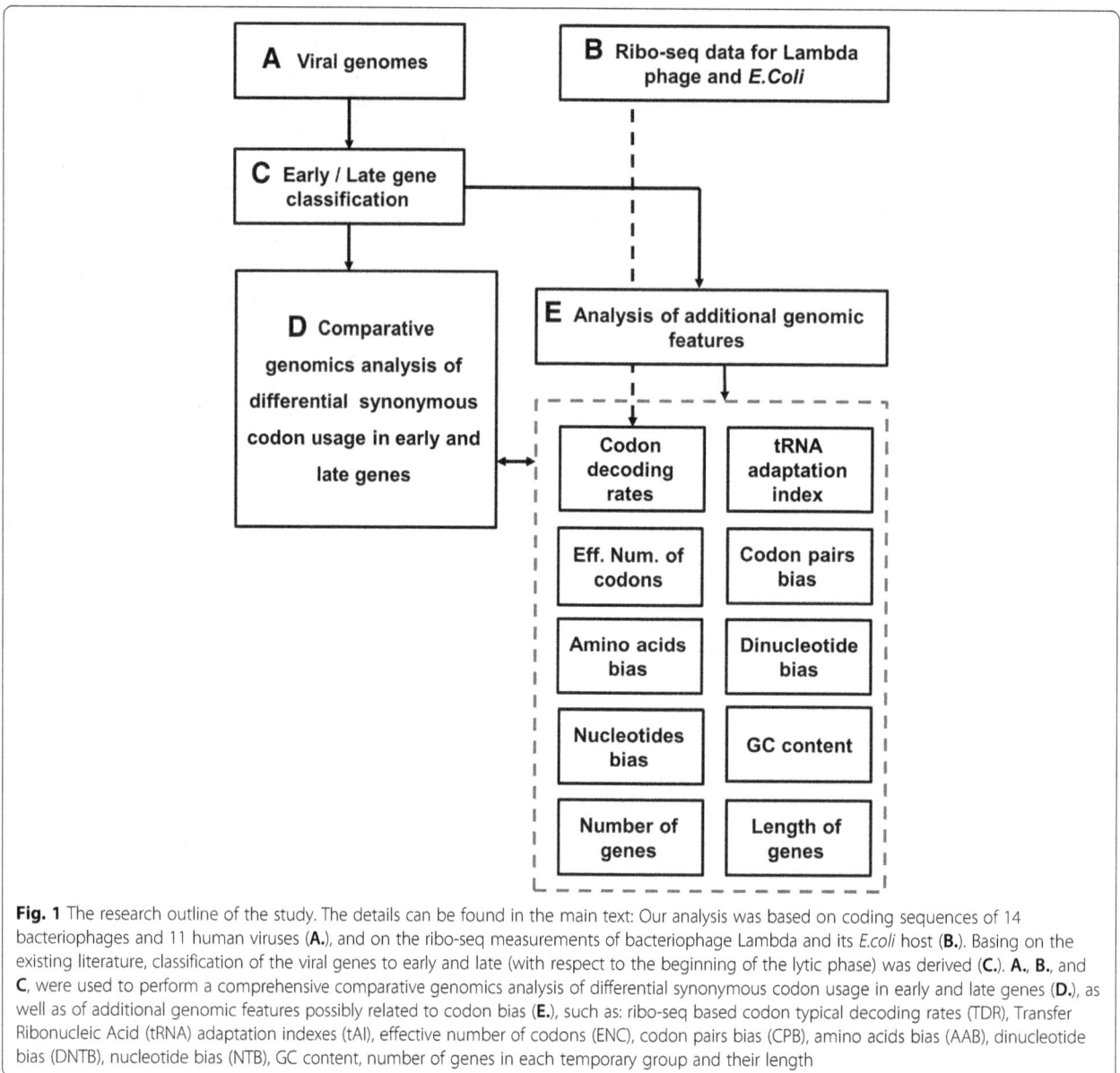

Fig. 1 The research outline of the study. The details can be found in the main text: Our analysis was based on coding sequences of 14 bacteriophages and 11 human viruses (**A.**), and on the ribo-seq measurements of bacteriophage Lambda and its *E.coli* host (**B.**). Basing on the existing literature, classification of the viral genes to early and late (with respect to the beginning of the lytic phase) was derived (**C.**). **A.**, **B.**, and **C**, were used to perform a comprehensive comparative genomics analysis of differential synonymous codon usage in early and late genes (**D.**), as well as of additional genomic features possibly related to codon bias (**E.**), such as: ribo-seq based codon typical decoding rates (TDR), Transfer Ribonucleic Acid (tRNA) adaptation indexes (tAI), effective number of codons (ENC), codon pairs bias (CPB), amino acids bias (AAB), dinucleotide bias (DNTB), nucleotide bias (NTB), GC content, number of genes in each temporary group and their length

(see details in Additional file 1: Section 1.3 and Figure S6), suggesting that the differential codon usage in early and late genes is a complex trait related to alternative determinants such as the bacterial niche, the specific phage proteins and their function/structure, etc.

Viruses undergo an extensive evolutionary selection for adaptation to their host's cell environment, and thus it can be assumed that their codon composition reflects an efficient adaptation of the viral genes to specific intracellular conditions (e.g. in terms of gene expression factors such as tRNA molecules, AA concentration, etc) that are prevalent in different gene expression stages, in accordance with the reported results.

Weaker separation between synonymous codon usage in early and late genes in human viruses

The results in the previous section suggest that bacteriophages undergo an extensive evolutionary selection on a synonymous level for temporal regulation of gene expression. Whether this also occurs in viruses of humans and other eukaryotic hosts is harder to ascertain. Human Immunodeficiency Virus 1 (HIV-1) was found to have a significant separation (p-value ≤ 0.05) of codon composition between early and late genes, while such separation was not statistically significant in the rest of the analyzed viruses (see Additional file 1: Table S2 in Section 1.4).

As evidenced in Additional file 1: Table S1 and Figure S1, human viruses tend to have fewer genes than

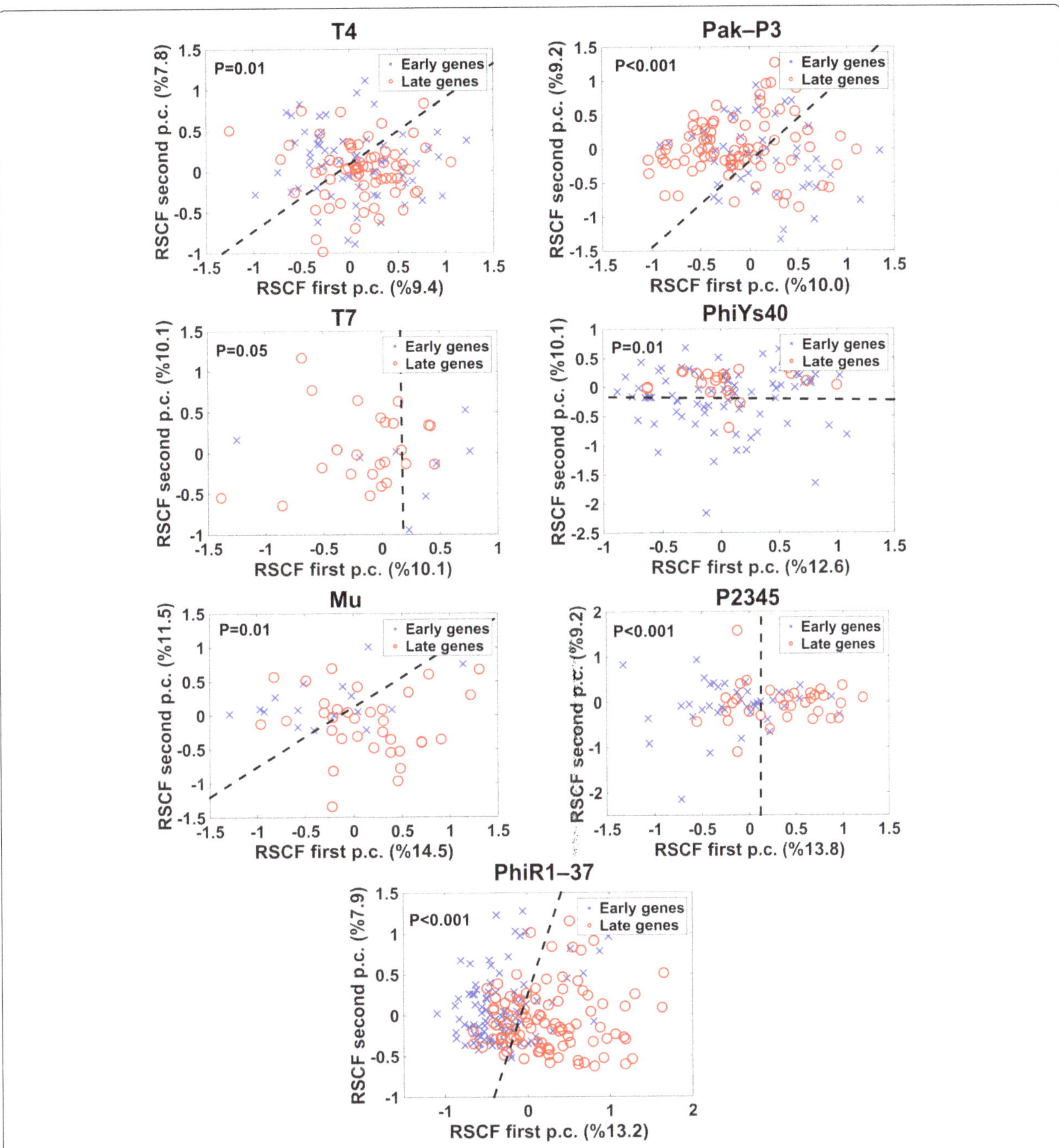

Fig. 2 Principal component analysis (PCA) of RSCF vectors for bacteriophages with significant separation in codon usage between early (blue circles) and late (red circles) genes. In order to visualize the clustering, PCA was applied to project the RSCF vectors to a plane spanned by their first two principal components. In order to visualize the separation between clusters a maximum margin separation line, a line for which the Euclidian distance between it and the nearest point from either of the groups is maximized, was calculated and plotted. The significance of cluster separation was assessed by comparing the Davies-Bouldin cluster score to the randomized scores obtained from 100 permutations of gene temporary (early or late) labels. The variances % of the first two principal components are mentioned in the figures axis

bacteriophages. Therefore, we were interested in checking whether this fact can explain the weaker signal for temporal separation in CUB, and if, in practice, human viruses may also behave as bacteriophages with respect to the differential usage of synonymous codons in their early and late genes. To this end we analyzed the 7 bacteriophages with temporary differential codon usage by sampling in each one of them a number of early and late genes that is typical to human viruses (average of 8 early genes and 14 late genes). We found that the

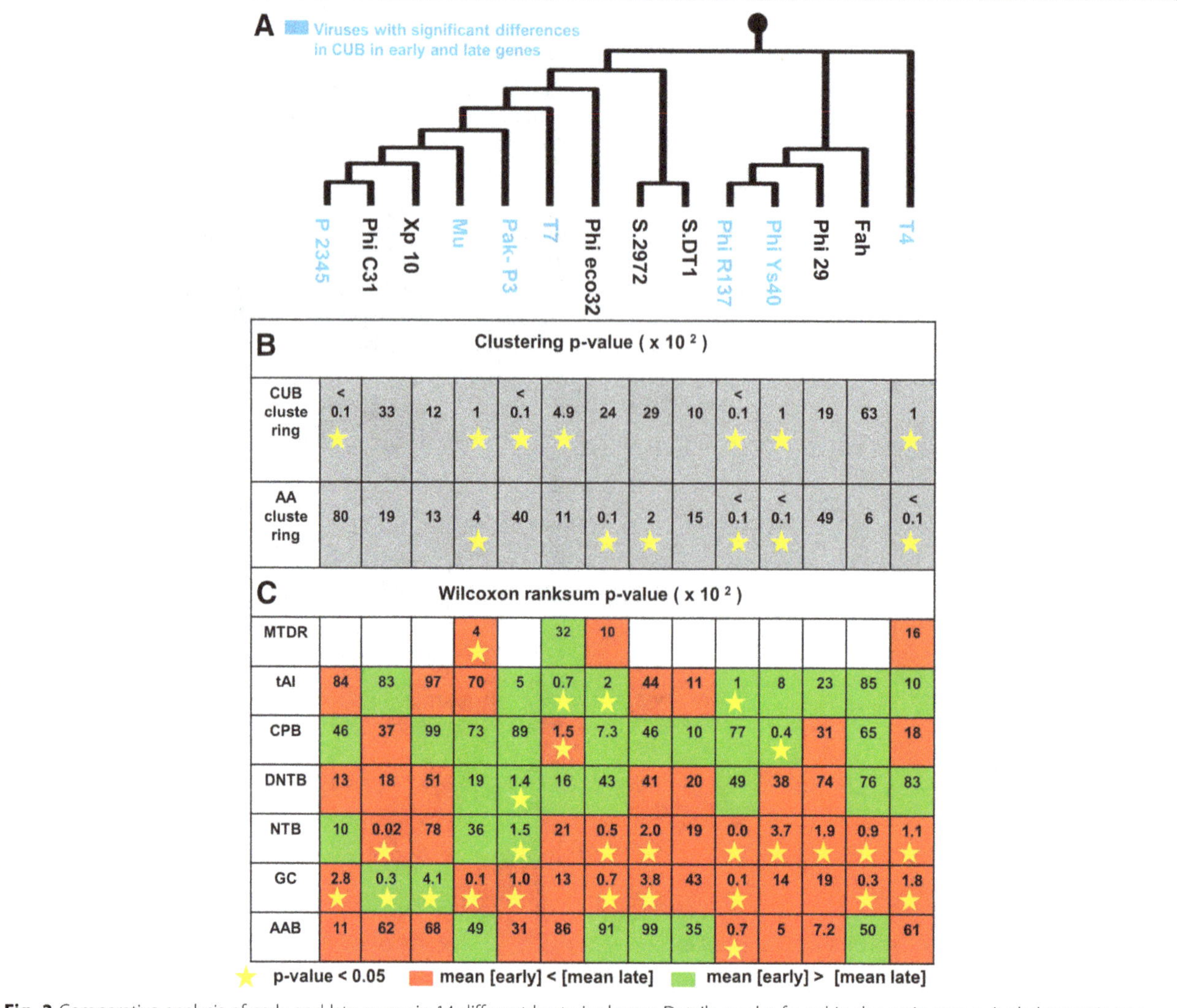

B	P 2345	Phi C31	Xp 10	Mu	Pak- P3	T7	Phi eco32	S.2972	S.DT1	Phi R137	Phi Ys40	Phi 29	Fah	T4
CUB clustering	< 0.1 ★	33	12	1 ★	< 0.1 ★	4.9 ★	24	29	10	< 0.1 ★	1 ★	19	63	1 ★
AA clustering	80	19	13	4 ★	40	11	0.1 ★	2 ★	15	< 0.1 ★	< 0.1 ★	49	6	< 0.1 ★

C	P 2345	Phi C31	Xp 10	Mu	Pak- P3	T7	Phi eco32	S.2972	S.DT1	Phi R137	Phi Ys40	Phi 29	Fah	T4
MTDR				4 ★		32	10							16
tAI	84	83	97	70	5	0.7 ★	2 ★	44	11	1 ★	8	23	85	10
CPB	46	37	99	73	89	1.5 ★	7.3	46	10	77	0.4 ★	31	65	18
DNTB	13	18	51	19	1.4 ★	16	43	41	20	49	38	74	76	83
NTB	10	0.02 ★	78	36	1.5 ★	21	0.5 ★	2.0 ★	19	0.0 ★	3.7 ★	1.9 ★	0.9 ★	1.1 ★
GC	2.8 ★	0.3 ★	4.1 ★	0.1 ★	1.0 ★	13	0.7 ★	3.8 ★	43	0.1 ★	14	19	0.3 ★	1.8 ★
AAB	11	62	68	49	31	86	91	99	35	0.7 ★	5	7.2	50	61

Fig. 3 Comparative analysis of early and late genes in 14 different bacteriophages. Details can be found in the main text. **a** A phylogenetic tree built from complete phage proteomes using ARS distance (see Materials and Methods). Phages with significant differences in temporary codon usage are marked by blue. **b** Viruses with significant (*p*-value <0.05) separation between early and late genes w.r.t synonymous codons or AA are marked by yellow stars. **c** Significance of separation between early and late genes w.r.t additional genomic features estimated by Wilcoxon ranksum p-value. Features/viruses with significant (p-value <0.05) separation between the two temporal groups are marked by yellow stars; green is related to higher mean in the case of the early genes and red is related to higher mean in the case of the late genes

temporal differences in codon usage remained significant even after randomly reducing the number of genes, indicating, among others, that these differences cannot be directly explained only by the genome size.

Comparison of early and late genes with respect to additional features of their coding regions

The signal of selection for temporarily regulated composition of synonymous codons in bacteriophages demonstrated in the previous subsection led us to analyze additional genomic features, such as: codon mean typical decoding rate (MTDR), tRNA adaptation index (tAI), codon pairs bias (CPB), dinucleotide bias (DNTB), nucleotide bias (NTB), GC content and amino acids bias (AAB).

Various studies related these features to different genomic mechanisms and biological processes involved in viral replication cycles and are related to the viral fitness.

For example, it was suggested that gene translation efficiency can be affected not only by single codons, but also by distribution of codon pairs [36]. In [37–39] it was argued that pairs of adjacent nucleotides may be an important genomic characteristic being under a significant evolutionary pressure in viruses and their hosts; specifically, it was suggested that CpG pairs are underrepresented in many Ribonucleic Acid (RNA) and in most small human DNA viruses, in correspondence to dinucleotide frequencies of their hosts. This phenomenon can be related, for example, to the contribution of the CpG stacking basepairs to RNA folding [40] and/or to the enhanced innate immune responses to viruses with elevated CpG [41]. The stability of the RNA secondary structures can be also affected by the genomic composition of nucleotides and in particular by GC content [42]. In addition, nucleotide compositions and AA usage bias may affect, among others, the synthesis of viral molecules, and the function and structure of the encoded proteins.

Consequently, we estimated the listed above features for all genes in all viruses, and evaluated the separation between early and late genes with respect to each one of them (see Materials and Methods). The results shown in Fig. 3c suggest that the differential usage of synonymous codons in early and late genes can be partially related to temporal differences in various characteristics of genomic sequences. Specifically, the features with the strongest temporal differences are the NTB and GC content which are significant (p-value <0.05) in most of the phages.

In addition, we wanted to check if the bacteriophages with a significant temporal separation with respect to synonymous codons tend also to be enriched with specific genomic features in comparison to the group of bacteriophages with non-significant temporal differences in synonymous codons. To this end, we compared the distribution of various genomic features in the two groups. Based on Wilcoxon rank-sum test we found no significant differences between the two groups of bacteriophages in terms of: genome length (p-value = 0.53), ENC (p-value = 0.4), CPB (p-value = 0.99), DNTB (p-value = 0.21), NTB (p-value = 0.9), GC content (p-value = 0.8) and AAB (p-value = 0.99). See also Additional file 1: Figure S7 in Section 1.5.

Discussion

In this study, we performed a comparative genomics analysis of viruses with annotations in literature regarding their genes division according to temporal expression. We examined 14 bacteriophages with different bacterial hosts and 11 human viruses in order to understand if there is a universal difference in synonymous codons usage as well as in additional genomic features (such as codon decoding rates, nucleotide/dinucleotide/AA biases, GC content and others) with respect to different temporal stages of viral life cycle.

Our results suggest that 50% of bacteriophages undergo an extensive evolutionary selection for distinct compositions of synonymous codons in early and late viral genes. This phenomenon was found to be weaker/less significant in human viruses, possibly because of the stronger selection pressure in bacteriophages / bacteria due to the larger size of their populations, and because of the fact that regulation processes in human gene expression are more 'complex' and thus may be mediated by additional aspects not necessary related to codons.

The differences between early and late genes, both with respect to the composition of synonymous codons and with respect to additional genomic features described in the previous sections, can be possibly influenced by various intracellular phenomena and processes related to the optimization of gene expression and to the overall fitness of the phage. To mention a few, these phenomena/processes include: adaptation of translation elongation efficiency in different phases of the viral lifecycle [17], Messenger Ribonucleic Acid (mRNA) folding [43, 44], adaptation of the viral genes to the (possibly altering) tRNA pool of their hosts [11, 12, 17, 31], mutation levels and biases [16, 20, 21, 23, 26, 27, 30, 32, 33], transcription regulation [45, 46], protein function and structure [47], cell metabolism [48], etc.

There can be various explanations to the fact that it seems that only 50% of the bacteriophages there is a significant difference in the codon usage in early vs. late genes:

First, it is possible that the effective population size (which is not easy to estimate) varies among the analyzed bacteriophages. The selection pressure is weaker in bacteriophages with smaller population size.

Second, this observation may be also related to the intracellular regimes during the development of the different bacteriophages. For example, it is possible that during the development of some bacteriophages the tRNA levels are modulated/changed, while in other cases the changes are minor. The changes in the tRNA levels may trigger evolution of different CUB in early/late genes in the bacteriophages that experience them.

Third, this result may be related to the nature of the protein encoded in the bacteriophages genome. The specific function and properties of the proteins in different bacteriophages may affect the observed levels of selection. For example, it is possible that only in some bacteriophages the early vs. late genes tend to have different structure with different co-translational folding constraints that eventually affect the codon bias. It is also possible that only in some bacteriophages the early vs. late genes tend to have different expression levels/patterns that eventually affect their codon bias.

It is possible that the results reported here have relevant practical applications. For example, vaccines, and their discovery, are topics of singular importance in present-day biomedical science; however, the discovery of vaccines has hitherto been primarily empirical in nature requiring considerable investments of time, efforts and resourced. To overcome the numerous pitfalls attributed to the classical vaccine design strategies, more efficient and robust rational approaches based on computer-based methods are highly desirable. One direction in designing in-silico vaccine candidates may be based on exploiting the temporally regulated synonymous information encoded in the genomes and investigated in this study for attenuating the viral replication cycle while retaining the wild type proteins. In particular, the result reported here suggest that viral genes can be designed with respect to phase specific temporary regulated gene expression constraints, and this design would result in controllable yields of the corresponding genetic products during a defined time period. To achieve this, codons would be selected with frequencies maximally dissimilar / similar to the set of early or late genes than a random set of genes. See Additional file 1: Section 2 and Figures S8, S9 for more details and examples.

Conclusions

The codon distribution in large fractions of bacteriophage genomes tend to be different in early and late genes. It seems that various additional genomic features (e.g. NTB and GC content) tend to be associated with this signal. This phenomenon seems to be related to various aspects of the viral life cycle, and to various intracellular processes. A similar signal may be observed in human viruses but it seems significantly less frequent. We believe that the reported results should contribute towards better understanding of viral evolution and may promote the development of relevant procedures in synthetic virology.

Material and methods

The research outline of the study is described in Fig. 1.

Viruses

Human Viruses analyzed in this study include Herpes viruses, papilloma viruses, Polyomavirus and HIV.

The analyzed bacteriophages include: bacteriophage Lambda, bacteriophage T4, bacteriophage Pak P3, bacteriophage phi29, bacteriophage T7, bacteriophage phiYs40, bacteriophage Fah, bacteriophage xp10, bacteriophage Streptococcus DT1, bacteriophage Streptococcus 2972, bacteriophage Mu, bacteriophage phiC31, bacteriophage phiEco32, bacteriophage p23–45 and bacteriophage phiR1–37.

These viruses were chosen since they have a known division to early and late genes annotated in the literature, as described in Additional file 1: Table S3.

Synonymous codon usage analysis

Codon composition of a coding sequence was represented by a 61-dimensional vector of RSCF of each one of 61 coding codons (stop codons are excluded).

Clustering analysis was performed on RSCF vectors of each viral coding sequence. Each viral sequence was assigned a group label corresponding to its temporal expression stage (early/late) (according to the classification known in the literature). The tendency of sequences to cluster according to the codons usage in two different clusters corresponding to their temporal expression stages (early/late) was measured using the Davies-Bouldin score (DBS) [49]. This score is based on a ratio of within-cluster and between-cluster distances. The optimal clustering solution has the smallest DBS value.

The significance of cluster separation was assessed by comparing the DBS of the wildtype sequences to the randomized scores obtained from 1000 permutations of gene group labels (early or late).

In addition, a similar analysis was performed on AA frequencies as well.

More details can be found in Additional file 1: Section 3.3.

We decided to use the RSCF, since in this study we are interested in comparing the frequencies of the codons without an a-priory assumption/focus on relative bias of codons; to this aim it is more natural to use the RSCF rather the widely used Relative Synonymous Codons Usage (RSCU) measure [50]. However, these measures are similar, and the same analysis performed with RSCU does not change the final conclusions.

Additional genomic features analyzed in this study

The tRNA adaptation index (tAI) quantifies the adaptation of a coding region to the tRNA pool with parameters describing the different tRNAs copy numbers and the selective constraints on the codon–anti-codon coupling efficiency. Since, currently, these parameters are based on gene expression measurements in a very limited number of organisms, and since the efficiencies of the different codon-tRNA interactions are expected to vary among different species, we used a novel approach proposed in [51] for adjusting the tAI weights to any target organism, without the need for gene expression measurements, basing on an optimization of the correlation between the tAI and a measure of codon usage bias. It is the first time, to our knowledge, that this approach is applied to study tAI in viruses with respect to their hosts. The resulting tAI values were computed by a standalone application [52]. See more details in Additional file 1: Section 3.4.

Effective number of codons (ENC) is a measure that quantifies how far the synonymous codon usage of a gene departs from what is expected under the assumption of uniformity [53]. See more details in Additional file 1: Section 3.5.

GC-content is the percentage of nitrogenous bases on a DNA or RNA molecule that are either guanine or cytosine. See more details in Additional file 1: Section 3.6.

Codon pair bias (CPB). To quantify the CPB, we follow [54] and define a codon pair score (CPS) as the log ratio of the observed over the expected number of occurrences of this codon pair in the coding sequence. The CPB of a virus is then defined as an average CPSs over all codon pairs comprising all viral coding sequences. See more details in Additional file 1: Section 3.7.

Dinucleotide bias (DNTB). We define a dinucleotide score (DNTS) for a pair of nucleotides as an observed over expected ratio of its occurrences in a sequence. The DNTB of a virus is defined as an average of DNTSs over all dinucleotides comprising all viral coding sequences. See more details in Additional file 1: Section 3.8.

Nucleotide (NTB) and amino acid (AAB) biases are defined as a normalized Shannon entropy over the frequencies of the nucleotides / AA in a genomic sequence. See more details in Additional file 1: Section 3.9.

Ribosome profiling analysis

Ribosome profiling (ribo-seq) data was taken from [55]. Ribosome profiles for bacteriophage Lambda and *Escherichia coli* (*E.Coli)* were reconstructed and normalized as in [17]. The normalization enables measuring the relative time a ribosome spends translating each codon in a specific gene relative to other codons, while considering the total number of codons in this gene, and results in codons normalized footprint count (NFC).

Codon typical decoding rate (TDR). Following [17], in order to estimate the typical decoding time of each codon based on the corresponding ribo-seq data, we used a novel statistical model [56] which takes into consideration the skewed nature of the NFC distribution and describes the NFC histogram of each codon as an output of a random variable which is a sum of a normally distributed and an exponentially distributed random variables called Exponentially Modified Gaussian (EMG). Maximum likelihood criterion was used to estimate the parameters of these distributions for each codon according to the ribo-seq data by fitting the suggested model to the NFC distribution. The mean of the normal distribution component of EMG was called μ, and $\frac{1}{\mu}$ was defined to be the TDR of a codon [17]. See more details in Additional file 1: Section 3.10.

Mean typical decoding rate (MTDR) is a measure which estimates the global translation elongation efficiency of the entire gene as a geometric average of TDRs of its codons. See more details in Additional file 1: Section 3.11.

Since bacteriophage Lambda is the only phage with publicly available ribo-seq data, a direct analysis of TDRs of other phages is currently impossible. Nevertheless, due to the adaptation of the viruses to the translation machinery of their hosts, a rough estimation of MTDR values for other *E.Coli* phages rather than Lambda may be obtained from the available ribose-seq of the *host* genes.

Phylogenetic reconstruction

Following [57], a phylogenetic reconstruction of bacteriophages was performed basing on an alignment-free distance that estimates the similarity of two sequences (in our case entire viral proteomes) according to the average length of subsequences that are repeated in both of them (the ARS). The tree was built using the neighbor joining algorithm as implemented in [58].

See more details in Additional file 1: Section 3.12.

Abbreviations

AA: Amino acid; AAB: Amino acid bias; ARS: Average repetitive subsequences; CPB: Codon pair bias; CPS: Codon pair score; CUB: Codon usage bias; DBS: Davies-bouldin score; DNA: Deoxy ribonucleic acid; DNTB: Dinucleotide bias; DNTS: Dinucleotide score; E.Coli: *Escherichia coli*; EMG: Exponentially modified gaussian; ENC: Effective number of codons; HIV: Human immunodeficiency virus; mRNA: Messenger ribonucleic acid; MTDR: Mean typical decoding rate; NFC: Normalized footprint count; NTB: Nucleotide bias; PCA: Principal component analysis; RNA: Ribonucleic acid; RSCF: Relative synonymous codons frequencies; RSCU: Relative synonymous codons usage; tAI: tRNA adaptation index; TDR: Typical decoding rate; tRNA: Transfer ribonucleic acid

Acknowledgements

None.

Funding

E.G. is supported, in part, by a fellowship from the Edmond J. Safra Center for Bioinformatics at Tel-Aviv University. T.T. is grateful to the Minerva ARCHES award.

Authors' contributions

OM, EG, TT analyzed the data and wrote the paper. All authors read and approved the final manuscript.

Competing interests

The authors declare that they have no competing interests.

Author details

[1]Department of Biomedical Engineering, Tel-Aviv University, Ramat Aviv, Israel. [2]SynVaccineLtd. Ramat Hachayal, Tel Aviv, Israel. [3]Sagol School of Neuroscience, Tel-Aviv University, Ramat Aviv, Israel.

References

1. Bonvicini F, Filippone C, Delbarba S, Manaresi E, Zerbini M, Musiani M, Gallinella G. Parvovirus B19 genome as a single, two-state replicative and transcriptional unit. Virology. 2006;347(2):447–54.
2. Fessler SP, Young CSH. Control of adenovirus early gene expression during the late phase of infection. J Virol. 1998;72(5):4049–56.
3. Gruffat H, Marchione R, Manet E. Herpesvirus late gene expression: a viral-specific pre-initiation complex is key. Front Microbiol. 2016;7:869.
4. Jia R, Liu XF, Tao MF, Kruhlak M, Guo M, Meyers C, Baker CC, Zheng ZM. Control of the Papillomavirus early-to-late Switch by differentially expressed SRp20. J Virol. 2009;83(1):167–80.

5. Liu Z, Carmichael GG. Polyoma-virus early-late switch - regulation of late rna accumulation by dna-replication. Proc Natl Acad Sci U S A. 1993;90(18):8494–8.
6. Nisole S, Saïb A. Early steps of retrovirus replicative cycle. Retrovirology. 2004;1(1):9.
7. Schiralli Lester GM, Henderson AJ. Mechanisms of HIV transcriptional regulation and their contribution to latency. Mol Biol Int. 2012;2012:11.
8. Yang H, Ma Y, Wang Y, Yang H, Shen W, Chen X. Transcription regulation mechanisms of bacteriophages. Bioengineered. 2014;5:300–4.
9. Levy JA, Fraenkel-Conrat H, Owens RA: Virology: prentice hall; 1994.
10. Saunders JBCaVA. Virology principles and applications. West Sussex: John Wiley & Sons Ltd; 2007.
11. Aragones L, Guix S, Ribes E, Bosch A, Pinto RM. Fine-tuning translation kinetics selection as the driving force of Codon usage bias in the hepatitis a virus Capsid. PLoS Pathog. 2010;6(3):e1000797.
12. Bahir I, Fromer M, Prat Y, Linial M. Viral adaptation to host: a proteome-based analysis of codon usage and amino acid preferences. Mol Syst Biol. 2009;5(1):311.
13. Bull JJ, Molineux IJ, Wilke CO. Slow fitness recovery in a Codon-modified viral genome. Mol Biol Evol. 2012;29(10):2997–3004.
14. Burns CC, Shaw J, Campagnoli R, Jorba J, Vincent A, Quay J, Kew O. Modulation of poliovirus replicative fitness in HeLa cells by deoptimization of synonymous codon usage in the capsid region. J Virol. 2006;80(7):3259–72.
15. Cai MS, Cheng AC, Wang MS, Zhao LC, Zhu DK, Luo QH, Liu F, Chen XY. Characterization of synonymous Codon usage bias in the duck plague virus UL35 gene. Intervirology. 2009;52(5):266–78.
16. Das S, Paul S, Dutta C. Synonymous codon usage in adenoviruses: influence of mutation, selection and protein hydropathy. Virus Res. 2006;117(2):227–36.
17. Goz E, Mioduser O, Diament A, Tuller T. Evidence of translation efficiency adaptation of the coding regions of the bacteriophage lambda. DNA Res. 2017;24(4):333–42.
18. Jia RY, Cheng AC, Wang MS, Xin HY, Guo YF, Zhu DK, Qi XF, Zhao LC, Ge H, Chen XY. Analysis of synonymous codon usage in the UL24 gene of duck enteritis virus. Virus Genes. 2009;38(1):96–103.
19. Liu YS, Zhou JH, Chen HT, Ma LN, Ding YZ, Wang M, Zhang J. Analysis of synonymous codon usage in porcine reproductive and respiratory syndrome virus. Infect Genet Evol. 2010;10(6):797–803.
20. Liu YS, Zhou JH, Chen HT, Ma LN, Pejsak Z, Ding YZ, Zhang J. The characteristics of the synonymous codon usage in enterovirus 71 virus and the effects of host on the virus in codon usage pattern. Infect Genet Evol. 2011;11(5):1168–73.
21. Ma MR, Ha XQ, Ling H, Wang ML, Zhang FX, Zhang SD, Li G, Yan W. The characteristics of the synonymous codon usage in hepatitis B virus and the effects of host on the virus in codon usage pattern. Virol J. 2011;8(1):544.
22. Michely S, Toulza E, Subirana L, John U, Cognat V, Marechal-Drouard L, Grimsley N, Moreau H, Piganeau G. Evolution of Codon usage in the smallest photosynthetic eukaryotes and their Giant viruses. Genome Biol Evol. 2013;5(5):848–59.
23. RoyChoudhury S, Mukherjee D. A detailed comparative analysis on the overall codon usage pattern in herpesviruses. Virus Res. 2010;148(1–2): 31–43.
24. Sau K, Gupta SK, Sau S, Ghosh TC. Synonymous codon usage bias in 16 Staphylococcus Aureus phages: implication in phage therapy. Virus Res. 2005;113(2):123–31.
25. Sharp PM, Bailes E, Grocock RJ, Peden JF, Sockett RE. Variation in the strength of selected codon usage bias among bacteria. Nucleic Acids Res. 2005;33(4):1141–53.
26. Su MW, Lin HM, Yuan HS, Chu WC. Categorizing host-dependent RNA viruses by principal component analysis of their Codon usage preferences. J Comput Biol. 2009;16(11):1539–47.
27. Tao P, Dai L, Luo MC, Tang FQ, Tien P, Pan ZS. Analysis of synonymous codon usage in classical swine fever virus. Virus Genes. 2009;38(1):104–12.
28. Tsai CT, Lin CH, Chang CY. Analysis of codon usage bias and base compositional constraints in iridovirus genomes. Virus Res. 2007;126(1–2): 196–206.
29. Wong EHM, Smith DK, Rabadan R, Peiris M, LLM P. Codon usage bias and the evolution of influenza a viruses. Codon usage biases of influenza virus. BMC Evol Biol. 2010;10(1):253.
30. Zhang ZC, Dai W, Wang Y, Lu CP, Fan HJ. Analysis of synonymous codon usage patterns in torque teno sus virus 1 (TTSuV1). Arch Virol. 2013;158(1): 145–54.
31. Zhao KN, Gru WY, Fang NX, Saunders NA, Frazer IH. Gene codon composition determines differentiation-dependent expression of a viral capsid gene in keratinocytes in vitro and in vivo. Mol Cell Biol. 2005;25(19): 8643–55.
32. Zhong J, Li Y, Zhao S, Liu S, Zhang Z. Mutation pressure shapes codon usage in the GC-rich genome of foot-and-mouth disease virus. Virus Genes. 2007;35(3):767–76.
33. Zhou JH, Zhang J, Chen HT, Ma LN, Liu YS. Analysis of synonymous codon usage in foot-and-mouth disease virus. Vet Res Commun. 2010; 34(4):393–404.
34. Novella IS, Zarate S, Metzgar D, Ebendick-Corpus BE. Positive selection of synonymous mutations in vesicular stomatitis virus. J Mol Biol. 2004;342(5): 1415–21.
35. Spencer PS, Barral JM. Genetic code redundancy and its influence on the encoded polypeptides. Comput Struct Biotechnol J. 2012;1(1):1–8.
36. Coleman JR, Papamichail D, Skiena S, Futcher B, Wimmer E, Mueller S. Virus attenuation by genome-scale changes in codon pair bias. Science. 2008; 320(5884):1784–7.
37. Greenbaum BD, Levine AJ, Bhanot G, Rabadan R. Patterns of evolution and host gene mimicry in influenza and other RNA viruses. PLoS Pathog. 2008; 4(6):e1000079.
38. Karlin S, Doerfler W, Cardon LR. Why is CpG suppressed in the genomes of virtually all small eukaryotic viruses but not in those of large eukaryotic viruses? J Virol. 1994;68(5):2889–97.
39. Rima BK, McFerran NV. Dinucleotide and stop codon frequencies in single-stranded RNA viruses. J Gen Virol. 1997;78(11):2859–70.
40. Yakovchuk P, Protozanova E, Frank-Kamenetskii MD. Base-stacking and base-pairing contributions into thermal stability of the DNA double helix. Nucleic Acids Res. 2006;34(2):564–74.
41. Cheng XF, Virk N, Chen W, Ji SQ, Ji SX, Sun YQ, Wu XY. CpG usage in RNA viruses: data and hypotheses. PLoS One. 2013;8(9):e74109.
42. Wang AHJ, Hakoshima T, Vandermarel G, Vanboom JH, Rich A. At base-pairs are less stable than Gc Base-pairs in Z-Dna - the crystal-structure of D(M5cgtam5cg). Cell. 1984;37(1):321–31.
43. Zur H, Tuller T. Strong association between mRNA folding strength and protein abundance in S. Cerevisiae. EMBO Rep. 2012;13(3):272–7.
44. Mortimer SA, Kidwell MA, Doudna JA. Insights into RNA structure and function from genome-wide studies. Nat Rev Genet. 2014;15(7):469.
45. Xia XH. Maximizing transcription efficiency causes codon usage bias. Genetics. 1996;144(3):1309–20.
46. Cohen E, Zafrir Z, and Tuller T. A Code for Transcription Elongation Speed. To appear in RNA Biology. 2017.
47. Zhang G, Ignatova Z. Folding at the birth of the nascent chain: coordinating translation with co-translational folding. Curr Opin Struct Biol. 2011;21(1):25–31.
48. Akashi H, Gojobori T. Metabolic efficiency and amino acid composition in the proteomes of Escherichia Coli and Bacillus Subtilis. Proc Natl Acad Sci U S A. 2002;99(6):3695–700.
49. Davies DL, Bouldin DW. A cluster separation measure. IEEE Trans Pattern Anal Mach Intell. 1979;1(2):224–7.
50. Sharp PM, Li WH. An evolutionary perspective on synonymous Codon usage in unicellular organisms. J Mol Evol. 1986;24(1–2):28–38.
51. Sabi R, Tuller T. Modelling the efficiency of Codon-tRNA interactions based on Codon usage bias. DNA Res. 2014;21(5):511–25.
52. Sabi R, Daniel RV, Tuller T. stAI(calc): tRNA adaptation index calculator based on species-specific weights. Bioinformatics. 2017;33(4):589–91.
53. Wright F. The effective number of Codons used in a gene. Gene. 1990; 87(1):23–9.
54. Karlin S. Global dinucleotide signatures and analysis of genomic heterogeneity. Curr Opin Microbiol. 1998;1(5):598–610.
55. Liu XQ, Jiang HF, Gu ZL, Roberts JW. High-resolution view of bacteriophage lambda gene expression by ribosome profiling. Proc Natl Acad Sci U S A. 2013;110(29):11928–33.
56. Dana A, Tuller T. The effect of tRNA levels on decoding times of mRNA codons. Nucleic Acids Res. 2014;42(14):9171–81.
57. Ulitsky I, Burstein D, Tuller T, Chor B. The average common substring approach to phylogenomic reconstruction. J Comput Biol. 2006;13(2):336–50.
58. Felsenstein J. PHYLIP - phylogeny inference package (version 3.2). Cladistics. 1989;5(2):163–6.

PERMISSIONS

All chapters in this book were first published in BMC GENOMICS, by BioMed Central; hereby published with permission under the Creative Commons Attribution License or equivalent. Every chapter published in this book has been scrutinized by our experts. Their significance has been extensively debated. The topics covered herein carry significant findings which will fuel the growth of the discipline. They may even be implemented as practical applications or may be referred to as a beginning point for another development.

The contributors of this book come from diverse backgrounds, making this book a truly international effort. This book will bring forth new frontiers with its revolutionizing research information and detailed analysis of the nascent developments around the world.

We would like to thank all the contributing authors for lending their expertise to make the book truly unique. They have played a crucial role in the development of this book. Without their invaluable contributions this book wouldn't have been possible. They have made vital efforts to compile up to date information on the varied aspects of this subject to make this book a valuable addition to the collection of many professionals and students.

This book was conceptualized with the vision of imparting up-to-date information and advanced data in this field. To ensure the same, a matchless editorial board was set up. Every individual on the board went through rigorous rounds of assessment to prove their worth. After which they invested a large part of their time researching and compiling the most relevant data for our readers.

The editorial board has been involved in producing this book since its inception. They have spent rigorous hours researching and exploring the diverse topics which have resulted in the successful publishing of this book. They have passed on their knowledge of decades through this book. To expedite this challenging task, the publisher supported the team at every step. A small team of assistant editors was also appointed to further simplify the editing procedure and attain best results for the readers.

Apart from the editorial board, the designing team has also invested a significant amount of their time in understanding the subject and creating the most relevant covers. They scrutinized every image to scout for the most suitable representation of the subject and create an appropriate cover for the book.

The publishing team has been an ardent support to the editorial, designing and production team. Their endless efforts to recruit the best for this project, has resulted in the accomplishment of this book. They are a veteran in the field of academics and their pool of knowledge is as vast as their experience in printing. Their expertise and guidance has proved useful at every step. Their uncompromising quality standards have made this book an exceptional effort. Their encouragement from time to time has been an inspiration for everyone.

The publisher and the editorial board hope that this book will prove to be a valuable piece of knowledge for researchers, students, practitioners and scholars across the globe.

LIST OF CONTRIBUTORS

Franck Cerutti, Ludovic Mallet, Claire Hoede, Annick Moisan, Christine Gaspin and Hélène Chiapello
Université de Toulouse, INRA, UR 875 Unité Mathématiques et Informatique Appliquées de Toulouse, Auzeville, 31326 Castanet-Tolosan, France

Anaïs Painset
Université de Toulouse, INRA, UR 875 Unité Mathématiques et Informatique Appliquées de Toulouse, uzeville, 31326 Castanet-Tolosan, France

Christophe Bécavin
Département de Biologie Cellulaire et Infection, Institut Pasteur, Unité des Interactions Bactéries-Cellules, F-75015 Paris, France
INSERM, U604,F-75015 Paris, France
INRA, USC2020, F-75015 Paris, France
Institut Pasteur - Bioinformatics and Biostatistics Hub - C3BI, USR 3756 IP CNRS, Paris, France

Mélodie Duval and Pascale Cossart
Département de Biologie Cellulaire et Infection, Institut Pasteur, Unité des Interactions Bactéries-Cellules, F-75015 Paris, France
INSERM, U604,F-75015 Paris, France
INRA, USC2020, F-75015 Paris, France

Olivier Dussurget
Département de Biologie Cellulaire et Infection, Institut Pasteur, Unité des Interactions Bactéries-Cellules, F-75015 Paris, France
INSERM, U604,F-75015 Paris, France
INRA, USC2020, F-75015 Paris, France
Université Paris Diderot, Sorbonne Paris Cité, F-75013 Paris, France

Steffen Grosse-Kock, Edward C. Schwalbe, Darren Smith, Iain C. Sutcliffe and Vartul Sangal
Faculty of Health and Life Sciences, Northumbria University, Newcastle upon Tyne, UK

Valentina Kolodkina and Leonid Titov
Republican Research and Practical Centre for Epidemiology and Microbiology, Minsk, Republic of Belarus

Jochen Blom
Justus-Liebig-Universität, Gießen, Germany

Andreas Burkovski
Friedrich-Alexander-Universität Erlangen-Nürnberg, Erlangen, Germany

Paul A. Hoskisson
Strathclyde Institute of Pharmacy and Biomedical Sciences, University of Strathclyde, Glasgow, UK

Sylvain Brisse
Institut Pasteur, Biodiversity and Epidemiology of Bacterial Pathogens, Paris, France

Aaron W. Kolb and Ralph Moeller Trane
Department of Ophthalmology and Visual Sciences, School of Medicine and Public Health, University of Wisconsin-Madison, 550 Bardeen Laboratories, 1300 University Ave., Madison, WI 53706, USA

Andrew C. Lewin
Department of Surgical Sciences, School of Veterinary Medicine, University of Wisconsin-Madison, Madison, WI, USA

Gillian J. McLellan
Department of Ophthalmology and Visual Sciences, School of Medicine and Public Health, University of Wisconsin-Madison, 550 Bardeen Laboratories, 1300 University Ave., Madison, WI 53706, USA
Department of Surgical Sciences, School of Veterinary Medicine, University of Wisconsin-Madison, Madison, WI, USA
McPherson Eye Research Institute, University of Wisconsin-Madison, Madison, WI, USA

Curtis R. Brandt
Department of Ophthalmology and Visual Sciences, School of Medicine and Public Health, University of Wisconsin-Madison, 550 Bardeen Laboratories, 1300 University Ave., Madison, WI 53706, USA
McPherson Eye Research Institute, University of Wisconsin-Madison, Madison, WI, USA
Medical Microbiology and Immunology, School of Medicine and Public Health, University of Wisconsin-Madison, Madison, WI, USA

Jun Qin, Ainong Shi, Yuejin Weng, Waltram Ravelombola, Gehendra Bhattarai, Lingdi Dong and Wei Yang
Department of Horticulture, University of Arkansas, Fayetteville, AR 72701, USA

Beiquan Mou
Crop Improvement and Protection Research Unit, USDA-ARS, Salinas, CA 93905, USA

Michael A. Grusak
USDA-ARS Red River Valley Agricultural Research Center, Fargo, ND 58102, USA

Yingchun Xu, Yanjie Wang and Qijiang Jin
College of Horticulture, Nanjing Agricultural University, Nanjing 210095, China

Neil Mattson
Horticulture Section, School of Integrative Plant Science, Cornell University, 134A Plant Science Bldg, Ithaca, NY 14853, USA

Liu Yang
Institute of Plant Protection, Jiangsu Academy of Agricultural Sciences, Nanjing 210095, China

Xinglin Zhang
College of Biosystems Engineering and Food Science, Zhejiang University, Hangzhou 310058, China
Department of Medical Microbiology, University Medical Center Utrecht, 3584CX Utrecht, The Netherlands

Vincent de Maat, Ana M. Guzmán Prieto, Jumamurat R. Bayjanov, Mark de Been, Malbert R. C. ogers, Marc J. M. Bonten and Rob J. L. Willems
Department of Medical Microbiology, University Medical Center Utrecht, 3584CX Utrecht, The Netherlands

Tomasz K. Prajsnar and Stéphane Mesnage
Krebs Institute, University of Sheffield, Sheffield S10 2TN, United Kingdom

Willem van Schaik
Department of Medical Microbiology, University Medical Center Utrecht, 3584CX Utrecht, The Netherlands
Institute of Microbiology and Infection, College of Medical and Dental Sciences, The University of Birmingham, Birmingham B15 2TT, United Kingdom

Vikram A. Misra and Michael P. Timko
Department of Biology, University of Virginia, Gilmer Hall 044, Charlottesville, VA 22904, USA

Yu Wang
Department of Biology, University of Virginia, Gilmer Hall 044, Charlottesville, VA 22904, USA
Center for Quantitative Sciences, Vanderbilt University, Nashville, TN 37232-6848, USA

Fengyan Zhou, Yong Zhang, Mei Wang and Tongchun Gao
Institute of Plant Protection and Agro-Products Safety, Anhui Academy of Agricultural Sciences, Hefei 230001, China

Wei Tang
State Key Laboratory of Rice Biology, China National Rice Research Institute, Hangzhou 311400, China

Weina Guo
Jiangsu Co-Innovation Center for Prevention and Control of Important Animal Infectious Diseases and Zoonoses, Yangzhou University College of Veterinary Medicine, Yangzhou, Jiangsu 225009, People's Republic of China
College of Animal Science, Anhui Science and Technology University, Maanshan, Anhui, China

Martina Jelocnik and Adam Polkinghorne
Centre for Animal Health Innovation, Faculty of Science, Health, Education and Engineering, University of the Sunshine Coast, Maroochydore, QLD, Australia

Jing Li and Jinfeng You
Jiangsu Co-Innovation Center for Prevention and Control of Important Animal Infectious Diseases and Zoonoses, Yangzhou University College of Veterinary Medicine, Yangzhou, Jiangsu 225009, People's Republic of China

Konrad Sachse
Institute of Bioinformatics, Friedrich-Schiller-Universität Jena, Jena, Germany

Yvonne Pannekoek
Department of Microbiology, University of Amsterdam, Amsterdam, The Netherlands

Bernhard Kaltenboeck
College of Veterinary Medicine, Auburn University, Auburn, AL, USA

Jiansen Gong
Poultry Institute, Chinese Academy of Agricultural Sciences, Jiangsu Co-Innovation Center for Prevention and Control of Important Animal Infectious Diseases and Zoonoses, Yangzhou, Jiangsu, China

Chengming Wang
Jiangsu Co-Innovation Center for Prevention and Control of Important Animal Infectious Diseases and Zoonoses, Yangzhou University College of Veterinary Medicine, Yangzhou, Jiangsu 225009, People's Republic of China
College of Veterinary Medicine, Auburn University, Auburn, AL, USA

Héctor Cordero
Fish Innate Immune System Group, Department of Cell Biology and Histology, Faculty of Biology, Regional Campus of International Excellence "Campus Mare Nostrum", University of Murcia, 30100 Murcia, Spain
Department of Microbiology and Immunology, Rosalind Franklin University of Medicine and Science, North Chicago, IL 60064, USA
Faculty of Biosciences and Aquaculture, Nord University, 8049 Bodø, Norway

Monica F. Brinchmann
Faculty of Biosciences and Aquaculture, Nord University, 8049 Bodø, Norway

Alberto Cuesta and María A. Esteban
Fish Innate Immune System Group, Department of Cell Biology and Histology, Faculty of Biology, Regional Campus of International Excellence"Campus Mare Nostrum", University of Murcia, 30100 Murcia, Spain

Ali Sharifi-Zarchi
Department of Bioinformatics, Institute of Biochemistry and Biophysics, University of Tehran, Tehran, Iran
Computer Science Department, Colorado State University, Fort Collins, CO, USA.
Department of Stem Cells and Developmental Biology, Cell Science Research Center, Royan Institute for Stem Cell Biology and Technology, ACECR, Tehran, Iran
Department of Computer Engineering, Sharif University of Technology, Tehran, Iran

Daniela Gerovska
Computational Biology and Systems Biomedicine, Biodonostia Health Research Institute, 20014 San Sebastián, Spain

Kenjiro Adachi
Department of Cell and Developmental Biology, Max Planck Institute for Molecular Biomedicine, Münster, Germany

Mehdi Totonchi
Department of Stem Cells and Developmental Biology, Cell Science Research Center, Royan Institute for Stem Cell Biology and Technology, ACECR, Tehran, Iran

Hamid Pezeshk
School of Mathematics, Statistics and Computer Science, College of Science, University of Tehran, Tehran, Iran
School of Biological Sciences, Institute for Research in Fundamental Sciences (IPM), Tehran, Iran

Ryan J. Taft
Illumina Inc., San Diego, USA

Hans R. Schöler
Department of Cell and Developmental Biology, Max Planck Institute for Molecular Biomedicine, Münster, Germany
Medical Faculty, University of Münster,Münster, Germany

Hamidreza Chitsaz
Computer Science Department, Colorado State University, Fort Collins, CO, USA

Mehdi Sadeghi
School of Biological Sciences, Institute for Research in Fundamental Sciences (IPM), Tehran, Iran.
National Institute of Genetic Engineering and Biotechnology (NIGEB), Tehran, Iran

Hossein Baharvand
Department of Stem Cells and Developmental Biology, Cell Science Research Center, Royan Institute for Stem Cell Biology and Technology, ACECR, Tehran, Iran
Department of Developmental Biology,University of Science and Culture, Tehran, Iran

Marcos J. Araúzo-Bravo
Computational Biology and Systems Biomedicine, Biodonostia Health Research Institute, 20014 San Sebastián, Spain
Computational Biology and Bioinformatics Group, Max Planck Institute for Molecular Biomedicine, Münster, Germany

IKERBASQUE, Basque Foundation for Science, 48011 Bilbao, Spain

Christine Marie George
Department of International Health, Program in Global Disease Epidemiology and Control, Johns Hopkins Bloomberg School of Public Health, 615 N. Wolfe Street, Room E5535, Baltimore, MD 21205-2103, USA

Mahamud Rashid, K. M. Saif-Ur-Rahman, Shirajum Monira and Munirul Alam
Md. Sazzadul Islam Bhuyian, Toslim T. Mahmud, Zillur Rahman, Munshi Mustafiz
International Centre for Diarrhoeal Disease Research, Shaheed Tajuddin Ahmed Ave, Dhaka 1213, Bangladesh

Mathieu Almeida, O. Colin Stine and Shan Li
University of Maryland School of Medicine, 655 W Baltimore S, Baltimore, MD 21201, USA

Khaled Hasan, Jessica Brubaker, Jamie Perin, David A. Sack and R. Bradley Sack
Johns Hopkins School of Public Health, 615 N. Wolfe Street, Baltimore, MD 21205-2103, USA

David S. Campo, Sumathi Ramachandran and Yury Khudyakov
Division of Viral Hepatitis, Molecular Epidemiology and Bioinformatics, Centers for Disease Control and Prevention, Atlanta, GA, USA

June Zhang
Division of Viral Hepatitis, Molecular Epidemiology and Bioinformatics, Centers for Disease Control and Prevention, Atlanta, GA, USA
Department of Electrical Engineering, University of Hawaii, Manoa, HI, USA

Pakeeza Akram and Li Liao
Department of Computer & Information Sciences, University of Delaware, Newark, DE, USA

Oriah Mioduser
Department of Biomedical Engineering, Tel-Aviv University, Ramat Aviv, Israel

Eli Goz
Department of Biomedical Engineering, Tel-Aviv University, Ramat Aviv, Israel
SynVaccineLtd. Ramat Hachayal, Tel Aviv, Israel

Tamir Tuller
Department of Biomedical Engineering, Tel-Aviv University, Ramat Aviv, Israel
SynVaccineLtd. Ramat Hachayal, Tel Aviv, Israel.
Sagol School of Neuroscience, Tel-Aviv University, Ramat Aviv, Israel

Index

www.ingramcontent.com/pod-product-compliance
Lightning Source LLC
Chambersburg PA
CBHW082028190326
41458CB00010B/3305
* 9 7 8 1 6 8 2 8 6 7 6 5 5 *